LONDON MATHEMATICAL SOCIETY STUDENT TEXTS

73 Groups, graphs and trees, JOHN MEIER
74 Representation theorems in Hardy spaces, JAVAD MASHREGHI
75 An introduction to the theory of graph spectra, DRAGOŠ CVETKOVIĆ, PETER ROWLINSON & SLOBODAN SIMIĆ
76 Number theory in the spirit of Liouville, KENNETH S. WILLIAMS
77 Lectures on profinite topics in group theory, BENJAMIN KLOPSCH, NIKOLAY NIKOLOV & CHRISTOPHER VOLL
78 Clifford algebras: An introduction, D. J. H. GARLING
79 Introduction to compact Riemann surfaces and dessins d'enfants, ERNESTO GIRONDO & GABINO GONZÁLEZ-DIEZ
80 The Riemann hypothesis for function fields, MACHIEL VAN FRANKENHUIJSEN
81 Number theory, Fourier analysis and geometric discrepancy, GIANCARLO TRAVAGLINI
82 Finite geometry and combinatorial applications, SIMEON BALL
83 The geometry of celestial mechanics, HANSJÖRG GEIGES
84 Random graphs, geometry and asymptotic structure, MICHAEL KRIVELEVICH *et al*
85 Fourier analysis: Part I – Theory, ADRIAN CONSTANTIN
86 Dispersive partial differential equations, M. BURAK ERDOĞAN & NIKOLAOS TZIRAKIS
87 Riemann surfaces and algebraic curves, R. CAVALIERI & E. MILES
88 Groups, languages and automata, DEREK F. HOLT, SARAH REES & CLAAS E. RÖVER
89 Analysis on Polish spaces and an introduction to optimal transportation, D. J. H. GARLING
90 The homotopy theory of $(\infty,1)$-categories, JULIA E. BERGNER
91 The block theory of finite group algebras I, MARKUS LINCKELMANN
92 The block theory of finite group algebras II, MARKUS LINCKELMANN
93 Semigroups of linear operators, DAVID APPLEBAUM
94 Introduction to approximate groups, MATTHEW C. H. TOINTON
95 Representations of finite groups of Lie type (2nd Edition), FRANÇOIS DIGNE & JEAN MICHEL
96 Tensor products of C^*-algebras and operator spaces, GILLES PISIER
97 Topics in cyclic theory, DANIEL G. QUILLEN & GORDON BLOWER
98 Fast track to forcing, MIRNA DŽAMONJA
99 A gentle introduction to homological mirror symmetry, RAF BOCKLANDT
100 The calculus of braids, PATRICK DEHORNOY
101 Classical and discrete functional analysis with measure theory, MARTIN BUNTINAS
102 Notes on Hamiltonian dynamical systems, ANTONIO GIORGILLI
103 A course in stochastic game theory, EILON SOLAN
104 Differential and low-dimensional topology, ANDRÁS JUHÁSZ
105 Lectures on Lagrangian torus fibrations, JONNY EVANS
106 Compact matrix quantum groups and their combinatorics, AMAURY FRESLON
107 Inverse problems and data assimilation, DANIEL SANZ-ALONSO, ANDREW STUART & ARMEEN TAEB
108 Künneth geometry, M.J.D. HAMILTON & D. KOTSCHICK
109 ADE: Patterns in mathematics, PETER J. CAMERON *et al*
110 Sets and transfinite algebra, THOMAS MÜLLER
111 Celestial mechanics, ANTONIO GIORGILLI, UGO LOCATELLI & MARCO SANSOTERRA
112 The Shrikhande graph, PETER J. CAMERON, APARNA LAKSHMANAN S. & AMBAT VIJAYAKUMAR

"The material of the book is comparable to that of Milnor's classic treatise, but Quick's exposition is much more detailed and concrete. Students with a basic knowledge of linear algebra and multidimensional calculus will have no difficulty understanding this material. Almost all exercises have detailed solutions, so, apart from being a great companion to lectures, the book is ideal for self-study."

Christopher Denniger, *University of Münster*

London Mathematical Society Student Texts 113

Differential Topology

GEREON QUICK
Norwegian University of Science and Technology, Trondheim

CAMBRIDGE
UNIVERSITY PRESS

CAMBRIDGE
UNIVERSITY PRESS

Shaftesbury Road, Cambridge CB2 8EA, United Kingdom

One Liberty Plaza, 20th Floor, New York, NY 10006, USA

477 Williamstown Road, Port Melbourne, VIC 3207, Australia

314-321, 3rd Floor, Plot 3, Splendor Forum, Jasola District Centre,
New Delhi – 110025, India

Cambridge University Press is part of Cambridge University Press & Assessment, a department of the University of Cambridge.

We share the University's mission to contribute to society through the pursuit of education, learning and research at the highest international levels of excellence.

www.cambridge.org
Information on this title: www.cambridge.org/9781009733878
DOI: 10.1017/9781009759847

When citing this work, please include a reference to the
DOI 10.1017/9781009759847

First published 2026

A catalogue record for this publication is available from the British Library

A Cataloging-in-Publication data record for this book is available from the Library of Congress

ISBN 978-1-009-73387-8 Hardback
ISBN 978-1-009-73386-1 Paperback

Contents

Preface

The purpose of this book is to guide readers with a background in multivariable calculus and linear algebra directly into the fascinating world of differential topology. To make the ideas and techniques as accessible as possible, the arguments are explained in great detail. Rather than aiming for the most elegant and shortest proofs, we often take a longer path, pausing to examine each step carefully. We hope that this level of detail will allow less experienced mathematicians to follow the development of the subject, appreciate the intuition behind the constructions, and witness how seemingly straightforward ideas can lead to deep results. This approach has the consequence that some arguments may appear lengthy to more experienced mathematicians. For concise expositions of the central ideas and theorems in differential topology, we warmly recommend the excellent texts by Milnor [35] and Guillemin–Pollack [14], which also served as foundations for this book. Almost all chapters are accompanied by exercises. We strongly encourage readers to work through them, as they are an integral part of the learning process. At the end of the book, suggestions for solutions to most exercises are provided.

Acknowledgements

This book grew out of lecture notes for a one-semester course taught over several years at the Norwegian University of Science and Technology in Trondheim. I am very grateful to colleagues and students whose feedback and insights helped improve the manuscript. In particular, I would like to thank Torgeir Aambø, Johanna Aigner, Anders Alexander Andersen, Daria Barjaktarevic, Denis Bergmann, Eivind Xu Djurhuus, Sigurd Gaukstad, Knut Bjarte Haus, Abigail Linton, Martin Löcsei, Trygve Poppe Oldervoll, Marius Thaule, and Glen Matthew Wilson. I am very grateful to Holly Paveling and Anna Scriven from Cambridge University Press for their kind and generous support throughout the preparation of this book.

Introduction

Manifolds

Differential topology studies smooth manifolds. The concept of a manifold is fundamental not only in topology but across mathematics, physics, and other sciences. Manifolds arise naturally in geometry and the theory of differential equations, in classical mechanics, general relativity, and string theory, and even in describing configuration spaces in fields such as robotics and cybernetics. There are several ways to arrive at the idea of a manifold. We begin with a concrete one. Suppose we would like to solve an equation of the form $f(x) = b$ where f is a collection of differentiable functions $\mathbb{R}^n \to \mathbb{R}$, i.e., we would like to find a simple and efficient way to describe the subset X of $\mathbb{R}^n$ given by $X = \{x \in \mathbb{R}^n : f(x) = b\}$. Sometimes there may be only finitely many solutions, and we can just list them all. More often, however, X is infinite, and we want to understand how its points are related to each other and how the set is situated inside $\mathbb{R}^n$. A natural approach is to describe points of X using a minimal number of parameters $t_1, \ldots, t_k$ via a map $\phi\colon \mathbb{R}^k \to X$. Such a parametrisation rarely covers all of X, often it describes only the points near some point x_0. We call such a map ϕ a *local parametrisation* around x_0. The idea is then to find enough such local parametrisations to describe every point of X. When this can be done using sufficiently nice, i.e., differentiable, maps ϕ, we say that $X \subset \mathbb{R}^n$ is a *smooth manifold.*

We now turn to a second, slightly more abstract approach to the notion of a manifold. It is based on the observation that calculus of functions on Euclidean space is very effective, and we would like to extend its tools to a wider class of spaces. Let X be a topological space. To perform calculus on X, we need coordinates, i.e., numerical descriptions of points on which differentiation makes sense. We would therefore like to have a map $\psi\colon X \to \mathbb{R}^k$, for some k, which allows us to describe points in X via coordinates in $\mathbb{R}^k$. However, such a map is rarely available globally without losing information. Instead, it is usually possible only locally, i.e., for each point $x \in X$ we may find an open neighbourhood $U \subset X$ and a map $\psi : U \to \mathbb{R}^k$ that provides coordinates on U. Moreover, whenever two such maps overlap, we require them to be compatible in an appropriate sense, i.e., the transition maps between the coordinate representations should be differentiable. If we can equip every point of X with such compatible local coordinate maps, then we call X a *manifold.* The maps ψ are called *local coordinate charts*, or simply *charts*. Thus, a manifold is a space in which every point can be equipped with compatible local coordinates such that we can perform calculus in a consistent way.

In the course of this book, we will learn that this abstract, coordinate-based viewpoint and the more concrete picture arising from level sets of smooth maps lead to the same class of spaces. We now summarise the common underlying idea in a first, informal definition, which will be made precise later.

Definition 1 (Working definition of a manifold) A k-dimensional *manifold* is a space which locally, i.e., in a small neighbourhood of every point, looks like a k-dimensional Euclidean space $\mathbb{R}^k$.

In this book, we will primarily adopt the first, more concrete viewpoint and regard manifolds as subsets of some Euclidean space $\mathbb{R}^n$. This perspective has several advantages, three of which we highlight here. First, the definition of a manifold becomes more transparent and requires far fewer technical details. Second, it allows us to describe differentiable maps between manifolds in a straightforward way, simply as restrictions of differentiable maps between Euclidean spaces. Third, it leads to an intuitive and easily visualised description of tangent spaces. At each point of the manifold, the tangent space naturally appears as a concrete vector subspace of $\mathbb{R}^n$. For example, the k-dimensional sphere

$$\mathbb{S}^k = \{(x_1, \ldots, x_{k+1}) \in \mathbb{R}^{k+1} : x_1^2 + \cdots + x_{k+1}^2 = 1\}$$

is a smooth manifold embedded in $\mathbb{R}^{k+1}$. The tangent space at a point $x \in \mathbb{S}^k$ is simply the k-dimensional subspace of $\mathbb{R}^{k+1}$ consisting of all vectors orthogonal to x. Other natural examples include the two-dimensional torus, Stiefel manifolds, and many matrix groups.

However, working exclusively with embedded manifolds has a limitation since some manifolds are most naturally described without reference to any ambient space. For this reason, we will also develop the more abstract, coordinate-based definition. This viewpoint will introduce new and important examples such as projective spaces, the Klein bottle, the Möbius band, and Grassmannians. A central result proved later in this book shows that every smooth manifold, whether defined concretely as a subset of $\mathbb{R}^n$ or abstractly via coordinate charts, can in fact be embedded into some Euclidean space. Thus, the two approaches ultimately describe the same class of spaces. Moreover, many key theorems in differential topology rely crucially on such embeddings into Euclidean space, making the concrete viewpoint not only intuitive but also technically useful.

Three Key Concepts

The existence of *tangent spaces* is a key feature of smooth manifolds. In fact, tangent spaces will allow us to formulate intuitive ideas about spaces in precise mathematical terms. For instance, it is easy to picture a nice and clear-cut intersection of two manifolds, e.g., two-dimensional spheres intersecting in $\mathbb{R}^3$, but it is through tangent spaces that we can give a precise definition of the ways manifolds intersect. This leads to two fundamental notions in differential topology: *regular values* and *transverse intersections*. Much of this book will be devoted to developing these ideas and exploring their consequences and generalisations.

Another central concept we will use extensively is *homotopy*. Homotopy plays a key role not only in differential topology but also in general topology and many other areas of mathematics. Roughly speaking, two maps are homotopic if one can be continuously deformed into the other. Homotopy defines an equivalence relation on the set of continuous maps between spaces. A crucial fact we will prove is that every continuous map between smooth manifolds is homotopic to a smooth map, i.e., a map that is infinitely differentiable. This allows us to prove statements about continuous maps using the techniques we develop for smooth maps. Studying manifolds up to homotopy is one of the features that distinguishes differential topology from differential geometry. Although this viewpoint sometimes ignores finer geometric properties of smooth manifolds, it allows us to construct powerful invariants that are essential for solving many deep mathematical problems. For instance, such invariants are, by design, a key tool in the classification of manifolds.

An *invariant* is a simple algebraic object; for us it will just be a number that we can assign to a smooth manifold or smooth map. The number is called an invariant since we construct it such that it does not change under the geometric transformations we consider. For smooth manifolds, these transformations are diffeomorphisms, i.e., smooth maps with smooth inverses, or smooth homotopies. We are familiar with the notion of an invariant from linear algebra in the form of the dimension of a vector space. In fact, the existence of tangent spaces, which are vector spaces, then shows that the dimension is an invariant for smooth manifolds as well, i.e., there cannot exist a diffeomorphism between smooth manifolds of different dimensions. One of the most important invariants in differential topology is the *Brouwer degree* of a smooth map $f\colon X \to Y$, where X and Y are smooth manifolds of the same dimension. The degree is based on the simple idea that, in a sense to be made precise, the number of points in the fibre $f^{-1}(y)$ is independent of the choice of $y \in Y$. Once we have put in the effort to show that this yields a well-defined invariant, we will see that the degree is a remarkably powerful tool. We will first define the Brouwer degree with values in $\mathbb{Z}/2\mathbb{Z}$. Later, after introducing orientations, we will refine the definition to obtain the degree with values in $\mathbb{Z}$. We will also study variations of the degree in the form of the *mod* 2 *intersection number*, the *Hopf invariant*, and the *index of a vector field*. Finally, we will see a far-reaching generalisation of the degree in terms of *framed cobordisms*.

Important Results

We will now describe five of the main results and applications of differential topology that we will prove in this book.

Theorem 2 (Brouwer Fixed Point Theorem) *Let $\mathbb{D}^n$ be the n-dimensional unit disc $\mathbb{D}^n = \{(x_1, \ldots, x_n) \in \mathbb{R}^n : x_1^2 + \cdots + x_n^2 \leq 1\}$. Then every continuous map $f\colon \mathbb{D}^n \to \mathbb{D}^n$ has a fixed point, i.e., there is an $x \in \mathbb{D}^n$ such that $f(x) = x$.*

This theorem has numerous applications. In fact, many problems in mathematics can be transformed into the question whether a certain map has a fixed point. Then Brouwer's theorem often provides the tool to show the existence of such a point. While the formulation of the theorem does not require any knowledge in differential topology, we will need a good amount of exciting theory and new skills to prove the theorem.

Using Sard's theorem and Whitney's approximation theorem, we will show that every continuous map $f\colon X \to \mathbb{S}^n$ from a manifold X of dimension k to the n-dimensional sphere $\mathbb{S}^n$ is homotopic to a constant map when $k < n$. For $k > n$, however, the story is very different. In fact, the Hopf invariant, constructed using the Brouwer degree, will enable us to prove the following deep result due to Hopf.

Theorem 3 (Hopf invariant one) *There is a smooth map $\hbar\colon \mathbb{S}^3 \to \mathbb{S}^2$ which is not homotopic to a constant map.*

The map $\hbar\colon \mathbb{S}^3 \to \mathbb{S}^2$ of the theorem is known as the *Hopf map*, a central object in the study of homotopy groups of spheres. We will use this important map as a recurring example throughout this book.

The story of the Hopf invariant and the Hopf map is related to the following purely algebraic result, which we will prove using mod 2 intersection theory on smooth manifolds and Brouwer's invariance of domain. Let $\mu\colon \mathbb{R}^n \times \mathbb{R}^n \to \mathbb{R}^n$, $(a,b) \mapsto a \cdot b$, be an $\mathbb{R}$-bilinear map with no zero-divisors, i.e., $a \cdot b = 0$ implies $a = 0$ or $b = 0$. If, in addition, $a \cdot b = b \cdot a$ for all $a, b \in \mathbb{R}^n$, μ is called a *commutative division algebra structure* on $\mathbb{R}^n$. Familiar examples of such structures occur in dimensions $n = 1$ and $n = 2$, provided by the fields of real and complex numbers, respectively. However, we will prove the surprising fact that there are no such structures in higher dimensions.

Theorem 4 *There is no commutative division algebra structure on $\mathbb{R}^n$ for $n > 2$.*

While invariants often help demonstrate that certain structures cannot exist, they can also show that a structure must exist. A famous example of such a case is the following theorem due to Hopf, which shows that the integer-valued Brouwer degree completely determines the homotopy class of continuous maps from manifolds into spheres of the same dimension.

Theorem 5 (Hopf Degree Theorem) *Let X be a compact, connected, oriented smooth n-dimensional manifold without boundary. Then two continuous maps $X \to \mathbb{S}^n$ are homotopic if and only if they have the same degree.*

Finally, we prove a first example of an index theorem which shows that analytical and purely topological properties are related to each other in a surprising way. A vector field on a smooth manifold X is a map which assigns to each point x a tangent vector at x. A zero of a vector field is a point where this vector is the zero vector. Using the Brouwer degree, we define the index of a zero, which captures the local behaviour of the vector field. The sum of the indices of the zeros of a vector field a priori depends on the smooth structure of the manifold. On the other hand, X is also equipped with another invariant, the Euler characteristic $\chi(X) \in \mathbb{Z}$, a purely topological invariant independent of any smooth structure. One of the highlights of this book is the proof of the following famous result due to Poincaré and Hopf, which relates the analytical properties of vector fields on a smooth manifold to its Euler characteristic.

Theorem 6 (Poincaré–Hopf Index Theorem) *Let X be a compact, oriented smooth manifold and let $\mathbf{v}$ be a vector field on X with only finitely many zeros. Then the sum of the indices of $\mathbf{v}$ equals the Euler characteristic of X. In particular, X possesses a nowhere-vanishing vector field if $\chi(X) = 0$. For example, the latter is the case when the dimension of X is odd. On the other hand, if $\chi(X) \neq 0$, then every vector field on X has a zero.*

1
A Brief Introduction to Topological Spaces

In this chapter, we recall some key concepts in topology. We begin with subspaces in $\mathbb{R}^n$ and will then consider more general topological spaces. We will then introduce some of the most relevant topological properties. Along the way, we will look at some basic examples. There will be some more interesting examples in the final section. We hope that this recollection is a good warm-up for the following chapters. The reader familiar with topological spaces may want to skip this chapter first and get back to the statements when needed. Other concepts and results in general topology will be recalled later in the book.

1.1 Open and Closed Subsets in Euclidean Space

We begin with a familiar example since it both helps motivate the general definition and is most relevant for us. Recall that the norm of a vector $x = (x_1, \dots, x_n) \in \mathbb{R}^n$ is defined as the non-negative real number

$$|x| := \sqrt{x_1^2 + x_2^2 + \cdots + x_n^2} \in \mathbb{R}^{\geq 0}.$$

The norm defines a distance between two points x, y in $\mathbb{R}^n$ by taking the norm $|x - y|$ of the difference of x and y. For every n, the space $\mathbb{R}^n$ together with this norm is called *n-dimensional Euclidean space.*

Definition 1.1 (Open sets in $\mathbb{R}^n$) Let x be a point in $\mathbb{R}^n$ and $r > 0$ a real number. We define the *n-dimensional open ball with radius r around x* to be

$$\mathbb{B}_r(x) = \mathbb{B}_r^n(x) = \{y \in \mathbb{R}^n : |x - y| < r\}.$$

We say that a subset $U \subseteq \mathbb{R}^n$ is *open* if, for every point $x \in U$, there exists a real number $\varepsilon > 0$ such that $\mathbb{B}_\varepsilon^n(x)$ is contained in U.

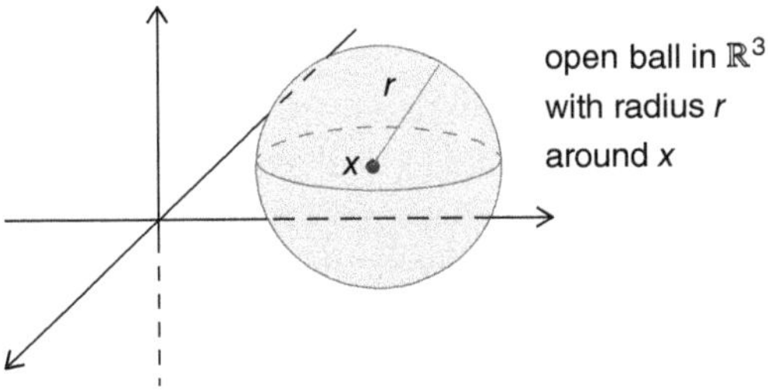

Figure 1.1 An open ball in $\mathbb{R}^3$.

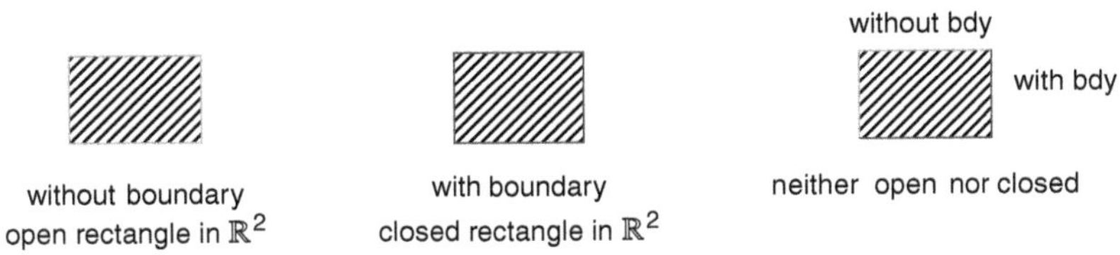

Figure 1.2 Examples of subsets in $\mathbb{R}^2$.

In particular, the whole space $\mathbb{R}^n$ is open since it contains all open balls (Figure 1.1). The empty set is open, as there is no condition to be checked. Familiar examples of open sets in $\mathbb{R}$ are open intervals such as $(-2, \sqrt{5})$ or $(1, \infty)$. However, there are also closed intervals in $\mathbb{R}$ such as $[-3, 1]$. Now we observe that we can write any closed interval as the complement of open sets in $\mathbb{R}$, e.g., $[-3, 1] = \mathbb{R} \setminus ((-\infty, -3) \cup (1, \infty))$. We will therefore say that a subset $Z \subset \mathbb{R}^n$ is *closed* if it is the complement of an open subset, i.e., if there is an open subset $U \subset \mathbb{R}^n$ such that $Z = \mathbb{R}^n \setminus U$. An important example of a closed set is the n-dimensional sphere $\mathbb{S}^n$ defined as $\mathbb{S}^n := \{x \in \mathbb{R}^{n+1} : |x| = 1\}$.

We note that the whole space $\mathbb{R}^n$ and the empty set are both open and closed at the same time. We should note, however, that not every subset of $\mathbb{R}^n$ is open or closed. In fact, there are a lot of subsets which are neither open nor closed. See Figure 1.2. For example, the interval $(0, 1]$ is neither open nor closed in $\mathbb{R}$.

1.2 Topological Spaces and Continuous Maps

We can think of the concept of an open subset in $\mathbb{R}^n$ as a generalisation of the distance function. We will now formulate the key features of open subsets in a new definition which does not require the existence of a distance function.

1.2.1 Topological Spaces

Definition 1.2 (Topological spaces) Let X be a set. A *topology* on X is a collection τ of subsets of X, called *open sets*, such that arbitrary unions of open sets are open, finite intersections of open sets are open, and the empty set $\emptyset$ and the whole space X are both open. A *topological space* is a pair (X, τ) consisting of a set X together with a topology τ. Note that we will often drop τ from the notation. A subset $Z \subseteq X$ is called *closed* if its complement $X \setminus Z$ is open.

We can turn any set into a topological space by using the discrete topology in which every subset is open. However, in most cases, this does not add anything interesting. More important examples of topological spaces are subspaces of $\mathbb{R}^n$ with the topology defined by the collection of open sets of Definition 1.1. Given two topological spaces, we can form their product, which we will always equip with the following topology.

Definition 1.3 (Product topology) Let X and Y be topological spaces. A subset $W \subset X \times Y$ is called open in the *product topology* if for each point $(x, y) \in W$ there are open subsets $U \subset X$ and $V \subset V$ such that $(x, y) \in U \times V \subset W$.

In analogy to the distance function on $\mathbb{R}^n$, we may think of open sets as a means to talk about things that happen close to a point. This is the idea behind the following terminology.

Definition 1.4 (Neighbourhoods) Let X be a topological space. We say that a subset $V \subseteq X$ containing a point $x \in X$ is a *neighbourhood of* x if there is an open subset $U \subseteq V$ with $x \in U$. If V itself is open, we call V an *open neighbourhood.*

When we refer to something that occurs in the neighbourhood of a point $x \in X$, we often say that it happens *locally at* x. A property of a space or a function that only needs to hold in a neighbourhood of each point is called a *local property*. In contrast, *global properties* describe the entire space. From now on, when we speak of a *space*, we mean a topological space, i.e., a set equipped with a specified topology. If this definition feels abstract, that is because it is indeed abstract. However, in most of our discussions, open sets will have a concrete geometric interpretation, typically arising from open balls in $\mathbb{R}^n$. Generalising the terminology will prove useful in many situations, so let us continue along this abstract path a little further.

Definition 1.5 (Subspace topology and relative open sets) Let X be a topological space and let $A \subset X$ be a subset. Then we say that $V \subset A$ is *open*

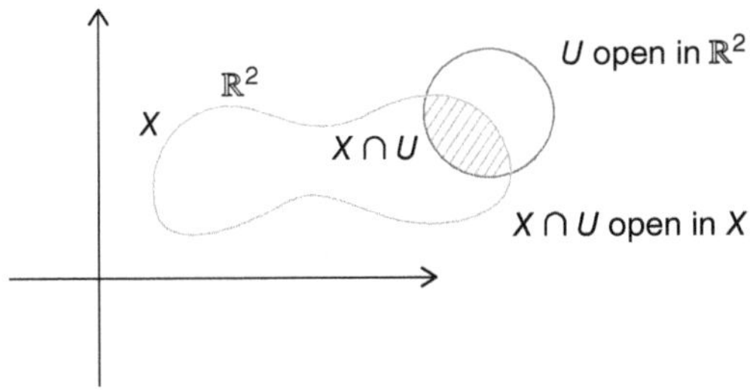

Figure 1.3 A relative open subset of X: $X \cap U$ is open in X.

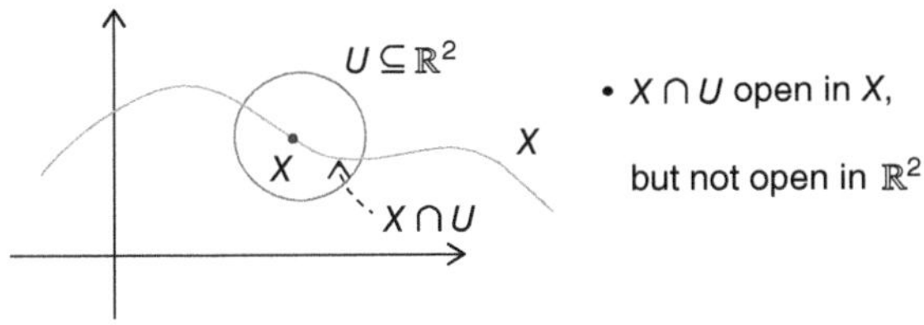

Figure 1.4 The relative open subset of $X \cap U$ is open in X, but it is not open in $\mathbb{R}^2$.

in A, or *relatively open*, if there is an open subset $U \subset X$ such that $V = U \cap A$. The collection of the relative open sets is called the *subset topology* on A.

When X is a topological space and A is a subset, then, unless otherwise stated, we consider A as a subspace of X with the subspace topology provided by relative open sets. Let us look at the case $X = \mathbb{R}^n$ and $A \subset \mathbb{R}^n$ an arbitrary subset. Then $V \subseteq A$ is open in A if and only if, for every point $x \in V$, there exists a real number $\varepsilon > 0$ such that $\mathbb{B}^n_\varepsilon(x) \cap A \subseteq V$. See Figure 1.3.

Definition 1.5 is particularly useful because it provides a method for constructing new topological spaces. At the same time, it highlights an important point: the notion of an open subset depends on the surrounding space. In other words, when we say a set is open, we always mean open in some specified ambient space. For example, given a subspace $X \subset \mathbb{R}^2$, then a set $U \subset X$ can be open in X, while it is not open in $\mathbb{R}^2$. See Figure 1.4.

Example 1.6 Let $\mathbb{S}^2$ be the two-dimensional sphere. We consider $\mathbb{S}^2$ as a subspace in $\mathbb{R}^3$ with the subspace topology. Let $x_0 \in \mathbb{S}^2$ be a point on $\mathbb{S}^2$. An example of a subset in $\mathbb{S}^2$ which is open in $\mathbb{S}^2$ and contains x_0 is the set $\mathbb{S}^2 \cap \mathbb{B}^3_{1/2}(x_0)$ with $\mathbb{B}^3_{1/2}(x_0) = \{y \in \mathbb{R}^3 : |y - x_0| < 1/2\}$. In fact, every subset which is open in $\mathbb{S}^2$ and contains x_0 contains a subset of the form $\mathbb{S}^2 \cap \mathbb{B}^3_\varepsilon(x_0)$ with $\varepsilon > 0$ sufficiently small. However, the set $\mathbb{S}^2 \cap \mathbb{B}^3_{1/2}(x_0)$ is not open in

$\mathbb{R}^3$, since there is no three-dimensional open ball $\mathbb{B}^3_\varepsilon(x_0)$ which is completely contained in $\mathbb{S}^2 \cap \mathbb{B}^3_{1/2}(x_0)$.

1.2.2 Continuous Maps

Maps between topological spaces should respect the topology, i.e., they should preserve open sets. We formulate what this means precisely in the following definition.

Definition 1.7 (Continuous maps) Let X and Y be topological spaces. A map $f: X \to Y$ is called *continuous* if, for every open subset $U \subseteq Y$, the subset $f^{-1}(U)$ is open in X. In the subspace topology in $\mathbb{R}^n$ this reads as follows: if A is a subset of $\mathbb{R}^n$ with the subset topology, then a map $f: A \to \mathbb{R}^m$ is continuous if and only if, for every open subset $U \subseteq \mathbb{R}^m$, there is some open subset $V \subseteq \mathbb{R}^n$ with $f^{-1}(U) = V \cap A$.

We are familiar with continuous maps from calculus. In our new terminology, the ε-δ-characterisation of continuity looks as follows.

Lemma 1.8 *Let $A \subset \mathbb{R}^n$ be a subset equipped with the subset topology. A map $f: A \to \mathbb{R}^m$ is continuous if, for every $\varepsilon > 0$, there is a $\delta > 0$ such that if $x \in \mathbb{B}^n_\delta(a) \cap A$, then $f(x) \in \mathbb{B}^m_\varepsilon(f(a))$.*

Proof: First, we assume f satisfies ε-δ-continuity. Let $U \subseteq \mathbb{R}^m$ be an open set in $\mathbb{R}^m$. If $f^{-1}(U)$ is empty, it is open by definition. So let $a \in f^{-1}(U)$ be a point in $f^{-1}(U)$. The fact that U is open means that there is an $\varepsilon > 0$ such that $\mathbb{B}^m_\varepsilon(f(a)) \subseteq U$. Given this ε, there is a $\delta > 0$ such that $x \in \mathbb{B}^n_\delta(a) \cap A$ implies $f(x) \in \mathbb{B}^m_\varepsilon(f(a))$. As $f(x) \in \mathbb{B}^m_\varepsilon(f(a)) \subseteq U$, we get $f(x) \in U$ and hence $x \in f^{-1}(U)$. Thus, for every $x \in \mathbb{B}^n_\delta(a) \cap A$, we have $x \in f^{-1}(U)$. In other words, $\mathbb{B}^n_\delta(a) \cap A \subseteq f^{-1}(U)$. Since a was an arbitrary point in $f^{-1}(U)$, this shows that $f^{-1}(U)$ is open in A. Now we assume $f^{-1}(U)$ is open in A for every open subset $U \subseteq \mathbb{R}^m$. Given $a \in A$ and $\varepsilon > 0$, let $\mathbb{B}^m_\varepsilon(f(a)) \subset \mathbb{R}^m$ be the open ball around $f(a)$ with radius ε. Since $\mathbb{B}^m_\varepsilon(f(a))$ is open in $\mathbb{R}^m$, our assumption tells us that $f^{-1}(\mathbb{B}^m_\varepsilon(f(a)))$ is open in A. Since $a \in f^{-1}(\mathbb{B}^m_\varepsilon(f(a)))$, this means that there is a $\delta > 0$ such that $\mathbb{B}^n_\delta(a) \cap A \subseteq f^{-1}(\mathbb{B}^m_\varepsilon(f(a)))$. In other words, $x \in \mathbb{B}^n_\delta(a) \cap A$ implies $f(x) \in \mathbb{B}^m_\varepsilon(f(a))$. □

Next, we introduce a name for the maps which have continuous inverses.

Definition 1.9 (Homeomorphisms) Let X and Y be topological spaces. A continuous map $f: X \to Y$ is called a *homeomorphism* if f is a bijection and its inverse f^{-1} is continuous as well.

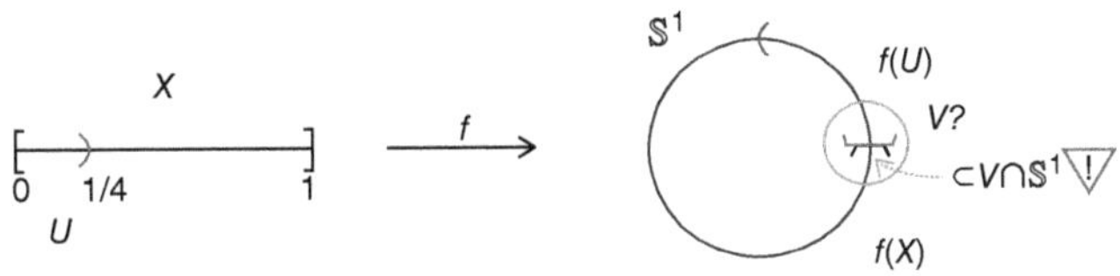

Figure 1.5 Wrapping the interval around the circle via f is a bijection but not a homeomorphism, since $f(U)$ may not be open in $\mathbb{S}^1$.

Homeomorphisms preserve the topology in the sense that if $f\colon X \to Y$ is a homeomorphism, then $U \subset X$ is open in X if and only if the image $f(U) \subset Y$ is open in Y.

Example 1.10 The map $\tan\colon (-\pi/2, \pi/2) \to \mathbb{R}$ is a homeomorphism. The map $f\colon \mathbb{R} \to \mathbb{R}$, $x \mapsto x^3$ is a homeomorphism. However, the map $g\colon \mathbb{R} \to \mathbb{R}$, $x \mapsto x^2$ is not a homeomorphism, since it is neither one-to-one nor onto.

Example 1.11 (A bijection which is not a homeomorphism) Consider the unit circle $\mathbb{S}^1 = \{(x, y) \in \mathbb{R}^2 : x^2 + y^2 = 1\} \subset \mathbb{R}^2$. We define a map $f\colon [0,1) \to \mathbb{S}^1$ on the interval $[0,1) \subset \mathbb{R}$ by $f(t) := (\cos(2\pi t), \sin(2\pi t))$. We know that f is bijective and continuous from trigonometry and calculus. However, the inverse function $f^{-1}\colon \mathbb{S}^1 \to [0,1)$ is not continuous. For example, we will now show that the image $f(U)$ of the open subset $U = [0, 1/4)$ is not open in $\mathbb{S}^1$. First, we observe that $U = [0, 1/4)$ is indeed open in $[0,1)$, since $U = [0,1) \cap (-1/4, 1/4)$. Now we look at the point $f(0) = (1,0) \in \mathbb{S}^1$. Since $0 \in U$, $f(0)$ is a point in $f(U)$. However, for every open ball $\mathbb{B}^2_\varepsilon((1,0)) \subset \mathbb{R}^2$, the intersection $\mathbb{B}^2_\varepsilon((1,0)) \cap \mathbb{S}^1$ contains points with strictly negative y-coordinate. In particular, $\mathbb{B}^2_\varepsilon((1,0)) \cap \mathbb{S}^1$ contains points which are not in $f(U)$, i.e., we get $\mathbb{B}^2_\varepsilon((1,0)) \cap \mathbb{S}^1) \not\subset f(U)$. Alternatively, we could observe that $f^{-1}(\mathbb{B}^2_\varepsilon((1,0)) \cap \mathbb{S}^1) \subset [0,1)$ contains points which are close to 1 in $[0,1)$ and therefore do not lie in U, i.e., $f^{-1}(\mathbb{B}^2_\varepsilon((1,0)) \cap \mathbb{S}^1) \not\subset U$. This shows that we cannot find an open subset V of $\mathbb{R}^2$ such that $V \cap \mathbb{S}^1 = f(U)$. Hence $f(U)$ is not open in $\mathbb{S}^1$ (Figure 1.5).

1.3 Topological Properties

A property of a topological space that is preserved under homeomorphisms is called a *topological property*. Such properties are often referred to as global properties, because they cannot be verified by examining only small neighbourhoods around each point. Many interesting phenomena in differential topology arise from the interplay between local and global properties. Next, we will review what it means for a space to be compact, Hausdorff, connected, and path-connected. These topological properties will play a central role in our study of smooth manifolds.

1.3.1 Compactness

Definition 1.12 (Compact space) A topological space X is called *compact* if every open cover $\{U_\alpha\}_\alpha$ of X, i.e., a collection of open subsets $U_\alpha \subset X$ such that $X = \bigcup_\alpha U_\alpha$ is the union of all the U_α, has a *finite* subcover. That is, among the $\{U_\alpha\}_\alpha$ it is always possible to pick finitely many $U_{\alpha_1}, \ldots, U_{\alpha_n}$ which already cover X, i.e., $X = U_{\alpha_1} \cup \ldots \cup U_{\alpha_n}$.

Compactness is a powerful concept because it allows us to control a space using only finitely many open sets. In fact, many proofs become significantly simpler when we can assume compactness. The definition itself may seem abstract and not immediately practical. The following observations will help us prove that a given space is compact. We will see a more concrete characterisation for subspaces in $\mathbb{R}^n$ in Theorem 1.16.

Lemma 1.13 *Let X and Y be topological spaces and let $f: X \to Y$ be a continuous map. Let $A \subset X$ be a compact subspace. Then $f(A)$ is compact. In particular, if X is compact and f is surjective, then Y is compact.*

Proof: See Exercise 1.1. □

Lemma 1.14 *Let X be a compact space and $A \subset X$ a closed subspace. Then A is compact.*

Proof: Let $\{U_\alpha\}_\alpha$ be a collection of open sets $U_\alpha \subset A$ such that $\bigcup_\alpha U_\alpha = A$. Since A is equipped with the subspace topology, for each α, there is an open subset $V_\alpha \subset X$ such that $U_\alpha = V_\alpha \cap A$. Since A is closed in X, its complement $X \setminus A$ is open in X. Thus, the collection $\{X \setminus A, V_\alpha\}_\alpha$ forms an open cover of X. Since X is compact, we can choose finitely many $\alpha_1, \ldots, \alpha_n$ such that $X = (X \setminus A) \cup V_{\alpha_1} \cup \cdots \cup V_{\alpha_n}$. Thus, $A = U_{\alpha_1} \cup \cdots \cup U_{\alpha_n}$, i.e., finitely many of the U_α suffice to cover A. □

Proposition 1.15 *Let X and Y be non-empty topological spaces. Then $X \times Y$ is compact if and only if both X and Y are compact.*

Proof: First, assume that $X \times Y$ is compact. Since the two projection maps $X \times Y \to X$ and $X \times Y \to Y$ are continuous and surjective, Lemma 1.13

implies that X and Y are compact. Now we assume that X and Y are compact. Let $\{W_\alpha\}_\alpha$ be an open cover of $X \times Y$. For every $(x, y) \in X \times Y$, we can choose an $\alpha(x, y)$ such that $(x, y) \in W_{\alpha(x,y)}$. Since $W_{\alpha(x,y)}$ is open in $X \times Y$, there are open subsets $U_{(x,y)} \subset X$ and $V_{(x,y)} \subset Y$ such that $U_{(x,y)} \times V_{(x,y)} \subset W_{\alpha(x,y)}$. For a fixed x, the collection $\{V_{(x,y)}\}_{y \in Y}$ is an open cover of Y. Since Y is compact, there are $y_1(x), \ldots, y_{n_x}(x)$ such that $Y = V_{(x,y_1(x))} \cup \cdots \cup V_{(x,y_{n_x}(x))}$. We set $U_x := U_{(x,y_1(x))} \cap \cdots \cap U_{(x,y_{n_x}(x))}$. Since $x \in U_x$, the collection of the U_x for all $x \in X$ is an open cover of X. Since X is compact, there are finitely many $x_1, \ldots, x_m$ such that $U_{x_1} \cup \cdots \cup U_{x_m}$. This shows that $X \times Y$ is covered by the finitely many $W_{\alpha(x_i, y_j(x_i))}$ for $1 \le i \le m$ and $1 \le j \le n_{x_i}$. □

Compact spaces in $\mathbb{R}^n$ have a much more concrete characterisation, which we now recall. A subset $Z \subset \mathbb{R}^n$ is called *bounded* if there exists $r > 0$ such that $Z \subset \mathbb{B}^n_r(0)$. For subspaces in Euclidean space, we then have the following important characterisation of compact subsets.

Theorem 1.16 (Heine–Borel) *A subset $Z \subset \mathbb{R}^n$ is compact if and only if it is closed and bounded.*

Theorem 1.16 provides us with plenty of examples of compact spaces. In particular, the n-dimensional sphere $\mathbb{S}^n$ is compact. On the other hand, any open subset of $\mathbb{R}^n$ is not compact. To illustrate the necessity of Z being closed, one may consider $Z = (0, 1) \subset \mathbb{R}$, which we can cover by the collection of open intervals $U_n = (1/n, 1)$ for all natural numbers n, but not with any finite subcollection. We postpone the proof of Theorem 1.16 to the end of Section 1.3.2, since we will use that Euclidean space is Hausdorff, a property we study in the next section. We will later make use of the following observation. Recall that a subspace $S \subset \mathbb{R}^n$ is called *discrete* if, for every point $s \in S$, there is an open ball $\mathbb{B}^n_\varepsilon(s)$ around s which does not contain any other point of S, i.e., such that $\mathbb{B}^n_\varepsilon(s)$ and $S \setminus \{s\}$ are disjoint.

Lemma 1.17 *Every compact and discrete subset S of $\mathbb{R}^n$ is finite.*

Proof: See Exercise 1.3. □

1.3.2 Hausdorff Spaces

We now turn to a property that may seem obvious at first glance. In fact, every space we will consider satisfies this property.

Definition 1.18 (Hausdorff spaces) A topological space X is called *Hausdorff* if, for any two distinct points $x, y \in X$, there are two disjoint open subsets $U, V \subset X$ such that $x \in U$ and $y \in V$.

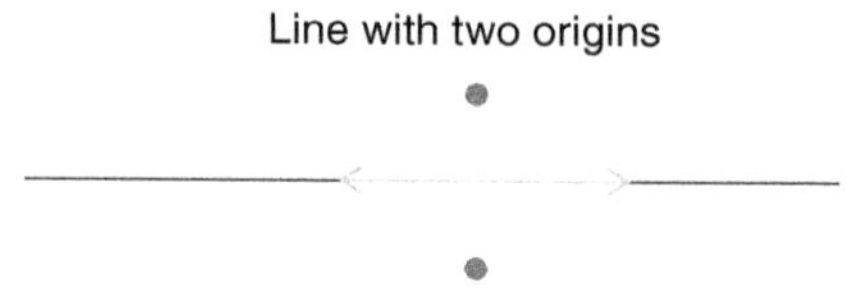

Figure 1.6 A space that looks like a line with a double point. The topology is such that we cannot separate the two points by open subsets.

In other words, in a Hausdorff space we can separate points by disjoint open neighbourhoods.

Example 1.19 Every subset of $\mathbb{R}^n$ with the subspace topology is a Hausdorff space since if $d = |x - y| > 0$ is the distance of two points x and y in $\mathbb{R}^n$, then the open balls $\mathbb{B}^n_{d/4}(x)$ and $\mathbb{B}^n_{d/4}(y)$ around x and y, respectively, are disjoint. In fact, when X is a Hausdorff space, then every subset A with the subspace topology is Hausdorff.

It may seem that every topological space should satisfy the Hausdorff condition. There are, however, spaces which are not Hausdorff.

Example 1.20 (A space which is not Hausdorff) For a typical example of a space which is not Hausdorff, consider two copies of the real line $Y_1 := \mathbb{R} \times \{1\}$ and $Y_2 := \mathbb{R} \times \{2\}$ as subspaces of $\mathbb{R}^2$. On $Y_1 \cup Y_2$, we define the equivalence relation $(x, 1) \sim (x, 2)$ for all $x \neq 0$. Let X be the set of equivalence classes. The topology on X is the quotient topology, i.e., a subset $W \subset X$ is open in X if and only if both its preimages in $\mathbb{R} \times \{1\}$ and $\mathbb{R} \times \{2\}$ are open. Then X looks like the real line except that the origin is replaced with two different copies of the origin. Every neighbourhood of the point $(0, 1)$ has a non-empty intersection in X with any neighbourhood of the point $(0, 2)$. Hence, we cannot separate the two origins by open subsets, and X is not Hausdorff. See Figure 1.6.

We will now look at some of the many useful consequences of the Hausdorff property. For example, in a Hausdorff space the following converse of Lemma 1.14 holds.

Proposition 1.21 *Let X be a Hausdorff topological space, and $A \subset X$ a compact subspace. Then A is closed in X.*

Remark 1.22 In the following proof of the proposition, we will use the following common argument. Let X be a topological space, and suppose we want to show that a subset $W \subset X$ is open. Then it suffices to show that, for every $x \in W$, there is a subset $U_x \subset W$ such that $x \in U_x$ and U_x is open in

X. Then W is the union of the open subsets U_x for all $x \in W$, and hence W is open in X by the definition of a topology in Definition 1.2. At first glance, this may seem like an unnecessary detour. However, it is often much easier to find a small open neighbourhood around each point than to describe the entire set at once. If we can do this by a general argument that applies to every point in W, the method becomes very efficient. We will encounter this type of reasoning frequently throughout the book. This observation illustrates that openness is a local property by definition: we can check it by examining neighbourhoods of individual points. This idea is familiar from calculus, where continuity of a real function f is defined at a point, and f is continuous if it is continuous at all points of its domain. Other properties, however, such as compactness, cannot be verified locally but require consideration of the entire space. This is why compactness is regarded as a global property. Not every condition in topology is purely local or global. For instance, many spaces are locally compact, such as all subspaces of $\mathbb{R}^n$, which combine local and global features.

Proof of Proposition 1.21: We prove the statement by showing that the complement $X \setminus A$ is open in X. This implies that A is closed in X. So, let $x \in X \setminus A$ be an arbitrary point in the complement. We will show that there is a subset $U_x \subset X \setminus A$ which is open in X and contains x. Since X is Hausdorff, for every $a \in A$, there is an open subset $V(a) \subset X$ such that $x \notin V(a)$. The collection of all the $V(a)$ then forms an open cover of A, i.e., $A \subset \cup_{a \in A} V(a)$. Now we use that A is compact. This means that we can pick finitely many points $a_1, \ldots, a_n$ in X and open subsets $V(a_1), \ldots, V(a_n)$ with $x \notin V(a_i)$ for all $i = 1, \ldots, n$ such that $A \subset V(a_1) \cup \cdots \cup V(a_n)$. Again, since X is Hausdorff, we can choose open subsets $U(a_i)$ in X for each i with $x \in U(a_i)$ and $U(a_i) \cap V(a_i) = \emptyset$. We set $U_x := U(a_1) \cap \cdots \cap U(a_n)$. Since this is a finite intersection of open subsets, U_x is open in X. Moreover, we have $x \in U_x$ since x is contained in each of the $U(a_i)$. It remains to show that U_x is contained in $X \setminus A$. So let $y \in U_x$ be another point. Then y is contained in each $U(a_i)$ by definition of U_x as the above intersection. Since $U(a_i) \cap V(a_i) = \emptyset$, this shows that $y \notin V(a_i)$ for all $i = 1, \ldots, n$. Since $A \subset V(a_1) \cup \cdots \cup V(a_n)$, this implies that $y \notin A$. This shows $U_x \subset X \setminus A$ and finishes the proof. □

The following lemma shows that the anomaly of Example 1.11 arises from the fact that the domain of the map is not compact.

Corollary 1.23 *Let $f\colon X \to Y$ be a continuous map. Assume that X is compact and Y is Hausdorff. Then f is a closed map, i.e., if $A \subset X$ is closed then $f(A) \subset Y$ is closed. Moreover, if f is a bijection then f is a homeomorphism.*

Proof: Let $A \subset Y$ be a closed subset. By Lemma 1.14 and since X is compact, this implies that A is compact. By Lemma 1.13, $f(A)$ is compact. By Proposition 1.21 and since Y is Hausdorff, this implies that $f(A)$ is closed in Y. This proves the first assertion. Now we assume in addition that f is bijective, and let $f^{-1}\colon Y \to X$ be its inverse. Let $U \subset X$ be an open subset. Since $f = (f^{-1})^{-1}$, we need to show that $f(U)$ is open in Y. Since $X \setminus U$ is closed in X, we know by the first assertion that $f(X \setminus U)$ is closed in Y. Thus, since $Y = f(X)$, we deduce that $f(U) = Y \setminus f(X \setminus U)$ is open in Y. □

Note that, in general, we cannot reverse the direction of the implication in Lemma 1.13, i.e., it is not true in general that the inverse image of a compact space is compact. The class of maps for which the latter implication does hold therefore gets a name.

Definition 1.24 A continuous map $f\colon X \to Y$ is called *proper* if the preimage of every compact subset is a compact subset, i.e., if $K \subset Y$ is compact then $f^{-1}(K)$ is compact.

Example 1.25 By Theorem 1.16, compact subsets in Euclidean space are closed and bounded. Hence, if a continuous map $f\colon \mathbb{R}^n \to \mathbb{R}^m$ is proper, then the preimage of a closed and bounded subset must be closed and bounded. This means, in particular, that f sends a sequence of points which are further and further away from the origin in $\mathbb{R}^n$ to a sequence of points which are further and further away from the origin in $\mathbb{R}^m$. Hence, intuitively we may characterise a proper map by saying that it *sends infinity to infinity*. An example of a proper map $\mathbb{R} \to \mathbb{R}$ is given by $x \mapsto x^2$. Some familiar maps like $\sin$ and $\cos$, however, are not proper. Both $\sin$ and $\cos$ send points far away from zero to points which are close to zero. More precisely, the preimage of $[-1, 1] \subset \mathbb{R}$, which is compact by Exercise 1.4, under $\sin$ or $\cos$ is $\mathbb{R}$ which is not compact by Theorem 1.16.

Since many spaces we consider in this book are compact and Hausdorff, the following lemma shows that many of the maps we meet are also proper.

Lemma 1.26 *Let $f\colon X \to Y$ be a continuous map. Assume that X is compact and that Y is Hausdorff. Then f is proper*

Proof: Let $K \subset Y$ be a compact subspace in Y. Then K is closed in Y by Proposition 1.21 since Y is Hausdorff. Since f is continuous, $f^{-1}(K)$ is closed in X. By Lemma 1.14, this implies that $f^{-1}(K)$ is compact since X is compact. □

We recall that a Hausdorff topological space Y is called *locally compact* if, for every $y \in Y$, we can find a compact neighbourhood K of y in Y, i.e., there is an open subset U and a compact subset K such that $y \in U \subset K$. All subspaces of Euclidean space are locally compact. For proper maps, we have the following stronger version of Corollary 1.23 where we can drop the assumption that X is compact.

Proposition 1.27 *Let X be a topological space and let Y be a Hausdorff topological space which is locally compact. Let $f: X \to Y$ be a continuous bijective map. Assume that f is proper. Then f is a homeomorphism.*

Proof: We have $Y = f(X)$ since f is surjective. Let Z be a closed subset in X. Let $y \in Y \setminus f(Z)$. We are going to show that this is an open subset of $Y = f(X)$ which then implies that $f(Z)$ is closed in Y. Since Y is locally compact, there is an open subset U and a compact subset K such that $y \in U \subset K$. Then $f^{-1}(K)$ is compact since f is proper. By Lemma 1.14, this implies that $f^{-1}(K) \cap Z$ is compact since it is a closed subset in $f^{-1}(K)$. Hence $f(f^{-1}(K) \cap Z) = K \cap f(Z)$ by Lemma 1.13. Since K is Hausdorff, this implies by Proposition 1.21 that $K \cap f(Z)$ is closed in K and hence also closed in Y. Thus, since U is open in Y, the subset $U \setminus (K \cap f(Z))$ is open in Y as well. We know $y \in U \setminus (K \cap f(Z))$ since $y \in U$ and $y \notin f(Z)$. This shows that y has an open neighbourhood $U \setminus (K \cap f(Z))$ in $Y \setminus f(Z)$. Thus, $Y \setminus f(Z)$ is an open subset of $Y = f(X)$. □

We can now provide the proof of the theorem of Heine–Borel.

Proof of Theorem 1.16: First assume that $K \subset \mathbb{R}^n$ is compact. By Proposition 1.21, we know that K is closed since $\mathbb{R}^n$ is Hausdorff by Example 1.19. Hence it remains to show that K is bounded. Let $\mathbb{B}^n_r(0) \subset \mathbb{R}^n$ be the open ball around the origin with radius r. Then the collection $\{\mathbb{B}^n_r(0)\}_{r>0}$ for all radii forms an open cover of $\mathbb{R}^n$. Hence the collection $\{U_r\}_{r>0}$ with $U_r := K \cap \mathbb{B}^n_r(0)$ forms an open cover of K. Since K is compact, there is a finite collection $U_{r_1}, \ldots, U_{r_n}$ such that $U_{r_1} \cup \cdots \cup U_{r_n} = K$. Let $R := \max\{r_1, \ldots, r_n\}$ be the maximum of the r_i. Since $U_r \subset U_{r'}$ if $r < r'$, we get $K \subset U_R$. Thus, K is bounded. Now we assume that $K \subset \mathbb{R}^n$ is closed and bounded. Since K is bounded, there is some real number r such that K is contained in the box $\mathcal{B} := [-r,r] \times \cdots \times [-r,r] \subset \mathbb{R}^n$ given by the n-fold product of the closed intervals $[-r,r] \subset \mathbb{R}$. By Lemma 1.14, it suffices to show that $\mathcal{B}$ is compact since K is closed in $\mathbb{R}^n$ and therefore K is also in $\mathcal{B}$ by definition of the subspace topology. By Exercise 1.4, the closed interval $[0,1]$ is compact. Since there is a homeomorphism between $[0,1]$ and any other closed interval in $\mathbb{R}$ defined by translating and multiplying with suitable real numbers, each closed interval $[-r,r]$ is compact. By Proposition 1.15, this implies by induction on n that $\mathcal{B}$ is compact. □

1.3.3 Connected Spaces

A quite intuitive topological property is connectedness. Roughly speaking, a space is connected if it consists of just one piece. However, making this precise requires some amount of abstraction. We will see soon that things are more intuitive for the spaces we study in this book.

Definition 1.28 (Connected spaces) A topological space X is called *connected* if X cannot be written as the union of two non-empty disjoint open subsets.

An equivalent formulation is that X is connected if X and $\emptyset$ are the only subsets which are both open and closed in X. Familiar examples of connected spaces are intervals in $\mathbb{R}$. For example, the interval $[0, 1]$ is connected.

Remark 1.29 If $X \neq \emptyset$ is not connected, we can consider the set of all connected subsets of X. We can order this set by inclusion $\subseteq$. The maximal connected subsets with respect to this ordering are called the *connected components* of X. Then X can be considered as the possibly infinite union of its connected components. The union of all connected subsets which contain a point x is called the *connected component of* x.

A frequent application of connectedness is the following. A continuous map $f : X \to S$ from a topological space X to any set S is called *locally constant* if for every $x \in X$ there is an open neighbourhood $U_x \subset X$ such that the restriction $f_{|U_x}$ is constant.

Lemma 1.30 *Let X be a connected topological space and $f : X \to S$ be a continuous map. Assume that f is locally constant. Then f is constant.*

Proof: Let $s \in S$ be a value of f, i.e., $s = f(x)$ for some $x \in X$. Let $A \subseteq X$ be the set $A = f^{-1}(s) = \{x \in X : f(x) = s\}$. Since f is locally constant, there is an open neighbourhood $U_a \subset A$ with $f(U_a) = \{s\}$. In other words, for every $a \in A$, there is a subset $U_a \subset A$ which is open in X and $a \in U_a$. This shows that A is open in X. Let $B = X \setminus A$ denote the complement of A, i.e., $B = \{x \in X : f(x) \neq s\}$. Since B is a union of preimages $f^{-1}(s')$ for all elements $s' \neq s$ in S, the previous argument shows that B is the union of open subsets and hence B is open. This implies that A and B are both open and closed. However, we can then write $X = A \sqcup B$ as the disjoint union of the open and closed subsets A and B. Since X is connected and $A \neq \emptyset$, we must have $A = X$, and f is constant. □

1.3.4 Path-Connected Spaces

The criterion for connectedness is easy to state but rather abstract. For example, it does not tell us if we can walk, i.e., draw a line without interruptions, from one point to another, as one would intuitively expect for a connected space. Let us formulate this more concrete property in a precise way.

Definition 1.31 (Path-connected spaces) Let X be a topological space, and let $x, y \in X$ be two points. A continuous map $\gamma\colon [0,1] \to X$ with $\gamma(0) = x$ and $\gamma(1) = y$ is called a *path* from x to y. The space X is called *path-connected* if for any two points $x, y \in X$ there is a path from x to y.

Example 1.32 Every open or closed ball $\mathbb{B}^n_r(y)$ in $\mathbb{R}^n$ is path-connected since there is a straight line-path $\gamma_{y\to z}\colon [0,1] \to \mathbb{B}^n_r(x)$ from the centre of the ball y to any point z in $\mathbb{B}^n_r(y)$, i.e., $\gamma_{y\to z}(t) = (1-t)\cdot y + t\cdot z$.

Path-connectedness is a stronger property than connectedness.

Lemma 1.33 *If a space is path-connected, then it is also connected.*

Proof: Let X be a path-connected space. If X was not connected, then there would be two disjoint non-empty open subsets A and B with $X = A \cup B$. Since A and B are non-empty, we can choose two points $a \in A$ and $b \in B$. Since X is path-connected, there is a continuous map $\gamma\colon [0,1] \to X$ with $\gamma(0) = a$ and $\gamma(1) = b$. Hence $0 \in \gamma^{-1}(A) \subset [0,1]$ and $1 \in \gamma^{-1}(B) \subset [0,1]$. Since A and B are disjoint and open, the subsets $\gamma^{-1}(A)$ and $\gamma^{-1}(B)$ are disjoint and open in $[0,1]$. Since $X = A \cup B$, we would have $[0,1] = \gamma^{-1}(A) \cup \gamma^{-1}(B)$, which contradicts the fact that $[0,1]$ is connected. Hence X must be connected. □

Remark 1.34 Be aware that there are connected spaces which are not path-connected. A standard example is the following subspace in $\mathbb{R}^2$ given by $X = \{(x, \sin(\log x)) \subset \mathbb{R}^2 : x > 0\} \cup (0 \times [-1,1])$.

Remark 1.35 Suppose we have a path γ from x to y in a space X. Then we can also get a path from y to x by reversing the direction, i.e., by considering the path $t \mapsto \gamma(1-t)$. Now let $x, y, z \in X$ be three points, and assume we have a path $\gamma_{x\to y}$ from x to y and a path $\gamma_{y\to z}$ from y to z. We can then define a new path $\gamma_{x\to z}\colon [0,1] \to X$ from x to z in X by

$$\gamma_{x\to z}(t) := \begin{cases} \gamma_{x\to y}(2t), & \text{for } 0 \le t \le \frac{1}{2} \\ \gamma_{y\to z}(2t-1), & \text{for } \frac{1}{2} \le t \le 1 \end{cases}.$$

We will show later that smooth manifolds, however, are connected if and only if they are path-connected. See Exercise 8.5. For open subsets in Euclidean space, we will prove the equivalence of both properties next.

Proposition 1.36 *Let $X \subset \mathbb{R}^n$ be a non-empty* open *subset. If X is connected, then X is path-connected.*

Proof: Since X is non-empty, we can pick a point $x \in X$. We define a subset U of $\mathbb{R}^n$ by $U := \{y \in X\colon$ there is a path in X from x to $y\}$. The set U is non-empty since it contains x. We now show that U is an open subset of X. Then we use the assumption that X is connected to prove that $U = X$. To show that U is open in X, let $y \in U$ be a point. Since X is open in $\mathbb{R}^n$, there is an open ball $\mathbb{B}^n_r(y) \subset \mathbb{R}^n$ with small enough radius r around y such that $\mathbb{B}^n_r(y) \subset X$. We learned in Example 1.32 that there is a path $\gamma_{z\to y}$ within $\mathbb{B}^n_r(y)$ from any point $z \in \mathbb{B}^n_r(y)$ to y. Since $y \in U$, we also have a path $\gamma_{x\to y}$ in X from x to y. We can then define a new path $\gamma_{x\to z}\colon [0,1] \to X$ from x to z as explained in Remark 1.35. The existence of the path $\gamma_{x\to z}$ shows that also z belongs to U. Since z was an arbitrary point in $\mathbb{B}^n_r(y)$, this proves that $\mathbb{B}^n_r(y) \subset U$. This shows that U is open in X. Now we look at the complement V of U in X, i.e., the set $V := \{y \in X\colon$ there is no path in X from x to $y\}$. If V is non-empty, there is a point $y \in V$, and we can find an open ball $\mathbb{B}^n_r(y)$ in $\mathbb{R}^n$ such that $\mathbb{B}^n_r(y) \subset X$. If there were a point $z \in \mathbb{B}^n_r(y) \cap U$, then there would be a path between z and y as both are in $\mathbb{B}^n_r(y)$, and also a path between z and x as z is in U. Hence we get a path between x and y. However, this contradicts our assumption $y \in V$. Hence, $\mathbb{B}^n_r(y)$ must be disjoint from U, i.e., we must have $\mathbb{B}^n_r(y) \subset V$. This shows that V is open. Thus, we have shown $X = U \cup V$ is a disjoint union of subsets which are both open and closed. Since X is connected, this can only happen if $V = \emptyset$ and $U = X$ as we know that $U \neq \emptyset$. Thus, there is a path from x to any point y in X. By composing paths, this shows that any two points in X can be connected via a path. □

1.3.5 Examples: Real and Complex Matrix Spaces

In this section, we will study some of the topological properties we have introduced so far for a familiar type of spaces. The results of this section will play a role later in the book. The reader preferring to learn about manifolds right away may skip this section for the moment and go back to the results when needed.

Let $\mathrm{M}(n) = \mathrm{M}(n, \mathbb{R})$ denote the set of $n \times n$-matrices with entries in $\mathbb{R}$. We consider $\mathrm{M}(n)$ as a topological space by identifying $\mathrm{M}(n)$ with the Euclidean space $\mathbb{R}^{n^2}$ of dimension n^2. In particular, $\mathrm{M}(n)$ and all the subspaces of $\mathrm{M}(n)$ are Hausdorff. Let

$$\mathrm{GL}(n) = \mathrm{GL}(n,\mathbb{R}) = \{A \in \mathrm{M}(n) : \det A \neq 0\}$$

denote the *real general linear group* consisting of all invertible $n \times n$-matrices with entries in $\mathbb{R}$. The group operation is given by matrix multiplication. The subset $\mathrm{GL}(n)$ is open in $\mathrm{M}(n) \cong \mathbb{R}^{n^2}$. To check that it is open, we look at its complement $\mathrm{M}(n) \setminus \mathrm{GL}(n)$, which is given by the set $\det^{-1}(0) = \{A \in \mathrm{M}(n) : \det A = 0\}$. Since $\det\colon \mathrm{M}(n) \to \mathbb{R}$ is continuous, being a polynomial in the entries of the matrix, and since $\{0\}$ is a closed subset of $\mathbb{R}$, the subset $\det^{-1}(0)$ is closed in $\mathrm{M}(n)$. Thus, $\mathrm{GL}(n)$ is open in $\mathrm{M}(n)$. By Theorem 1.16, neither $\mathrm{M}(n)$ nor $\mathrm{GL}(n,\mathbb{R})$ is compact, since they are unbounded and $\mathrm{GL}(n,\mathbb{R})$ is not closed. While $\mathrm{M}(n)$ is path-connected, we will show in Exercise 1.9 that $\mathrm{GL}(n,\mathbb{R})$ is not connected.

Matrices with complex entries, however, behave better with respect to connectedness as we will now show. Let $\mathrm{M}(n,\mathbb{C})$ denote the set of $n \times n$-matrices with entries in the field of complex numbers $\mathbb{C}$. We consider $\mathbb{C}$ as a topological space by identifying it with $\mathbb{R}^2$, and we then consider $\mathrm{M}(n,\mathbb{C})$ as a topological space by identifying it with $\mathbb{R}^{n^4}$. Let $\mathrm{GL}(n,\mathbb{C}) = \{A \in \mathrm{M}(n,\mathbb{C}) : \det A \neq 0\}$ denote the *complex general linear group* consisting of all invertible $n \times n$-matrices with entries in $\mathbb{C}$. Again, $\mathrm{GL}(n,\mathbb{C})$ is an open subspace of $\mathrm{M}(n,\mathbb{C})$.

Proposition 1.37 *The space* $\mathrm{GL}(n,\mathbb{C})$ *is path-connected.*

Proof: Let $A \in \mathrm{GL}(n,\mathbb{C})$. It suffices to show that there is a path from A to the identity matrix I in $\mathrm{GL}(n,\mathbb{C})$. This will prove that there is a path between any two points in $\mathrm{GL}(n,\mathbb{C})$ since we can connect paths and reverse their direction as explained in Remark 1.35. First, we define the map

$$P\colon \mathbb{C} \to \mathbb{C},\ z \mapsto \det(A + z(I - A)).$$

Since P is a polynomial of degree n, it has only finitely many zeros. We also note that neither $0 \in \mathbb{C}$ nor $1 \in \mathbb{C}$ are zeros of P, since $P(0) = \det A \neq 0$ and $P(1) = \det I = 1 \neq 0$. Since $\mathbb{C} \setminus \{\text{finitely many points}\}$ is path-connected (see Figure 1.7), we can find a path $\gamma\colon [0,1] \to \mathbb{C}$ with $\gamma(0) = 0$, $\gamma(1) = 1$ which avoids the zeros of P, i.e., $P(\gamma(t)) \neq 0$ for all t. Then the continuous map

$$[0,1] \to \mathrm{GL}(n,\mathbb{C}),\ t \mapsto A + \gamma(t)(I - A)$$

is the desired path from A to I. □

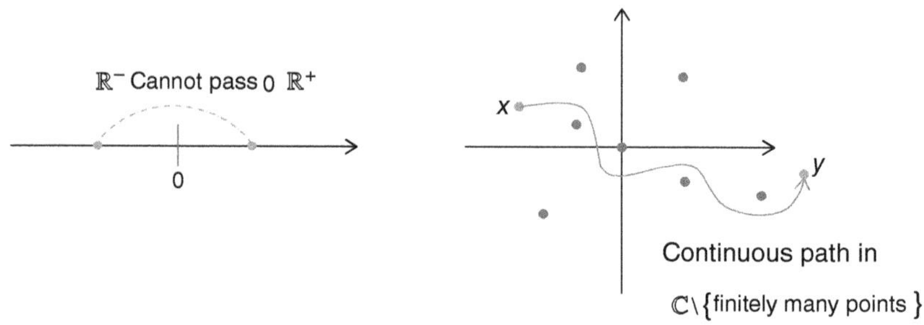

Figure 1.7 Two points in $\mathrm{GL}(n,\mathbb{R})$ may not be connected. In $\mathrm{GL}(n,\mathbb{C})$, however, we can always find a path between two matrices. This reduces, via the determinant function, to the fact that the plane remains path-connected after removing finitely many points.

We see the difference between $\mathrm{GL}(n,\mathbb{R})$ and $\mathrm{GL}(n,\mathbb{C})$ most clearly for the case $n = 1$. While $\mathrm{GL}(1,\mathbb{R}) = \mathbb{R}^{\times}$ is not path-connected since we cannot cross 0, the space $\mathrm{GL}(1,\mathbb{C}) = \mathbb{C}^{\times}$ is indeed path-connected since we can walk around 0 in the plane. See Figure 1.7.

Remark 1.38 The fact that $\mathrm{GL}(n,\mathbb{C})$ is path-connected, while $\mathrm{GL}(n,\mathbb{R})$ is not, plays a crucial role for orientations of vector spaces. In fact, every complex vector space has a natural orientation. We will get back to this in Chapter 14.

We will now show that $\mathrm{GL}(n,\mathbb{R})$ has two connected components, one of which is an open subgroup consisting of all matrices A with $\det A > 0$, and the other is the open subset consisting of all matrices A with $\det A < 0$. We first show that a certain subspace of the former is path-connected. Let $\mathrm{O}(n)$ denote the subspace of matrices A of $\mathrm{GL}(n)$ which satisfy $AA^{\top} = I$ where $A^{\top}$ denotes the transpose of A and I is the $n \times n$-identity matrix. Elements in $\mathrm{O}(n)$ are called *orthogonal matrices* and $\mathrm{O}(n)$ is called the *orthogonal group*, since the defining condition means that their columns are orthogonal to each other with respect to the standard inner product on $\mathbb{R}^n$. Note that $\mathrm{O}(n)$ is the subgroup of invertible matrices which preserve the inner product of $\mathbb{R}^n$. In particular, matrices in $\mathrm{O}(n)$ preserve distances and angles in $\mathbb{R}^n$. In Exercise 1.9, we show that $\mathrm{O}(n)$ is a compact space. If $A \in \mathrm{O}(n)$, we know that $\det A = \pm 1$. The matrices in $\mathrm{O}(n)$ with determinant $+1$ have a special name as we will now recall. For $n \geq 1$, the subspace $\mathrm{SO}(n) = \{A \in \mathrm{O}(n) \colon \det A = 1\}$ is called the *special orthogonal group*. We will show that $\mathrm{SO}(n)$ is path-connected. We first consider a special case to illustrate one of the main ideas of the argument. For $n = 1$, the only orthogonal map $\mathbb{R} \to \mathbb{R}$ with determinant $+1$ is the identity map, i.e., we have $\mathrm{SO}(1) = \{I\}$. So let us look at $n = 2$.

Lemma 1.39 *The space* SO(2) *is path-connected.*

Proof: Let $A = \begin{pmatrix} a & b \\ c & d \end{pmatrix} \in \mathrm{SO}(2)$, i.e., we have $\det A = 1$ and

$$AA^\top = \begin{pmatrix} a & b \\ c & d \end{pmatrix}\begin{pmatrix} a & c \\ b & d \end{pmatrix} = \begin{pmatrix} a^2+b^2 & ab+cd \\ ab+cd & c^2+d^2 \end{pmatrix} = \begin{pmatrix} 1 & 0 \\ 0 & 1 \end{pmatrix}.$$

Hence the entries in A satisfy the equations $\det A = ad - bc = 1$, $a^2 + b^2 = c^2 + d^2 = 1$, and $ab + cd = 0$. This shows that we can find an $\alpha \in [0, 2\pi)$ such that $a = d = \cos\alpha$ and $c = -b = \sin\alpha$ and $A = \begin{pmatrix} \cos\alpha & -\sin\alpha \\ \sin\alpha & \cos\alpha \end{pmatrix}$. In other words, A acts on $\mathbb{R}^2$ as a rotation by the angle α. This gives us a simple way to define a continuous path $\gamma\colon [0,1] \to \mathrm{SO}(2)$, which starts at the identity matrix and ends at A by gradually increasing the angle of the rotation, i.e., we define γ by $\gamma(t) := \begin{pmatrix} \cos(t\alpha) & -\sin(t\alpha) \\ \sin(t\alpha) & \cos(t\alpha) \end{pmatrix}$. □

Now, we look at the general case.

Proposition 1.40 *The space* SO(n) *is path-connected.*

Proof: Let $A \in \mathrm{SO}(n)$. We will show that there is a continuous path γ in $\mathrm{SO}(n)$ which starts at A and ends at the identity matrix $I \in \mathrm{SO}(n)$. The key point is that we can describe A using rotations as we did in the proof of Lemma 1.39. We recall from linear algebra that there is an $n \times n$-matrix P such that $P^\top P = I$ and $A = P^\top Q P$ with

$$Q = \begin{pmatrix} R_1 & & & & \\ & \ddots & & & \\ & & R_k & & \\ & & & 1 & \\ & & & & \ddots & \\ & & & & & 1 \end{pmatrix} \quad \text{where } R_i = \begin{pmatrix} \cos\alpha_i & -\sin\alpha_i \\ \sin\alpha_i & \cos\alpha_i \end{pmatrix}.$$

The key is that each R_i is a rotation matrix. We can then define a continuous path $\gamma\colon [0,1] \to \mathrm{SO}(n)$ by $\gamma(t) := P^\top Q(t) P$ where we obtain $Q(t)$ from Q by replacing R_i by $R_i(t) = \begin{pmatrix} \cos(t\alpha_i) & -\sin(t\alpha_i) \\ \sin(t\alpha_i) & \cos(t\alpha_i) \end{pmatrix}$. We then get $\gamma(0) = I_n$ and $\gamma(1) = A$. □

The final step is to use the Gram–Schmidt process to find a path from an arbitrary invertible matrix to a matrix with orthogonal columns.

Proposition 1.41 *The space* $\mathrm{GL}(n,\mathbb{R})^+$ *of matrices which have strictly positive determinant is path-connected.*

Proof: Let $A \in \mathrm{GL}(n,\mathbb{R})^+$. By Remark 1.35 and Proposition 1.40, it suffices to show that there is a continuous path $\gamma\colon [0,1] \to \mathrm{GL}(n,\mathbb{R})^+$ with $\gamma(0) = A$ and $\gamma(1) = A'$ for some $A' \in \mathrm{SO}(n)$. For then we find a path from A' to the identity matrix I_n within $\mathrm{SO}(n) \subset \mathrm{GL}(n,\mathbb{R})^+$. Let $\{e_1, \ldots, e_n\}$ denote the standard basis of $\mathbb{R}^n$, and let $v_i := A(e_i)$ for $i = 1, \ldots, n$. Then $\{v_1, \ldots, v_n\}$ is again a basis of $\mathbb{R}^n$ since A is invertible. Now, we recall that we can use the Gram–Schmidt process to transform $\{v_1, \ldots, v_n\}$ into an orthogonal basis $\{w_1, \ldots, w_n\}$ as follows. We set $w_1 := v_1$, and, for $2 \le k \le n$, we set $w_k := v_k - \sum_{i=1}^{k-1} \frac{\langle v_k, w_i \rangle}{\langle w_i, w_i \rangle} w_i$. We turn this construction into a continuous path of vectors $w_k\colon [0,1] \to \mathbb{R}^n$ by setting $w_1(t) := v_1$, and

$$w_k(t) := v_k - t \cdot \sum_{i=1}^{k-1} \frac{\langle v_k, w_i \rangle}{\langle w_i, w_i \rangle} w_i$$

for $2 \le k \le n$, where $\langle v, w \rangle$ denotes the standard inner product on $\mathbb{R}^n$. We then have $w_k(0) = v_k$ and $w_k(1) = w_k$. Now we normalise each $w_k(t)$ and set $u_k(t) := w_k(t)/|w_k(t)|^t$ for $1 \le k \le n$. Note that we need the exponent t in the denominator to make sure that $u_k(0) = v_k$. For each k, the assignment $t \mapsto u_k(t)$ defines a continuous map $[0,1] \to \mathbb{R}^n$. We let $A'(t)$ denote the $(n \times n)$-matrix with kth column given by $u_k(t)$. For each t and k, we have $v_k \in \mathrm{Span}\{u_1(t), \ldots, u_k(t)\}$. This shows that, for every $t \in [0,1]$ and each $1 \le k \le n$, we have $\mathrm{Span}\{u_1(t), \ldots, u_k(t)\} = \mathrm{Span}\{v_1, \ldots, v_k\}$. This implies, in particular, that $\{u_1(t), \ldots, u_n(t)\}$ is a basis of $\mathbb{R}^n$ and hence $\det A'(t) \neq 0$ for every $t \in [0,1]$. Since $\det A > 0$ by assumption and since both $t \mapsto A'(t)$ and $\det\colon M(n) \to \mathbb{R}$ are continuous maps, we must have $\det A'(t) > 0$ for all $t \in [0,1]$ by the Intermediate Value Theorem in calculus. Thus, the assignment $t \mapsto \gamma(t) := A'(t)$ defines a continuous map $\gamma\colon [0,1] \to \mathrm{GL}(n,\mathbb{R})^+$. Finally, by construction, $\{u_1(1), \ldots, u_n(1)\}$ is an orthonormal basis of $\mathbb{R}^n$. Thus, $A' := A'(1) \in \mathrm{SO}(n)$ is the matrix we were looking for, together with the continuous path γ in $\mathrm{GL}(n,\mathbb{R})^+$ such that $\gamma(0) = A$ and $\gamma(1) = A' \in \mathrm{SO}(n)$. □

1.4 Exercises

1.1 Prove the assertion of Lemma 1.13, i.e., show that if $f\colon X \to Y$ is a continuous map and $A \subset X$ is a compact subspace, then $f(A)$ is compact. Hint: Begin with an arbitrary open covering of $f(A)$, apply f^{-1} to the open sets and use the assumption on A.

1.2 Let $f\colon X \to Y$ be a continuous map between topological spaces. Show that if f is closed, i.e., f sends closed subsets in X to closed subsets in

Y, then, for all $y \in Y$ and open subset $W \subseteq X$ with $f^{-1}(y) \subseteq W$, there is an open neighbourhood U of y with $f^{-1}(U) \subseteq W$.

1.3 Prove the assertion of Lemma 1.17, i.e., show that every compact and discrete subset S of $\mathbb{R}^n$ is finite. Hint: Assume S is infinite. Use Theorem 1.16 and divide a big box containing S into smaller boxes one of which then contains infinitely many points. Continue to get a Cauchy sequence of points. Use that every convergent series in a closed subspace A in $\mathbb{R}^n$ has a limit in A.

1.4 Prove Lebesgue Number Lemma for $[0,1] \subset \mathbb{R}$: Show that for every open cover $\{I_\alpha\}_\alpha$ of $[0,1]$ there is a *Lebesgue number*, i.e., a real number $\lambda > 0$ such that every subinterval of length λ lies in one of the I_α. Deduce that $[0,1]$ is compact. Hint: Prove the existence of a Lebesgue number by assuming there is no such number and deriving a contradiction. You may want to use that Cauchy sequences in $[0,1]$ have a limit in $[0,1]$.

1.5 Now, we prove a more general form of Lebesgue Number Lemma. Let $X \subset \mathbb{R}^n$ be a compact subset. Show that for every open cover $\{U_\alpha\}_\alpha$ of X, there is a *Lebesgue number*, i.e., a real number $\lambda > 0$ such that every subset C of X with diameter less than λ, i.e., $|x-y| < \lambda$ for all $x, y \in C$, satisfies $C \subset U_\alpha$ for some α. Hint: Use that the U_α are open and define a new cover with open balls around all points in X. Use the compactness to reduce to a finite cover with open balls. Then consider the minimum of the finitely many radii of these balls.

1.6 Let X be a topological space and let $\Delta := \{(x,x) \in X \times X : x \in X\}$ be the diagonal. We equip $X \times X$ with the product topology. Show that X is Hausdorff if and only if Δ is closed in $X \times X$.

1.7 Let X be a topological space and let R be an equivalence relation on X. Let X/R denote the set of equivalence classes and let $q : X \to X/R$ be the quotient map. We equip X/R with the quotient topology, i.e., a subset U in X/R is open if and only if $q^{-1}(U)$ is open in X. Let $\Gamma(R) := \{(x,y) \in X \times X : x \sim_R y\}$. Assume that q is an open map and that $\Gamma(R)$ is closed. Show that X/R is Hausdorff. Hint: Consider the map $\widetilde{q} := q \times q : X \times X \to (X/R) \times (X/R)$ and use Exercise 1.6.

1.8 The subspace $\mathrm{SL}(n)$ of $(n \times n)$-matrices with determinant $+1$ forms a subspace of $\mathrm{GL}(n,\mathbb{R})$. Show that $\mathrm{SL}(n)$ is a closed subspace of $\mathrm{GL}(n,\mathbb{R})$. Is $\mathrm{SL}(n)$ also closed in $\mathrm{M}(n)$?

1.9 Show that $\mathrm{O}(n)$ is compact. Is $\mathrm{SO}(n)$ compact? Hint: Show that if $A = (a_{ij})$ lies in $\mathrm{O}(n)$, then for each i, $\sum_j a_{ij}^2 = 1$. Consider the continuous map $\mathrm{M}(n) \to \mathrm{M}(n)$ sending A to $AA^\top$.

1.10 Show that $\mathrm{GL}(n,\mathbb{R})$ is not connected. Hint: Use Proposition 1.36 and show that $\mathrm{GL}(n,\mathbb{R})$ is not path-connected by using the Intermediate Value Theorem in calculus.

2
Smooth Manifolds

We begin this chapter by introducing smooth maps between subspaces in Euclidean space. We will then define smooth manifolds and smooth maps between manifolds and look at some first examples. Then we will introduce tangent spaces and the induced derivative of smooth maps. We end the chapter with a longer discussion of an example of a manifold, the tangent bundle.

2.1 Smooth Maps

In this section we will introduce smooth, i.e., infinitely often differentiable, maps between subspaces in Euclidean space. This will prepare us for the definition of smooth manifolds. We first recall the familiar case of maps which are differentiable on an open subset. Then we will explain how we extend the definition to arbitrary subspaces.

2.1.1 Maps on Open Domains

In differential topology, we typically require maps to be not just once differentiable, but infinitely differentiable. This condition is extremely useful, though at first it may seem rather strict. However, as we will see later, it is not as restrictive as it might appear.

Definition 2.1 (Smooth maps on open subsets) Let $U \subseteq \mathbb{R}^n$ and $V \subseteq \mathbb{R}^m$ be open sets, and $k \geq 0$ be an integer. A continuous map $f: U \to V$ is said to be $\mathcal{C}^0$, and f is said to be $\mathcal{C}^k$ if, at every point $a \in U$, the partial derivatives $\frac{\partial^k f_i}{\partial x_{j_1} \dots \partial x_{j_k}}(a)$ of f of order k exist and are continuous. A map $f: U \to V$ is said to be $\mathcal{C}^\infty$ and henceforth called *smooth* if, at every point $a \in U$, the partial derivatives of f of all orders exist and are continuous, i.e., all the partial derivatives $\frac{\partial^k f_i}{\partial x_{j_1}\dots\partial x_{j_k}}(a)$ exist and are continuous for all $k \geq 1$.

Examples 2.2 Familiar maps like exp, cos, sin, and all polynomials are smooth maps from $\mathbb{R}$ to $\mathbb{R}$. More generally, let $p_1, p_2, \dots, p_m$ be polynomials with coefficients in $\mathbb{R}$ in the variables $x_1, x_2, \dots, x_n$. They induce a smooth map $P\colon \mathbb{R}^n \to \mathbb{R}^m$ by evaluating $p_1, \dots, p_m$ on the coordinates of $\mathbb{R}^n$. The partial derivative $\frac{\partial^k P_i}{\partial x_{j_1}\dots\partial x_{j_k}}(a)$ is just the partial derivative of the polynomial p_i. The latter always exists and is continuous. For example, the map $\mathbb{R}^2 \to \mathbb{R}^2$ defined by $(x, y) \mapsto (x^2 - y^2, 2xy)$ is smooth. Another example is the smooth map

$$F\colon \mathbb{R}^4 \to \mathbb{R}^3, \quad \begin{pmatrix} x_1 \\ x_2 \\ x_3 \\ x_4 \end{pmatrix} \mapsto \begin{pmatrix} 2x_1x_3 + 2x_2x_4 \\ 2x_2x_3 - 2x_1x_4 \\ x_1^2 + x_2^2 - x_3^2 - x_4^2 \end{pmatrix}. \tag{2.1}$$

2.1.2 Extension to Arbitrary Subsets

We now want to extend the notion of smoothness to maps between arbitrary subsets of Euclidean spaces. This requires some care. In calculus, we learned what it means for a function $f\colon (a-\varepsilon, a+\varepsilon) \to \mathbb{R}$ defined on an open interval to be differentiable at the point a. However, this definition only makes sense if there is room on both sides of a, i.e., if $\varepsilon > 0$. For example, we cannot speak of differentiability for a function $f\colon [a, a+\varepsilon) \to \mathbb{R}$ at a, because the limit requires approaching a from both directions. This is why we usually require maps to be defined on an open neighbourhood of the point a. Fortunately, there is a way to overcome this limitation. Let $X \subset \mathbb{R}^n$ be an arbitrary subset of $\mathbb{R}^n$, and let $f\colon X \to \mathbb{R}^m$ be a map. To say f is *smooth at* $a \in X$, we require that f is the restriction of a map defined on an open ball $\mathbb{B}^n_\varepsilon(a)$ around a in $\mathbb{R}^n$. This simple idea allows the entire machinery of smoothness to work nicely.

Definition 2.3 (Smooth maps on arbitrary subsets) Let $X \subseteq \mathbb{R}^n$ be an arbitrary subset. A map $f\colon X \to \mathbb{R}^m$ is called *smooth* if, for each $x \in X$, there exists an open subset $U \subseteq \mathbb{R}^n$ containing x and a smooth map $F\colon U \to \mathbb{R}^m$ that coincides with f on all of $X \cap U$, i.e., $F_{X\cap U} = f_{X\cap U}$.

Note that smoothness at a point x is by definition a local property, i.e., we need to check it only in a small neighbourhood of x.

Examples 2.4 The identity map of any set X is smooth. If $f\colon X \to Y$ and $g\colon Y \to Z$ are smooth, then the composition $g \circ f$ is also smooth. The projection map $\mathbb{S}^1 \to \mathbb{R}$, $(x, y) \mapsto x$, is smooth since it can be extended to the projection $\mathbb{R}^2 \to \mathbb{R}$ onto the first coordinate which is smooth. Let $\hbar \colon \mathbb{S}^3 \to \mathbb{S}^2$ be the *Hopf map* defined by sending (x_1, x_2, x_3, x_4) to

$$\begin{pmatrix} 2x_1x_3 + 2x_2x_4 \\ 2x_2x_3 - 2x_1x_4 \\ x_1^2 + x_2^2 - x_3^2 - x_4^2 \end{pmatrix}.$$

This map is the restriction of the map F defined in (2.1) to $\mathbb{S}^3 \subset \mathbb{R}^4$. We verify in Exercise 2.10 that, for $x \in \mathbb{S}^3$, we have $F(x) = \hbar(x) \in \mathbb{S}^2 \subset \mathbb{R}^3$. Since F is smooth, we deduce that $\hbar$ is smooth.

2.1.3 Diffeomorphisms

A bijective smooth map has an inverse as a map of sets. We are interested in the maps for which this inverse is smooth as well.

Definition 2.5 (Diffeomorphism) Let $X \subseteq \mathbb{R}^n$ and $Y \subseteq \mathbb{R}^m$ be arbitrary subsets. A smooth map $f \colon X \to Y$ is called a *diffeomorphism* if f is bijective and its inverse f^{-1} is smooth as well. We say that X and Y are *diffeomorphic* if there exists a diffeomorphism $f \colon X \to Y$.

Note that every diffeomorphism is a homeomorphism, but not the other way around. Here are some concrete examples. For the first example, we note that the Inverse Function Theorem from calculus provides a simple way to recognise a diffeomorphism. We will discuss the Inverse Function Theorem in a more general form in Section 3.1.

Example 2.6 The map $f \colon \mathbb{R} \to \mathbb{R}$, $f(x) = x^3 + x$, is a diffeomorphism since $f'(x) = 3x^2 + 1 > 0$ for all $x \in \mathbb{R}$. However, $f \colon \mathbb{R} \to \mathbb{R}$, $x \mapsto x^3$, is a homeomorphism but not a diffeomorphism, since the inverse map is not differentiable and therefore not smooth at the origin.

Example 2.7 The map $\mathbb{R}^2 \setminus \{(0,0)\} \to \mathbb{R}^2 \setminus \{(0,0)\}$, $(x,y) \mapsto (x^2 - y^2, 2xy)$, is not a diffeomorphism even though its derivative is invertible everywhere because it is not one-to-one.

Example 2.8 Let $\mathbb{S}^2 = \{(x,y,z) \in \mathbb{R}^3 : x^2 + y^2 + z^2 = 1\}$ be the two-dimensional sphere. The map $f \colon \mathbb{S}^2 \setminus \{(0,0,1)\} \to \mathbb{R}^2$ defined by $(x,y,z) \mapsto \frac{1}{1-z}(x,y)$ is a diffeomorphism.

Example 2.9 The argument of the proof of Lemma 1.39 provides a diffeomorphism $\mathbb{S}^1 \to \mathrm{SO}(2)$ defined by $(\cos t, \sin t) \mapsto \begin{pmatrix} \cos t & -\sin t \\ \sin t & \cos t \end{pmatrix}$.

Example 2.10 The Hopf map $\hbar\colon \mathbb{S}^3 \to \mathbb{S}^2$ defined by sending (x_1, x_2, x_3, x_4) to

$$\begin{pmatrix} 2x_1x_3 + 2x_2x_4 \\ 2x_2x_3 - 2x_1x_4 \\ x_1^2 + x_2^2 - x_3^2 - x_4^2 \end{pmatrix}$$

is not a diffeomorphism, since it is not one-to-one. For example, the unit circle in the x_1-x_2-plane on $\mathbb{S}^3$, i.e., points on $\mathbb{S}^3$ with $x_3 = x_4 = 0$, is mapped to the north pole $(0, 0, 1)$ on $\mathbb{S}^2$. The unit circle in the x_3-x_4-plane on $\mathbb{S}^3$, i.e., points on $\mathbb{S}^3$ with $x_1 = x_2 = 0$, is mapped to the south pole $(0, 0, -1)$ on $\mathbb{S}^2$. In fact, we will see in Exercise 2.10 that each fibre of $\hbar$, i.e., the preimage $\hbar^{-1}(y)$ of a $y \in \mathbb{S}^2$, is diffeomorphic to a circle on $\mathbb{S}^3$. We will meet the Hopf map $\hbar$ many times in this book. In particular, we will show later that the fibres of $\hbar$ are all linked to each other. This is a fascinating and very rare phenomenon. More on this later. For the moment, we conclude this example with the observation that after defining tangent spaces for manifolds we will see that, in fact, there cannot exist a diffeomorphism between $\mathbb{S}^3$ and $\mathbb{S}^2$.

2.2 Smooth Manifolds

We are now ready to introduce smooth manifolds. To motivate the definition, we begin with a guiding question.

2.2.1 How to Describe a Space of Solutions?

Many interesting spaces are given as the set of solutions of an equation of the form $f(x) = b$ where $f\colon \mathbb{R}^n \to \mathbb{R}^m$ is a map and $b \in \mathbb{R}^m$ is some point.

Question Can we describe the space of solutions $X = \{x \in \mathbb{R}^n : f(x) = b\}$ in a way that is both simple and efficient, while still capturing all of its essential and interesting properties?

Let us look at a familiar situation and consider the linear map $A\colon \mathbb{R}^4 \to \mathbb{R}^3$ which sends $x = (x_1, x_2, x_3, x_4)$ to

$$\begin{pmatrix} 2x_2 + 2x_4 \\ 2x_3 - 2x_1 \\ x_1 + x_2 - x_3 - x_4 \end{pmatrix}.$$

Suppose we would like to solve the equation $A(0) = 0$. We can, for example, use Gauss elimination to get the set of solutions $L = \{x \in \mathbb{R}^4 : x_2 = 0, x_3 = x_1, x_4 = -x_2\} \subset \mathbb{R}^4$. This is a one-dimensional vector subspace of $\mathbb{R}^4$ and we can describe all the points in L by using just one variable, say t: $L = \{x \in \mathbb{R}^4 : x = t \cdot (1,0,1,0)\}$. We think of the variable t as a parameter and would like to say that t parametrises the set of solutions L. In a more formal way, we have the linear isomorphism $\psi : \mathbb{R} \to L$ defined by $t \mapsto t \cdot (1,0,1,0)$. We may think of ψ as a way to express the one-dimensionality of L, and that L has a linear structure. We call the map ψ a *parametrisation of* L.

Now we look at the map $f : \mathbb{R}^4 \to \mathbb{R}^3$ defined by sending (x_1, x_2, x_3, x_4) to

$$\begin{pmatrix} 2x_1x_3 + 2x_2x_4 \\ 2x_2x_3 - 2x_1x_4 \\ x_1^2 + x_2^2 - x_3^2 - x_4^2 \end{pmatrix}.$$

This map is not linear, since we multiply variables. However, f is a smooth map since each of its coordinates is a polynomial. The equation $f(x) = 0$ has only a single solution, namely the zero vector. So, let us rather determine the solutions of the equation $f(x) = b = (0,1,0)$. After some calculation we obtain the set of solutions

$$X = \{x \in \mathbb{R}^4 \colon x_3 = x_2, x_4 = -x_1, x_1^2 + x_2^2 = 1/2\}.$$

The set X is not a straight line in $\mathbb{R}^4$. However, it looks like something one-dimensional since knowing one of the variables, say x_1, should determine x_2, x_3, and x_4, and thereby x. Let us follow this idea a bit further and fix a value $x_1 = t$. Then we get $1/2 = x_1^2 + x_2^2 = t^2 + x_2^2$, i.e., $x_2 = \pm\sqrt{1/2 - t^2}$. Hence x_2 is only determined up to a choice. To remedy this defect, let us focus on points $x \in X$ with $x_2 > 0$; then $x_1 = t$ determines x completely. We write $X_{x_2>0}$ for the subset of such points. In addition, we need to make sure that the square root is defined, i.e., we need that t only varies in the range $t \in (-\sqrt{1/2}, \sqrt{1/2})$. Hence we can use the map $\phi_+ \colon (-\sqrt{1/2}, \sqrt{1/2}) \to X_{x_2 \geq 0} \subset \mathbb{R}^4$ defined by $\phi_+(t) := (t, \sqrt{1/2 - t^2}, \sqrt{1/2 - t^2}, -t)$ to describe one part of the set of solutions X. We can check that this map is a bijection. This is very similar to the parametrisation we used to describe L. However, this map is far from linear. There is no way to fix this, since f was not linear in the first place. However, ϕ_+ is smooth. In fact, ϕ_+ is a diffeomorphism, and this is the best we can hope for. Hence ϕ_+ does a similar job as the parametrisation ψ above. For, ϕ_+ expresses the one-dimensionality of X for points in $X_{x_2>0}$, and that X has a smooth structure[1] for points in $X_{x_2>0}$. However, since ϕ_+ describes only some part of X, we call the map ϕ_+ a *local parametrisation of* X. Now we should take care of the points of X that ϕ_+ is not describing, in particular where $x_2 \leq 0$. We do this by defining further local parametrisations similar to

[1] This is not yet a well-defined term. For the moment we may think of it as saying that X does not have any corners or any other nasty points.

ϕ_+ to cover all of X. The collection of such local parametrisations will then express X as a one-dimensional smooth subspace of $\mathbb{R}^4$. In fact, these maps give X the structure of a one-dimensional smooth manifold, a notion we will now define rigorously based on what we learned from this example.

2.2.2 Smooth Manifolds and Smooth Maps

We recall our working definition of a k-dimensional smooth manifold X as a space that locally resembles k-dimensional Euclidean space. The preceding discussion now shows how we can give this idea a precise mathematical meaning.

Definition 2.11 (Smooth manifolds) Let $X \subset \mathbb{R}^N$ be a subset of Euclidean space for some N. Then X is called a *smooth manifold of dimension k* if, for every point $x \in X$, there is an open subset $U \subseteq \mathbb{R}^k$, an open subset $V \subset X$ containing x and a diffeomorphism $\phi\colon U \to V$. Any such diffeomorphism $\phi\colon U \to V$ is called a *local parametrisation around x*.

We note that, since $V \subset X \subset \mathbb{R}^N$ and $U \subset \mathbb{R}^k$, we know from Definition 2.3 what it means that $\phi\colon U \to V$ and $\phi^{-1}\colon V \to U$ are smooth.

Remark 2.12 When $\phi\colon U \to V$ is a local parametrisation, we call the inverse diffeomorphism $\phi^{-1}\colon V \to U$ a *local coordinate system.* We may then talk about the coordinates $(\phi_1^{-1}(x), \phi_2^{-1}(x), \ldots, \phi_k^{-1}(x))$ of a point $x \in V \subset X$. Writing $u_i(x) = \phi_i^{-1}(x)$ for $i = 1, \ldots, k$, we sometimes omit mentioning ϕ^{-1} and just talk about the coordinates $(u_1(x), u_2(x), \ldots, u_k(x))$ of x. Here we need to remember that the $u_1, \ldots, u_k$ are smooth functions which depend on x.

The definition of a smooth manifold as a subspace of $\mathbb{R}^N$ makes it easy to talk about smooth maps.

Definition 2.13 (Smooth maps) Let $X \subset \mathbb{R}^N$ and $Y \subset \mathbb{R}^M$ be two smooth manifolds. Then a *smooth map* $f\colon X \to Y$ between the manifolds X and Y is just a smooth map in the sense of Definition 2.3.

Notation 2.14 Let $X \subset \mathbb{R}^N$ be a smooth manifold of dimension k and let $x \in X$ be a point. Let $\phi\colon U \to V$ be a local parametrisation around x, i.e., $U \subseteq \mathbb{R}^k$ and $V \subseteq X$ are open subsets with $x \in V$ and $\phi\colon U \to V$ is a diffeomorphism. Then we also write $\phi\colon U \to X$ for the composite $U \xrightarrow{\phi} V \hookrightarrow X$. We often assume that ϕ is adjusted such that $\phi(0) = x$.

We note that, in general, there are many different ways to choose parametrisations for a smooth manifold X. The definition only requires that it is possible to find sufficiently many local parametrisations to cover all of X. The natural number N in Definition 2.11 is not important for the definition either. We just assume that there is some $\mathbb{R}^N$ big enough to fit X into it. In particular, if $X \subset \mathbb{R}^N$, then also have $X \subset \mathbb{R}^{N+n}$ for every $n \geq 0$. It is, however, an interesting problem to find the minimal N for a given X. We will get back to this question in Section 7.2.

2.2.3 First Examples

We are already familiar with some examples of smooth manifolds.

Examples 2.15 Euclidean space $\mathbb{R}^k$ is a k-dimensional smooth manifold. An open subset $U \subseteq \mathbb{R}^k$ is a k-dimensional manifold. The identity map $U \to U$ is a parametrisation for all of U. For instance, every k-dimensional open ball $\mathbb{B}_r^k(x)$ around some point $x \in \mathbb{R}^k$ is a smooth manifold of dimension k. A zero-dimensional manifold M just consists of a collection of discrete points. Given $x \in M$, the set $\{x\} \subset M$ consisting of x alone is open in M and is diffeomorphic to the one-point set $\mathbb{R}^0 = \{0\}$.

Example 2.16 (The unit circle) Let $\mathbb{S}^1 := \{(x, y) \in \mathbb{R}^2 : x^2 + y^2 = 1\} \subset \mathbb{R}^2$ be the *unit circle*. We will now show that $\mathbb{S}^1$ is a one-dimensional smooth manifold. First, suppose that (x, y) lies in the upper semicircle where $y > 0$. Then $\phi_1(x) = (x, \sqrt{1 - x^2})$ maps the open interval $W = (-1, 1) \subset \mathbb{R}$ bijectively onto the upper semicircle, and it is a smooth map. For the latter, it is important that we do not include the endpoints of the interval $(-1, 1)$. Its inverse is the projection map $\phi_1^{-1}(x, y) = x$ defined on the upper semicircle. This ϕ_1^{-1} is smooth, since it extends to a smooth map of all of $\mathbb{R}^2$ to $\mathbb{R}^1$. Therefore, ϕ_1 is a local parametrisation. A local parametrisation of the lower semicircle where $y < 0$ is similarly defined by $\phi_2(x) = (x, -\sqrt{1 - x^2})$ with inverse $\phi_2^{-1}(x, y) = x$. These two maps give local parametrisations of $\mathbb{S}^1$ around all points except the two points $(1, 0)$ and $(-1, 0)$. To cover these points, we can use the maps $\phi_3(y) = (\sqrt{1 - y^2}, y)$ and $\phi_4(y) = (-\sqrt{1 - y^2}, y)$ which map W to the right and left semicircles, respectively. This shows that $\mathbb{S}^1$ is a one-dimensional manifold.

Example 2.17 (n-sphere) More generally, we will show in the exercises that the *n-dimensional sphere* $\mathbb{S}^n := \{x \in \mathbb{R}^{n+1} : |x| = 1\} \subset \mathbb{R}^{n+1}$ is an n-dimensional smooth manifold.

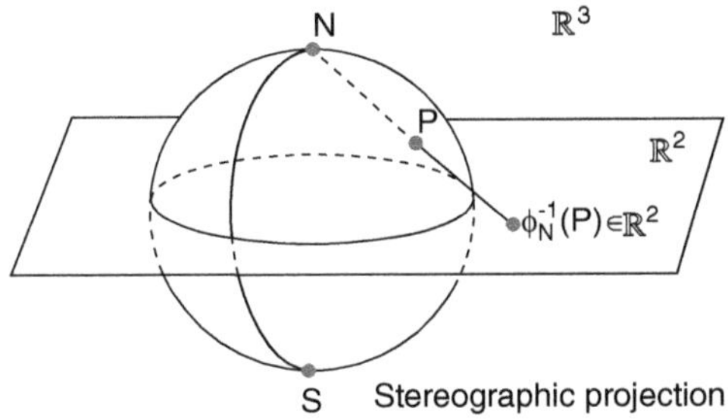

Figure 2.1 The stereographic projection is a diffeomorphism between $\mathbb{S}^2 \setminus \{N\}$ and $\mathbb{R}^2$.

Another example is the set of solutions $X \subset \mathbb{R}^4$ of the equation $f(x) = b$ that we have seen at the beginning of the chapter. We can check that the map ϕ_+ we defined is a local parametrisation of X. Using what we learned from the local parametrisations of $\mathbb{S}^1$ we can write down the missing local parametrisations for X by adjusting for how X sits inside $\mathbb{R}^4$. We will see in Chapter 4 that, in fact, many smooth manifolds arise as the set of solutions of a suitable equation involving a smooth map.

Remark 2.18 The definition of a manifold requires local parametrisations that together cover the entire space. A natural question is: what is the minimal number of parametrisations needed? The answer, of course, depends on the manifold in question. So, let X be a smooth k-dimensional manifold. If only one local parametrisation suffices, then X is diffeomorphic to an open subset of $\mathbb{R}^k$. We can then conclude that if X is compact then we cannot cover X by just one local parametrisation, since there is no homeomorphism between a compact space and an open subset of $\mathbb{R}^k$ by Theorem 1.16. For example, we cannot cover the n-sphere with just one local parametrisation. However, stereographic projection yields a cover of the n-sphere with only two local parametrisations, see Figure 2.1. In Exercise 2.9, we find the formulae for the corresponding diffeomorphisms.

Remark 2.19 (Product of manifolds) Let $X \subset \mathbb{R}^N$ and $Y \subset \mathbb{R}^M$ be smooth manifolds of dimensions k and n, respectively. Then the *product* $X \times Y \subseteq \mathbb{R}^{N+M}$ is a manifold of dimension $k+n$. For, let $\phi\colon W \to X$ a local parametrisation around $x \in X$, and $\psi : U \to Y$ a local parametrisation around $y \in Y$. Then we can use the map $\phi \times \psi : W \times U \to X \times Y$ from the open subset $W \times U$ in $\mathbb{R}^k \times \mathbb{R}^n = \mathbb{R}^{k+n}$ to $X \times Y$ defined by $(\phi \times \psi)(w, u) = (\phi(w), \psi(u))$ as a local parametrisation around (x, y).

Example 2.20 (Torus) The two-dimensional torus is the product $\mathbb{T}^2 = \mathbb{S}^1 \times \mathbb{S}^1$. By Remark 2.19, $\mathbb{T}^2$ is a smooth manifold of dimension two. Here, we consider $\mathbb{T}^2$ as a subset of $\mathbb{R}^4$ since $\mathbb{S}^1$ being a subspace of $\mathbb{R}^2$ forces us to take the product of $\mathbb{R}^2$ with itself. We may, however, be more used to visualising the torus as a subspace in three dimensions, and we will find a way to describe $\mathbb{T}^2$ as a subspace in $\mathbb{R}^3$ in Exercise 2.8.

2.2.4 Submanifolds

A natural way to discover new manifolds is to look at subsets of a given manifold. We are interested in those subsets which are themselves.

Definition 2.21 (Submanifolds) Let $X \subset \mathbb{R}^N$ be a smooth manifold, and let $Z \subset X$ be a subset considered as a topological space with the subspace topology induced from X and hence from $\mathbb{R}^N$. If $Z \subset \mathbb{R}^N$ is a smooth manifold, possibly of lower dimension, then Z is called a *submanifold of* X.

Examples 2.22 Every open subset of a smooth manifold is a submanifold. We can consider the equator in $\mathbb{S}^2$ as a copy of $\mathbb{S}^1$ and hence as a submanifold of $\mathbb{S}^2$. More generally, we can consider copies of $\mathbb{S}^1, \mathbb{S}^2, \ldots, \mathbb{S}^{n-1}$ as submanifolds of $\mathbb{S}^n$. Similarly, we have at least two ways of considering a copy of the circle on the two-dimensional torus: once as a horizontal circle, once as a vertical circle.

Let $\mathrm{M}(p,n)$ denote the space of real $(p \times n)$-matrices which we identify with $\mathbb{R}^{pn}$ as before. Let $\mathrm{M}(p,n;k)$ denote the subspace of matrices of rank k.

We will now show that $\mathrm{M}(p,n;k)$ is a smooth manifold. See also Exercise 2.6.

Theorem 2.23 (Matrices of fixed rank form a manifold) *For* $k \leq \min(p,n)$, $\mathrm{M}(p,n;k)$ *is a smooth submanifold of* $\mathrm{M}(n,p)$ *of dimension* $k \cdot (p+n-k)$. *In particular, for* $p \geq n$, $\mathrm{M}(p,n;n)$ *is an open subset of* $\mathrm{M}(p,n)$ *and hence a smooth manifold of dimension* $p \cdot n$.

Proof: Let $E_0 \in M(p,n;k)$. After performing row operations if necessary, we may assume that E_0 is of the form $E_0 = \begin{pmatrix} A_0 & B_0 \\ C_0 & D_0 \end{pmatrix}$, where A_0 is an invertible $k \times k$ matrix. Since $\det A_0 \neq 0$ and $\det\colon \mathrm{M}(k,k) \to \mathbb{R}$ is a function which depends continuously on the entries in a matrix, there is an $\varepsilon > 0$ such that, if the absolute values of all the entries of the matrix $A - A_0$ are less than ε, then $\det A \neq 0$ as well. Let $U \subset \mathrm{M}(p,n)$ denote the open subset consisting of all matrices of the form $E = \begin{pmatrix} A & B \\ C & D \end{pmatrix}$ with the absolute values of all the entries of $A - A_0$ being less than ε.

- Claim: E is in $\mathrm{M}(p,n;k)$ if and only if $D = CA^{-1}B$.

Proof of the claim: The matrix $\begin{pmatrix} A & B \\ XA + C & XB + D \end{pmatrix} = \begin{pmatrix} I_k & 0 \\ X & I_{p-k} \end{pmatrix} \cdot \begin{pmatrix} A & B \\ C & D \end{pmatrix}$ has the same rank as E since we multiply E by a $(p \times p)$-matrix with identity matrices on the diagonal. For $X = -CA^{-1}$, this matrix becomes $\widetilde{E} = \begin{pmatrix} A & B \\ 0 & -CA^{-1}B + D \end{pmatrix}$. If $D = CA^{-1}B$, then the matrix $\widetilde{E}$ has the same rank as A, i.e., it has rank k. This proves one direction of the claim. For the other direction, assume that $D \neq CA^{-1}B$. Then at least one entry in $-CA^{-1}B + D$ is non-zero. This implies that $\widetilde{E}$ has rank strictly bigger than k. Hence neither $\widetilde{E}$ nor E is in $M(p,n;k)$. This proves the claim. □

Let W be the open subset in $\mathrm{M}(p,n)$ consisting of matrices $\begin{pmatrix} A & B \\ C & 0 \end{pmatrix}$ with the absolute values of all the entries of $A - A_0$ being less than ε. We can consider W as an open subset in Euclidean space of dimension $(pn - (p-k)(n-k)) = k(p+n-k)$. By the claim above and since matrix multiplication defines a smooth map, the map $\begin{pmatrix} A & B \\ C & 0 \end{pmatrix} \mapsto \begin{pmatrix} A & B \\ C & CA^{-1}B \end{pmatrix}$ is a diffeomorphism from W onto the open neighbourhood $U \cap \mathrm{M}(p,n;k)$ of E_0. This shows that every point in $\mathrm{M}(p,n;k)$ has an open neighbourhood which is diffeomorphic to an open subset of Euclidean space of dimension $k(p+n-k)$. □

Example 2.24 (A non-example) Finally, we will now discuss the case of a space which is not a manifold. Let X denote the union of the x- and the y-axis in $\mathbb{R}^2$, i.e., $X := \{(x,y) \in \mathbb{R}^2 : xy = 0\} \subset \mathbb{R}^2$. Every other point on X, except possibly from the origin $O := (0,0)$, has an open neighbourhood which is diffeomorphic to an open interval in $\mathbb{R}$. Since no point in $\mathbb{R}^d$ with $d \geq 2$ has an open neighbourhood in $\mathbb{R}^d$ diffeomorphic to an open interval in $\mathbb{R}$,[2] X could only be a one-dimensional manifold. Now let us check the point $O = (0,0)$. If X was a manifold of dimension one, there would be an open subset $V \subseteq X$ around O diffeomorphic to an open interval in $\mathbb{R}$. By definition of open sets in a subset of $\mathbb{R}^2$, there must be an open ball $\mathbb{B}^2_\varepsilon(O)$ such that $\mathbb{B}^2_\varepsilon(O) \cap X$ is

[2] We can prove this fact once we have introduced tangent spaces.

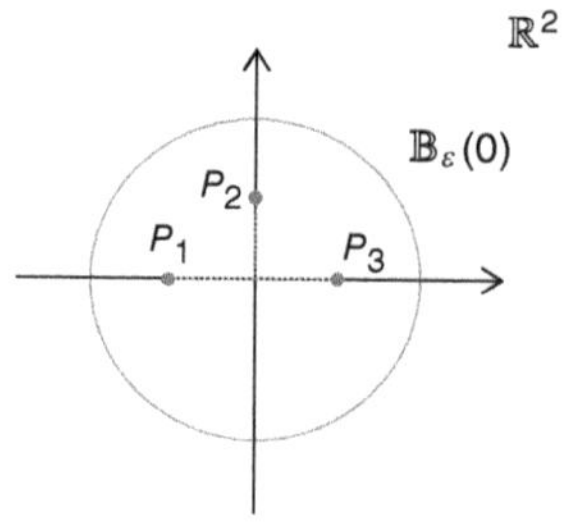

Figure 2.2 Three points on the coordinate axes, two of which can be connected without passing through the other one.

contained in V. Let I be the open interval in $\mathbb{R}$ homeomorphic to $\mathbb{B}^2_\varepsilon(O) \cap X$. The subset $\mathbb{B}_\varepsilon(O) \cap X$ looks like a cross and contains, in particular, the points $P_1 = (-\varepsilon/2, 0)$, $P_2 = (0, \varepsilon/2)$, and $P_3 = (\varepsilon/2, 0)$. In $\mathbb{B}^2_\varepsilon(O) \cap X$, there are paths, i.e., continuous maps $\gamma\colon [0,1] \to \mathbb{B}^2_\varepsilon(O) \cap X$, of the form γ_1 from P_2 to P_3 not passing through P_1, γ_2 from P_1 to P_3 not passing through P_2, and γ_3 from P_1 to P_2 not passing through P_3. However, there is no triple of distinct points with this property in the open interval $I \subset \mathbb{R}$. In more detail, we can argue as follows. Since P_1, P_2, and P_3 are pairwise distinct points, their images under ϕ must be pairwise distinct as well. Since $\mathbb{R}$ is a totally ordered set, we can order these three points. Assume first $\phi(P_1) < \phi(P_2) < \phi(P_3)$. Then the Intermediate Value Theorem of calculus implies that for the path $\phi \circ \gamma_2\colon [0,1] \to \mathbb{B}^2_\varepsilon(O) \cap X$ with $\gamma_2(0) = P_1$ and $\gamma_2(1) = P_3$, there is an $s \in (0,1)$ such that $(\phi \circ \gamma_2)(s) = \phi(P_2)$. This would imply $\gamma_2(s) = P_2$, contradicting the choice of γ_2. Thus the diffeomorphism ϕ with the assumed ordering $\phi(P_1) < \phi(P_2) < \phi(P_3)$ cannot exist. Now we can adjust and repeat this argument for any ordering of the three points $\phi(P_1)$, $\phi(P_2)$, and $\phi(P_3)$ and get contradictions to the choices of paths γ_1, γ_2, and γ_3. Hence the homeomorphism $\phi\colon \mathbb{B}^2_\varepsilon(O) \cap X \to I$ cannot exist. We conclude that O does not have a neighbourhood homeomorphic to an open interval in $\mathbb{R}$, and X is not a smooth manifold (Figure 2.2).

2.3 Tangent Spaces and Derivatives

We now introduce one of the key tools for studying smooth manifolds. To motivate the definition, consider the following. Let $X \subset \mathbb{R}^N$ be a smooth manifold of dimension k, and let x be a point. By definition, we can choose a local parametrisation $\phi\colon U \to X$ around x, where $U \subset \mathbb{R}^k$ is open, so that locally near x, X is the image of U under the diffeomorphism ϕ. Images under diffeomorphisms are convenient, but images under linear maps are even better, since linear spaces are vector spaces, and for these it is much easier to describe concepts such as intersections and orthogonal complements. This raises a natural question: how can we describe X, at least locally near x, using a vector space? Although ϕ itself is not a linear map, we can approximate ϕ by a linear map near x. Then we can use this map to approximate X linearly at x. This leads us to the concept of the tangent space at x.

In order to motivate the construction of the tangent space further, we begin with a familiar situation. Let $f: \mathbb{R} \to \mathbb{R}$ be a smooth function and let $f'(x)$ be the derivative of f at x. In calculus, we think of the derivative often as the slope of the tangent line at the graph of f at the point $(x, f(x))$. In fact, the subset $\Gamma(f) := \{(x, f(x)) \in \mathbb{R}^2 : x \in \mathbb{R}\} \subset \mathbb{R}^2$, called the graph of f, is an example of a smooth manifold as we will now explain. See also Exercise 2.14. We have the natural map $\phi\colon \mathbb{R} \to \Gamma(f)$ given by $x \mapsto (x, f(x))$. The map ϕ is smooth since f is smooth, ϕ is injective because of the first coordinate, and it is surjective by definition of $\Gamma(f)$. Moreover, the projection $(x, f(x)) \mapsto x$ defines a smooth inverse. Thus, ϕ is a diffeomorphism and yields a parametrisation for all points of $\Gamma(f)$. Now, the tangent line at the point $(x, f(x))$ is the prototype of an example of a tangent space. More precisely, we prefer to consider the parallel translate L_x of the tangent line through the origin since we want the tangent space to be a vector space. How can we describe the line L_x passing through the origin? It is determined by its slope, which is given by the derivative of f at x. This means that L_x consists of all points in $\mathbb{R}^2$ of the form $t \cdot \begin{pmatrix} 1 \\ f'(x) \end{pmatrix}$ for all $t \in \mathbb{R}$. Now we observe that $\begin{pmatrix} 1 \\ f'(x) \end{pmatrix}$ is the derivative at x of the map ϕ we defined above, i.e., $d\phi_x = \begin{pmatrix} 1 \\ f'(x) \end{pmatrix}$. Thus, the line L_x is the image in $\mathbb{R}^2$ of the linear map $d\phi_x : \mathbb{R} \to \mathbb{R}^2$.

2.3.1 The Tangent Space

Our goal is to generalise this observation to an arbitrary smooth manifold by following the same recipe, i.e., the tangent space at a point should be the image of the derivative of a local parametrisation at that point. To make this precise, we recall some facts about the derivative of a smooth map. Let $\phi\colon U \to V$ where $U \subset \mathbb{R}^n$ and $V \subset \mathbb{R}^N$ are open subsets. Let $u \in U$ and $h \in \mathbb{R}^n$ be a vector in $\mathbb{R}^n$. Then the derivative of ϕ in the direction h can be defined as the limit

$$d\phi_u(h) = \lim_{t \to 0} \frac{\phi(u+th) - \phi(u)}{t}.$$

For a fixed u, the derivative is a map $d\phi_u \colon \mathbb{R}^n \to \mathbb{R}^N$ sending a vector $h \in \mathbb{R}^n$ to the vector $d\phi_u(h) \in \mathbb{R}^N$. Moreover, we can show or recall from calculus that this map is $\mathbb{R}$-linear, i.e., $d\phi_u(h+g) = d\phi_u(h) + d\phi_u(g)$ and $d\phi_u(\lambda h) = \lambda d\phi_u(h)$ for all $h, g \in \mathbb{R}^n$, and $\lambda \in \mathbb{R}$. In particular, the derivative of ϕ is a map on its own, which is defined on all of $\mathbb{R}^n$ even when ϕ may not be. Recall that we can calculate $d\phi_u$ in the standard bases of Euclidean spaces as the Jacobian matrix. Its entry in row i and column j is the partial derivative $\frac{\partial \phi_i}{\partial x_j}(u)$. The derivative gives us a simple approximation to ϕ at u, i.e., knowing $\phi(u)$ and $d\phi_u$ gives us a good guess for where $\phi(u+h)$ might be, namely a point close to $\phi(u) + d\phi_u(h)$. Now we are ready to define tangent spaces.

Definition 2.25 (Tangent space) Let $X \subset \mathbb{R}^N$ be a k-dimensional smooth manifold and $x \in X$ a point. Let $\phi \colon U \to V$ be a local parametrisation around x with $\phi(u) = x$. We define the *tangent space of X at x*, denoted by $T_x(X)$, to be the vector subspace of $\mathbb{R}^N$ given by the image of the linear map $d\phi_u \colon \mathbb{R}^k \to \mathbb{R}^N$.

By definition, a *tangent vector* at x is a vector $v \in \mathbb{R}^N$ that lies in the subspace $T_x(X) \subset \mathbb{R}^N$. However, we usually imagine v as the arrow pointing from x to $x + v$ in the translate $x + T_x(X)$. To define $T_x(X)$ we made a choice of a parametrisation ϕ. We have to check what happens if we choose a different parametrisation. In more mathematical terms, we need to show that $T_x(X)$ is *well defined*. That is, we need to show that, when we use another local parametrisation at x, we get the same vector subspace in $\mathbb{R}^N$.

Proposition 2.26 *Let $X \subset \mathbb{R}^N$ be a smooth manifold, and $x \in X$. Then the vector space $T_x(X) \subset \mathbb{R}^N$ is independent of the choice of local parametrisation.*

Proof: So let $\psi \colon W \to X$ be another local parametrisation around x with $\psi(0) = x$. After shrinking U and W, i.e., after replacing U with $\phi^{-1}(\phi(U) \cap \psi(W)) \subset U$ and W with $\psi^{-1}(\phi(U) \cap \psi(W)) \subset W$, we can assume that $\phi(U) = \psi(W)$. Then the map $\theta := \psi^{-1} \circ \phi \colon U \to W$ is a diffeomorphism since it is the composite of two diffeomorphisms, and the chain rule in calculus implies that this yields a diffeomorphism as well. By definition of θ, we have $\phi = \psi \circ \theta$. Taking derivatives yields $d\phi_0 = d\psi_0 \circ d\theta_0$ according to the chain rule in calculus. This implies that the image of $d\phi_0$ is contained in the image of $d\psi_0$, i.e., $d\phi_0(\mathbb{R}^k) \subseteq d\psi_0(\mathbb{R}^k)$ in $\mathbb{R}^N$. By switching the roles of ϕ and ψ in the argument, we also get $d\psi_0(\mathbb{R}^k) \subseteq d\phi_0(\mathbb{R}^k)$ in $\mathbb{R}^N$. Hence $d\phi_0(\mathbb{R}^k) = d\psi_0(\mathbb{R}^k)$

in $\mathbb{R}^N$. This shows that whatever local parametrisation around x we start with, the vector subspace $T_x(X) \subseteq \mathbb{R}^N$ is always the same. □

Lemma 2.27 *The dimension of $T_x(X)$ as a vector space over $\mathbb{R}$ is k.*

Proof: Since a local parametrisation ϕ is a diffeomorphism onto its image, its derivative $d\phi_u$ is injective. Hence, by definition of $T_x(X)$ as the image of $\mathbb{R}^k$ under $d\phi_u$, the dimension of $T_x(X)$ is k. □

We now look at some first examples. There will be more examples in the exercise section.

Example 2.28 (Tangent space of the circle) Let $p = (a,b) \in \mathbb{S}^1$ be a point with $b > 0$. A local parametrisation around p with $\phi(0) = p$ is given by

$$\phi\colon (-\varepsilon,\varepsilon) \to \mathbb{S}^1,\ t \mapsto \left(t+a, \sqrt{1-(t+a)^2}\right)$$

for some small enough real number $\varepsilon > 0$. The derivative at t is the linear map

$$d\phi_t\colon \mathbb{R} \to \mathbb{R}^2,\ d\phi_t = \begin{pmatrix} 1 \\ -\frac{t+a}{\sqrt{1-(t+a)^2}} \end{pmatrix}.$$

Hence the image of $\mathbb{R}$ under $d\phi_0$ in $\mathbb{R}^2$ is the line spanned by $(-b,a)$ where we use $b = \sqrt{1-a^2}$. For points $(a,b) \in \mathbb{S}^1$ with $b \leq 0$, we use other local parametrisations with a suitable square root. The tangent space $T_{(a,b)}\mathbb{S}^1$ is always orthogonal to the vector (a,b), i.e., it is spanned by $(-b,a)$ in $\mathbb{R}^2$.

Example 2.29 (Tangent space of the two-sphere) Let $p = (a,b,c)$ be a point on $\mathbb{S}^2$ which is not the north pole $(0,0,1)$. Then we can use the stereographic projection $\phi_N\colon \mathbb{R}^2 \to \mathbb{S}^2 \setminus \{N\}$ as a local parametrisation. Recall that

$$\phi_N(x,y) = \frac{1}{1+x^2+y^2}\left(2x, 2y, x^2+y^2-1\right).$$

The derivative at (x,y) is the linear map $d\phi_N\colon \mathbb{R}^2 \to \mathbb{R}^3$ which, with respect to the standard bases of $\mathbb{R}^2$ and $\mathbb{R}^3$, is given by the matrix

$$d(\phi_N)_{(x,y)} = \frac{2}{(1+x^2+y^2)^2}\begin{pmatrix} 1-x^2+y^2 & -2xy \\ -2xy & 1+x^2-y^2 \\ 2x & 2y \end{pmatrix}.$$

The image of $\mathbb{R}^2$ under the linear map $d(\phi_N)_{(x,y)}$ is the tangent space $T_{\phi_N(x,y)}\mathbb{S}^2$. This image is spanned by the two column vectors of the matrix $d(\phi_N)_{(x,y)}$. Let us check that we get the space we would expect, i.e., the plane which is orthogonal to the vector $\phi_N(x, y)$. We do this by checking that the inner product of the vectors is zero:

$$\begin{aligned}
&\left(2x, 2y, x^2 + y^2 - 1\right) \cdot \begin{pmatrix} 1 - x^2 + y^2 \\ -2xy \\ 2x \end{pmatrix} \\
&\quad = 2x(1 - x^2 + y^2) - 4xy^2 + 2x(x^2 + y^2 - 1) \\
&\quad = 2x - 2x^3 + 2xy^2 - 4xy^2 + 2x^3 + 2xy^2 - 2x = 0,
\end{aligned}$$

and similarly

$$\begin{aligned}
&\left(2x, 2y, x^2 + y^2 - 1\right) \cdot \begin{pmatrix} -2xy \\ 1 + x^2 - y^2 \\ 2y \end{pmatrix} \\
&\quad = -4x^2y + 2y(1 + x^2 - y^2) + 2y(x^2 + y^2 - 1) \\
&\quad = -4x^2y + 2y + 2x^2y - 2y^3 + 2x^2y + 2y^3 - 2y = 0.
\end{aligned}$$

Hence the plane spanned by the column vectors is orthogonal to $\phi_N(x, y)$.

Let $X \subset \mathbb{R}^N$ be a k-dimensional manifold and W be an *open* subset. Then W is also a k-dimensional manifold, since we can restrict all local parametrisations to the intersection with W.

Lemma 2.30 *Let X be a smooth manifold and W be an open subset. At every point $x \in W$, we have $T_x(X) = T_x(W)$.*

Proof: For $x \in W$, let $\phi\colon U \to V \subset X$ be a local parametrisation around x of X. We can assume $\phi(0) = x$. Then $\phi_{|U_W} : U_W \to W \cap V$ is a local parametrisation around x in W with $U_W := \phi^{-1}(W \cap V) \subset U$. Since the derivative only depends on an open neighbourhood around a point, we get $d\phi_0 = d(\phi_{|U_W})_0$. In particular, for the tangent spaces at x, we get

$$T_x(X) = d\phi_0(\mathbb{R}^k) = d(\phi_{|U_W})_0(\mathbb{R}^k) = T_x(W)$$

as vector subspaces of $\mathbb{R}^N$. □

Tangent spaces behave well under taking products.

Lemma 2.31 *Let $X \subseteq \mathbb{R}^N$ and $Y \subseteq \mathbb{R}^M$ be smooth manifolds and $x \in X$, $y \in Y$. Then the tangent space of the product X and Y is the product of the tangent spaces, i.e., $T_{(x,y)}(X \times Y) = T_x(X) \times T_y(Y)$.*

Proof: This follows from the fact that we can choose neighbourhoods in $X \times Y$ by taking the product of neighbourhoods in X and Y, respectively, as explained in Remark 2.19. □

2.3.2 The Derivative

Now we will study the effect of smooth maps on tangent spaces. In fact, we will show that every smooth map f between manifolds induces a linear map between tangent spaces which we may think of as a linear approximation of f. Let $f\colon X \to Y$ be a smooth map from a k-dimensional smooth manifold $X \subseteq \mathbb{R}^N$ to an n-dimensional smooth manifold $Y \subseteq \mathbb{R}^M$. Let $\phi\colon U \to W \subset X$ be a local parametrisation around x with $U \subseteq \mathbb{R}^k$, and let $\psi\colon V \to \widetilde{W} \subset Y$ be a local parametrisation around $f(x)$ with $V \subseteq \mathbb{R}^n$. We can assume $\phi(0) = x$ and $\psi(0) = f(x)$. Moreover, we can choose U and V small enough such that $f(W) \subset \widetilde{W}$. We then obtain the following diagram

$$\begin{array}{ccc} W & \xrightarrow{\quad f_{|W} \quad} & \widetilde{W} \\ {\scriptstyle \phi}\uparrow & & \uparrow{\scriptstyle \psi} \ \ \downarrow{\scriptstyle \psi^{-1}} \\ U & \underset{\theta := \psi^{-1} \circ f_{|W} \circ \phi}{\dashrightarrow} & V \end{array} \tag{2.2}$$

in which we define the map $\theta\colon U \to V$ as the composition $\theta := \psi^{-1} \circ f_{|W} \circ \phi$. The map θ is now a smooth map defined on the open subset $U \subseteq \mathbb{R}^k$. Hence the derivatives $d\phi_0$, $d\psi_0$, and $d\theta_0$ exist. Moreover, since $d\phi_0$ is an isomorphism, we can form the linear map $d\psi_0 \circ d\theta_0 \circ (d\phi_0)^{-1}$. By definition of the tangent spaces as $T_x(X) = \operatorname{Im}(d\phi_0)$ and $T_{f(x)}(Y) = \operatorname{Im}(d\psi_0)$, this yields a map $T_x(X) \to T_{f(x)}(Y)$, and this is the desired derivative of f at x.

Definition 2.32 (The derivative df_x) With the above notation, we define the *derivative of f at x* as the linear map $df_x\colon T_x(X) \to T_{f(x)}(Y)$ given by the composition $df_x := d\psi_0 \circ d\theta_0 \circ (d\phi_0)^{-1}$. That is, df_x is defined to be the linear map which makes the following diagram commutative:

$$\begin{array}{ccc} T_x(X) & \overset{df_x}{\dashrightarrow} & T_{f(x)}(Y) \\ {\scriptstyle d\phi_0}\uparrow \ \downarrow{\scriptstyle (d\phi_0)^{-1}} & & \uparrow{\scriptstyle d\psi_0} \\ \mathbb{R}^k & \xrightarrow[d\theta_0]{\qquad} & \mathbb{R}^n \end{array} . \tag{2.3}$$

Lemma 2.33 *The map df_x does not depend on the choice of local parametrisations around x and $f(x)$.*

Proof: So let $\phi'\colon U' \to X$ and $\psi'\colon V' \to Y$ be another choice of local parametrisations around x and $f(x)$, respectively. Again after possibly shrinking U, U', V, and V' we can assume that $\phi(U) = \phi'(U') \subseteq X$ and $\psi(V) = \psi'(V') \subseteq Y$, and $f_{|\phi'(U')}(\phi'(U')) \subset \psi'(V')$. Then the linear maps $d\phi_0, d\phi'_0 \colon \mathbb{R}^k \to Tx(X)$ differ by a linear isomorphism of $\mathbb{R}^k$, say α: $d\phi_0 = d\phi'_0 \circ \alpha$. Similarly, there is a linear isomorphism β of $\mathbb{R}^n$ such that $d\psi_0 = d\psi'_0 \circ \beta$ as linear maps $\mathbb{R}^n \to T_{f(X)}(Y)$. Let $\theta' \colon U \to V$ be defined similarly to θ, i.e., we set $\theta' = \psi'^{-1} \circ f_{|\phi'(U')} \circ \phi'$. This gives us the following diagram in which each square commutes.

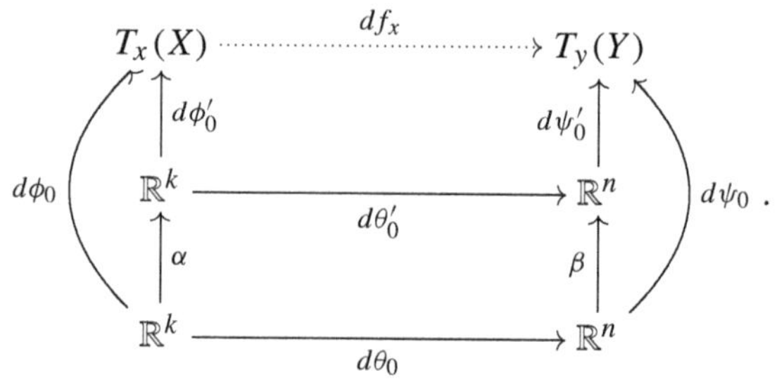

(2.4)

The relation between $d\theta_0$ and $d\theta'_0$ is given by $d\theta'_0 = \beta \circ d\theta_0 \circ \alpha^{-1}$. Putting all relations together, we get

$$\begin{aligned} d\psi'_0 \circ d\theta'_0 \circ d(\phi'^{-1})_x &= d\psi'_0 \circ (\beta \circ d\theta_0 \circ \alpha^{-1}) \circ d(\phi'^{-1})_x \\ &= (d\psi'_0 \circ \beta) \circ d\theta_0 \circ (\alpha^{-1}) \circ d(\phi'^{-1})_x) \\ &= d\psi_0 \circ d\theta_0 \circ d(\phi^{-1})_x. \end{aligned}$$

This implies the desired identity for the a priori different constructions of the derivative. □

Remark 2.34 Let $f \colon X \to Y$ be a smooth map between manifolds. If $X \subset \mathbb{R}^k$ is an open subset of $\mathbb{R}^k$, then the derivative df_x at any point $x \in X$ coincides with the usual derivative from calculus, as recalled at the beginning of Section 2.3.1. In general, however, defining df_x requires additional work, even though we have already defined smooth maps between arbitrary subsets of Euclidean space. One reason is that the classical derivative presupposes that f is defined on an open neighbourhood of x. This forces us to use local parametrisations. Moreover, we must ensure that df_x is a linear map with domain $T_x(X)$ and codomain $T_{f(x)}(Y)$. The construction of df_x may seem elaborate, but it is precisely the linear algebra needed to guarantee that the derivative is well defined between the appropriate tangent spaces.

The derivative is compatible with composition of smooth maps.

Proposition 2.35 (Chain Rule) *Let* $X \xrightarrow{f} Y \xrightarrow{g} Z$ *be smooth maps of manifolds. Then* $d(g \circ f)_x = dg_{f(x)} \circ df_x$.

Proof: Let $\eta\colon W \to Z$ be a local parametrisation around $z = g(f(x))$ with an open subset $W \subseteq \mathbb{R}^m$ and $\eta(0) = z$. Then we have a commutative diagram

$$\begin{array}{ccccc} X & \xrightarrow{\quad f \quad} & Y & \xrightarrow{\quad g \quad} & Z \\ \phi\big\uparrow & & \big\uparrow\psi & & \big\uparrow\eta \\ U & \xrightarrow[\theta=\psi^{-1}\circ f\circ\phi]{} & V & \xrightarrow[\iota=\eta^{-1}\circ g\circ\psi]{} & W \end{array}. \tag{2.5}$$

Thus, by definition, $d(g \circ f)_x = d\eta_0 \circ d(\iota \circ \theta)_0 \circ (d\phi_0)^{-1}$. The chain rule from calculus for maps of open sets of Euclidean spaces then gives $d(\iota \circ \theta)_0 = (d\iota_0) \circ (d\theta_0)$. Thus,

$$d(g \circ f)_x = (d\eta_0 \circ d\iota_0 \circ (d\psi_0)^{-1}) \circ (d\psi_0 \circ d\theta_0 \circ (d\phi_0)^{-1}) = dg_y \circ df_x.$$

□

2.4 The Hopf Map

We now look at an important example of a smooth map and will compute its derivative concretely. Recall from Example 2.10 the Hopf map $\hbar\colon \mathbb{S}^3 \to \mathbb{S}^2$ defined by

$$\hbar\colon \mathbb{S}^3 \to \mathbb{S}^2, \quad \begin{pmatrix} x_1 \\ x_2 \\ x_3 \\ x_4 \end{pmatrix} \mapsto \begin{pmatrix} 2x_1x_3 + 2x_2x_4 \\ 2x_2x_3 - 2x_1x_4 \\ x_1^2 + x_2^2 - x_3^2 - x_4^2 \end{pmatrix}.$$

See also Exercise 2.10 for an alternative formula for $\hbar$. We now compute the derivative of $\hbar$ at two points in $\mathbb{S}^3$. Beyond serving as an illustration of the theory developed so far, the importance of these computations, and of our choice of points, will become evident later. First, we pick a point which is mapped to the south pole $\mathbf{s}_2 = (0, 0, -1) \in \mathbb{S}^2$. The formula for $\hbar$ shows that all points $(x_1, x_2, x_3, x_4) \in \mathbb{S}^3$ with $x_1 = x_2 = 0$ are mapped to $\mathbf{s}_2$. So let us look at the north pole $\mathbf{n}_3 = (0, 0, 0, 1) \in \mathbb{S}^3$ in $\hbar^{-1}(\mathbf{s}_2)$. We choose the open ball $\mathbb{B}^3(\mathbf{0}_3)$ around the origin $\mathbf{0}_3$ in $\mathbb{R}^3$ of radius $1/\sqrt{2}$ and use the local parametrisation $\phi\colon U = \mathbb{B}^3_{1/\sqrt{2}}(\mathbf{0}_3) \to W_3$ around $\mathbf{n}_3$ defined by $\phi(x, y, z) = \left(x, y, z, \sqrt{1 - (x^2 + y^2 + z^2)}\right)$ where $W_3 \subset \mathbb{S}^3_{x_4>0} \subset \mathbb{S}^3$ denotes the open subset of $\mathbb{S}^3$ consisting of points with coordinates $x_4 > 0$ and $2x_1^2 + 2x_2^2 < 1$. We note $\phi(\mathbf{0}_3) = \mathbf{n}_3$. Similarly, we choose the open ball $\mathbb{B}^2_1(\mathbf{0}_2)$ around the origin $\mathbf{0}_2$ in $\mathbb{R}^2$ of radius 1 and use the local parametrisation

$\psi\colon V = \mathbb{B}^2_1(\mathbf{0}_2) \to W_2$ defined by $\psi(x,y) = \left(x, y, -\sqrt{1-(x^2+y^2)}\right)$ where $W_2 = \mathbb{S}^2_{x_3<0} \subset \mathbb{S}^2$ denotes the open subset of $\mathbb{S}^2$ consisting of points with coordinate $x_3 < 0$. We note $\psi(\mathbf{0}_2) = \mathbf{s}_2$. Moreover, we have $\hbar(W_3) \subset W_2$. Now we need to calculate the induced map $\theta\colon \mathbb{R}^3 \to \mathbb{R}^2$ such that the diagram commutes

$$\begin{array}{ccc} W_3 & \xrightarrow{\quad \hbar \quad} & W_2 \\ \uparrow \phi & & \uparrow \psi \ \ \vdots\, \psi^{-1} \\ \mathbb{B}^3_{1/\sqrt{2}}(\mathbf{0}_3) & \xrightarrow[\theta=\psi^{-1}\circ\hbar\circ\phi]{} & \mathbb{B}^2_1(\mathbf{0}_2) \end{array} . \tag{2.6}$$

Recall that we write $|\mathbf{x}|$ for the norm of points $\mathbf{x} \in \mathbb{R}^k$. First we apply the composite $\hbar \circ \phi$ to a point $\mathbf{x} = (x,y,z)$ in $\mathbb{B}^3(\mathbf{0}_3)$:

$$\hbar \circ \phi(\mathbf{x}) = \begin{pmatrix} 2xz + 2y\sqrt{1-|\mathbf{x}|^2} \\ 2yz - 2x\sqrt{1-|\mathbf{x}|^2} \\ x^2+y^2-z^2-1+|\mathbf{x}|^2 \end{pmatrix}.$$

Since ψ^{-1} is just the projection onto the first two coordinates, we get

$$(\psi^{-1}\circ\hbar\circ\phi)(\mathbf{x}) = \begin{pmatrix} 2xz + 2y\sqrt{1-|\mathbf{x}|^2} \\ 2yz - 2x\sqrt{1-|\mathbf{x}|^2} \end{pmatrix}.$$

Now we need to apply this result to compute the linear map $d\hbar_{\mathbf{n}_3}$ given by

$$d\hbar_{\mathbf{n}_3}\colon T_{\mathbf{n}_3}(\mathbb{S}^3) \xrightarrow{(d\phi_{\mathbf{0}_3})^{-1}} \mathbb{R}^3 \xrightarrow{d\theta_{\mathbf{0}_3}} \mathbb{R}^2 \xrightarrow{d\psi_{\mathbf{0}_2}} T_{\mathbf{s}_2}(\mathbb{S}^2).$$

In particular, we need to calculate the derivative $d\theta_{\mathbf{0}_3}\colon \mathbb{R}^3 \to \mathbb{R}^2$ at the origin $\mathbf{0}_3$. We do this by first computing the Jacobian matrix J_θ of θ:

$$J_\theta = \begin{pmatrix} 2z - \frac{2xy}{\sqrt{1-|\mathbf{x}|^2}} & 2\sqrt{1-|\mathbf{x}|^2} - \frac{2y^2}{\sqrt{1-|\mathbf{x}|^2}} & 2x - \frac{2yz}{\sqrt{1-|\mathbf{x}|^2}} \\ -2\sqrt{1-|\mathbf{x}|^2} + \frac{2x^2}{\sqrt{1-|\mathbf{x}|^2}} & 2z + \frac{2xy}{\sqrt{1-|\mathbf{x}|^2}} & 2y + \frac{2xz}{\sqrt{1-|\mathbf{x}|^2}} \end{pmatrix}.$$

Since we want to calculate the matrix representing $d\theta_{\mathbf{0}_3}$ at the origin, we set $x = y = z = 0$ and see that $d\theta_{\mathbf{0}_3}\colon \mathbb{R}^3 \to \mathbb{R}^2$ is given by the matrix $d\theta_{\mathbf{0}_3} = \begin{pmatrix} 0 & 2 & 0 \\ -2 & 0 & 0 \end{pmatrix}$ with respect to the standard bases of $\mathbb{R}^3$ and $\mathbb{R}^2$. By definition of $T_{\mathbf{n}_3}(\mathbb{S}^3)$ as the image of $d\phi_{\mathbf{0}_3}\colon \mathbb{R}^3 \to \mathbb{R}^4$, we have a canonical choice of a basis of $T_{\mathbf{n}_3}(\mathbb{S}^3)$ as the image of the standard basis $\mathbf{e}^3_1, \mathbf{e}^3_2, \mathbf{e}^3_3$ of $\mathbb{R}^3$ under $d\phi_{\mathbf{0}_3}$. With respect this basis $d\phi_{\mathbf{0}_3}(\mathbf{e}^3_1), d\phi_{\mathbf{0}_3}(\mathbf{e}^3_2), d\phi_{\mathbf{0}_3}(\mathbf{e}^3_3)$ for $T_{\mathbf{n}_3}(\mathbb{S}^3)$ and the standard basis for $\mathbb{R}^3$, the derivative $(d\phi_{\mathbf{0}_3})^{-1}\colon T_{\mathbf{n}_3}(\mathbb{S}^3) \to \mathbb{R}^3$ is represented by the (3×3)-identity matrix. Similarly, for $T_{\mathbf{s}_2}(\mathbb{S}^2)$ we can choose the image of the standard basis $\mathbf{e}_1^2, \mathbf{e}_2^2$ of $\mathbb{R}^2$ under $d\psi_{\mathbf{0}_2}$. With respect to the standard basis for $\mathbb{R}^2$ and the basis $d\psi_{\mathbf{0}_2}(\mathbf{e}_1^2), d\psi_{\mathbf{0}_2}(\mathbf{e}_2^2)$ for $T_{\mathbf{s}_2}(\mathbb{S}^2)$, the derivative $d\psi_{\mathbf{0}_2}\colon \mathbb{R}^2 \to T_{\mathbf{s}_2}(\mathbb{S}^2)$ is represented by the (2×2)-identity matrix. Hence, with respect to these bases for $T_{\mathbf{n}_3}(\mathbb{S}^3)$ and $T_{\mathbf{s}_2}(\mathbb{S}^2)$, we see that $d\hbar_{\mathbf{n}_3} = d\psi_{\mathbf{0}_2} \circ d\theta_{\mathbf{0}_3} \circ (d\phi_{\mathbf{0}_3})^{-1}$ is represented by the matrix $d\hbar_{\mathbf{n}_3} = \begin{pmatrix} 0 & 2 & 0 \\ -2 & 0 & 0 \end{pmatrix}$. In particular, $d\hbar_{\mathbf{n}_3}(d\phi_{\mathbf{0}_3}(\mathbf{e}_1^3)) = -2 \cdot d\psi_{\mathbf{0}_2}(\mathbf{e}_2^2)$, $d\hbar_{\mathbf{n}_3}(d\phi_{\mathbf{0}_3}(\mathbf{e}_2^3)) = 2 \cdot d\psi_{\mathbf{0}_2}(\mathbf{e}_1^2)$, and $d\hbar_{\mathbf{n}_3}(d\phi_{\mathbf{0}_3}(\mathbf{e}_3^3)) = \mathbf{0}$. Note that our choices of bases made it very easy to compute the matrix for $d\hbar_{\mathbf{n}_3}$. So in the end, there is not as much to compute as one might expect.

Remark 2.36 (Something we learn from this example) Among other things, we learn from this computation how we can find a matrix which represents the derivative. Let $f\colon X \to Y$ be a smooth map, and let $x \in X$ and $y = f(x) \in Y$. Then, once we have computed θ, there is a straight forward way to compute a matrix which describes df_x. For, there is a *canonical choice* for the bases of $T_x(X)$ and $T_y(Y)$, namely the images of the standard basis vectors of $\mathbb{R}^n$ and $\mathbb{R}^m$ under the isomorphisms $d\phi_0$ and $d\psi_0$, respectively. Then we can compute the matrix which represents the linear map $df_x\colon T_x(X) \to T_y(Y)$ with respect to these bases as the matrix which represents $d\theta_0$ with respect to the standard bases of $\mathbb{R}^n$ and $\mathbb{R}^m$. We get this matrix as the Jacobian matrix of θ.

Now we look at a point on $\mathbb{S}^3$ which is mapped to the north pole $\mathbf{n}_2 = (0,0,1) \in \mathbb{S}^2$. The formula for $\hbar$ shows that all points $(x_1,x_2,x_3,x_4) \in \mathbb{S}^3$ with $x_3 = x_4 = 0$ are mapped to $\mathbf{n}_2$. Hence, in particular, the point $\mathbf{q}_3 = (1,0,0,0) \in \mathbb{S}^3$ is mapped to $\mathbf{n}_2 = (0,0,1) \in \mathbb{S}^2$, and our next goal is to compute $d\hbar_{\mathbf{q}_3}$. Much of the calculation is the same as in the previous case. However, there are some interesting changes, in particular, of some signs. Again, we will be able to appreciate our effort even more later. We choose the open ball $\mathbb{B}^3(\mathbf{0}_3)$ around the origin $\mathbf{0}_3$ in $\mathbb{R}^3$ of radius $1/\sqrt{2}$ and use the local parametrisation $\phi\colon U = \mathbb{B}^3_{1/\sqrt{2}}(\mathbf{0}_3) \to W_3$ defined by $\phi(x,y,z) = \left(\sqrt{1-(x^2+y^2+z^2)}, x, y, z\right)$ where $W_3 \subset \mathbb{S}^3_{x_1>0} \subset \mathbb{S}^3$ denotes the open subset of $\mathbb{S}^3$ consisting of points with coordinate $x_1 > 0$ and $2x_3^2 + 2x_4^2 < 1$. We choose the open ball $\mathbb{B}^2_1(\mathbf{0}_2)$ around the origin $\mathbf{0}_2$ in $\mathbb{R}^2$ of radius 1 and use the local parametrisation $\psi\colon V = \mathbb{B}^2_1(\mathbf{0}_2) \to W_2$ defined by $\psi(x,y) = \left(x, y, \sqrt{1-(x^2+y^2)}\right)$ where $W_2 \subset \mathbb{S}^2_{x_3>0}$ denotes the open subset of $\mathbb{S}^2$ consisting of points with coordinate $x_3 > 0$. The composite $\hbar \circ \phi$ evaluated at $\mathbf{x} = (x,y,z) \in \mathbb{B}^3_{1/\sqrt{2}}(\mathbf{0}_3)$ is then

$$\hbar \circ \phi(\mathbf{x}) = \begin{pmatrix} 2y\sqrt{1-|\mathbf{x}|^2} + 2xz \\ 2xy - 2z\sqrt{1-|\mathbf{x}|^2} \\ 1 - |\mathbf{x}|^2 + x^2 - y^2 - z^2 \end{pmatrix},$$

and applying ψ^{-1} yields $(\psi^{-1} \circ \hbar \circ \phi)(\mathbf{x}) = \begin{pmatrix} 2xz + 2y\sqrt{1-|\mathbf{x}|^2} \\ 2xy - 2z\sqrt{1-|\mathbf{x}|^2} \end{pmatrix}$. Next, we compute the derivative $d\theta_{\mathbf{0}_3} : \mathbb{R}^3 \to \mathbb{R}^2$ at the origin $\mathbf{0}_3$. We do this again by calculating the Jacobian matrix J_θ of θ:

$$J_\theta = \begin{pmatrix} 2z - \frac{2xy}{\sqrt{1-|\mathbf{x}|^2}} & 2\sqrt{1-|\mathbf{x}|^2} - \frac{2y^2}{\sqrt{1-|\mathbf{x}|^2}} & 2x - \frac{2yz}{\sqrt{1-|\mathbf{x}|^2}} \\ 2y + \frac{2xz}{\sqrt{1-|\mathbf{x}|^2}} & 2x + \frac{2yz}{\sqrt{1-|\mathbf{x}|^2}} & \frac{2z^2}{\sqrt{1-|\mathbf{x}|^2}} - 2\sqrt{1-\mathbf{x}|^2} \end{pmatrix}.$$

At the origin we set $x = y = z = 0$ and see that $d\theta_{\mathbf{0}_3} : \mathbb{R}^3 \to \mathbb{R}^2$ is represented by the matrix $d\theta_{\mathbf{0}_3} = \begin{pmatrix} 0 & 2 & 0 \\ 0 & 0 & -2 \end{pmatrix}$ with respect to the standard bases of $\mathbb{R}^3$ and $\mathbb{R}^2$, respectively. We make our canonical choice of bases of $T_{\mathbf{q}_3}(\mathbb{S}^3)$, i.e., $d\phi_{\mathbf{0}_3}(\mathbf{e}_1^3), d\phi_{\mathbf{0}_3}(\mathbf{e}_2^3), d\phi_{\mathbf{0}_3}(\mathbf{e}_3^3)$, and of $T_{\mathbf{n}_2}(\mathbb{S}^2)$, i.e., $d\psi_{\mathbf{0}_2}(\mathbf{e}_1^2), d\psi_{\mathbf{0}_2}(\mathbf{e}_2^2)$. Then, with respect to these bases, the composition $d\hbar_{\mathbf{q}_3} = d\psi_{\mathbf{0}_2} \circ d\theta_{\mathbf{0}_3} \circ (d\phi_{\mathbf{0}_3})^{-1}$ is represented by the matrix $d\hbar_{\mathbf{q}_3} = \begin{pmatrix} 0 & 2 & 0 \\ 0 & 0 & -2 \end{pmatrix}$. That is, $d\hbar_{\mathbf{q}_3}(d\phi_{\mathbf{0}_3}(\mathbf{e}_1^3)) = \mathbf{0}$, $d\hbar_{\mathbf{q}_3}(d\phi_{\mathbf{0}_3}(\mathbf{e}_2^3)) = 2 \cdot d\psi_{\mathbf{0}_2}(\mathbf{e}_1^2)$, and $d\hbar_{\mathbf{q}_3}(d\phi_{\mathbf{0}_3}(\mathbf{e}_3^3)) = -2 \cdot d\psi_{\mathbf{0}_2}(\mathbf{e}_2^2)$.

2.5 Tangent Bundle

In this section, we introduce another important example of a smooth manifold: the tangent bundle, constructed from the collection of all tangent spaces of a manifold. The tangent bundle will serve as a powerful geometric tool in many later constructions. At first glance, it may seem counterintuitive that tangent spaces, objects defined using a manifold, can themselves be assembled into a new manifold. If this feels confusing, you may choose to skip this section for now and return to it later when we begin using the tangent bundle in applications.

2.5.1 Tangent Bundles Are Smooth Manifolds

Let $X \subset \mathbb{R}^N$ be a smooth manifold. For every $x \in X$, the tangent space $T_x(X)$ to X at x is a vector subspace of $\mathbb{R}^N$. If we let x vary, these tangent spaces will overlap or intersect in $\mathbb{R}^N$. Hence, we need a trick to keep these spaces apart.

Definition 2.37 (Tangent bundle) The *tangent bundle*, denoted $T(X)$, is the subspace of $X \times \mathbb{R}^N \subset \mathbb{R}^N \times \mathbb{R}^N$ defined by

$$T(X) := \left\{(x, v) \in X \times \mathbb{R}^N : v \in T_x(X)\right\}$$

together with the natural projection map $\pi : T(X) \to X$ given by $\pi(x, v) = x$.

Let $Y \subset \mathbb{R}^M$ be another smooth manifold. Then a smooth map $f : X \to Y$ induces a global derivative map

$$df : T(X) \to T(Y),\ (x, v) \mapsto (f(x), df_x(v)).$$

Proposition 2.38 *The map $df : T(X) \to T(Y)$ is smooth as a map between subsets of $\mathbb{R}^{2N}$ and $\mathbb{R}^{2M}$, respectively.*

Proof: Since $f : X \to \mathbb{R}^M$ is smooth, it extends by definition around any point $x \in X$ to a smooth map $F : U \to \mathbb{R}^M$, where U is an open set of $\mathbb{R}^N$. Since $U \subset \mathbb{R}^N$ is open and hence $T_u(U) = \mathbb{R}^N$ for every $u \in U$, we have $T(U) = U \times \mathbb{R}^N$ by Lemma 2.30. Since $U \times \mathbb{R}^N$ is an open set in $\mathbb{R}^{2N}$, the linear map dF is smoothly defined on an open subset of $\mathbb{R}^{2N}$. Since dF extends df on $(U \times \mathbb{R}^N) \cap T(X)$, we conclude that df is smooth. □

Remark 2.39 (Tangent bundles are intrinsic) Given smooth maps $f : X \to Y$ and $g : Y \to Z$, it follows from the chain rule that the global derivative of the composite is equal to the composite of global derivatives, i.e., $d(g \circ f) = dg \circ df : T(X) \to T(Z)$. In particular, this shows that, if $f : X \to Y$ is a diffeomorphism, then $df : T(X) \to T(Y)$ is a diffeomorphism. Thus, diffeomorphic manifolds have diffeomorphic tangent bundles. As a result, $T(X)$ is an object intrinsically associated to X, i.e., it does not depend on the ambient Euclidean space.

We will now show that $T(X)$ is in fact itself a smooth manifold. We will use this fact, for example, in the proof of Whitney's theorem in Section 7.1.

Proposition 2.40 (Tangent bundles are manifolds) *Let X be a smooth manifold of dimension k. The tangent bundle $T(X)$ is a smooth manifold of dimension $2k$.*

Proof: Let W be an open set of X. Then W is also a smooth manifold, and its tangent bundle $T(W)$ is given by

$$T(W) = \{(x, v) \in T(X) : x \in W\} = T(X) \cap (W \times \mathbb{R}^N) \subset T(X).$$

Since $W \times \mathbb{R}^N$ is open in $X \times \mathbb{R}^N$, $T(W)$ is open in $T(X)$. Now suppose that W is the image of a local parametrisation $\phi : U \to W$, where U is an open set in $\mathbb{R}^k$. Then the global derivative $d\phi\colon T(U) \to T(W)$ is a diffeomorphism. Since $T(U) = U \times \mathbb{R}^k$ is an open subset of $\mathbb{R}^{2k}$, $d\phi$ is a local parametrisation of the open set $T(W)$ in $T(X)$. Since every point of $T(X)$ sits in such an open neighbourhood, we have proved the assertion. □

2.5.2 Tangent Bundles Are Vector Bundles

Tangent bundles are examples of a more general class of spaces, called smooth vector bundles. Roughly speaking, an n-dimensional vector bundle E consists of an assignment of an n-dimensional vector space V to each point $x \in X$, and a rule for how to glue all these vector spaces together in a nice way.

Definition 2.41 Let X be a smooth manifold. An *n-dimensional smooth vector bundle E over X* consists of a topological space E together with a continuous map $\pi : E \to X$, and, for each $x \in X$, the structure of a vector space over the real numbers on the set $\pi^{-1}(x)$, which is called the *fibre over* x. Moreover, these data are required to satisfy the following local triviality condition: for every point $x \in X$, there is an open subset $U \subset X$ around x and a diffeomorphism $h\colon U \times \mathbb{R}^n \to \pi^{-1}(U)$ such that, for every $y \in X$, the map $v \mapsto h(y, v)$ defines a linear isomorphism between the vector spaces $\mathbb{R}^n$ and $\pi^{-1}(y)$.

The tangent bundle of a smooth manifold is an important example of a smooth vector bundle. Vector bundles have a very interesting theory, and many problems can be formulated in terms of vector bundles. For example, we will see in Section 6.3 that there is a nice classifying space for vector bundles, called the Grassmannian. If it is possible to choose the open subset around x in the above condition to be all of X, then we call $E \to X$ a *trivial bundle*. The following lemma gives us a useful test for the triviality of a bundle.

Lemma 2.42 *Let $\pi : E \to X$ be a trivial bundle. Then π has a nowhere-zero section, i.e., there is a smooth map $\sigma : X \to E$ with $\pi \circ \sigma = \mathrm{id}_X$ and, for all $x \in X$, $\sigma(x)$ is not the zero vector in $\pi^{-1}(x)$.*

Proof: Let n denote the dimension of the fibres of π. Since $\pi : E \to X$ is assumed to be trivial, there is a diffeomorphism $h\colon X \times \mathbb{R}^n \to E$ such that, for every $x \in X$, the map $v \mapsto h(x, v)$ defines a linear isomorphism between the vector spaces $\mathbb{R}^n$ and $\pi^{-1}(x)$. We fix a non-zero vector $w \in \mathbb{R}^n$. Since the map $v \mapsto h(x, v)$ is a linear isomorphism, we have $h(x, w)$ is not the zero vector in $\pi^{-1}(x)$. Thus we can define the section σ by setting $\sigma(x) := h(x, w)$. □

Remark 2.43 (Parallelisable spheres) A smooth manifold for which the tangent bundle is a trivial vector bundle is called *parallelisable*. Examples of manifolds which are parallelisable are $\mathbb{S}^1$, $\mathbb{S}^3$, and $\mathbb{S}^7$, whereas $\mathbb{S}^2$ is not parallelisable. In fact, it is a famous result of Bott–Milnor [5] and Kervaire [25] that $\mathbb{S}^n$ is parallelisable if and only if $n = 0, 1, 3$, or 7. This is a consequence of a fundamental result on the possible multiplicative structures on $\mathbb{R}^n$. Let $\mathbb{R}^n \times \mathbb{R}^n \to \mathbb{R}^n$ be a map with a two-sided identity element and no zero-divisors. Then n must be either 1, 2, 4, or 8. See also Theorem 13.17 and Remark 13.18. We will show in Example 6.42 in Section 6.3.1 that the canonical line bundle over real projective space is non-trivial.

2.6 Exercises

2.1 Let $f: \mathbb{R}^2 \to \mathbb{R}^2$, $f(x, y) := (x^2 - y^2, 2xy)$, and $U := \mathbb{R}^2 \setminus \{(0,0)\}$.
   (a) Show that the Jacobian matrix of the restriction $f_{|U}: U \to U$ is invertible for every point in U.
   (b) Is $f_{|U}$ a diffeomorphism?

2.2 Let $X \subseteq \mathbb{R}^N$, $Y \subseteq \mathbb{R}^M$ and $Z \subseteq \mathbb{R}^L$, and let $f: X \to Y$, $g: Y \to Z$ be smooth maps with $f(X) \subseteq Y$.
   (a) Show the composite $g \circ f: X \to Z$ is smooth if g and f are smooth. Hint: If all subsets are open, this is just the chain rule from calculus.
   (b) Show that if g and f are diffeomorphisms, so is $g \circ f$.

2.3 For a real number $r > 0$, let $\mathbb{B}_r = \mathbb{B}_r^k(0) = \{x \in \mathbb{R}^k : |x| < r\}$ be the open ball around the origin with radius r in $\mathbb{R}^k$. Let X be a k-dimensional smooth manifold.
   (a) Show that the map $f: \mathbb{B}_r \to \mathbb{R}^k$ given by $f(x) = \frac{rx}{\sqrt{r^2 - |x|^2}}$ is a diffeomorphism from $\mathbb{B}_r$ to $\mathbb{R}^k$. Hint: Compute the inverse directly.
   (b) Show that every point in X has an open neighbourhood diffeomorphic to an open ball in $\mathbb{R}^k$ around the origin.
   (c) Show that every point in X has a neighbourhood diffeomorphic to all of $\mathbb{R}^k$.

2.4 The *special unitary group* $\mathrm{SU}(n)$ is defined as $\mathrm{SU}(n) := \{A \in \mathrm{M}(n, \mathbb{C}) : \bar{A}^\top A = \mathrm{Id} \text{ and } \det(A) = 1\}$, where $\bar{A}$ denotes the complex conjugate of A, i.e., the matrix obtained from A by conjugating every entry of A. Show that there is a diffeomorphism between $\mathrm{SU}(2)$ and $\mathbb{S}^3$.

2.5 Show that every k-dimensional vector subspace V of $\mathbb{R}^N$ is a smooth manifold diffeomorphic to $\mathbb{R}^k$, and show that every linear map $V \to \mathbb{R}^m$ is smooth.

2.6 Let $\mathrm{M}(p,n)$ denote the space of real $(p \times n)$-matrices, which we consider as a subspace of $\mathbb{R}^{pn}$. Let $\mathrm{M}(p,n;k)$ denote the subspace of matrices of rank k. Prove the following statement: For $p \geq n$, $\mathrm{M}(p,n;n)$ is an open subset of $\mathrm{M}(p,n)$ and hence a smooth manifold of dimension $p \cdot n$. Hint: Use that the determinant function is continuous.

2.7 In this exercise we study spaces of the form $H_a = \{(x,y,z) \in \mathbb{R}^3 : x^2 + y^2 - z^2 = a\}$ for $a \in \mathbb{R}^{\geq 0}$.

(a) Prove that the hyperboloid H_a for $a > 0$ is a smooth manifold of dimension two.

(b) Explain why H_0 is not a smooth manifold.

2.8 The torus $\mathbb{T}(a,b)$ is the set of points in $\mathbb{R}^3$ at distance b from the circle of radius a in the xy-plane, where $0 < b < a$. Prove that each $\mathbb{T}(a,b)$ is diffeomorphic to $\mathbb{S}^1 \times \mathbb{S}^1 \subset \mathbb{R}^4$. What happens when $b = a$?

2.9 Let $N = (0,\ldots,0,1) \in \mathbb{S}^k$ be the *north pole* on the k-dimensional sphere. The stereographic projection ϕ_N^{-1} from $\mathbb{S}^k \setminus \{N\}$ onto $\mathbb{R}^k$ is the map which sends a point p to the point at which the line through N and p intersects the subspace in $\mathbb{R}^{k+1}$ defined by $x_{k+1} = 0$. See Figure 2.1 for the case $k = 2$.

(a) Show that ϕ_N^{-1} is given by $\phi_N^{-1}(x_1,\ldots,x_{k+1}) = \frac{1}{1-x_{k+1}}(x_1,\ldots,x_k)$.

(b) Find a formula for the inverse ϕ_N of ϕ_N^{-1}, and explain why both maps are smooth.

(c) Let $S = (0,\ldots,0,-1) \in \mathbb{S}^k$ be the south pole. Describe the parametrisation using the stereographic projection starting in S instead of N, and conclude that $\mathbb{S}^k$ is a k-dimensional smooth manifold.

2.10 We consider $\mathbb{S}^3$ as a subset of $\mathbb{C}^2$, i.e., $\mathbb{S}^3 = \{(z_0,z_1) \in \mathbb{C}^2 : |z_0|^2 + |z_1|^2 = 1\}$, and $\mathbb{S}^2$ as a subset of $\mathbb{C} \times \mathbb{R}$, i.e., $\mathbb{S}^2 = \{(z,x) \in \mathbb{C} \times \mathbb{R} : |z|^2 + x^2 = 1\}$. Then the map $\hbar \colon \mathbb{S}^3 \to \mathbb{S}^2$ defined by

$$\hbar(z_0,z_1) := \left(2z_0\bar{z}_1, |z_0|^2 - |z_1|^2\right).$$

equals the map from Example 2.10 and Section 2.4 by identifying $\mathbb{C}^2 = \mathbb{R}^4$ and $\mathbb{C} \times \mathbb{R} = \mathbb{R}^3$. The map $\hbar$ is called the *Hopf map*.

(a) Check that this actually defines a map $\mathbb{S}^3 \to \mathbb{S}^2$.

(b) Show that $\hbar(z_0,z_1) = \hbar(w_0,w_1)$ if and only if there is a complex number α with $|\alpha|^2 = \alpha\bar{\alpha} = 1$ such that $(w_0,w_1) = (\alpha z_0, \alpha z_1)$.

(c) Show that, for every point $p \in \mathbb{S}^2$, the fibre $\hbar^{-1}(p)$ is diffeomorphic to $\mathbb{S}^1$. We conclude that the Hopf map $\hbar$ realises $\mathbb{S}^3$ as a disjoint union of fibres which each look like $\mathbb{S}^1$.

2.11 Let V be a finite-dimensional vector subspace of $\mathbb{R}^N$ which we consider as a smooth manifold. Show that $T_x(V) = V$ for $x \in V$.

2.12 Determine the tangent space to the torus $\mathbb{S}^1 \times \mathbb{S}^1 \subset \mathbb{R}^4$ at an arbitrary point p. Recall the description of the torus $\mathbb{T}(a,b) \subset \mathbb{R}^3$ from the previous exercise set. Can you describe the tangent space at a point in $\mathbb{T}(a,b)$?

2.13 Determine the tangent space to $H_a = \{(x,y,z) \in \mathbb{R}^3 \colon x^2 + y^2 - z^2 = a\}$ at $\left(\sqrt{a}, 0, 0\right)$ for $a > 0$.

2.14 The graph of a map $f\colon X \to Y$ is the subset of $X \times Y$ defined by

$$\Gamma(f) = \{(x, f(x)) \in X \times Y : x \in X\}.$$

Define $F\colon X \to \Gamma(f)$ by $F(x) = (x, f(x))$. We now assume that X and Y are smooth manifolds and f is a smooth map.

(a) Show that F is a diffeomorphism, and conclude that $\Gamma(f)$ is a smooth manifold.

(b) Consider F as a smooth map $F\colon X \to X \times Y$. Show that the derivative dF_x is given by $dF_x(v) = (v, df_x(v))$.

(c) Show that the tangent space to $\Gamma(f)$ at the point $(x, f(x))$ is the graph of $df_x\colon T_x(X) \to T_{f(x)}(Y)$.

2.15 Show that the derivative of the determinant map $\det\colon \mathrm{GL}(n,\mathbb{R}) \to \mathbb{R}$ at $A \in \mathrm{GL}(n,\mathbb{R})$ is given by $d(\det)_A(B) = (\det A) \cdot (\operatorname{tr} A^{-1}B)$ for all $B \in \mathrm{M}(n)$. In particular, $d(\det)_A(AB) = (\det A) \cdot (\operatorname{tr} B)$ for all $B \in \mathrm{M}(n)$.

2.16 A *smooth curve* on a smooth manifold X is a smooth map $t \mapsto c(t)$ of an open interval of $\mathbb{R}$ into X. The *velocity vector* of a curve c at time t_0 in $x_0 = c(t_0)$, denoted $\frac{dc}{dt}(t_0)$, is defined to be the vector $dc_{t_0}(1) \in T_{x_0}(X)$, where $dc_{t_0}\colon \mathbb{R}^1 \to T_{x_0}(X)$.

(a) For $X = \mathbb{R}^k$ and $c(t) = (c_1(t), \ldots, c_k(t))$, show that

$$\frac{dc}{dt}(t_0) = dc_{t_0}(1) = (c_1'(t_0), \ldots, c_k'(t_0)) \in T_{x_0}\mathbb{R}^k.$$

(b) Let X be a k-dimensional smooth manifold. Show that every vector in $T_{x_0}(X)$ is the velocity vector of some curve in X. Hint: Use point (a) and a local parametrisation.

Aside: Note that two curves $c_1\colon I \to X$ and $c_2\colon J \to X$, with I and J open in $\mathbb{R}$, have the same velocity vector in $c_1(t_1) = x_0 = c_2(t_2)$ if $d(c_1)_{t_1}(1) = d(c_2)_{t_2}(1) \in T_{x_0}(X)$. One can show that having the same velocity vector at a point of X is an equivalence relation on the set of curves through x_0 in X. Using this relation, we have shown that there is a unique correspondence between tangent vectors at X in x_0 and equivalence classes of smooth curves through x_0 in X.

3

The Inverse Function Theorem, Immersions, and Embeddings

We now study smooth maps between manifolds in more detail. Let $f: X \to Y$ be a smooth map between smooth manifolds. We learned that the derivative $df_x: T_xX \to T_{f(x)}Y$ at a point $x \in X$ is a linear map between vector spaces. Since it is easier to understand linear maps, we will study smooth maps according to the behaviour of their derivatives. We then ask:

Question How much does the derivative df_x tell us about the map f?

3.1 The Inverse Function Theorem and Local Diffeomorphisms

Manifolds are characterised by the way they look in a neighbourhood around any point. So let us think locally and introduce the following terminology.

Definition 3.1 (Local diffeomorphism) Let $f: X \to Y$ be a smooth map between smooth manifolds. Then f is called a *local diffeomorphism at x* if there is an open subset $U \subset X$ containing x such that $f(U) \subset Y$ is open in Y and $f_{|U}: U \to f(U)$ is a diffeomorphism. We say that f is a *local diffeomorphism* if it is a local diffeomorphism at every $x \in X$.

We note that if the existence of a local diffeomorphism implies $\dim X = \dim Y$. Moreover, if $f_{|U}$ is a diffeomorphism $U \to V$ between neighbourhoods U around $x \in X$ and $y = f(x) \in Y$, respectively, let $f_{|U}^{-1}$ be its smooth inverse. Then we have $f_{|U}^{-1} \circ f_{|U} = \mathrm{id}_U$ and $f_{|U} \circ f_{|U}^{-1} = \mathrm{id}_V$. The chain rule implies $d(\mathrm{id}_U)_x = d(f_{|U}^{-1})_y \circ df_x$ and $d(\mathrm{id}_V)_y = df_x \circ d(f_{|U}^{-1})_y$. We have $d(\mathrm{id}_X) = \mathrm{Id}_{T_x(X)}$ for any manifold X and every point $x \in X$. Hence df_x is an isomorphism with inverse $d(f_{|U}^{-1})_{f(x)}$. Thus, a necessary condition for f to be a local diffeomorphism at x is that its derivative $df_x\colon T_x(X) \to T_y(Y)$ is an isomorphism. We now set out to prove that this is actually a sufficient condition. In order to prove this, we recall the corresponding result for Euclidean space from calculus.

Theorem 3.2 (The Inverse Function Theorem in calculus) *Let $U \subseteq \mathbb{R}^n$ be an open subset, and let $f\colon U \to \mathbb{R}^n$ be a smooth map. Assume that there is a point $a \in U$ such that df_a is an invertible linear map $\mathbb{R}^n \to \mathbb{R}^n$. Then f is a local diffeomorphism at a.*

We will use the following lemma in the proof[1] of Theorem 3.2.

Lemma 3.3 *For $r > 0$, let $f\colon \mathbb{B}^n_r \to \mathbb{R}^n$ be a smooth map which satisfies $\left|\frac{\partial f_i}{\partial x_j}(x)\right| \le b$ for all i, j. Then $|f(x) - f(y)| \le bn^2|(x-y)|$ for all $x, y \in \mathbb{B}^n_r$.*

Proof: We can rewrite the difference $f_i(y) - f_i(x)$ as a telescoping sum

$$f_i(y) - f_i(x) = \sum_{j=1}^{n} \Big[f_i(y_1, \dots, y_j, x_{j+1}, \dots, x_n) - f_i(y_1, \dots, y_{j-1}, x_j, \dots, x_n) \Big].$$

By the Mean Value Theorem in calculus, there are points $p_{i,j} \in \mathbb{B}^n_r$ such that

$$f_i(y_1, \dots, y_j, x_{j+1}, \dots, x_n) - f_i(y_1, \dots, y_{j-1}, x_j, \dots, x_n) = (y_j - x_j) \cdot \frac{\partial f_i}{\partial x_j}(p_{i,j}).$$

By our assumption, we have $\left|(y_j - x_j) \cdot \frac{\partial f_i}{\partial x_j}(p_{i,j})\right| < b \cdot |y_j - x_j|$, and since $|y_j - x_j| \le |y - x|$, this implies

$$f_i(y) - f_i(x) \le \sum_{j=1}^{n} |y_j - x_j| \cdot b \le n \cdot |y - x| \cdot b.$$

Thus, we get $|f(x) - f(y)| \le \sum_{j=1}^{n} |f_i(y) - f_i(x)| \le n^2 b |y - x|$. □

Proof of Theorem 3.2: To simplify the notation, we first observe the following. Since the translation which sends $f(x)$ to $f(0)$ is a diffeomorphism, we may assume that $a = 0$ and $f(0) = 0$. Moreover, since a linear isomorphism is a diffeomorphism, we can also assume that $f(0) = 0$ and $df_0 = \mathrm{Id}$ after replacing $f(x)$ with $(df_0)^{-1} \circ (f(x) - f(0))$ and using the chain rule. We then define a new map $g\colon U \to \mathbb{R}^n$ by setting $g(x) := f(x) - x$. The

[1] There are many different ways to prove the Inverse Function Theorem. Here we follow [43].

linear transformation dg_0 is given by the zero matrix. Since $\det(df_x)$ depends continuously on x in an open set containing a, we can choose an $r > 0$ with $\mathbb{B}_r^n \subseteq U$ and such that df_x is an isomorphism, i.e., $\det(df_x) \neq 0$ for all x with $|x| < r$. Moreover, since the partial derivatives of f and thereby of g depend continuously on x; we can, after possibly shrinking r, assume that the partial derivatives satisfy

$$\left|\frac{\partial g_i}{\partial x_j}(x)\right| \leq \frac{1}{2n^2} \text{ for all } x \in \mathbb{B}_r. \tag{3.1}$$

- Claim: For $y \in \mathbb{B}_{r/2}^n$, there is a unique $x \in \mathbb{B}_r^n$ such that $f(x) = y$.

Proof of the Claim: By Lemma 3.3 and the inequality in (3.1), we have

$$|g(x) - g(0)| \leq \frac{1}{2}|x| \text{ on } \mathbb{B}_r^n. \tag{3.2}$$

Now we define a sequence of points $\{x_i\}$ inductively by setting $x_0 = 0$, $x_1 = y$, $x_{i+1} = y - g(x_i)$. For each i, we have $x_i - x_{i+1} = g(x_{i-2}) - g(x_{i-1})$ and thus

$$|x_i - x_{i-1}| \leq \frac{1}{2}|x_{i-2} - x_{i-1}| \leq \frac{|y|}{2^{n-1}} \text{ on } \mathbb{B}_r^n. \tag{3.3}$$

Since $y \in \mathbb{B}_{r/2}^n$, this implies $x_i \in \mathbb{B}_r^n$ for all i. Moreover, (3.3) also implies that the sequence $\{x_i\}$ converges. By (3.3), we know $|x_i| \leq |y|$ for all i, and hence the sequence $\{x_i\}$ converges to a point $\tilde{x}$ with $|\tilde{x}| \leq 2|y|$. Again, since $y \in \mathbb{B}_{r/2}^n$, this shows $\tilde{x} \in \mathbb{B}_r^n$. As the limit of the sequence $\{x_i\}$, $\tilde{x}$ also satisfies the defining formula, i.e., we have $\tilde{x} = y - g(\tilde{x})$ and hence $f(\tilde{x}) = y$. This proves the existence part of the claim. To show uniqueness, assume there was another point $\tilde{x}' \in \mathbb{B}_r^n$ with $f(\tilde{x}') = y$. Then we would have $g(\tilde{x}') - g(\tilde{x}) = f(\tilde{x}') - \tilde{x}' - (f(\tilde{x}) - \tilde{x}) = \tilde{x}' - \tilde{x}$. By the inequality in (3.1), this contradicts Lemma 3.3 Hence we must have had $\tilde{x}' = \tilde{x}$. This proves the claim. □

This shows that the inverse map $f^{-1}\colon \mathbb{B}_{r/2}^n \to \mathbb{B}_r^n$ exists.

- Claim: The inverse $f^{-1}\colon \mathbb{B}_{r/2}^n \to \mathbb{B}_r^n$ is continuous.

Proof of the Claim: By definition of g, the triangle inequality, and by (3.2), we have $|f(x) - f(x')| \geq |x - x'| - |g(x) - g(x')| \geq \frac{1}{2}|x - x'|$ for all $x, x' \in \mathbb{B}_r^n$. Thus, for $y = f(x)$ and $y' = f(x')$, we get

$$|f^{-1}(y) - f^{-1}(y')| \leq 2|y - y'|. \tag{3.4}$$

This proves that f^{-1} is continuous. □

- Claim: The inverse $f^{-1}\colon \mathbb{B}^n_{r/2} \to \mathbb{B}^n_r$ is differentiable with $d(f^{-1})_y = (df_x)^{-1}$ at $y = f(x)$.

Proof of the Claim: For $x \in \mathbb{B}^n_r$, we write $A = df_x$, and let $y = f(x)$. We are going to show $d(f^{-1})_y = A^{-1}$. To do so, we set $\varphi(h) := f(x+h) - f(x) - A(h)$. By definition of differentiability, we then have

$$\lim_{h\to 0} \frac{|\varphi(h)|}{|h|} = 0. \tag{3.5}$$

By assumption, A is invertible. Let A^{-1} be its linear inverse. Then we get

$$\begin{aligned} A^{-1}(f(x+h) - f(x)) &= A^{-1}(A(h) + \varphi(h)) \\ &= h + A^{-1}(\varphi(h)) \\ &= [(x+h) - x] + A^{-1}(\varphi((x+h) - x)). \end{aligned}$$

With $y = f(x)$ and $y' = f(x+h)$, we can rewrite this observation as

$$A^{-1}(y' - y) = [f^{-1}(y') - f^{-1}(y)] + A^{-1}(\varphi(f^{-1}(y') - f^{-1}(y))),$$

and hence

$$-A^{-1}(\varphi(f^{-1}(y') - f^{-1}(y))) = [f^{-1}(y') - f^{-1}(y)] - A^{-1}(y' - y). \tag{3.6}$$

To prove that f^{-1} is differentiable at y with $d(f^{-1})_y = A^{-1}$, we need to show

$$\lim_{|y'-y|\to 0} \frac{|f^{-1}(y') - f^{-1}(y) - A^{-1}(y' - y)|}{|y' - y|} = 0.$$

To compute this limit, we rewrite the quotient first using (3.6) to get

$$\frac{|f^{-1}(y') - f^{-1}(y) - A^{-1}(y' - y)|}{|y' - y|} = \frac{|A^{-1}(\varphi(f^{-1}(y') - f^{-1}(y)))|}{|y' - y|}.$$

Now we recall that, for a linear transformation $T\colon \mathbb{R}^n \to \mathbb{R}^n$, we can find a real number M such that $|T(z)| \le M|z|$ for all $z \in \mathbb{R}^n$. Since A^{-1} is a linear map, it suffices to show

$$\lim_{|y'-y|\to 0} \frac{|\varphi(f^{-1}(y') - f^{-1}(y))|}{|y' - y|} = 0.$$

By factoring the quotient we get

$$\frac{|\varphi(f^{-1}(y') - f^{-1}(y))|}{|y' - y|} = \frac{|\varphi(f^{-1}(y') - f^{-1}(y))|}{|f^{-1}(y') - f^{-1}(y)|} \cdot \frac{|f^{-1}(y') - f^{-1}(y)|}{|y' - y|}.$$

By definition of φ and (3.5), the continuity of f^{-1} implies that the first factor tends to 0 when $|y' - y| \to 0$. By (3.4), we have $\frac{|f^{-1}(y')-f^{-1}(y)|}{|y'-y|} \leq 2$, i.e., the second factor is bounded. Thus, the limit of the product on the right-hand side is 0 when $|y' - y| \to 0$. This proves that f^{-1} is differentiable with $d(f^{-1})_y = A^{-1}$. This proves the claim. □

- Claim: The map $\mathbb{B}^n_{r/2} \to \mathrm{GL}(n)$ defined by $y \mapsto d(f^{-1})_y$ is smooth.

Proof of the Claim: We have proven that, for $y \in \mathbb{B}^n_{r/2}$, we can write the matrix $d(f^{-1})_y$ as $d(f^{-1})_y = (df_{f^{-1}(y)})^{-1}$. Thus, the map $y \mapsto d(f^{-1})_y$ is the composition of the maps f^{-1}, then taking the derivation of f at $f^{-1}(y)$ followed by taking the inverse of the matrix $df_{f^{-1}(y)}$. Since f is smooth, the map $x \mapsto df_x$ is smooth. Matrix inversion is smooth as well. Thus, since we proved that f^{-1} is continuous, $y \mapsto d(f^{-1})_y$ is continuously differentiable. Moreover, by induction we get that the map $y \mapsto d(f^{-1})_y$ is $\mathcal{C}^k$ if f^{-1} is $\mathcal{C}^k$. (See Definition 2.1 for what it means that a map is $\mathcal{C}^k$.) Thus, by induction, we get that f^{-1} is $\mathcal{C}^\infty$, or in other words, f^{-1} is smooth. This proves the claim. □

This proves that f has a local smooth inverse, which finishes the proof of Theorem 3.2. □

We will now generalise the theorem to maps between smooth manifolds.

Theorem 3.4 (Inverse Function Theorem) *Let X and Y be smooth manifolds, and let $f : X \to Y$ be a smooth map. Assume $df_x : T_x(X) \to T_{f(x)}(Y)$ is an isomorphism at $x \in X$. Then f is a local diffeomorphism at x.*

Proof: We can assume that both X and Y are subsets in $\mathbb{R}^N$. Let $\phi : U \to W \subset X$ be a local parametrisation around x with $U \subseteq \mathbb{R}^k$, and let $\psi : V \to \widetilde{W} \subset Y$ be a local parametrisation around $f(x)$ with $V \subseteq \mathbb{R}^k$. Note that the dimension k has to be the same when the tangent spaces are isomorphic. We can assume $\phi(0) = x$ and $\psi(0) = f(x)$, and choose U and V small enough such that $f(W) \subset \widetilde{W}$. Then we obtain the diagram

$$\begin{array}{ccc} W & \xrightarrow{\quad f_{|W}\quad} & \widetilde{W} \\ \phi \big\uparrow \; \big\downarrow \phi^{-1} & & \psi \big\uparrow \; \big\downarrow \psi^{-1} \\ U & \xrightarrow[\theta := \psi^{-1} \circ f_{|W} \circ \phi]{} & V \end{array} . \tag{3.7}$$

By the assumption, $df_x = d\psi_0 \circ d\theta_0 \circ (d\phi_0)^{-1}$ is an isomorphism. This implies that $d\theta_0$ is an isomorphism. By Theorem 3.2, this implies that there is an open neighbourhood $\widetilde{U} \subseteq U$ around 0 and there is an open neighbourhood $\widetilde{V} \subseteq V$ around 0 such that $\theta_{|\widetilde{U}} : \widetilde{U} \to \widetilde{V}$ is a diffeomorphism. Since ϕ and ψ are diffeomorphisms, $\phi(\widetilde{U}) \subseteq X$ and $\psi(\widetilde{V}) \subseteq Y$ are open neighbourhoods of x and y, respectively, and $f_{|\phi(\widetilde{U})} : \phi(\widetilde{U}) \to \psi(\widetilde{V})$ is a diffeomorphism. □

Theorem 3.4 tells us that in order to check that f is a diffeomorphism in a neighbourhood of a point x, we just need to check that a single number, the determinant of df_x, is non-zero.

Example 3.5 (Global diffeomorphisms) The map $f(t) = \tan t$ is a diffeomorphism $(-\pi/2, \pi/2) \to \mathbb{R}$. There is a diffeomorphism between $\mathbb{S}^1$ and SO(2). We have shown in the exercises to Chapter 2 that $f : \mathbb{B}_r^k \to \mathbb{R}^k$ given by $f(x) = \frac{rx}{\sqrt{r^2 - |x|^2}}$ is a diffeomorphism, and, by considering $\mathbb{S}^3 \subset \mathbb{C}^2$, that the map $\mathbb{S}^3 \to \mathrm{SU}(2)$ defined by $(a, b) \mapsto \begin{pmatrix} a & b \\ -\bar{b} & \bar{a} \end{pmatrix}$ is a diffeomorphism.

However, we have to remember that Theorem 3.4 makes a local statement, i.e., if df_x is invertible, it only tells us that f is invertible in a neighbourhood of x. Even if df_x is invertible for every $x \in X$, $f : X \to Y$ may not be a diffeomorphism.

Example 3.6 (A local but not global diffeomorphism) In this example, we show that the map $f : \mathbb{R}^1 \to \mathbb{S}^1 \subset \mathbb{R}^2$ defined by $f(t) = (\cos t, \sin t)$ is a local diffeomorphism but not a global diffeomorphism. First, f is not a global diffeomorphism because it is not injective. And we have seen that f is not a homeomorphism even when we restrict it to $[0, 2\pi) \to \mathbb{S}^1$. We could also argue that $\mathbb{S}^1$ is compact and $\mathbb{R}$ is not, so we cannot find a diffeomorphism between $\mathbb{S}^1$ and $\mathbb{R}$ by Lemma 1.13 and Exercise 1.1. However, Theorem 3.4 tells us that f is indeed a local diffeomorphism since df_t is an isomorphism for every $t \in \mathbb{R}$. For, let $t_0 \in \mathbb{R}$ such that $\cos(t_0) > 0$. For other points, the argument is similar. Consider the local parametrisation $\psi : V = (-1, 1) \to \widetilde{W} \subset \mathbb{R}^2$ defined by $\psi(y) = (\sqrt{1 - y^2}, y)$ of $\mathbb{S}^1$ around $f(t_0)$ with $V = \{(x, y) \in \mathbb{S}^1 : x < 0\}$. The inverse is given by projecting onto the second coordinate $\psi^{-1}(x, y) = y$. Let $\varepsilon > 0$ be such that both $\cos(t_0 - \varepsilon) > 0$ and $\cos(t_0 + \varepsilon) > 0$. We let $\phi : U = (t_0 - \varepsilon, t_0 + \varepsilon) \to \mathbb{R}$ be the local parametrisation around t_0 to be the inclusion. Then the map $\theta : U \to V$ is $\theta = \psi^{-1} \circ f \circ \phi$ given by $\theta(t) = \sin t$. Then we get $d\theta_t : \mathbb{R} \to \mathbb{R}$, $z \mapsto (\cos t) \cdot z$, at any point t in U. We have $d\phi_t = \mathrm{id}$ for $t \in U$. To calculate the derivative of ψ, we consider it first as a map $V \to \mathbb{R}^2$. Then we get

$$d\psi_y : \mathbb{R} \to \mathbb{R}^2,\ z \mapsto \left(-\frac{y}{\sqrt{1 - y^2}}, 1 \right) \cdot z.$$

Remember that the tangent space $T_{f(t_0)}\mathbb{S}^1$ is by definition the image of $d\psi_{y_0}\colon \mathbb{R} \to \mathbb{R}^2$ where y_0 is the point in W which maps to $f(t_0)$. Thus we have

$$T_{f(t_0)}\mathbb{S}^1 = d\psi_{t_0}(\mathbb{R}) = \mathrm{Span}\{(-\sin(t_0), \cos(t_0))\} \subset \mathbb{R}^2.$$

We have $df_{t_0} = d\psi_{\sin t_0} \circ d\theta_{t_0}$ and hence

$$df_{t_0}(z) = \left(-\frac{\sin t_0}{\cos t_0}, 1\right)(\cos t_0) \cdot z = (-\sin t_0, \cos t_0) \cdot z.$$

This shows that $df_{t_0}\colon T_{t_0}\mathbb{R} \to T_{f(t_0)}\mathbb{S}^1$ is an isomorphism.

Lemma 3.7 *A bijective local diffeomorphism is a global diffeomorphism.*

Proof: See Exercise 3.2. □

3.2 Immersions

Now, we study smooth maps for which the derivative is injective.

3.2.1 Definition and Examples

First, we introduce some terminology.

Definition 3.8 (Immersion) Let $f\colon X \to Y$ be a smooth map. If df_x is injective, we say that f is an *immersion at* x. If f is an immersion at every point, we say that f is an *immersion*.

Examples 3.9 Every linear injective map $L\colon \mathbb{R}^k \to \mathbb{R}^n$ is an immersion, since L equals its own derivative dL_x at every point $x \in \mathbb{R}^k$. In Exercise 3.5, we show that the map $f\colon \mathbb{R} \to \mathbb{R}^2$ defined by $f(t) := \left(\frac{e^t+e^{-t}}{2}, \frac{e^t-e^{-t}}{2}\right)$ is an immersion. Every local diffeomorphism is also an immersion. Example 3.6 therefore shows that an immersion does not have to be injective itself, only its derivative is injective at every point.

Example 3.10 Let $f\colon \mathbb{R}^2 \to \mathbb{R}^3$ be the smooth map defined by

$$f(s,t) := ((1 + 2\cos s)\cos t, (1 + 2\cos s)\sin t, 2\sin s)\,.$$

The map f is not injective, but we will now show that it is an immersion. The derivative of f at a point (s,t) can be described in the standard bases of $\mathbb{R}^2$ and $\mathbb{R}^3$, respectively, by the matrix

$$\begin{pmatrix} -2\sin s\cos t & -(2\cos s + 1)\sin t \\ -2\sin s\sin t & (2\cos s + 1)\cos t \\ 2\cos s & 0 \end{pmatrix}.$$

We claim that this matrix has rank 2 for all (s,t). From the third row, we get that the two columns can only be linearly dependent if $\cos s = 0$, i.e., if $s \in \{n\frac{\pi}{2} | n \in \mathbb{Z}\}$. However, the determinant of the submatrix of the two top rows is $-2\sin s(2\cos s + 1)$, which is $\neq 0$ for all $s \notin \{n\pi : n \in \mathbb{Z}\}$. Thus the two columns are always linearly independent, which proves the claim. Hence f is an immersion. The image of f is the torus $\mathbb{T}(1,2)$ consisting of all points in $\mathbb{R}^3$ at distance 2 from the circle of radius 1 in the xy-plane centred at the origin.

3.2.2 The Local Immersion Theorem

Before we study more examples and interesting phenomena, we pause for a moment and show a technical, but useful result about immersions. Among all the injective maps $\mathbb{R}^k \to \mathbb{R}^n$ with $k \leq n$, there is a very simple one, namely the map

$$\iota_n^k \colon \mathbb{R}^k \hookrightarrow \mathbb{R}^n,\ (x_1, \dots, x_k) \mapsto (x_1, \dots, x_k, 0, \dots, 0),$$

which sends the coordinates of a point in $\mathbb{R}^k$ to the first k coordinates in $\mathbb{R}^n$ and adds 0 at the remaining $n - k$ positions. This corresponds to the inclusion $\mathbb{R}^k \times \{\mathbf{0}\} \subset \mathbb{R}^n$ where $\mathbf{0}$ denotes the $(n-k)$-tuple of zeros. The map ι_n^k is called the *canonical immersion from* $\mathbb{R}^k$ *into* $\mathbb{R}^n$. The matrix which represents ι_n^k with respect to the standard bases of $\mathbb{R}^k$ and $\mathbb{R}^n$, respectively, is the $(n \times k)$-matrix J_n^k which has the $(k \times k)$-identity matrix in the first k rows and only zeros in the remaining $n-k$ rows at the bottom. However, not all injective maps $\mathbb{R}^k \to \mathbb{R}^n$ are as simple as ι_n^k. To motivate the next theorem, let $L \colon \mathbb{R}^k \hookrightarrow \mathbb{R}^n$ be a linear injective map. Then we can write L as the composition

$$L \colon \mathbb{R}^k \xrightarrow{\iota_n^k} \mathbb{R}^n \xrightarrow{\alpha} \mathbb{R}^n,$$

where α is a linear isomorphism. The canonical immersion ι_n^k places a copy of $\mathbb{R}^k$ into $\mathbb{R}^n$ as the k-dimensional plane containing the origin. Then we need to move this plane inside $\mathbb{R}^n$ to the position of $L(\mathbb{R}^k)$ via α. Thus, α is the linear isomorphism which sends the first k standard basis vectors $\mathbf{e}_1^n, \dots, \mathbf{e}_k^n$ of $\mathbb{R}^n$ to the images $\mathbf{v}_1 = L(\mathbf{e}_1^k), \dots, \mathbf{v}_k = L(\mathbf{e}_k^k)$ in $\mathbb{R}^n$ of the standard basis vectors $\mathbf{e}_1^k, \dots, \mathbf{e}_k^k$ of $\mathbb{R}^k$, and α sends the remaining basis vectors $\mathbf{e}_{k+1}^n, \dots, \mathbf{e}_n^n$ of $\mathbb{R}^n$ to a basis $\{\mathbf{w}_{k+1}, \dots, \mathbf{w}_n\}$ of the orthogonal complement of the image of L in $\mathbb{R}^n$. Since L is injective, this complement has dimension $n - k$. Actually, we are used to such a manoeuvre. For what α does is changing the basis of $\mathbb{R}^n$ from the standard basis to the new basis which consists of the vectors $\mathbf{v}_1, \ldots, \mathbf{v}_k, \mathbf{w}_{k+1}, \ldots, \mathbf{w}_n$. With respect to this new basis for $\mathbb{R}^n$ and the standard basis for $\mathbb{R}^k$, the matrix which represents L is the $n \times k$-matrix J_n^k described above. Now we can translate this idea to immersions between smooth manifolds.

Theorem 3.11 (Local Immersion Theorem) *Let X and Y be smooth manifolds of dimensions k and n, respectively, with $k \leq n$. Suppose that $f: X \to Y$ is an immersion at x, and write $y = f(x)$. Then we can choose local parametrisations $\phi: U \to W$ around x with $U \subset \mathbb{R}^k$ and $\psi: V \to \widetilde{W}$ around y with $V \subset \mathbb{R}^n$ such that in the commutative diagram*

$$
\begin{array}{ccc}
W & \xrightarrow{\quad f_{|W} \quad} & \widetilde{W} \\
\phi \big\uparrow & & \big\uparrow \psi \ \ \psi^{-1} \\
U & \xrightarrow[\theta := \psi^{-1} \circ f_{|W} \circ \phi]{\text{canonical immersion}} & V
\end{array}
\tag{3.8}
$$

the map $\theta: U \to V$ is the canonical immersion restricted to U.

Proof: We start by choosing any local parametrisations $\phi': U' \to W'$ with $\phi'(0) = x$ and $\psi': V' \to \widetilde{W}'$ with $\psi'(0) = y$. We set $\theta' = (\psi')^{-1} \circ f_{|W'} \circ \phi'$. By the assumption, df_x is injective, and hence $d\theta_0: \mathbb{R}^k \to \mathbb{R}^n$ is injective. After choosing a suitable basis for $\mathbb{R}^n$ as explained above, we can assume that $d\theta_0$ is given by the $n \times k$-matrix J_n^k which has the $k \times k$-identity matrix in the first k rows and only zeros in the remaining $n - k$ rows. Now we define a new map $\Theta: U' \times \mathbb{R}^{n-k} \to \mathbb{R}^n$ by $\Theta(x, z) := \theta(x) + (0, z)$. It is related to θ' by the commutative diagram

$$
\begin{array}{ccc}
U' & \xrightarrow[\text{immersion}]{\text{canonical}} & U' \times \mathbb{R}^{n-k} \\
& {}_{\theta}\searrow \quad \swarrow_{\Theta} & \\
& \mathbb{R}^n &
\end{array}
\quad .
\tag{3.9}
$$

Since f is an immersion at x, we can choose U' and V' small enough such that θ sends open sets to open sets. By the assumption on $d\theta_0$ and the construction of Θ, $d\Theta_0$ is represented by the $(n \times n)$-identity matrix. By the Inverse Function Theorem 3.4, this implies that $\Theta: U' \times \mathbb{R}^{n-k} \to \mathbb{R}^n$ is a local diffeomorphism at 0. Hence we can find open subsets $V \subset U' \times \mathbb{R}^{n-k}$ and $\widetilde{V} \subset \mathbb{R}^n$ containing the origins such that $\Theta_{|V}: V \to \widetilde{V}$ is a diffeomorphism. After possibly shrinking U' to an open subset $U \subset U'$ such that $U \times \{\mathbf{0}\} \subset V$, we get the commutative diagram

$$\begin{array}{ccc} \phi'(U) & \xrightarrow{\quad f_{|\phi'(U)} \quad} & \psi'(\widetilde{V}) \\ {\scriptstyle \phi'_{|U}}\uparrow & & \uparrow{\scriptstyle \psi'_{|\widetilde{V}}} \\ U & \xrightarrow{\quad \theta' \quad} & \widetilde{V} \\ & \searrow{\scriptstyle \text{canonical immersion}} & \uparrow{\scriptstyle \Theta} \\ & & V \end{array} \quad . \tag{3.10}$$

Since ψ' and Θ are local diffeomorphisms at 0, so is the composition $\psi'_{|\widetilde{V}} \circ \Theta$. Hence we can use $\psi'_{|\widetilde{V}} \circ \Theta$ as a local parametrisation around y. Then we set $\phi := \phi'_{|U'}$, $W := \phi'(U)$, and $\psi := \psi'_{|\widetilde{V}} \circ \Theta$, $\widetilde{W} := \psi'(\widetilde{V})$ to get the desired local paramatrisations $\phi\colon U \to W$ and $\psi\colon V \to \widetilde{W}$ such that $\theta := \psi^{-1} \circ f_{|W} \circ \phi$ is the canonical immersion. □

We observe from the proof that to be an immersion is a local condition.

Corollary 3.12 *Let $f\colon X \to Y$ be an immersion at x_0. Then there is an open neighbourhood $W \subset X$ around x_0 such that f is an immersion at all $x \in W$.*

Proof: Let $\phi\colon U \to W$ and $\psi\colon V \to \widetilde{W}$ be local parametrisations as in Theorem 3.11 such that $\theta := \psi^{-1} \circ f_{|W} \circ \phi$ is the canonical immersion. Then df_x is injective for all $x \in W$. □

3.2.3 The Image of an Immersion

Let $f\colon X \to Y$ be an immersion. We now set out to study the subspace $f(X) \subset Y$ and would like to find answers to the following two questions.

Question Is $f(X)$ always a submanifold, and is f a diffeomorphism onto its image?

We will see below that if f is a diffeomorphism onto $f(X)$ then this provides the subspace $f(X) \subset Y$ with the structure of a smooth manifold and hence $f(X)$ is a submanifold in Y. However, in principle, $f(X)$ can be a smooth manifold without f being a diffeomorphism onto $f(X)$. In Example 3.10, the image of the map $f\colon \mathbb{R}^2 \to \mathbb{R}^3$ is the two-dimensional torus and hence a submanifold. Nevertheless, f is not a diffeomorphism onto its image, since f is not one-to-one. We will now look at two examples of an immersion for which the image is not a smooth manifold.

Example 3.13 (Figure eight immersion as wrap) Consider the map

$$f\colon (-\pi, \pi) \to \mathbb{R}^2,\ t \mapsto (\sin 2t, \sin t).$$

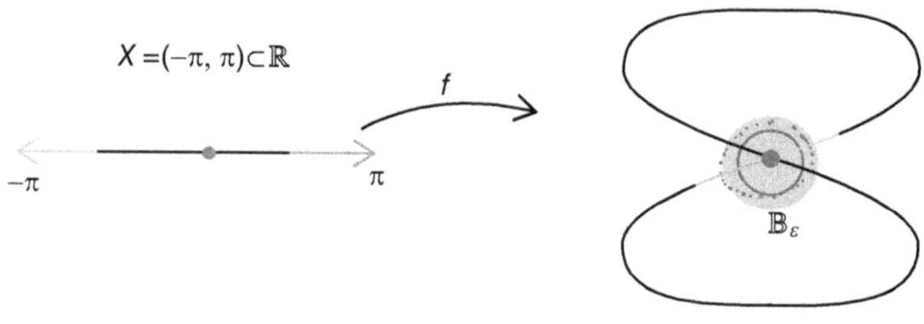

Figure 3.1 The map f is an immersion and one-to-one. However, the image of f, considered as a subspace of $\mathbb{R}^2$, is still not a submanifold.

We note that the image of f is a simple example of a *Lissajous curve* which plays an important role in, for example, the theory of complex harmonic motion. It is also an example of a *lemniscate*, as its points (x, y) satisfy the equation $x^2 = 4y^2(1 - y^2)$. See Figure 3.1. The map f is smooth and one-to-one. To show that it is an immersion, we note that df_t can be represented by the 2×1-matrix $J_f(t) = \begin{pmatrix} 2\cos 2t \\ \cos t \end{pmatrix}$ which is never zero for $t \in (-\pi, \pi)$ and hence df_t is injective for all t. Nevertheless, f is not a diffeomorphism onto its image. In fact, we see that f is not a homeomorphism onto its image $\mathrm{Im}(f)$. For, if f was a homeomorphism onto $\mathrm{Im}(f)$, then $f(-\delta, \delta)$ is open in $\mathrm{Im}(f)$ for every $\delta > 0$. However, every open ball $\mathbb{B}^2_\varepsilon$ of radius $\varepsilon > 0$ around $f(0) = (0, 0)$ contains points of $\mathrm{Im}(f)$ which are in the image of the intervals $(-\pi, -\pi + \delta')$ and $(\pi - \delta', \pi)$ for some small $\delta' > 0$. This shows that $f(-\delta, \delta)$ is not open in the subspace topology on $\mathrm{Im}(f)$, and f is not a homeomorphism onto its image. Moreover, the subset $\mathrm{Im}(f)$ in $\mathbb{R}^2$ is not a smooth manifold, since the point $(0, 0) \in \mathrm{Im}(f)$ does not admit a local parametrisation from an open subset in $\mathbb{R}$ while all other points do. See also the argument in Example 2.24.

Example 3.14 (Image of a line with irrational slope) We now look at a classical example of another case of a map which is a one-to-one immersion but not a homeomorphism, and hence not a diffeomorphism, onto its image See Figure 3.2. Let $g \colon \mathbb{R} \to \mathbb{S}^1 \subset \mathbb{C}$ be the local diffeomorphism $t \mapsto e^{2\pi i t}$. We define the map $G \colon \mathbb{R}^2 \to \mathbb{T}^2 = \mathbb{S}^1 \times \mathbb{S}^1 \subset \mathbb{C}^2$ by $G(s, t) := (g(s), g(t))$. The map G is a local diffeomorphism from the plane onto the two-dimensional torus $\mathbb{T}^2$. Let $\alpha \in \mathbb{R}$ be an *irrational* number. We define the map $\gamma : \mathbb{R} \to \mathbb{T}^2$ by

$$\gamma(t) := (g(t), g(\alpha t)).$$

The map γ is an immersion because $d\gamma_t$ is non-zero for every t, and a non-zero linear map from a one-dimensional vector space to a four-dimensional vector

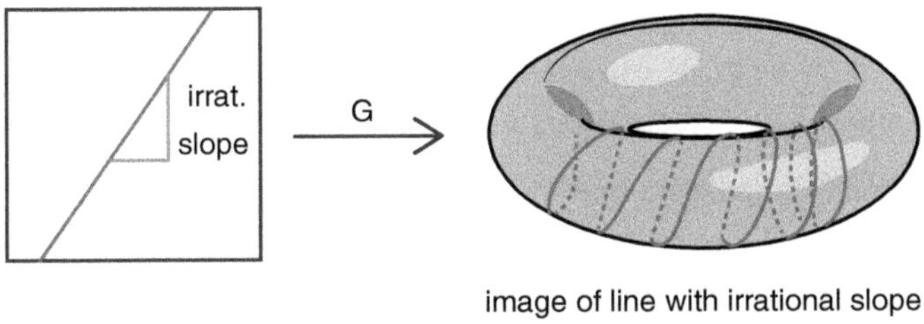

Figure 3.2 The map γ wraps a line around the torus. If the slope is irrational, then the image will never meet itself. However, the image meets every open subset of $\mathbb{T}^2$.

space is injective. Moreover, γ itself is injective, since $\gamma(t_1) = \gamma(t_2)$ implies $g(t_1) = g(t_2)$ and $g(\alpha t_1) = g(\alpha t_2)$, i.e., $e^{2\pi i t_1} = e^{2\pi i t_2}$ and $e^{2\pi i \alpha t_1} = e^{2\pi i \alpha t_2}$. Thus, we get $t_1 - t_2 \in \mathbb{Z}$ and $\alpha(t_1 - t_2) \in \mathbb{Z}$. Since α is irrational, this implies $t_1 = t_2$. However, γ is not a diffeomorphism onto its image, since it is not even a homeomorphism onto its image as we will now explain. Consider the set $\gamma(\mathbb{Z}) := \{\gamma(n) : n \in \mathbb{Z}\}$. By Dirichlet's Approximation Theorem, for every $\varepsilon > 0$, there are integers n and m such that $|\alpha n - m| < \varepsilon$. Thus, since the line segment in the plane between two points $e^{2\pi i t_1}$ and $e^{2\pi i t_2}$ on the unit circle is shorter than the circular arc of length $2\pi|t_1 - t_2|$, we can, for every $\varepsilon > 0$, find integers n and m such that $|e^{2\pi i \alpha n} - e^{2\pi i m}| \leq 2\pi|\alpha n - m| < 2\pi\varepsilon$. Now, with coordinates in $\mathbb{C}^2$, we have $|\gamma(n) - \gamma(0)| = |(e^{2\pi i n}, e^{2\pi i \alpha n}) - (1,1)| = |e^{2\pi i \alpha n} - 1|$. Since $e^{2\pi i m} = 1$ for all $m \in \mathbb{Z}$, we can, for every $\varepsilon > 0$, find integers n and m such that $|\gamma(n) - \gamma(0)| \leq 2\pi|\alpha n - m| < 2\pi\varepsilon$. In particular, for every $\varepsilon > 0$, find an integer n such that $|\gamma(n) - \gamma(0)| < 2\pi\varepsilon$. This shows that that $\gamma(\mathbb{Z})$ is not a discrete subset of $\mathbb{C}^2 = \mathbb{R}^4$, since we cannot find an open ball $\mathbb{B}^n_\varepsilon(\gamma(0))$ around $\gamma(0)$ which does not contain any other point of $\gamma(\mathbb{Z})$, and $\gamma(\mathbb{Z})$ is therefore not discrete in $\gamma(\mathbb{R})$ either. If γ was a homeomorphism onto its image, then it would have to send discrete subsets in $\mathbb{R}$ to discrete subsets. Thus, since $\mathbb{Z}$ is discrete in $\mathbb{R}$, γ is not a homeomorphism onto its image. See also Exercise 3.7 for the case when α is rational.

3.3 Embeddings

We would now like to understand which additional conditions ensure that an immersion is a diffeomorphism onto its image and that the image is a submanifold. First, we introduce a new terminology for this case.

Definition 3.15 (Embedding) An immersion $f: X \to Y$ is called an *embedding* if $f(X) \subset Y$ considered with the subspace topology is a smooth manifold and f is a diffeomorphism onto its image.

An embedding must be one-to-one, since it is a diffeomorphism onto its image. So, let $f: X \to Y$ be an immersion which is also one-to-one. We know from Examples 3.13 and 3.14 that this does not imply that f is an embedding.

Question Under which circumstances is f an embedding?

By the Local Immersion Theorem 3.11, we can choose local parametrisations $\phi: U \to W \subset X$ around x and $\psi: V \to \widetilde{W} \subset Y$ around $y = f(x)$ such that the induced map $\theta: U \to V$ is the canonical immersion. This provides the commutative diagram

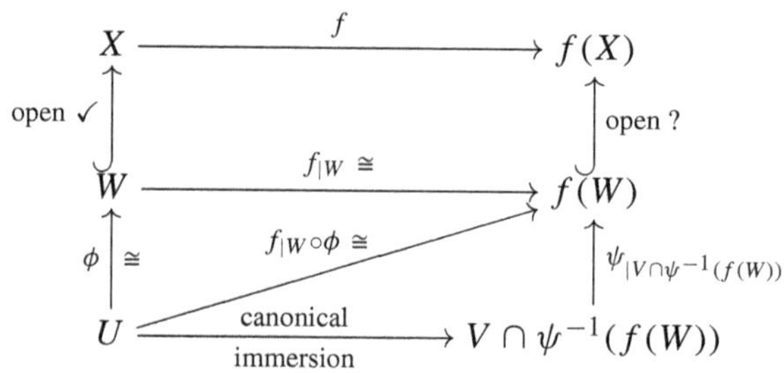

. (3.11)

Since the bottom horizontal map is the canonical immersion, the restriction $f_{|W}$ is a diffeomorphism. As ϕ is a diffeomorphism, the composite $f_{|W} \circ \phi: U \to f(W)$ is a diffeomorphism, too. Thus $f(W)$ is diffeomorphic to an open subset in $\mathbb{R}^n$. Since we can do this for every point $y \in f(X)$, we would like to say that the collection of diffeomorphisms $f_{|W} \circ \phi$ for varying ϕ provides local parametrisations for $f(X)$. However, we do not know whether $f(W)$ is open in $f(X)$. This does not follow from the given assumptions on f. Let us assume for a moment that $f(W)$ was open for each open subset $W \subset X$. Then f would be a bijective and a local diffeomorphism since, for every point $x \in X$, there would be an open subset $x \in W \subset X$ and an open subset $f(W) \subset f(X)$ such that $f_{|W}$ is a diffeomorphism $W \to f(W)$. Then f is a bijective local diffeomorphism, and hence f a diffeomorphism by Lemma 3.7 and Exercise 3.2.

We learn from this discussion that the missing condition is a purely topological one. Namely, if an immersion $f: X \to Y$ is a homeomorphism onto its image, then f is a diffeomorphism onto its image and $f(X)$ is a smooth manifold. In fact, if we know that f is one-to-one, then, by Corollary 1.23, we only need to check that f is an open or a closed map as Y is Hausdorff. Note that this fits well with Examples 3.13 and 3.14 since we showed that the maps of the examples are not open. Finally, we recall that if f is one-to-one and proper, i.e., if the preimages of compact subsets under f are again compact, then, by Proposition 1.27, we know that f is a homeomorphism onto its image as Y is Hausdorff. Thus, we have proven the following theorem.

Theorem 3.16 (Embedding Theorem) *Let $f: X \to Y$ be a one-to-one immersion which satisfies one of the following conditions: f is either open, or closed, or proper. Then f is an embedding.*

Corollary 3.17 *Let X be a compact smooth manifold. Then every one-to-one immersion $f: X \to Y$ is an embedding.*

Proof: Since X is compact and Y is Hausdorff, f is proper by Lemma 1.26. Hence the assertion follows from Theorem 3.16. □

3.4 Exercises

3.1 Let $A: \mathbb{R}^n \to \mathbb{R}^n$ be a linear map, and $b \in \mathbb{R}^n$. Show that the map $f: \mathbb{R}^n \to \mathbb{R}^n$ given by $f(x) := Ax + b$ is a diffeomorphism of $\mathbb{R}^n$ if and only if A is invertible.

3.2 Let $f: X \to Y$ be a smooth map. Assume that f is a local diffeomorphism, and assume that f is bijective as a map of sets. Show that f is a diffeomorphism.

3.3 Show that a local diffeomorphism $f: X \to Y$ which is one-to-one is a diffeomorphism of X onto an open subset of Y. Hint: Use Exercise 3.2.

3.4 Show that $f: \mathbb{R}^3 \to \mathbb{R}^3$, $f(x, y, z) := (e^{2y} + e^{2z}, e^{2x} - e^{2z}, x - y)$, is an immersion. Is f a local diffeomorphism?

3.5 Show that $f: \mathbb{R} \to \mathbb{R}^2$, $f(t) := \left(\frac{e^t + e^{-t}}{2}, \frac{e^t - e^{-t}}{2}\right)$, is an embedding.

3.6 Show that the map $f: \mathbb{R}^2 \to \mathbb{R}^3$ defined by

$$(s, t) \mapsto ((2 + \cos(2\pi s)) \cos(2\pi t), (2 + \cos(2\pi s)) \sin(2\pi t), \sin(2\pi s))$$

is an immersion. Is it an embedding?

3.7 Let $\gamma_{a,b}: \mathbb{R} \to \mathbb{S}^1 \times \mathbb{S}^1$ be the curve on the torus defined by $\gamma_{a,b}(t) := (e^{2\pi i a t}, e^{2\pi i b t})$ where we consider $\mathbb{S}^1$ as a subset of $\mathbb{C} \cong \mathbb{R}^2$. Let a and b are integers with $a \neq 0$. Show that $\gamma_{a,b}$ factors through an embedding $\mathbb{S}^1 \to \mathbb{S}^1 \times \mathbb{S}^1$, i.e., find an embedding $g_{a,b}: \mathbb{S}^1 \to \mathbb{S}^1 \times \mathbb{S}^1$ such that $\gamma_{a,b}$ is the composite of the map $\mathbb{R} \to \mathbb{S}^1, t \mapsto e^{2\pi i t}$, followed by $g_{a,b}$.

3.8 Consider the map $f: (0, 3\pi/4) \to \mathbb{R}^2$, $t \mapsto \sin(2t)(\cos t, \sin t)$.

(a) Show that f is an immersion.

(b) Let $\mathrm{Im}(f) = f((0, 3\pi/4)) \subset \mathbb{R}^2$ be the image of f considered as a subspace in $\mathbb{R}^2$. Show that $f\colon (0, 3\pi/4) \to \mathrm{Im}(f)$ is not a homeomorphism. (Draw a picture of the image of f.)

(c) Answer the following questions:

- What is the difference between $\mathrm{Im}(f)$ and the graph $\Gamma(f)$?
- Is the map $F\colon (0, 3\pi/4) \to (0, 3\pi/4) \times \mathbb{R}^2$, $t \mapsto (t, f(t))$, an embedding? Hint: Remember Exercise 2.14.
- Would f be an embedding if it were defined on the closed interval $[0, 3\pi/4]$? Hint: Check what happens at the endpoints.
- Is the map $g\colon (0, 3\pi/4) \to \mathbb{R}^3$, $t \mapsto \sin(2t)(\cos t, \sin t, t)$ an embedding?
- Is the map $h\colon [0, 3\pi/4] \to \mathbb{R}^3$, $t \mapsto (\sin(2t)\cos t, \sin(2t)\sin t, 2t)$ an embedding?

3.9 In this exercise, we are going to prove the first part of the Constant Rank Theorem. The remaining parts will be proven in Exercise 4.6. Let $f\colon X \to Y$ be a smooth map. We say that f has *constant rank* if the linear map df_x has rank r for all $x \in X$. Show that an injective smooth map $f\colon X \to Y$ of constant rank is an immersion.

3.10 Let X be an n-dimensional smooth manifold, Z be a k-dimensional smooth submanifold of X, and let $z \in Z$. Show that there exists a local coordinates $(x_1, \ldots, x_n)$ defined in a neighbourhood U of z in X such that $Z \cap U$ is defined by the equations $x_{k+1} = 0, \ldots, x_n = 0$, i.e., $Z \cap U$ is the subset of points in U for which the functions $x_{k+1}, \ldots, x_n$ all vanish.

4
Submersions and Regular Values

We will now turn to the question of how much we know about a smooth map between manifolds if the derivative is surjective. This case will lead to the notion of regular values, one of the key concepts in differential topology, and to the Preimage Theorem on the fibres of smooth maps at regular values. As an application, we will discuss Milnor's proof of the Fundamental Theorem of Algebra using the techniques we have learned so far.

4.1 Submersions

We begin with some terminology.

Definition 4.1 (Submersion) Let X and Y be smooth manifolds and let $f\colon X \to Y$ be a smooth map. If df_x is surjective, we say that f is a *submersion at* x. If f is a submersion at every point, we say that f is a *submersion*.

Example 4.2 Every linear surjective map $L\colon \mathbb{R}^k \to \mathbb{R}^n$ is a submersion. This follows from the fact that L equals its own derivative dL_x at every point $x \in \mathbb{R}^k$.

Example 4.3 Every local diffeomorphism $f\colon X \to Y$ is also a submersion. Considering our examples for local diffeomorphisms, this also shows that a submersion f does not have to be surjective itself, only its derivative df_x is surjective at every x.

Example 4.4 Let $f\colon \mathbb{R}^4 \to \mathbb{R}$ be given by $f(x_1,x_2,x_3,x_4) := x_1 + x_2^2 + x_3^3 + x_4^4$. The derivative df_x at a point $x = (x_1,x_2,x_3,x_4)$ is a linear map $\mathbb{R}^4 \to \mathbb{R}$ given in the standard basis by the 1×4-matrix $df_x = \begin{pmatrix} 1 & 2x_2 & 3x_3^2 & 4x_4^3 \end{pmatrix}$. Since the codomain of the linear map df_x has dimension one, it suffices to observe that df_x is not the zero map to conclude that df_x is surjective for all $x \in \mathbb{R}^4$. Hence f is a submersion.

Example 4.5 We show in Exercise 4.11 that the determinant $\det\colon \mathrm{M}(n) \to \mathbb{R}$ is a submersion at every matrix $A \in \mathrm{GL}(n)$.

Example 4.6 Recall the Hopf map $\hbar\colon \mathbb{S}^3 \to \mathbb{S}^2$ from Example 2.10 and Exercise 2.10. We show in Exercise 4.12 that $\hbar$ is a submersion. So far, our previous calculations show that the derivative of f is surjective at $\mathbf{n}_3 = (0,0,0,1) \in \mathbb{S}^3$ and at $\mathbf{q}_3 = (1,0,0,0) \in \mathbb{S}^3$. For we showed that $d\hbar_{\mathbf{n}_3}$ can be represented by the matrix $\begin{pmatrix} 0 & 2 & 0 \\ -2 & 0 & 0 \end{pmatrix}$, and $d\hbar_{\mathbf{q}_3}$ can be represented by the matrix $\begin{pmatrix} 0 & 2 & 0 \\ 0 & 0 & -2 \end{pmatrix}$. Both these matrices have rank two, and the corresponding linear maps are surjective. Hence $\hbar$ is a submersion at $\mathbf{n}_3$ and $\mathbf{q}_3$.

The possibly simplest surjective map $\mathbb{R}^n \to \mathbb{R}^m$ with $n \geq m$ is the map

$$\sigma_m^n\colon \mathbb{R}^n \to \mathbb{R}^m,\ (x_1,\ldots,x_m,x_{m+1},\ldots,x_n) \mapsto (x_1,\ldots,x_m),$$

which sends the coordinates of a point in $\mathbb{R}^n$ to the point in $\mathbb{R}^m$ with the first m coordinates and forgets the remaining $n-m$ ones. The map σ_m^n is called the *canonical submersion* from $\mathbb{R}^n$ onto $\mathbb{R}^m$. The matrix which represents σ_m^n with respect to the standard bases of $\mathbb{R}^n$ and $\mathbb{R}^m$, respectively, is the $m \times n$-matrix S_m^n which has the $m \times m$-identity matrix in the first m columns and only zeros in the remaining $n-m$ columns on the right-hand side. Starting with a linear surjective map $L\colon \mathbb{R}^n \to \mathbb{R}^m$, we can split L into a composition of the form

$$\mathbb{R}^n \xrightarrow{\alpha} \mathbb{R}^n \xrightarrow{\sigma_m^n} \mathbb{R}^m,$$

where α is a linear isomorphism defined as follows. Since L is surjective, the kernel of L has dimension $n-m$, and the quotient vector space $\mathbb{R}^n/\operatorname{Ker} L$ has dimension m. Let $\mathbf{v}_1,\ldots,\mathbf{v}_m$ be vectors in $\mathbb{R}^n$ which map to a basis of $\mathbb{R}^n/\operatorname{Ker} L$ and such that $L(\mathbf{v}_1) = \mathbf{e}_1^n,\ldots,L(\mathbf{v}_m) = \mathbf{e}_m^n$ in $\mathbb{R}^n$. Let $\mathbf{w}_1,\ldots,\mathbf{w}_{n-m}$ be a basis of $\operatorname{Ker} L$. We define α by sending the first m standard basis vectors $\mathbf{e}_1^n,\ldots,\mathbf{e}_m^n$ of $\mathbb{R}^m$ to $\mathbf{v}_1,\ldots,\mathbf{v}_m$ and the remaining $n-m$ basis vectors $\mathbf{e}_{m+1}^n,\ldots,\mathbf{e}_n^n$ of $\mathbb{R}^n$ to $\mathbf{w}_1,\ldots,\mathbf{w}_{n-m}$. That is, first we move all the action into $\mathbb{R}^n$, while the second step, where the dimension drops, consists of one fixed map which just forgets coordinates. Again, this is a familiar manoeuvre. For α changes the basis of $\mathbb{R}^n$ from the standard basis to the basis which consists of the vectors $\mathbf{v}_1,\ldots,\mathbf{v}_m,\mathbf{w}_1,\ldots,\mathbf{w}_{n-m}$. With respect to this new basis for $\mathbb{R}^n$ and the standard basis for $\mathbb{R}^m$, the matrix which represents L is the $m \times n$-matrix S^n_m described above.

Theorem 4.7 (Local Submersion Theorem) *Let X and Y be smooth manifolds of dimensions n and m, respectively, with $n \geq m$. Let $f : X \to Y$ be a submersion at x, and write $y = f(x)$. Then we can choose local parametrisations $\phi : U \to W$ around x with $U \subset \mathbb{R}^n$ and $\psi : V \to \widetilde{W}$ around y with $V \subset \mathbb{R}^m$ such that in the commutative diagram*

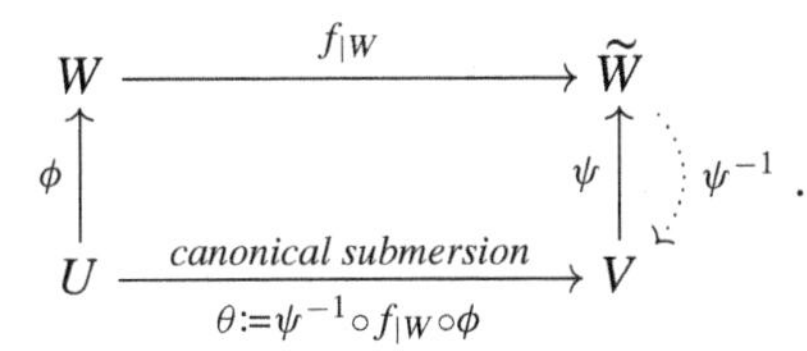

(4.1)

the map $\theta : U \to V$ is the canonical submersion restricted to U.

Proof: We start by choosing any local parametrisations $\phi' : U' \to W'$ with $\phi'(0) = x$ and $\psi' : V' \to \widetilde{W}'$ with $\psi'(0) = y$. We set $\theta' = (\psi')^{-1} \circ f_{|W'} \circ \phi'$. By the assumption, df_x is surjective, and hence $d\theta'_0 : \mathbb{R}^n \to \mathbb{R}^m$ is surjective. Hence, after choosing a suitable basis for $\mathbb{R}^n$, we can assume that $d\theta_0$ is given by the $m \times n$-matrix S^n_m which has the $m \times m$-identity matrix in the first m columns and only zeros in the remaining $n - m$ columns.[1] We define a new map $\Theta : U \to \mathbb{R}^n$ by $\Theta(\mathbf{x}) := (\theta'(\mathbf{x}), x_{m+1}, \ldots, x_n)$ for a point $\mathbf{x} = (x_1, \ldots, x_n)$. It is related to θ' by the commutative diagram

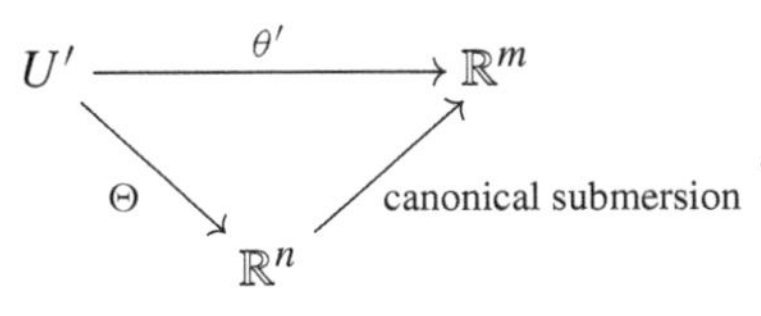

(4.2)

By the construction, the derivative $d\Theta_0$ at 0 is given by the $(n \times n)$-identity matrix with respect to our suitably chosen basis. By the Inverse Function Theorem 3.4, this implies that $\Theta : U' \to \mathbb{R}^n$ is a local diffeomorphism at 0. Thus, we can find open subsets $U \subset \mathbb{R}^n$ and $\widetilde{U} \subset U'$ such that $\Theta_{|\widetilde{U}} : \widetilde{U} \to U$ is a diffeomorphism. i.e., we have $\theta \circ \Theta^{-1} = \sigma^n_m$ on U'. This gives us the commutative diagram

[1] The reason we make this change of bases is to make sure that the map Θ we are about to define is a local diffeomorphism, which we know since $d\Theta_0$ is the identity matrix. Without the change of bases, we would not be sure that $d\Theta_0$ is an isomorphism.

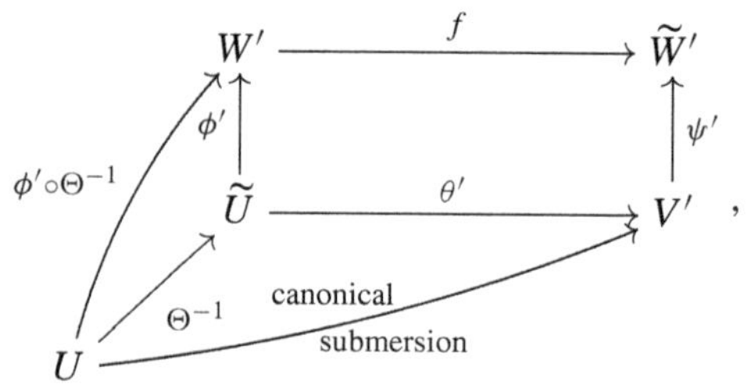

(4.3)

in which we omit the respective restrictions to simplify the notation. Then we set $\phi := \left(\phi' \circ \Theta^{-1}_{|\widetilde{U}}\right)$, $W := \phi'(\widetilde{U})$, and $V := \theta'(\widetilde{U})$ and $\psi := \psi'_{|V}$ to get the desired local paramatrisations $\phi\colon U \to W$ and $\psi\colon V \to \widetilde{W}$ such that the composition $\theta := \psi^{-1} \circ f_{|W} \circ \phi$ is the canonical submersion. □

We observe from the proof that to be an immersion is a local condition.

Corollary 4.8 *Let $f\colon X \to Y$ be a submersion at x_0. Then there is an open neighbourhood $W \subset X$ around x_0 such that f is a submersion at all $x \in W$.*

Proof: Let $\phi\colon U \to W$ and $\psi\colon V \to \widetilde{W}$ be local parametrisations as in Theorem 4.7 such that $\theta := \psi^{-1} \circ f_{|W} \circ \phi$ is the canonical submersion. Then df_x is surjective for all $x \in W$. □

As another consequence, we get the following useful observation.

Proposition 4.9 *Let $f\colon X \to Y$ be a smooth map. Then the set R of regular values for f is open in Y.*

Proof: By Corollary 4.8, the set of regular points for f is open in X. This implies that its complement, the set C of critical points for f, is closed in X. Since X is compact, this implies that C is a compact subset by Lemma 1.14. Since f is continuous, $f(C)$ is a compact subset of Y by Lemma 1.13. Since Y is Hausdorff, this implies that $f(C)$ is closed in Y by Proposition 1.21. The set $f(C)$ is the set of critical values of f, and its complement is the set R of regular values. Hence R is open in Y. □

4.2 Regular Values and the Preimage Theorem

Let $f\colon X \to Y$ be a smooth map and $y \in Y$. Our next goal is to understand the structure of the fibre of f over y, i.e., the set

$$f^{-1}(y) = \{x \in X\colon f(x) = y\} \subset X.$$

In general, there is no reason for the set $f^{-1}(y)$ to be a manifold. So let us make a convenient assumption, and assume that $f\colon X \to Y$ is a submersion at a point $x_0 \in X$. We denote $n = \dim X \geq \dim Y = m$ and assume that $f^{-1}(y)$ is non-empty. Let x_0 be a point with $f(x_0) = y$. By the Local Submersion Theorem 4.7, we choose local parametrisations $\phi\colon U \to W$ around x_0 with $U \subset \mathbb{R}^n$ and $\psi\colon V \to \widetilde{W}$ around y with $V \subset \mathbb{R}^m$ such that $\theta = \psi^{-1} \circ f \circ \phi$ is the canonical submersion $\mathbb{R}^n \to \mathbb{R}^m$ restricted to U:

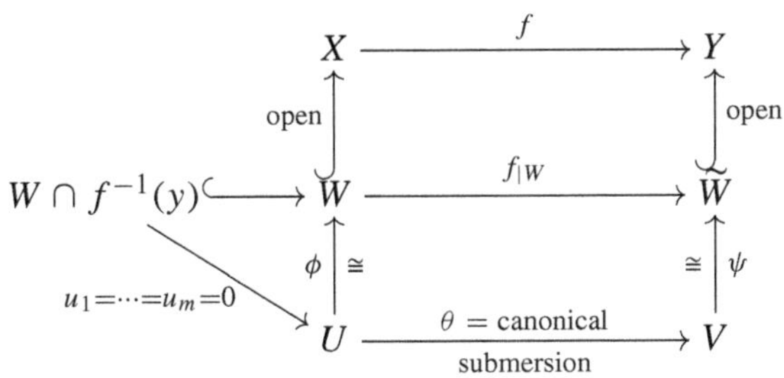

(4.4)

Since $\psi^{-1}(y) = 0$, we have $x \in V \cap f^{-1}(y)$ if and only if $\theta(\phi^{-1}(x)) = 0$. We now write $\phi^{-1}(x) =: (u_1(x), u_2(x), \ldots, u_n(x))$ for smooth functions $u_i\colon U \to \mathbb{R}$. With this notation, we can rewrite the above equivalence as $x \in V \cap f^{-1}(y)$ if and only if $\theta\,(u_1(x), u_2(x), \ldots, u_n(x)) = 0$. Now, since θ is the canonical submersion, we can rewrite this yet again as $x \in V \cap f^{-1}(y)$ if and only if $u_i(x) = 0$ for $i = 1, \ldots, m$. Hence, in terms of sets, we get

$$f^{-1}(y) \cap V = \{x \in V\colon u_1(x) = \cdots = u_m(x) = 0\}.$$

On $f^{-1}(y) \cap V$, the remaining coordinate functions $u_{m+1}(x), \ldots, u_n(x)$ do not have to satisfy any additional condition. Hence we can use them to define a local coordinate system on $\mathbb{R}^{n-m}$ for points $x \in f^{-1}(y) \cap V$. In fact, the subset $f^{-1}(y) \cap V \subset f^{-1}(y)$ is open in $f^{-1}(y)$ since V is open in X and $f^{-1}(y)$ has the subspace topology. Similarly, the subset $(\{0\} \times \mathbb{R}^{n-m}) \cap U$ is open in $\{0\} \times \mathbb{R}^{n-m}$ since U is open in $\mathbb{R}^n$. Thus, we obtain a local coordinate system

$$f^{-1}(y) \cap V \to (\{0\} \times \mathbb{R}^{n-m}) \cap U,\ x \mapsto (u_{m+1}(x), \ldots, u_n(x))$$

for $f^{-1}(y)$ around x_0. Now let $U' \subset \mathbb{R}^{n-m}$ be the open subset such that $\{0\} \times U' = (\{0\} \times \mathbb{R}^{n-m}) \cap U$. We have shown that the restriction of ϕ to U' defines a diffeomorphism $\phi'\colon U' \to f^{-1}(y) \cap V$ which we would like to use a local parametrisation for the points in $f^{-1}(y) \cap V$. If we could do this for every x with $f(x) = y$ we would obtain that $f^{-1}(y)$ is a smooth manifold. In general, however, df_x may not be surjective. Hence we introduce new terminology for the case in which df_x is surjective at all points in $f^{-1}(y)$.

Definition 4.10 (Regular points and regular values) Let X and Y be smooth manifolds, and let $f\colon X \to Y$ be a smooth map. A point $x \in X$ is called a *regular point for* f if $df_x\colon T_x(X) \to T_y(Y)$ is surjective, i.e., f is a submersion at x. An element $y \in Y$ is called a *regular value for* f if all points $x \in f^{-1}(y)$ are regular, i.e., $y \in Y$ is a regular value if $df_x\colon T_x(X) \to T_y(Y)$ is surjective at every point $x \in X$ such that $f(x) = y$.

Remark 4.11 We note that, in particular, all values y which are *not* in the image of f are regular values for f. For, if $f^{-1}(y)$ is the empty set, then there is no condition to be satisfied.

The above argument shows the following important result.

Theorem 4.12 (Preimage Theorem) *Let X and Y be smooth manifolds, and let $f\colon X \to Y$ be a smooth map. Let y be a regular value for $f\colon X \to Y$ with $f^{-1}(y) \neq \emptyset$. Then $f^{-1}(y)$ is a submanifold of X with* $\dim f^{-1}(y) = \dim X - \dim Y$.

Theorem 4.12 is one of the main tools to show that a space is a smooth manifold. It is a milestone on our journey, and we should have it in mind whenever we are asked to show that a space has the structure of a smooth manifold.

Proposition 4.13 (Tangent space of a regular fibre) *Let $f\colon X \to Y$ be a smooth map of manifolds, and let y be a regular value with $f^{-1}(y) \neq \emptyset$. Let $Z = f^{-1}(y)$ be the fibre $y \in Y$. Then the tangent space $T_x(Z)$ at a point $x \in Z$ is the kernel of the derivative $df_x\colon T_x(X) \to T_y(Y)$, i.e.,* $T_x(Z) = \mathrm{Ker}\,(df_x)$.

Proof: With the notation of the argument to prove Theorem 4.12, we have the following commutative diagram:

$$\begin{array}{ccccc} T_x(Z) & \hookrightarrow & T_x(X) & \xrightarrow{\;df_x\;} & T_y(Y) \\ \uparrow{\scriptstyle d\phi_0'} & & \uparrow{\scriptstyle d\phi_0} & & \uparrow{\scriptstyle d\psi_0} \\ \mathbb{R}^{n-m} & \hookrightarrow & \mathbb{R}^n & \xrightarrow{\;d\theta_0=\sigma^n_m\;} & \mathbb{R}^m \end{array} \qquad (4.5)$$

Since σ^n_m is the canonical submersion, the composition of the lower horizontal map is zero. Since $T_x(Z)$ is defined as the image of $d\phi_0'$ and the diagram commutes, this shows that df_x sends all of $T_x(Z)$ to 0. This proves $T_x(Z) \subseteq \mathrm{Ker}\, df_x$. On the other hand, since y is a regular value, df_x is surjective. Hence the dimension of the kernel of df_x is $\dim T_x(X) - \dim T_y(Y) =$

$\dim X - \dim Y = \dim Z$. This shows that $T_x(Z)$ is a subspace of the kernel of df_x of the same dimension as $\operatorname{Ker} df_x$. Thus $T_x(Z) = \operatorname{Ker} df_x$. □

Definition 4.14 (Critical values) Let X and Y be smooth manifolds, and let $f: X \to Y$ be a smooth map. A point $x \in X$ is called a *critical point* for f if df_x is not surjective, i.e., if x is not a regular point. An element $y \in Y$ is called a *critical value* for f if it is not a regular value for f, i.e., if there is at least one critical point x in $f^{-1}(y)$.

4.3 Examples

Now we look at some first examples. We will see many more applications of Theorem 4.12 in this book. See also the following subsections and exercises.

Example 4.15 Let $f: \mathbb{R}^{n+1} \to \mathbb{R}$ be the map $f(x) = |x|^2$. The derivative df_x at the point $x = (x_1, \ldots, x_{n+1})$ is the linear map given by the matrix $(2x_1 \ldots 2x_{n+1})$ expressed in the standard basis. Thus $df_x: \mathbb{R}^{n+1} \to \mathbb{R}$ is surjective unless $f(x) = 0$, and every non-zero real number is a regular value of f. In particular, we see again that the sphere $\mathbb{S}^n = f^{-1}(1)$ is a smooth manifold of dimension n.

Example 4.16 We will show in Exercise 4.11 that $1 \in \mathbb{R}$ is a regular value of the determinant function $\det: \mathrm{GL}(n) \to \mathbb{R}$. This implies that the subgroup $\mathrm{SL}(n)$ of matrices with determinant one is a smooth manifold. It is called the *special linear group*. Exercise 4.11 also shows that the only critical value of the determinant map $\det: \mathrm{M}(n) \to \mathbb{R}$ is 0.

Example 4.17 Recall the map $f: \mathbb{R}^4 \to \mathbb{R}$, $f(x_1,x_2,x_3,x_4) = x_1+x_2^2+x_3^3+x_4^4$ from Example 4.4, where we showed that f is a submersion. In particular, this implies that 0 is a regular value of f. Hence the preimage $f^{-1}(0)$ is a three-dimensional smooth manifold.

Example 4.18 Let $p_1, \ldots, p_n$ be polynomials with real coefficients in variables $t_1, \ldots, t_m$. We can consider the collection $(p_1, \ldots, p_n)$ as a smooth map $\mathbf{p}: \mathbb{R}^m \to \mathbb{R}^n$. Then the set of common zeros of $p_1, \ldots, p_n$ is a smooth submanifold of $\mathbb{R}^m$ if $\mathbf{0} = (0, \ldots, 0) \in \mathbb{R}^n$ is a regular value of p. This is the case if the Jacobian matrix $J_{\mathbf{s}}$ at a point $\mathbf{s} = (s_1, \ldots, s_m)$, i.e., the matrix with (i,j)-entry the partial derivative $\frac{\partial p_i}{\partial t_j}(\mathbf{s})$, has maximal rank whenever $\mathbf{p}(\mathbf{s}) = \mathbf{0}$.

For example, if $n = m$, this requires that $\det J_s \neq 0$. For $n = 1$ and $m > 1$, this only requires that the partial derivatives do not vanish simultaneously. Such a set of zeros of polynomials is often referred to as an *algebraic variety*. See also Exercise 4.10.

Remark 4.19 The study of the zeros of polynomials is a central theme in algebraic geometry. In the past decades, fascinating connections between algebraic geometry and homotopy theory emerged, mostly in the field of motivic homotopy theory developed by Morel and Voevodsky in [38]. See also [37]. Motivic homotopy theory has recently led to surprising new developments in the classical field of enumerative geometry based on motivic versions of the invariants in differential topology. See, for example, the pioneering work of Kass and Wickelgren in [24].

4.3.1 The Orthogonal Group Is a Manifold

Recall that $\mathrm{O}(n)$ denotes the subspace of matrices A of $\mathrm{GL}(n)$ which satisfy $AA^\top = I$ where $A^\top$ denotes the transpose of A and I is the $(n \times n)$-identity matrix. It is a group under matrix multiplication.

Theorem 4.20 *The space* $\mathrm{O}(n)$ *of orthogonal matrices is a smooth manifold of dimension* $n(n-1)/2$.

Proof: First, we note that $AA^\top$ is a symmetric matrix since we have $(AA^\top)^\top = (A^\top)^\top A^\top = AA^\top$. The subspace $S(n)$ of symmetric matrices is a smooth submanifold of $\mathrm{M}(n)$ diffeomorphic to $\mathbb{R}^k$ with $k = n(n+1)/2$. The dimension is determined by the fact that the entries below the diagonal are determined by the entries above the diagonal, such that there are $n(n+1)/2$ free choices. We consider the map $f\colon \mathrm{M}(n) \to S(n)$ defined by $A \mapsto AA^\top$. The map f is smooth, since multiplication of matrices is smooth and taking transposes is smooth as well, as we just reshuffle the entries in the matrix. Now we observe $\mathrm{O}(n) = f^{-1}(I)$. Hence, by Theorem 4.12, it suffices to show that I is a regular value for f. Thus, we need to show that derivative

$$df_A\colon T_A(\mathrm{M}(n)) \to T_{f(A)}(S(n))$$

of f at a matrix A is surjective for all $A \in \mathrm{O}(n)$. Since $\mathrm{M}(n)$ and $S(n)$ are vector spaces, we have $T_A(\mathrm{M}(n)) = \mathrm{M}(n)$ and $T_{f(A)}(S(n)) = S(n)$. Thus, tangent vectors are given by matrices. We can then compute df_A at a matrix $A \in \mathrm{O}(n)$ by determining its effect on a matrix $B \in T_A(\mathrm{M}(n)) = \mathrm{M}(n)$:

$$\begin{aligned}
df_A(B) &= \lim_{s\to 0}\frac{f(A+sB)-f(A)}{s} = \lim_{s\to 0}\frac{(A+sB)(A+sB)^\top - AA^\top}{s}\\
&= \lim_{s\to 0}\frac{(A+sB)(A^\top + sB^\top) - AA^\top}{s}\\
&= \lim_{s\to 0}\frac{AA^\top + sBA^\top + sAB^\top + s^2BB^\top - AA^\top}{s}\\
&= \lim_{s\to 0}\frac{sBA^\top + sAB^\top + s^2BB^\top}{s} = \lim_{s\to 0} BA^\top + AB^\top + sBB^\top\\
&= AB^\top + BA^\top.
\end{aligned}$$

Hence, given a matrix $C \in S(n)$, we need to show that there is a matrix $B \in \mathrm{M}(n)$ with $df_A(B) = BA^\top + AB^\top = C$. Since C is symmetric by assumption, we have $C = 1/2(C+C) = 1/2(C+C^\top)$. Now we set $B = \frac{1}{2}CA$ and compute $df_A(B) == 1/2(CAA^\top + AA^\top C^\top) = 1/2(C + C^\top) = C$, where we used that $AA^\top = I$ since $A \in \mathrm{O}(n)$. Thus, I is a regular value, and $\mathrm{O}(n)$ is a submanifold of $\mathrm{M}(n)$ by Theorem 4.12. The dimension of $\mathrm{O}(n)$ is given by the difference $\dim \mathrm{M}(n) - \dim S(n)$, i.e., $\dim \mathrm{O}(n) = n^2 - \frac{n(n+1)}{2} = \frac{n(n-1)}{2}$. □

We end this section with an outlook to an exciting field which has ramifications to many areas of mathematics.

Remark 4.21 (Lie groups) The manifold $\mathrm{O}(n)$ is an example of an important class of smooth manifolds. We have shown that $\mathrm{O}(n)$ is a smooth manifold. Moreover, $\mathrm{O}(n)$ is a group, and the group operations in $\mathrm{O}(n)$ are smooth since both the multiplication map $\mathrm{O}(n) \times \mathrm{O}(n) \to \mathrm{O}(n)$, $(A, B) \mapsto AB$, and the map of forming the inverse $\mathrm{O}(n) \to \mathrm{O}(n)$, $A \mapsto A^{-1}$, are smooth. For the latter note $A^{-1} = A^\top$ for $A \in \mathrm{O}(n)$, though taking inverses is also a smooth map for other matrix groups. In general, a group which is a smooth manifold such that the group operation and taking inverses are smooth maps is called a *Lie group*. Lie groups have a rich and exciting theory. For example, the tangent space at a Lie group at the identity element is a *Lie algebra*, a vector space with a certain additional operation. Such Lie algebras can be classified completely. Lie groups and Lie algebras play an important role in many different areas of mathematics and physics.

4.3.2 Stiefel Manifolds

As another application of the Preimage Theorem 4.12 and the computations we made in the previous section, we consider the following space.

Definition 4.22 (Stiefel manifold) A *k-frame* in $\mathbb{R}^{n+k}$ is a k-tuple $[v_1, \ldots, v_k]$ of orthonormal vectors in $\mathbb{R}^{n+k}$. The *Stiefel manifold* $V_k(\mathbb{R}^{n+k})$ is defined as the set $V_k(\mathbb{R}^{n+k}) := \{k\text{-frames in } \mathbb{R}^{n+k}\} \subset \mathbb{R}^{(n+k)k}$.

Remark 4.23 The set $V_k(\mathbb{R}^{n+k})$ can be identified with the quotient $\mathrm{O}(n+k)/\mathrm{O}(n)$. For, if $[v_1, \ldots, v_k]$ and $T \in \mathrm{O}(n+k)$, then multiplying each v_i with T yields another k-frame $[Tv_1, \ldots, Tv_k]$. In fact, any two k-frames are connected in this way, i.e., if $[v_1, \ldots, v_k]$ and $[v_1', \ldots, v_k']$ are two k-frames then there is a $T \in \mathrm{O}(n+k)$ such that $[v_1', \ldots, v_k'] = [Tv_1, \ldots, Tv_k]$. In other words, $\mathrm{O}(n+k)$ acts transitively on the set of k-frames. Moreover, the stabiliser subgroup of a given frame is the subgroup isomorphic to $\mathrm{O}(n)$, which acts non-trivially on the orthogonal complement of the space spanned by that frame.

The topology on $V_k(\mathbb{R}^{n+k})$ is given as follows. We consider $V_k(\mathbb{R}^{n+k})$ as the subspace of $\mathbb{S}^{n+k-1} \times \ldots \times \mathbb{S}^{n+k-1}$ of k copies of spheres given by all orthonormal k-tuples. We equip $V_k(\mathbb{R}^{n+k})$ with the subspace topology induced by the product topology on $\mathbb{S}^{n+k-1} \times \ldots \times \mathbb{S}^{n+k-1}$. Note that we have already met a Stiefel manifold since $V_1(\mathbb{R}^{n+1})$ may be identified with $\mathbb{S}^n$.

Theorem 4.24 *The space $V_k(\mathbb{R}^{n+k})$ is a compact smooth manifold of dimension $nk + \frac{k(k-1)}{2}$.*

Proof: Since orthogonality of two vectors can be expressed by an algebraic equation, $V_k(\mathbb{R}^{n+k})$ is a closed subset of $\mathbb{S}^{n+k-1} \times \ldots \times \mathbb{S}^{n+k-1}$. Since the product of spheres is compact by Proposition 1.15 and since $V_k(\mathbb{R}^{n+k})$ is a closed subset, $V_k(\mathbb{R}^{n+k})$ is compact by Lemma 1.14. It remains to show that $V_k(\mathbb{R}^{n+k})$ is a smooth manifold. Note that a k-frame $[v_1, \ldots, v_k]$ corresponds to an $(n+k) \times k$-matrix A with v_i as ith column. That the column vectors are orthonormal is equivalent to that the product of the transpose of the ith column with the jth column is 1 if $i = j$ and 0 otherwise, i.e., $v_i^\top \cdot v_j = 1$ if $i = j$ and $v_i^\top \cdot v_j = 0$ if $i \neq j$. In other words, A represents a k-frame in $\mathbb{R}^{n+k}$ if and only if $A^\top A = I_k$, where I_k denotes the $(k \times k)$-identity matrix. Let $\mathrm{M}(n+k,k)$ be the space of $(n+k) \times k$-matrices. We can thus consider $V_k(\mathbb{R}^{n+k})$ as a subspace in $\mathrm{M}(n+k,k)$. Let $S(k)$ denote the space of symmetric $k \times k$-matrices. We define the map $f\colon \mathrm{M}(n+k,k) \to S(k)$ by $f(A) := A^\top A$, and we observe $V_k(\mathbb{R}^{n+k}) = f^{-1}(I_k)$. Hence, by Theorem 4.12, it suffices to show that I_k is a regular value for f. That is, we need to show that the derivative $df_A\colon T_A(\mathrm{M}(n+k,k)) = \mathrm{M}(n+k,k) \to S(k) = T_{f(A)}(S(k))$ is surjective for all $A \in V_k(\mathbb{R}^{n+k})$. As in the proof of Theorem 4.20, we compute the derivative of f at a matrix $A \in V_k(\mathbb{R}^{n+k})$ by $df_A(B) = A^\top B + B^\top A$ as a linear map $df_A\colon T_A(\mathrm{M}(n+k,k)) = \mathrm{M}(n+k,k) \to S(k) = T_{f(A)}(S(k))$. For $C \in S(k)$, we set $B = \frac{1}{2}AC \in \mathrm{M}(n+k,k)$ and get $df_A(B) = C$. Thus, df_A is surjective for all $A \in V_k(\mathbb{R}^{n+k})$. By Theorem 4.12, this shows that $V_k(\mathbb{R}^{n+k})$ is a smooth manifold of dimension given by the difference $\dim \mathrm{M}(n+k,k) - \dim S(k)$, i.e., $\dim V_k(\mathbb{R}^{n+k}) = (n+k)k - \frac{k(k+1)}{2} = nk + \frac{k(k-1)}{2}$. □

4.4 The Stack of Records Theorem

We will now take a closer look at a special, though important, case of regular values. In this section, We assume that X and Y are smooth manifolds, that X is compact, and that $\dim X = \dim Y$. Let $f\colon X \to Y$ be a smooth map. Our goal is to understand the fibres of the regular values of f. We do this in a series of observations which will also be useful later in the book. First, let $y \in Y$ be a regular value for f such that $f^{-1}(y)$ is non-empty. Let x be a point in $f^{-1}(y)$. Since y is a regular value, x is a regular point, i.e., df_x is surjective and hence an isomorphism as $\dim X = \dim Y$. Hence f is a local diffeomorphism at x by the Inverse Function Theorem 3.4. Let $V \subset X$ and $U \subset Y$ be open neighbourhoods around x and y, respectively, such that $f_{|V}\colon V \to U$ is a diffeomorphism. Let x' be another point in $f^{-1}(y)$ with $x \neq x'$. Then $df_{x'}$ is an isomorphism as well, and we can choose an open neighbourhood $V' \subset X$ around x' such that $f_{|V'}\colon V' \to U'$ is a diffeomorphism onto an open subset $U' \subset Y$ containing y.

Lemma 4.25 *The subsets V and V' are disjoint.*

Proof: If $V \cap V' \neq \emptyset$, then f restricts to a diffeomorphism from the open subset $V \cap V'$ onto $U \cap U'$. Since $y \in U \cap U'$ and $f(x) = y = f(x')$, this would imply $x = x' \in V \cap V'$, since x and x' are the unique points in V and V', respectively, which map to y. Thus, if $x \neq x'$, we must have $V \cap V' = \emptyset$. □

Lemma 4.26 *The set $f^{-1}(y)$ is a finite discrete subset of X.*

Proof: By Lemma 4.25, all the points in $f^{-1}(y)$ lie in pairwise disjoint open subsets of X. This shows that $f^{-1}(y)$ is a discrete subset of X and therefore of Euclidean space. Since the subset $\{y\}$ is closed in Y and f is continuous, the fibre $f^{-1}(y)$ is a closed subset of X. Since X is compact and closed subsets in compact spaces are compact by Lemma 1.13 and Exercise 1.1, this implies that $f^{-1}(y)$ is compact as well. Hence $f^{-1}(y)$ is a compact and discrete subset of Euclidean space. By Lemma 1.17, this implies that $f^{-1}(y)$ is a finite set. □

Let $f^{-1}(y) = \{x_1, \ldots, x_n\}$. We can thus pick pairwise disjoint open subsets $W_1, \ldots, W_n$ in X with $x_i \in W_i$ and open subsets $U_1, \ldots, U_n$ in Y, each containing y, such that f maps W_i diffeomorphically onto the open subset U_i.

Lemma 4.27 *The subset* $U := (U_1 \cap \cdots \cap U_n) \setminus f(X \setminus (W_1 \cup \ldots \cup W_n))$ *is open in* Y.

Proof: First, the finite intersection $U_1 \cap \cdots \cap U_n$ is open in Y. Second, the set $W := W_1 \cup \ldots \cup W_n$ is open in X and its complement $X - W$ in X is closed. Since X is compact, $X \setminus W$ is compact by Lemma 1.14 and hence $f(X \setminus W)$ is compact by Lemma 1.13. Since Y is Hausdorff, this implies that $f(X \setminus W)$ is a closed subset of Y by Proposition 1.21. Hence the subset $f(X \setminus W)$ is closed in $U_1 \cap \cdots \cap U_n$. This shows that U is open. □

We know $y \in U$ since $y \in U_1 \cap \cdots \cap U_n$ and $y \notin f(X \setminus W)$ by our choices of the U_i and W_i. We set $V_i := W_i \cap f^{-1}(U)$. Then $f_{|V_i} : V_i \to U$ is a diffeomorphism which maps x_i to y. Thus, the inverse image $f^{-1}(U)$ is a disjoint union of open subsets $V_1, \ldots, V_n$ with $x_i \in V_i$ and the restriction of f to any of the V_i is a diffeomorphism onto U. Hence we have shown the following useful result, illustrated in Figure 4.1.

Theorem 4.28 (Stack of Records Theorem) *Let X and Y be smooth manifolds with* $\dim X = \dim Y$ *and X compact. Let $f: X \to Y$ be a smooth map, and let $y \in Y$ be a regular value such that $f^{-1}(y) \neq \emptyset$. Then the set $f^{-1}(y)$ is a discrete finite subset $\{x_1, \ldots, x_n\}$ of X, and we can choose an open neighbourhood $U \subset Y$ around y such that $f^{-1}(U) \subset X$ is the disjoint union $V_1 \cup \cdots \cup V_n$ of open subsets of X with $x_i \in V_i$ and the restriction of f to any of the V_i is a diffeomorphism onto U.*

Remark 4.29 (Covering spaces) The above situation is a particular instance of a more general phenomenon. In Topology, a continuous map $f: X \to Y$ is an *unramified covering* if every point in Y has an open neighbourhood U such that $f^{-1}(U)$ is the disjoint union of open sets V_i such that f maps each V_i homeomorphically onto U. Suppose that in addition to the assumptions of Theorem 4.28, f is surjective and all values in Y are regular for f. Then $X \to Y$ is an unramified covering.

We keep the assumptions from the beginning of this section. Given a regular value y, the fibre $f^{-1}(y)$ is either empty or finite. Hence it makes sense to talk about the number of elements in the set $f^{-1}(y)$. We denote this number by $\# f^{-1}(y)$.

Proposition 4.30 (Locally constant fibre) *Let $R \subset Y$ denote the set of regular values for f, considered as a topological space with the subspace topology. The function $R \to \mathbb{Z}$ defined by $y \mapsto \# f^{-1}(y)$ is locally constant, i.e., for every regular value y_0, there is an open neighbourhood $U \subset Y$ of y_0 such that $\# f^{-1}(y) = \# f^{-1}(y_0)$ for all $y \in U$.*

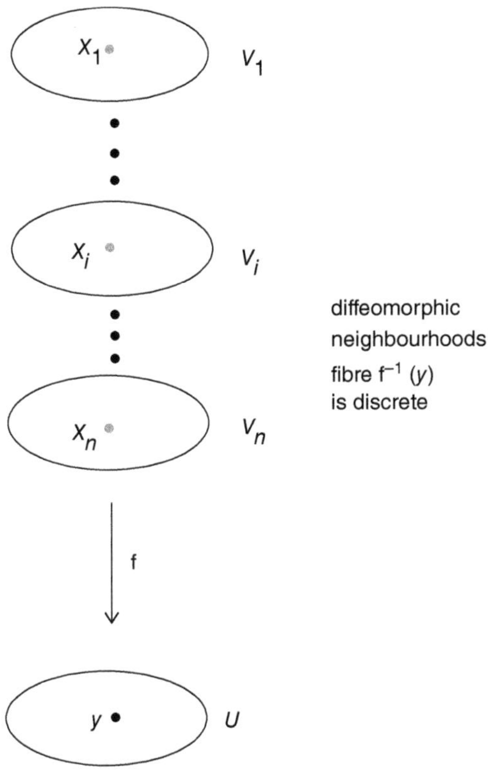

Figure 4.1 Every point y in Y has an open neighbourhood U such that $f^{-1}(U)$ consists of disjoint copies of open subsets V_i in X which are diffeomorphic to U. In three-dimensional space, this may be drawn as several old-school records stacked on top of each other. One key consequence is that the fibre over y is a finite and discrete set.

Proof: First, we assume $f^{-1}(y_0) \neq \emptyset$. By Theorem 4.28, we can choose the open neighbourhood U of y_0 in Y such that $f^{-1}(U) = V_1 \cup \cdots \cup V_n$ is the pairwise disjoint union of open neighbourhoods V_i of x_i which are all mapped diffeomorphically onto the open subset U. Thus, for every point $y \in U$, there is exactly one point in V_i which maps to y, and these are the only points in X which map onto y by the choice of U. Thus, $\#f^{-1}(y) = \#f^{-1}(y_0)$ for all $y \in U$, and the function $y \mapsto \#f^{-1}(y)$ is constant on the open subset U. Now assume $f^{-1}(y_0) = \emptyset$. Since X is compact, $f(X)$ is compact by Lemma 1.13. Hence, since Y is Hausdorff, $f(X)$ is closed in Y by Proposition 1.21. Thus $U := Y \setminus f(X)$ is open in Y. Since by our assumption $y_0 \in Y \setminus f(X)$, the subset U is an open neighbourhood around y_0 such that $\#f^{-1}(y) = \#f^{-1}(y_0)$ for all $y \in U$. This proves the claim. □

4.5 Milnor's Proof of the Fundamental Theorem of Algebra

As an application of Theorem 4.28 and of what we learned about regular values, we will now study Milnor's proof in [35, pages 8–9] of the following important theorem. Let $P(z) = a_n z^n + a_{n-1} z^{n-1} + \cdots + a_1 z + a_0$ denote a non-constant complex polynomial, i.e., $n \geq 1$, $a_0, \ldots, a_n \in \mathbb{C}$ and $a_n \neq 0$. We say that P *has a zero* if there is a $z_0 \in \mathbb{C}$ such that $P(z_0) = 0$. Our goal is to prove the following important theorem.

Theorem 4.31 (Fundamental Theorem of Algebra) *Every non-constant complex polynomial P has a zero.*

In fact, by induction on the degree of the polynomial, we can conclude that P has exactly n zeros when we count them with multiplicities. We will deduce Theorem 4.31 from the following stronger result.

Theorem 4.32 *Every non-constant complex polynomial P defines a surjective smooth map $\mathbb{C} \to \mathbb{C}$.*

Theorem 4.32 implies Theorem 4.31 since the former shows $P^{-1}(0) \neq \emptyset$.

Proof of Theorem 4.31: As before, we identify the complex numbers $\mathbb{C}$ with the points in the real plane $\mathbb{R}^2$, but we keep in mind that $\mathbb{C}$ is a field, i.e., we can multiply and form inverses for points in $\mathbb{C}$, and we will use both points of view on $\mathbb{C}$ in this proof. Since P is a polynomial, it defines a smooth map $\mathbb{C} \to \mathbb{C}$. The assertion we need to prove is that it is a surjective map. To prove this, we will extend the map $P \colon \mathbb{C} \to \mathbb{C}$ to a smooth map on a compact space. The idea is then to apply the Stack of Records Theorem and its consequences.

We therefore define a new map

$$f \colon \mathbb{S}^2 \to \mathbb{S}^2, \quad \begin{cases} f(x) := \phi_N \circ P \circ \phi_N^{-1}(x) & \text{for all } x \in \mathbb{S}^2 \setminus \{N\} \\ f(N) := N & \text{for } x = N \end{cases}, \tag{4.6}$$

where $\phi_N \colon \mathbb{R}^2 \to \mathbb{S}^2 \setminus \{N\}$ and $\phi_N^{-1} \colon \mathbb{S}^2 \setminus \{N\} \to \mathbb{R}^2$ denote the diffeomorphisms given by the stereographic projection from the north pole N on $\mathbb{S}^2$. See Exercise 2.9 for the formulas. Since ϕ_N and ϕ_N^{-1} are diffeomorphisms, f is surjective if and only if P is surjective. Hence we reduced our task to showing that f is surjective. Assuming that f is surjective, we get a zero for P as follows. Let S denote the south pole on $\mathbb{S}^2$. Then there is an $x \in \mathbb{S}^2$ with $f(x) = S$. Since ϕ_N is a diffeomorphism and $\phi_N(0) = S$, we have $P(\phi_N^{-1}(x)) = 0$ and $z := \phi_N^{-1}(x) \in \mathbb{C}$ is a zero of P. We now show that f is surjective in a sequence of lemmas.

Lemma 4.33 *The map $f \colon \mathbb{S}^2 \to \mathbb{S}^2$ is smooth.*

Proof: Since ϕ_N and ϕ_N^{-1} are diffeomorphisms, it is clear that f is smooth at all points which are not the north pole. It remains to show that it is also smooth on an open neighbourhood of N. Now let $\phi_S\colon \mathbb{R}^2 \to \mathbb{S}^2 \setminus \{S\}$ and $\phi_S^{-1}\colon \mathbb{S}^2 \setminus \{S\} \to \mathbb{R}^2$ denote the diffeomorphisms given by the stereographic projection from the south pole N on $\mathbb{S}2$. Again, see Exercise 2.9 for the formulas. We define the map $Q\colon \mathbb{C} \to \mathbb{C}$ by $Q := \phi_S^{-1} \circ f \circ \phi_S$. Note that $Q(0) = 0$ since $\phi_S(0) = N$, $f(N) = N$, and $\phi_S^{-1}(N) = 0$. To compute Q for other values, we need to calculate the composite $\phi_N^{-1} \circ \phi_S$. For a point $x = (x_1, x_2) \neq (0,0)$ we get

$$\begin{aligned}
\phi_N^{-1} \circ \phi_S(x_1,x_2) &= \phi_N^{-1}\left(\frac{1}{1+|x|^2}(2x_1, 2x_2, 1-|x|^2)\right) \\
&= \frac{1}{1-\frac{1-|x|^2}{1+|x|^2}}\left(\frac{2x_1}{1+|x|^2}, \frac{2x_2}{1+|x|^2}\right) \\
&= \frac{1+|x|^2}{2|x|^2}\left(\frac{2x_1}{1+|x|^2}, \frac{2x_2}{1+|x|^2}\right) = \frac{1}{|x|^2}(x_1, x_2).
\end{aligned}$$

Remembering complex conjugation $z \mapsto \bar{z}$ on $\mathbb{C}$ and $|z|^2 = z\bar{z}$, we can rewrite this as $\phi_N^{-1} \circ \phi_S(z) = \frac{z}{|z|^2} = 1/\bar{z}$ for all $z \in \mathbb{C} \setminus \{0\}$. Similarly, we get $\phi_S^{-1} \circ \phi_N(z) = \frac{z}{|z|^2} = 1/\bar{z}$ for all $z \in \mathbb{C} \setminus \{0\}$. Thus, for $z \neq 0$, we have

$$\begin{aligned}
Q(z) &= \phi_S^{-1} \circ \phi_N \circ P \circ \phi_N^{-1} \circ \phi_S(z) = \phi_S^{-1} \circ \phi_N(P(1/\bar{z})) \\
&= \phi_S^{-1} \circ \phi_N(a_n \bar{z}^{-n} + a_{n-1}\bar{z}^{-(n-1)} + \cdots + a_1\bar{z}^{-1} + a_0) \\
&= 1/\left(\bar{a}_n z^{-n} + \bar{a}_{n-1} z^{-(n-1)} + \cdots + \bar{a}_1 z^{-1} + \bar{a}_0\right) \\
&= z^n/\left(\bar{a}_n + \bar{a}_{n-1}z + \cdots + \bar{a}_1 z^{n-1} + \bar{a}_0 z^n\right).
\end{aligned}$$

Note that $z = 0$ is not a zero of $D(z) := \bar{a}_n + \bar{a}_{n-1}z + \cdots + \bar{a}_1 z^{n-1} + \bar{a}_0 z^n$, since $\bar{a}_n \neq 0$. Thus in a small neighbourhood of $z = 0$ in $\mathbb{C}$, the denominator in the expression for $Q(z)$ is bounded below, i.e., there exist $\varepsilon > 0$ and $c > 0$ such that $|D(z)| \geq c$ for all $z \in \mathbb{B}^2_\varepsilon(0) \subset \mathbb{C}$. Since $Q(z) = z^n/D(z)$, this shows that Q is smooth at $z = 0$. See also Remark 4.34 below. This implies that Q is smooth in a small open neighbourhood of 0. Since ϕ_S and ϕ_S^{-1} are diffeomorphisms and since ϕ_S sends an open neighbourhood of N in $\mathbb{S}^2$ to an open neighbourhood of 0 in $\mathbb{C}$, this implies that $f = \phi_S^{-1} \circ Q \circ \phi_S$ is smooth in an open neighbourhood of N. This proves the lemma. □

Remark 4.34 (More on the derivative of Q) We can also calculate the derivative of Q at 0. Since we checked above that $Q(0) = 0$, we get

$$dQ_0 = \lim_{h\to 0} \frac{Q(h) - Q(0)}{h} = \lim_{h\to 0} \frac{h^n/(\bar{a}_n + \bar{a}_{n-1}h + \cdots + \bar{a}_0 h^n) - 0}{h}$$
$$= \lim_{h\to 0} \frac{h^{n-1}}{\bar{a}_n + h(\bar{a}_{n-1} + \cdots + \bar{a}_0 h^{n-1})} = \begin{cases} 1/\bar{a}_n & \text{if } n = 1 \\ 0 & \text{if } n \geq 2. \end{cases}$$

Now we use the fact that P is a complex polynomial. This will give us a simple way to calculate its derivative. For the derivative of the real function $P\colon \mathbb{R}^2 \to \mathbb{R}^2$ is the same as the underlying derivative of $P\colon \mathbb{C} \to \mathbb{C}$ as a complex function. In addition, we know that the derivative is $\mathbb{C}$-linear, not just $\mathbb{R}$-linear. More importantly, the derivative of a polynomial has at most finitely many zeros. Recall that a point $x \in \mathbb{S}^2$ is called a *critical point for f* if the derivative df_x is not surjective.

Lemma 4.35 *The map $f\colon \mathbb{S}^2 \to \mathbb{S}^2$ has at most n many critical points.*

Proof: Since ϕ_N and ϕ_N^{-1} are diffeomorphisms, the only points that might be critical for f are the points where P is not a local diffeomorphism, and possibly N. The derivative of P, as a linear map $dP_z\colon \mathbb{C} \to \mathbb{C}$, $w \mapsto dP_z \cdot w$, is given by $dP_z = P'(z) = \sum_{j=1}^{n} ja_j z^{j-1}$. Hence dP_z fails to be an isomorphism, only if it is the zero map, i.e., $dP_z = 0$. However, as a function of z, dP_z is a polynomial of degree $n-1$, and it is not identically zero, since $a_n \neq 0$. This shows that there are at most $n-1$ complex numbers z such that $dP_z = 0$. Hence there are at most n points on $\mathbb{S}^2$ where df_z is not surjective. □

Remark 4.36 The computation of the derivative of Q at 0 in Remark 4.34 and the chain rule imply that N is, in fact, a critical point of f when $n \geq 2$. Hence, for example, if P is a polynomial of degree two, then f has two critical points since $P'(z)$ is of degree one and therefore has a zero. This observation is relevant in case we would like to apply the argument to a real polynomial P.

Lemma 4.37 *The set R of regular values for f is connected, and the function $R \to \mathbb{Z}$ given by $y \mapsto \#f^{-1}(y)$ is constant.*

Proof: By Lemma 4.35, f has at most n critical values. Thus, the set R of regular values for f is $\mathbb{S}^2$ with only finitely many points removed. This shows that R is connected. Now we use the fact that $\mathbb{S}^2$ is compact. The map f satisfies the assumption of the Stack of Records Theorem 4.28, and we can apply Proposition 4.30 to see that the function $R = \mathbb{S}^2 \setminus \{\text{critical values}\} \to \mathbb{Z}$, $y \to \#f^{-1}(y)$, is locally constant. Since it is defined on a connected space, it must be constant by Lemma 1.30. □

Lemma 4.38 *The map f has infinitely many different values.*

Proof: Assume that f has only finitely many different values $y_1, \dots, y_k \in \mathbb{S}^2$. Then we can write $\mathbb{S}^2$ as the union $\mathbb{S}^2 = f^{-1}(y_1) \cup \cdots \cup f^{-1}(y_k)$. For $y_i \neq y_j$, we have $f^{-1}(y_i) \cap f^{-1}(y_j) = \emptyset$. Each $f^{-1}(y_i)$ is a closed subset of $\mathbb{S}^2$ since $\{y_i\}$ is a closed subset of $\mathbb{S}^2$, as the latter is Hausdorff, and f is continuous. Moreover, each $f^{-1}(y_i)$ is also open since the complement is a finite union of closed subsets and is therefore closed. Thus, we could write $\mathbb{S}^2$ as the union of non-empty, open, and closed subsets. Since $\mathbb{S}^2$ is connected, this implies that $k = 1$, i.e., f is constant. However, P is not constant, and ϕ_N and ϕ_N^{-1} are diffeomorphisms. Thus f cannot be constant, and our initial assumption had to be wrong. □

Lemma 4.39 *The map f is surjective.*

Proof: Assume that there is a $y_0 \in \mathbb{S}^2$ with $f^{-1}(y_0) = \emptyset$, i.e., $\#f^{-1}(y_0) = 0$. Then y_0 is a regular value for f by definition. Since the function $y \mapsto \#f^{-1}(y)$ is constant on the set of regular values, it would have to be zero for every regular value. Hence $\#f^{-1}(y) \neq 0$ only for critical values y. But this would mean that f had only finitely many values, as f has only finitely many critical values. This contradicts the Lemma 4.38. Thus f must be surjective. □

As we pointed out before, this implies that P is surjective as well and completes the proof of Theorem 4.31. □

Remark 4.40 ($\mathbb{R}$ is too one-dimensional) If we try the above strategy for a real polynomial $P: \mathbb{R} \to \mathbb{R}$, we will replace P with a smooth map $f: \mathbb{S}^1 \to \mathbb{S}^1$. Then f is smooth and defined on a compact domain, and it has only finitely many critical values. But the conclusion that the function $y \mapsto \#f^{-1}(y)$ is constant on the set of regular values on $\mathbb{S}^1$ would not work anymore. For, after removing at least two different points from $\mathbb{S}^1$ we get a space which is not connected. And for a polynomial of degree at least two, we cannot exclude the possibility that f may have two critical values. See also Remark 4.36.

4.6 Exercises

4.1 Let $f: X \to Y$ be a submersion and U an open subset of X. Show that $f(U)$ is open in Y. In other words, submersions are open maps.

4.2 For $g: \mathbb{R}^2 \to \mathbb{R}$, $g(x, y) := x^2 - y^2$. Determine the set of regular values of g, and determine the set of critical values of g. Is g a submersion?

4.3 Let $f: X \to Y$ be a nontrivial submersion. Show that if X is compact and Y is connected, then f is surjective.

4.4 Show that there exist no submersions from compact manifolds to $\mathbb{R}^n$ for any n. Hint: Use Exercise 4.3.

4.5 Let Z be the subset of points (x, y, z) in $\mathbb{R}^3$ which satisfy the two equations $x^3 + y^3 + z^3 = 1$ and $x + y + z = 0$. Show that $Z \subset \mathbb{R}^3$ is a smooth manifold. What is the dimension of Z?

4.6 In this exercise, we are going to prove the remaining cases of the Constant Rank Theorem. Recall that we proved the first case in Exercise 3.9. Let $f: X \to Y$ be a smooth map. Recall that we say that f has *constant rank* if the linear map df_x has rank r for all $x \in X$.
   (a) Assume that X is compact. Show that a surjective smooth map $f: X \to Y$ of constant rank is a submersion.
   (b) Using Baire's Category Theorem show that a surjective smooth map $f: X \to Y$ of constant rank is a submersion.
   (c) Assume that X is compact. Show that a bijective smooth map $f: X \to Y$ of constant rank is a diffeomorphism.

4.7 Show that the tangent space to $\mathrm{O}(n)$ at the identity matrix I is the vector space of *skew symmetric* $n \times n$-matrices, i.e., matrices B with $B^\top = -B$.

4.8 Let $\mathrm{M}(2)$ denote the set of all real (2×2)-matrices.
   (a) Show that the restriction of the determinant function $\det: \mathrm{M}(2) \setminus \{0\} \to \mathbb{R}$ to the open submanifold of non-zero matrices is a submersion.
   (b) Conclude that the set R_1 of all (2×2)-matrices of rank 1 is a three-dimensional submanifold of $\mathbb{R}^4 = \mathrm{M}(2)$.

4.9 In this exercise, we prove Euler's identity for homogeneous polynomials: Let $P(x_1, \ldots, x_k)$ be a homogeneous polynomial of degree m in k variables, i.e., $P(tx_1, \ldots, tx_k) = t^k P(x_1, \ldots, x_k)$ for all t. Show Euler's identity

$$\sum_i x_i \partial P / \partial x_i = mP. \tag{4.7}$$

Hint: Define a new function Q by setting $Q(x_1, \ldots, x_k, t)$ equal to $P(tx_1, \ldots, tx_k) - t^m P(x_1, \ldots, x_k)$ and compute its derivative.

4.10 In this exercise, we show that the fibres of homogeneous polynomials form manifolds. Let $P(x_1, \ldots, x_k)$ be a homogeneous polynomial of degree m in k variables. We consider P as a map

$$\mathbb{R}^k \to \mathbb{R}, \ (x_1, \ldots, x_k) \mapsto P(x_1, \ldots, x_k).$$

   (a) Show that 0 is the only critical value of P. Conclude that $P^{-1}(a)$ is a $k-1$-dimensional submanifold of $\mathbb{R}^k$ for all $a \neq 0$. Hint: You may want to use Euler's identity for homogeneous polynomials.

(b) For two positive real numbers $a, b > 0$, show that $P^{-1}(a)$ is diffeomorphic to $P^{-1}(b)$. Similarly, for two negative real numbers $a, b < 0$, show that $P^{-1}(a)$ is diffeomorphic to $P^{-1}(b)$.

4.11 The set $\mathrm{SL}(n)$ of $n \times n$- matrices with determinant $+1$ form a subspace of $\mathrm{GL}(n)$. In this exercise, we show that $\mathrm{SL}(n)$ is a smooth manifold of dimension $n^2 - 1$ and compute its tangent space.

(a) Show that 0 is the only critical value of $\det\colon \mathrm{M}(n) \to \mathbb{R}$. Hint: You may either want to think of det as a homogeneous polynomial given by Leibniz' formula

$$\det(A) = \sum_{\sigma} \left(\operatorname{sgn}(\sigma) \prod_{i=1}^{n} a_{i\sigma(i)} \right), \tag{4.8}$$

where the sum runs over all permutations of the set $\{1, \ldots, n\}$ and $\operatorname{sgn}(\sigma)$ denotes the sign of the permutation σ. Then use a previous exercise. Or you use the formula $\det A = \sum_{i=1}^{n} (-1)^{i+j} a_{ij} \det A_{ij}$ where A_{ij} denotes the $(n-1) \times (n-1)$-matrix defined by removing the ith row and jth column from A.

(b) Conclude that $\mathrm{SL}(n)$ is a submanifold of $\mathrm{M}(n)$ of dimension $n^2 - 1$.

(c) Show that the tangent space to $\mathrm{SL}(n)$ at the identity matrix consists of all matrices with trace equal to zero. Hint: You may want to use Proposition 4.13 and Leibniz' formula for det.

4.12 Recall the Hopf map $\hbar\colon \mathbb{S}^3 \to \mathbb{S}^2$ from Example 2.10 and Exercise 2.10 with $\mathbb{S}^3 \subset \mathbb{C}^2$ and $\mathbb{S}^2 \subset \mathbb{C} \times \mathbb{R}$ given by

$$\hbar(z_0, z_1) = \left(2z_0\bar{z}_1, |z_0|^2 - |z_1|^2\right).$$

(a) First we consider $\hbar$ as a map $\tilde{\hbar}\colon \mathbb{R}^4 \cong \mathbb{C}^2 \to \mathbb{C} \times \mathbb{R} \cong \mathbb{R}^3$ using the same formula as for $\hbar$, i.e., $\hbar = \tilde{\hbar}_{|\mathbb{S}^3}$. Compute the derivative of $\tilde{\hbar}$ at a point $q = (z_0, z_1)$.

(b) Let $q \in \mathbb{S}^3$. Explain why the restriction of $d\tilde{\hbar}_q$ to $T_q\mathbb{S}^3$ has an image contained in $T_{\hbar(q)}\mathbb{S}^2$.

(c) Consider the points $a = (0,0,1)$ and $b = (0,1,0)$ on $\mathbb{S}^2 \subset \mathbb{R}^3 \cong \mathbb{C} \times \mathbb{R}$. Determine the fibres $\hbar^{-1}(a)$ and $\hbar^{-1}(b)$ and show that a and b are regular values for $\hbar$.

(d) Extend your computations from the previous points to show that actually each point in $\mathbb{S}^2$ is a regular value for $\hbar$.

5
Transversality

In this chapter, we generalise the idea of regular values and introduce a fundamental geometric concept: transverse intersections. As an application, we prove Sard's Theorem, a very important and useful result in differential topology. In the final section, we discuss Morse functions as an application of Sard's Theorem.

5.1 Transversality and Preimages

We would like to understand what happens when we take the preimage of not a point, but of a submanifold. More precisely, let X and Y be smooth manifolds and let $f: X \to Y$ be a smooth map. Let $Z \subset Y$ be a submanifold. We now ask:

Question When is the subset $f^{-1}(Z) \subseteq X$ a smooth manifold?

Our strategy to answer this question is to work locally and to apply our previous results. Let us denote $n = \dim X$, $m = \dim Y$, and $k = \dim Z$. Let $x_0 \in X$ be a point with $f(x_0) \in Z \subset Y$. The inclusion of subsets $Z \hookrightarrow Y$ is an immersion. Hence we can apply the Local Immersion Theorem 3.11. We choose a local parametrisation $\psi: U \to V$ with $\psi(0) = f(x_0)$ with $V \subset Y$ and $U \subset \mathbb{R}^m$ open subsets. Since Z inherits the subspace topology from Y, $Z \cap V$ is open in Z. Since Z is a smooth manifold, we can choose V small enough such that there is a local parametrisation $V' \to Z \cap V$ around $f(x_0)$ with $V' \subset \mathbb{R}^k$ an open subset such that

$$\begin{array}{ccc} Z \cap V & \xhookrightarrow{\text{inclusion}} & V \\ \uparrow & & \uparrow \psi \\ V' & \xrightarrow[\text{immersion}]{\text{canonical}} & U \end{array} \tag{5.1}$$

commutes. Let $\pi^m_{m-k} \colon \mathbb{R}^m \to \mathbb{R}^{m-k}$ denote the projection onto the last $m-k$ coordinates. Since ϕ and ψ are diffeomorphisms and the bottom map in (5.1) is the canonical immersion, the subspace $Z \cap V$ is exactly the preimage of $0 \in \mathbb{R}^{m-k}$ under the composition $V \xrightarrow{\psi^{-1}} U \xrightarrow{\pi^m_{m-k}} \mathbb{R}^{m-k}$, i.e., we have $Z \cap V = (\pi^m_{m-k} \circ \psi^{-1})^{-1}(0)$. To simplify the notation, we set $g := \pi^m_{m-k} \circ \psi^{-1}$. We then have $f^{-1}(Z \cap V) = (g \circ f)^{-1}(0)$. By the Preimage Theorem 4.12, we know that $f^{-1}(Z \cap V)$ is a smooth manifold if 0 is a regular value for $g \circ f$. Hence we would like to have $d(g \circ f)_x \colon T_x(X) \to T_0(\mathbb{R}^{m-k}) = \mathbb{R}^{m-k}$ is surjective for all $x \in (g \circ f)^{-1}(0)$. By the chain rule, $d(g \circ f)_x = dg_{f(x)} \circ df_x$ and hence $d(g \circ f)_x$ is surjective if and only if the restriction of $dg_{f(x)}$ to the image of df_x surjects onto $\mathbb{R}^{m-k}$. Since π^m_{m-k} is a linear map and ψ^{-1} is the inverse of the diffeomorphism, we use to define $T_{f(x)}(Y)$, $dg_{f(x)} \colon T_{f(x)}(Y) \to \mathbb{R}^{m-k}$ is a surjective linear map with kernel $T_{f(x)}(Z)$. This means that $(dg_{f(x)})_{|\mathrm{Im}\,(df_x)}$ is surjective if and only if $\mathrm{Im}\,(df_x)$ and $T_{f(x)}(Z)$ together span all of $T_{f(x)}(Y)$. We conclude that $g \circ f$ is a submersion at $x \in f^{-1}(Z)$ if and only if $\mathrm{Im}\,(df_x) + T_{f(x)}(Z) = T_{f(x)}(Y)$. We note that this condition is independent of the choice of local parametrisations, and we now give it a name.

Definition 5.1 (Transversality) Let X and Y be smooth manifolds, $f \colon X \to Y$ a smooth map, and $Z \subseteq Y$ a submanifold. Then f is said to be *transverse to* Z, denoted $f \pitchfork Z$, if $\mathrm{Im}\,(df_x) + T_{f(x)}(Z) = T_{f(x)}(Y)$ for every $x \in f^{-1}(Z)$.

Now we can prove the following important result.

Theorem 5.2 *Let $f \colon X \to Y$ be a smooth map, and let $Z \subseteq Y$ be a submanifold. Assume that f is transverse to Z. Then $f^{-1}(Z)$ is either empty or it is a submanifold of X such that the codimension of $f^{-1}(Z)$ in X equals the codimension of Z in Y, i.e., we have* $\dim X - \dim f^{-1}(Z) = \dim Y - \dim Z$.

Proof: Let $x_0 \in X$ be a point with $f(x_0) \in Z$. As explained above, we can choose an open subset $V \subset Y$ with $f(x_0) \in V$ such that $f^{-1}(Z \cap V)$ is a smooth manifold of dimension $m-k$ since f and Z are transverse. Then, for every point $x \in f^{-1}(Z \cap V)$ has a local parametrisation $\phi \colon U \to W$ with $U \subset \mathbb{R}^{m-k}$ and $W \subset f^{-1}(Z \cap V)$ open subsets with $x \in W$. Since f is continuous and $Z \cap V$ is open in Z, $f^{-1}(Z \cap V)$ is an open subset in $f^{-1}(Z)$. This implies that W is open in $f^{-1}(Z)$, too. Since we started with an arbitrary point in $f^{-1}(Z)$, this shows that every point $f^{-1}(Z)$ has a local parametrisation, and $f^{-1}(Z)$ is a smooth manifold of dimension $m - k$. □

We can simplify the assertion on the dimensions by using the following terminology. For a submanifold $Z \subset Y$, we call the number $\operatorname{codim}_Y Z := \dim Y - \dim Z$ the *codimension of Z in Y*. The theorem then says that the codimension of $f^{-1}(Z)$ in X is equal to the codimension of Z in Y. If $Z = \{y\}$ consists of a single point, then $T_y(Z)$ is the vector space consisting only of the zero vector. Hence, in this case, transversality means $\operatorname{Im} df_x = T_y(Y)$ at all points with $f(x) = y$, i.e., y is a regular value of f. It is useful to note that any smooth map $f: X \to Y$ whose image does not meet a submanifold Z of Y, i.e., $f^{-1}(Z) = \emptyset$, is transverse to Z since no condition needs to be satisfied. On the other hand, if $f: X \to Y$ is a submersion, then f is transverse to every submanifold Z of Y since then $\operatorname{Im}(df_x) = T_{f(x)}(Y)$ for every x. Geometrically, transversality tells us something about how the image of f and Z meet in Y. Tangent spaces are the key tool that allows us to make precise how f and Z meet at a point. We will give further geometric intuition for transversality soon. We first look at some simple examples.

Example 5.3 We consider $Y = \mathbb{R}^2$ with the submanifold Z being the x-axis. Then the map $f: \mathbb{R}^1 \to \mathbb{R}^2$ defined by $f(t) = (0, t)$ is transverse to Z, with $f^{-1}(Z) = \{(0,0)\}$. The map $f: \mathbb{R}^1 \to \mathbb{R}^2$ defined by $f(t) = (t, t^2)$, however, is not transverse to Z, with $f^{-1}(Z) = \{(0,0)\}$. See Figure 5.1. The map $f: \mathbb{R}^1 \to \mathbb{R}^2$ defined by $f(t) = (t, t^2 - 1)$ is transverse to Z, with $f^{-1}(Z) = \{(-1,0), (1,0)\}$. The map $f: \mathbb{R}^1 \to \mathbb{R}^2$ defined by $f(t) = (t, \cos t - 1)$ is not transverse to Z, with $f^{-1}(Z) = \{(0,0)\}$. See Figure 5.2.

Example 5.4 We can generalise the observations from Example 5.3. Let $p(t) = t^n + a_{n-1}t^{n-1} + \cdots + a_0$ be a polynomial with real coefficients. We can consider $p(t)$ as a smooth map $\mathbb{R} \to \mathbb{R}$. The map $f: \mathbb{R}^1 \to \mathbb{R}^2$

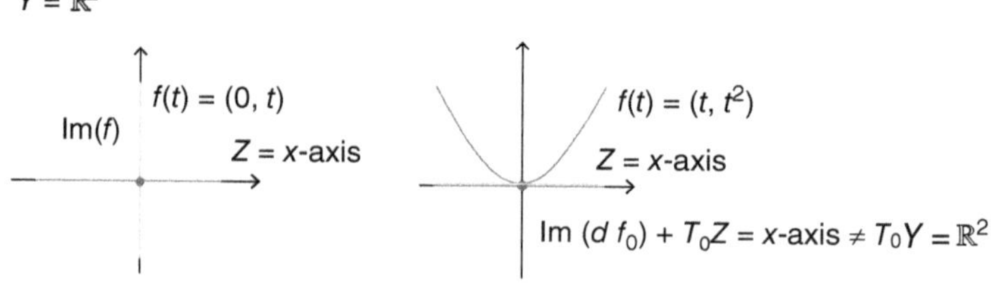

Figure 5.1 (a) f is transverse to the x-axis, since $\operatorname{Im}(df_0)$ and $T_0 Z$ span all of $T_0 Y = \mathbb{R}^2$. (b) However, f is not transverse to Z, since both $\operatorname{Im}(df_0)$ and $T_0 Z$ span the same one-dimensional subspace.

Figure 5.2 We begin to see a pattern in the plane: f is transverse to the x-axis unless the x-axis is tangent to the graph of f.

defined by $f(t) = (t, p(t))$ is transverse to the x-axis Z if and only if all zeros of p are simple, i.e., if and only if we can write $p(t)$ as a product $p(t) = (t - r_1) \cdot (t - r_2) \cdot \ldots \cdot (t - r_n)$ with $r_i \neq r_j$ if $i \neq j$. The reason is the following. The tangent space of Z at a point $f(t_0) \in \mathbb{R}^2 = Y$ is the x-axis $T_{f(t_0)}Z = \mathbb{R} \times \{0\} \subset \mathbb{R}^2$, and the derivative df_{t_0} of f at t_0 is the map $df_{t_0} \colon \mathbb{R} \to \mathbb{R}^2$, $s \mapsto s \cdot \begin{pmatrix} 1 \\ p'(t_0) \end{pmatrix}$. Now we observe that the vectors $\begin{pmatrix} 1 \\ 0 \end{pmatrix}$ and $\begin{pmatrix} 1 \\ p'(t_0) \end{pmatrix}$ are linearly independent in $\mathbb{R}^2$ if and only if $p'(t_0) \neq 0$. Hence $T_{f(t_0)}(Z)$ and $\operatorname{Im} d_{f(t_0)}$ span $T_{f(t_0)}(Y)$ if and only if $p'(t_0) \neq 0$, i.e., t_0 is a simple zero of p.

Proposition 5.5 (Tangent space of the preimage) *Let $f \colon X \to Y$ be a smooth map and Z be a submanifold of Y. Assume that f is transverse to Z with $f^{-1}(Z) \neq \emptyset$. Then $T_x(f^{-1}(Z))$ is the preimage of $T_{f(x)}(Z)$ under the linear map $df_x \colon T_x(X) \to T_{f(x)}(Y)$, i.e., $T_x(f^{-1}(Z)) = (df_x)^{-1}(T_{f(x)}(Z))$.*

Proof: Let $x \in f^{-1}(Z)$. As we explained in the discussion leading to Definition 5.1, we can choose a local parametrisation $\psi \colon U \to V$ around $f(x)$ with $V \subset Y$ open and $Z \cap V$ non-empty, and define a smooth map $g \colon V \to \mathbb{R}^{m-k}$ such that $Z \cap V = g^{-1}(0)$ with 0 a regular value for both g and $g \circ f$. By Proposition 4.13, and since tangent spaces are determined locally, we then have $T_z(Z) = \operatorname{Ker} dg_z$ and $T_x(f^{-1}(Z)) = \operatorname{Ker}(d(g \circ f)_x)$. By the chain rule, we then get $T_x(f^{-1}(Z)) = (df_x)^{-1}(\operatorname{Ker}(dg_{f(x)})) = (df_x)^{-1}(T_{f(x)}(Z))$. □

5.2 Transverse Intersections

We now study the more geometric situation, in which two submanifolds X and Z in a given bigger manifold Y intersect. We would like the intersection $X \cap Z$ to be a submanifold in Y. In order to get sufficient conditions for when this is the case, we can apply Definition 5.1 and Theorem 5.2 to the special case that

f is the inclusion map $i: X \hookrightarrow Y$. Since $x \in X$ belongs to the intersection $X \cap Z$ if and only if $x \in i^{-1}(Z)$ and since the derivative $di_x: T_x(X) \to T_x(Y)$ is the inclusion of $T_x(X)$ into $T_x(Y)$ as a subspace, we make the following definition.

Definition 5.6 (Transverse submanifolds) Let Y be a smooth manifold. Two submanifolds $X \subset Y$ and $Z \subset Y$ are *transverse in* Y, denoted $X \pitchfork Z$, if $T_x(X) + T_x(Z) = T_x(Y)$ for all $x \in X \cap Z$.

Note that the condition is symmetric in X and Z. Moreover, two submanifolds which do not intersect at all are transverse. We also note that, for the equality $T_x(X) + T_x(Z) = T_x(Y)$, it is not sufficient that $\dim T_x(X) + \dim T_x(Z) = \dim T_x(Y)$. The two subspaces together must span all of $T_x(Y)$. A simple example is the intersection of Z with itself, i.e., $X = Z$, with $\dim Z = \frac{1}{2} \dim Y$. Then we have $\dim T_z(X) + \dim T_z(Z) = \dim T_z(Y)$, but the tangent spaces $T_z(Z)$ and $T_z(X)$ do not span $T_z(Y)$. However, we will see later a way to make self-intersections interesting using Thom's Transversality Theorem and invariance under homotopy. See Chapters 12 and 13.

Remark 5.7 (Tangent spaces are the key) A simple way to think of transversality is that two manifolds are transverse if they intersect as little as possible at every point. The degree of intersection is measured using tangent spaces. Tangent spaces translate the local behaviour of smooth manifolds into linear algebra, and in linear algebra we have a rather simple way to check how much two subspaces intersect by determining their linear span.

Remark 5.8 (The ambient space matters) Whether two submanifolds are transverse or not depends on the ambient space. For example, the two coordinate axes are transverse in $\mathbb{R}^2$, but are not transverse when considered as submanifolds of $\mathbb{R}^3$. In general, if the sum of the dimensions of X and Z is strictly less than the dimension of Y, then X and Z can be transverse only if they do not intersect. See Figure 5.3.

Theorem 5.9 (Intersection of transverse submanifolds) *Let Y be a smooth manifold and let X and Z be submanifolds of Y. We assume that $X \cap Z \neq \emptyset$ and that X and Z are transverse in Y. Then $X \cap Z$ is a smooth submanifold of Y with* $\operatorname{codim}_Y(X \cap Z) = \operatorname{codim}_Y X + \operatorname{codim}_Y Z$.

Proof: The assertion is the special case of Theorem 5.2 when f is the inclusion map $i: X \hookrightarrow Y$ since $X \cap Z = i^{-1}(Z)$. □

Note that the converse of the theorem is not true. When $X = Z$, so that X and Z are the same submanifold, then they are not transverse but $X \cap Z$

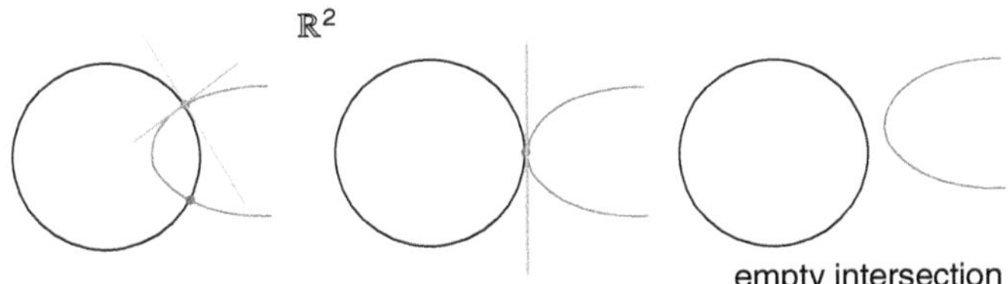

Figure 5.3 The intersection of two circles in the plane is transverse unless there is exactly one intersection point.

is a smooth manifold. For another simple case, we recall Example 5.3. The submanifolds $X = \{(x, y) \in \mathbb{R}^2 : y = x^2\}$ and $Z = \{(x, y) \in \mathbb{R}^2 : y = 0\}$ are transverse at 0 in $Y = \mathbb{R}^2$, but their intersection $X \cap Z = \{0\}$ is a zero-dimensional manifold. There do, however, exist intersections which are not transverse and where the intersection is not a manifold. See Example 5.13.

Proposition 5.10 (Tangent space of intersections) *Let $X \subset Y$ and $Z \subset Y$ be submanifolds such that $X \pitchfork Z$ in Y. Then the tangent space to $X \cap Z$ is the intersection of the tangent spaces, i.e., $T_x(X \cap Z) = T_x(X) \cap T_x(Z)$ for all $x \in X \cap Z$.*

Proof: This is the special case of Proposition 5.5 when f is the inclusion map $i : X \hookrightarrow Y$ since $T_x(X) \cap T_x(Z) = (di_x)^{-1}(T_x(Z))$. □

Example 5.11 (A primer to Brieskorn manifolds) Recall the map $f : \mathbb{R}^4 \to \mathbb{R}$ defined by $f(x_1, x_2, x_3, x_4) = x_1 + x_2^2 + x_3^3 + x_4^4$. We showed in Example 4.4 that f is a submersion and hence $Z = f^{-1}(0)$ is a smooth manifold in $\mathbb{R}^4$ by Theorem 4.12. Now we show that Z and $\mathbb{S}^3$ are transverse in $\mathbb{R}^4$ where $\mathbb{S}^3 \subset \mathbb{R}^4$. We need to check that $T_z(Z) + T_z(\mathbb{S}^3) = T_z(\mathbb{R}^4) = \mathbb{R}^4$ for all $z \in Z \cap \mathbb{S}^3$. Since $T_z(Z)$ and $T_z(\mathbb{S}^3)$ are both three-dimensional subspaces of $\mathbb{R}^4$, it suffices to show that, for every $z \in Z \cap \mathbb{S}^3$, there is at least one vector $\mathbf{v}$ in $T_z(Z)$ which is not contained in $T_z(\mathbb{S}^3)$. The tangent space to Z at a point $z \in Z$ is the subspace of $\mathbb{R}^4$ given by the kernel of the derivative df_z which in the standard bases is given by the 1×4-matrix $df_z = \begin{pmatrix} 1 & 2x_2 & 3x_3^2 & 4x_4^3 \end{pmatrix}$. Let $z = (x_1, x_2, x_3, x_4)$ be any point in $Z \cap \mathbb{S}^3$. Then the vector $\mathbf{v} := (12x_1, 6x_2, 4x_3, 2x_4)$ lies in $T_z(Z)$, since $df_z(\mathbf{v}) = 12x_1 + 12x_2^2 + 12x_3^3 + 12x_4^4 = 12f(z) = 0$. We will now show that $\mathbf{v}$ is not an element in $T_z(\mathbb{S}^3)$. Recall that $T_z(\mathbb{S}^3)$ is the subspace in $\mathbb{R}^4$ which is orthogonal to the vector z, i.e., $T_z(\mathbb{S}^3) = \{\mathbf{w} \in \mathbb{R}^4 : z \perp \mathbf{w} = 0\}$. We can check orthogonality via the scalar product in $\mathbb{R}^4$, i.e., $z \perp \mathbf{w}$ if and only if $z \cdot \mathbf{w} = 0$. We then compute $z \cdot \mathbf{v} = 12x_1^2 + 6x_2^2 + 4x_3^2 + 3x_4^2 > 0$. Thus, $\mathbf{v}$ is not an element in $T_z(\mathbb{S}^3)$. Hence the intersection of Z and $\mathbb{S}^3$ is transverse in $\mathbb{R}^4$. By Theorem 5.9, the codimension of $Z \cap \mathbb{S}^3$ in $\mathbb{S}^3$ equals the codimension of Z in $\mathbb{R}^4$. Thus $\dim Z \cap \mathbb{S}^3 = 2$.

Example 5.12 (Hyperboloid meets a sphere) In $Y = \mathbb{R}^3$, we consider the submanifolds $X = \{(x, y, z) \in \mathbb{R}^3 : x^2 + y^2 - z^2 = 1\}$ and the sphere $Z_a = \{(x, y, z) \in \mathbb{R}^3 : x^2 + y^2 + z^2 = a\}$ of radius $\sqrt{a}$ with $a > 0$. We show in Exercise 5.4 that the intersection of X and Z_a is transverse in $Y = \mathbb{R}^3$ if and only if $a \neq 1$. Nevertheless, $X \cap Z_1$ is a circle and thus a smooth manifold.

Example 5.13 (A non-transverse intersection) Let $Y = \mathbb{R}^3$ and let Z be the hyperplane defined by $Z = \{(x, y, z) \in \mathbb{R}^3 : x = 1\}$, and let X be the hyperboloid defined by $X = \{(x, y, z) \in \mathbb{R}^3 : x^2 + y^2 - z^2 = 1\}$. The intersection of X and Z is given by the points satisfying $x = 1$ and $x^2 + y^2 - z^2 = 1$, i.e., all points such that $x = 1$ and $y^2 = z^2$. This means $X \cap Z = \{(x, y, z) \in \mathbb{R}^3 : x = 1,\ y = \pm z\}$, i.e., $X \cap Z$ is the intersection of the two lines in the plane $x = 1$ in $\mathbb{R}^3$ given by the lines $y = z$ and $y = -z$. We have seen in Example 2.24 that a space consisting of two lines crossing each other is not a manifold. The intersection point, i.e., the point $p = (1, 0, 0)$, does not have a neighbourhood in $X \cap Z$ which is diffeomorphic to an open subset in one-dimensional Euclidean space. Thus $X \cap Z$ is not a manifold.

Remark 5.14 (Exotic Spheres) We let $\mathbb{S}^7_k$ denote the following intersection in $\mathbb{C}^5 \setminus \{0\}$: $\mathbb{S}^7_k := \{z_1^2 + z_2^2 + z_3^2 + z_4^3 + z_5^{6k-1} = 0\} \cap \{|z_1|^2 + \cdots + |z_5|^2 = 1\}$. In Exercise 5.8, we show that this is a transverse intersection. One can show that, for each value $k = 1, \ldots, 28$, the space $\mathbb{S}^7_k$ is a smooth manifold which is homeomorphic to the seven-sphere $\mathbb{S}^7$. But none of these manifolds are diffeomorphic. These are so-called exotic seven-spheres. They were constructed by Brieskorn [7] and represent each of the 28 diffeomorphism classes on the space $\mathbb{S}^7$. The existence of such exotic seven-spheres is a famous and groundbreaking result of Milnor [33]. Milnor's work started an exciting story about the diffeomorphic structures on spheres, see also [26] and [27], which culminated in the solution of the Kervaire invariant problem by Barratt–Jones–Mahowald [3, 4], Hill–Hopkins–Ravenel [16, 17], and Lin–Wang–Xu [30]. See also [32].

Remark 5.15 (Lefschetz maps) Let $f : X \to X$ be a map, and let x be a fixed point of f, i.e., $f(x) = x$. If $+1$ is not an eigenvalue of $df_x : T_x(X) \to T_x(X)$, then x is called a *Lefschetz fixed point* of f. The map f is called a *Lefschetz map* if all its fixed points are Lefschetz. In Exercise 5.7, we will prove the following interesting application of transversality: if X is compact and f is a Lefschetz map, then f has only finitely many fixed points.

5.3 The Theorem of Brown and Sard

Let $f: X \to Y$ be a smooth map. We have seen how useful regular values are. This motivates us to ask:

Question How many regular values are there in Y?

If f is a submersion, for example, then all $y \in Y$ are regular values. In Section 4.5, we studied a smooth map $f: \mathbb{S}^2 \to \mathbb{S}^2$ for which all but a finite number of $y \in Y$ are regular, i.e., f has finitely many critical values. To formulate a precise general statement, we need to recall some terminology in general topology.

Definition 5.16 (Dense subsets) Let X be a topological space, and $A \subset X$ be a subset. Then A is called a *dense subset* of X if its intersection with every non-empty open subset of X is non-empty.

Now we can formulate the theorem we are going to prove in this section.

Theorem 5.17 (Sard's Theorem) *Let X and Y be smooth manifolds and let $f: X \to Y$ be a smooth map. Then the set of regular values for f is a dense subset of Y, i.e., every open subset of Y contains a regular value.*

This theorem is a key result in differential topology, and we will apply it in many situations. Our proof will follow the outline of Holm in [19]. There are many other proofs of the theorem. See, for example, also [35, chapter 3]. We will first reformulate and simplify the assertion. To do so, we are going to use the following terminology.

Definition 5.18 (The interior of a set) Let X be a topological space, and S be a subset of X. Then the *interior* of S, denoted $\text{int}(S)$, is the union of all open subsets of X contained in S.

By definition, the interior of S is an open subset of X. In fact, it is the largest open subset of X which is contained in S. See Figure 5.4. If $S \subset \mathbb{R}^N$, then $\text{int}(S)$ is the set of all points $s \in S$ such that there is a small open ball centred at x which is contained in S. If U is an open subset of X, then $\text{int}(U) = U$. In general, however, $\text{int}(S)$ is a proper subset of X.

We recall the following facts from general topology. A subset $A \subset X$ has empty interior if and only if its complement $X \setminus A$ is dense in X. The countable intersection of open dense subsets is a dense subset in $\mathbb{R}^p$. Equivalently, the countable union of compact subsets with empty interior in $\mathbb{R}^p$ is a subset with empty interior. The analogous statements hold in every smooth p-dimensional

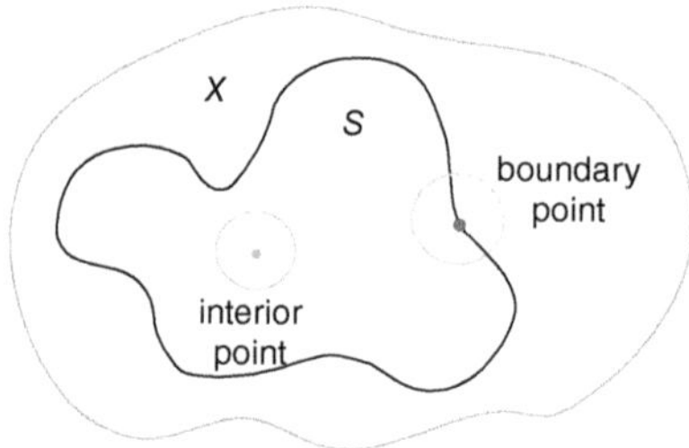

Figure 5.4 The interior of a subset S in the topological space X is the union of all open subsets of X which are contained in S.

manifold, i.e., $\mathbb{R}^p$ and smooth manifolds in general are Baire spaces. We will show that Theorem 5.17 is a consequence of the following result.

Theorem 5.19 (Euclidean case) *Let $f\colon \mathbb{R}^n \to \mathbb{R}^p$ be a smooth map. Then the set of regular values of f is a dense subset of $\mathbb{R}^p$. Equivalently, the set of critical values of f is a set with empty interior.*

Before we start proving Theorem 5.19, we show that it implies Theorem 5.17. We will use the following notation.

Notation 5.20 For a smooth map $f\colon X \to Y$, we denote by C_f the set of critical points of f, by D_f the set of critical values, and denote the set of regular values by R_f.

Proof that Theorem 5.19 implies Theorem 5.17: First assume that $Y = \mathbb{R}^p$. Let $x \in X$ be a point and let $\phi\colon \mathbb{R}^n \to X$ be a local parametrisation where, by Exercise 2.3, we can choose the domain of ϕ to be all of $\mathbb{R}^n$. We set $W := \phi(\mathbb{R}^n)$ and $K = \phi(\bar{\mathbb{B}}^n_1(0))$, where $\bar{\mathbb{B}}^n_1(0)$ is the closed unit ball in $\mathbb{R}^n$. In particular, K is a compact subspace in X. By Theorem 5.19, the set $D_{f\circ\phi} = f(C_f \cap W)$ of critical values of $f \circ \phi$ has empty interior. By Proposition 4.9, C_f is a closed subset. Thus, since C_f is closed and hence $C_f \cap K$ is compact, $f(C_f \cap K)$ is a compact subset of $\mathbb{R}^p$ with empty interior. Since X can be covered by countably many neighbourhoods of the form K by covering points with rational coordinates, $D_f = f(C_f)$ is a countable union of compact subsets with empty interior. This implies that D_f has empty interior. Thus, the complement $Y \setminus D_f$, i.e., the set of regular values of f, is a dense subset. Now let Y be an arbitrary smooth p-dimensional manifold. Then we can find countably many local parametrisations ψ_i of Y such that the union of the compact sets $K_i := \psi_i(\bar{\mathbb{B}}^p_1(0))$ covers Y. By the first case, the intersection of the D_f with any of these sets K_i is a compact subset with empty interior. Hence D_f is a countable union of compact subsets with empty interior and has itself empty interior. □

Proof of Theorem 5.19: The proof will proceed in two steps: first, we prove the result for a smooth function $\mathbb{R}^n \to \mathbb{R}$. Then, we use this to reduce the case $\mathbb{R}^p$ to $\mathbb{R}^{p-1}$ and will conclude by induction.

- The case $p = 1$, i.e., $f\colon \mathbb{R}^n \to \mathbb{R}$ is a smooth function.

The assertion is true for $n = 0$ as there is nothing to prove. Now assume that the assertion is true for $n - 1$. Let $C_i \subset \mathbb{R}^n$ be the closed set of points where all partial derivatives of f of order $\leq i$ vanish. We then have $C_f = C_1$ and $C_1 \supset C_2 \supset C_3 \supset \cdots$. We can therefore write C_f as the union of subsets $C_f = (C_1 \setminus C_2) \cup \cdots \cup (C_{n-1} \setminus C_n) \cup C_n$. This implies that the set of critical values for f can be written as the union $D_f = f(C_1 \setminus C_2) \cup \cdots \cup f(C_{n-1} \setminus C_n) \cup f(C_n)$. Hence the theorem follows for n if we can show that each of the sets $f(C_i \setminus C_{i+1})$ for all $i \geq 1$ and $f(C_i)$ for $i \geq n$ is a countable union of closed subsets with empty interior. This will be shown in the following lemmas.

Lemma 5.21 *For each $u \in C_i \setminus C_{i+1}$, there is a compact neighbourhood K disjoint from C_{i+1} and an $(n-1)$-dimensional submanifold $Z \subset \mathbb{R}^n$ such that $C_i \cap K \subset Z$.*

Proof: Let $u \in C_i \setminus C_{i+1}$. Then there is some i-th order partial derivative of f whose first-order partial derivatives do not all vanish at u. We denote this partial derivative by g. Then u is a regular point for g. By definition of C_i, we have $C_i \subseteq g^{-1}(0)$, and, in particular, $g(u) = 0$. By Corollary 4.8, we can then find an open neighbourhood U of u such that 0 is a regular value for $g_{|U}$. Hence, by the Preimage Theorem 4.12, $g^{-1}(0) \cap U$ is an $(n-1)$-dimensional submanifold in U. We set $Z := g^{-1}(0) \cap U$ and choose a sufficiently small compact neighbourhood K of u in U. By definition of g, we also know that K is disjoint from C_{i+1}. □

Lemma 5.22 *The set $f(C_i \setminus C_{i+1})$ is a countable union of closed subsets with empty interior for all $i \geq 1$.*

Proof: Let K and Z be as in Lemma 5.21. Then every point of $C_i \cap K$ is critical for $f_{|Z}$, since it is critical for f. By our induction hypothesis, $f(C_i \cap K)$ is a closed subset in $\mathbb{R}$ without interior points. Since $C_i \setminus C_{i+1}$ can be covered by countably many arbitrarily small compact neighbourhoods K. □

Lemma 5.23 *For $i \geq n$, the set $f(C_i)$ is a countable union of closed subsets with empty interior.*

Proof: By Taylor's Theorem, we have

$$f(x+u) = f(x) + \partial_u f(x) + \cdots + \frac{1}{i!}\partial_u^i f(x) + \frac{1}{(i+1)!}\partial_u^{i+1} f(x+\lambda u)$$

for any two points x and u in $\mathbb{R}^n$, where $0 \leq \lambda \leq 1$ is a real number and ∂_u is the differential operator $\partial_u = u_1\partial/\partial x_1 + \cdots + u_n\partial/\partial x_n$. Thus, if $x \in C_i$, then by definition of C_i, $f(x+u) - f(x) = \frac{1}{(i+1)!}\partial_u^{i+1} f(x+\lambda u)$. If in addition x and $y := x + u$ lie in a convex set K, then $x + \lambda u$ is also in K, and we get the inequality $|f(y) - f(x)|_{\max} \leq c|y - x|_{\max}^{i+1}$ where $|w|_{\max} = \max\{|w_1|, \ldots, |w_n|\}$ and c is a constant depending on K and f only. Now we let K be the unit cube in $\mathbb{R}^n$ and consider the subdivision of K into k^n subcubes with sides of length $1/k$. Let K' be one of these subcubes and suppose $x \in C_i \cap K$ and $y \in K'$. Then $|y - x|_{\max} \leq 1/k$. This implies that $f(C_i \cap K')$ is contained in an interval of length c/k^{i+1}. Thus, $f(C_i \cap K)$ is contained in a union of k^n intervals of joint length $k^n c/k^{i+1} \leq c/k$ where we use that $i \geq n$ by our assumption. Since k is any positive integer, this length can be arbitrarily small. Hence the set $f(C_i \cap K)$ has empty interior. Finally, $\mathbb{R}^n$ and therefore C_i is contained in a countable union of unit cubes K. This proves the lemma. □

- The induction step for $p \geq 2$, i.e., a smooth map $f\colon \mathbb{R}^n \to \mathbb{R}^p$.

We assume that the assertion holds for every smooth map $f\colon \mathbb{R}^n \to \mathbb{R}^{p-1}$. Let $U \subset \mathbb{R}^p$ be a non-empty open subset of $\mathbb{R}^p$. We will show that f has regular values in U. This implies that the set of regular values for f is dense in $\mathbb{R}^p$. If U contains points which are not in $f(\mathbb{R}^n)$, then these points are regular values for f, and we are done. Hence assume $U \subseteq f(\mathbb{R}^n)$. Let $\pi\colon \mathbb{R}^p \to \mathbb{R}^{p-1}$ be the projection onto the first $p-1$ coordinates. Then $\pi(U)$ is open in $\mathbb{R}^{p-1}$, since π is a submersion and hence an open map by Exercise 4.1. By our induction hypothesis, the map $\pi \circ f\colon \mathbb{R}^n \to \mathbb{R}^{p-1}$ has a regular value $y' \in \pi(U)$. In other words, f is transverse to the line $Y' = \pi^{-1}(y')$ in $\mathbb{R}^p$. Let $f'\colon X' \to Y'$ be the induced map. Since $U \subseteq f(\mathbb{R}^n)$, the open set U meets Y', i.e., $Y' \cap U \neq \emptyset$. Since Y' is diffeomorphic to $\mathbb{R}$, f' has a regular value $y'' \in Y' \cap U$ by the case $p = 1$. Thus, we have shown that f is transverse to Y' in $\mathbb{R}^p$, and $f'\colon f^{-1}(Y') \to Y'$ has a regular value y'' in $Y' \cap U$. By the chain rule and what we learned about transversality, this implies that $y'' \in Y' \cap U$ is a regular value for f. This finishes the proof of Theorem 5.19. □

Remark 5.24 At the end of the proof we used a special case of a fact we proved in Exercise 5.5: if $f: X \to \mathbb{R}^p$ is transverse to a submanifold $Y' \subset \mathbb{R}^p$ and the induced map $f': f^{-1}(Y') \to Y'$ is transverse to $Y'' \subset Y'$, then f is transverse to Y''. We may therefore think of the argument for the general case as *splitting off a transverse component* of lower dimension to use induction.

We now look at some first applications of Sard's theorem. Recall that $\mathrm{M}(p,n)$ denotes the space of real $(p \times n)$-matrices, which we consider as a subspace of $\mathbb{R}^{pn}$, and $\mathrm{M}(p,n;k)$ denotes the subspace of matrices of rank k.

Theorem 5.25 (Making the origin a regular value) *Let $U \subset \mathbb{R}^n$ be an open subset, and let $f: U \to \mathbb{R}^p$ be a smooth map. Then, for every $\varepsilon > 0$, there is a $(p \times n)$-matrix A with entries a_{ij} satisfying $|a_{ij}| < \varepsilon$ and a $y \in \mathbb{R}^p$ with entries y_i satisfying $|y_i| < \varepsilon$ such that the smooth map $g: U \to \mathbb{R}^p$ defined by $g(x) = f(x) + A \cdot x + y$ has the origin as a regular value.*

Proof: If $p > n$, then the only regular values of f are the points in $\mathbb{R}^p$ which are not in the image of f. By Sard's Theorem 5.17, the latter set is dense in $\mathbb{R}^p$. Hence we can choose $A = 0$ and a point $y \in \mathbb{R}^p$ with small enough entries such that the origin is not in the image of $g: x \mapsto f(x) + y$. We assume now that $p \leq n$. The origin is a regular value for g if $dg_x = df_x + A$ has rank p at all points x such that $0 = g(x) = f(x) + A \cdot x + y$. To show that we can find the desired matrix $A = dg_x - df_x$ and point $y = -f(x) - (dg_x - df_x) \cdot x$, we define, for $0 \leq k \leq p$, the map $F_k: \mathrm{M}(p,n;k) \times U \to \mathrm{M}(p,n) \times \mathbb{R}^p$ by

$$F_k(Q,x) := (Q - df_x, \; - f(x) - (Q - df_x) \cdot x).$$

Since f is smooth, the map F_k is smooth. By Theorem 2.23, the domain of F_k is a smooth manifold of dimension $k(p+n-k)+n$. For $k \leq k' < n$, we have $k(p+n-k)+n \leq k'(p+n-k')+n$. Hence for $k = 0, 1, \ldots, p-1$, we have $\dim_{\mathbb{R}}(\mathrm{M}(p,n;k) \times U) = k(p+n-k)+n \leq (p-1)(p+n-(p-1))+n$, and hence $\dim_{\mathbb{R}}(\mathrm{M}(p,n;k) \times U) \leq p + pn - 1$. Since we assume $p > n$, we get $\dim_{\mathbb{R}}(\mathrm{M}(p,n;k) \times U) < pn + p = \dim_{\mathbb{R}}(M(p,n) \times \mathbb{R}^p)$ for $0 \leq k < p$. This means that the derivative of F_k cannot be surjective for $0 \leq k < p$, and the only regular values for F_k with $0 \leq k < p$ are pairs in $\mathrm{M}(p,n) \times \mathbb{R}^p$ which are not in the image of F_k. By Theorem 5.17, the set of regular values for F_k is dense in $\mathrm{M}(p,n) \times \mathbb{R}^p$. Since the finite intersection of dense subsets is again a dense subset, the set of pairs which are not in the image of F_k for $0 \leq k < p$ is dense in $\mathrm{M}(p,n) \times \mathbb{R}^p$. Hence there is a pair (A, y) in $\mathrm{M}(p,n) \times \mathbb{R}^p$ arbitrarily close to the zero pair which is not in the image of the F_k for $0 \leq k < p$. Thus, the matrix $df_x + A$ has rank p since otherwise (A, y) had been in the image of one of the F_k with $k < p$. Hence we have found the desired matrix A and point y such that $dg_x = df_x + A \cdot x + y$ has rank p for all x in the fibre $g^{-1}(0)$, and thus such that the origin is a regular value for g. □

Theorem 5.26 (An immersion via matrices) *Let $U \subset \mathbb{R}^n$ be an open subset, and let $f: U \to \mathbb{R}^p$ be a smooth map with $p \geq 2n$. Then, for every $\varepsilon > 0$, there is a $(p \times n)$-matrix A with entries a_{ij} satisfying $|a_{ij}| < \varepsilon$ such that the smooth map $g: U \to \mathbb{R}^p$ defined by $g(x) = f(x) + A \cdot x$ is an immersion.*

Proof: See Exercise 5.9. □

We remark that Sard's Theorem is often formulated using the following terminology. A rectangular solid in $\mathbb{R}^n$ is a Cartesian product of n intervals in $\mathbb{R}^n$, and its volume is the product of the lengths of the n intervals. An arbitrary set A in $\mathbb{R}^n$ is said to have *(Lebesgue) measure zero* if, for every $\varepsilon > 0$, there exists a countable collection $\{S_1, S_2, \ldots\}$ of rectangular solids in $\mathbb{R}^n$, such that A is contained in the union of the S_i, and $\sum_{i=1}^{\infty} \text{vol}(S_i) < \varepsilon$. In a manifold X, an arbitrary subset $C \subset X$ has *measure zero* if, for every local parametrisation ϕ of X, the preimage $\phi^{-1}(C)$ has measure zero in Euclidean space. Note that measures and volumes depend on the ambient space. An example of a measure zero subset is given by the set of rational numbers in $\mathbb{R}$. We note that every non-empty open subset of a manifold X has measure strictly bigger than zero. Hence no subset of measure zero in a manifold X can contain a non-empty open subset of X. While Theorem 5.17 suffices for our purposes, there is the following stronger version of Sard's Theorem.

Theorem 5.27 (Sard's Theorem revisited) *Let $f: X \to Y$ be a smooth map of manifolds. Then the subset of critical values has measure zero in Y.*

5.4 Morse Functions

We will now study an interesting application of Sard's Theorem. Morse theory, due to Marston Morse [39], has many applications. We will see one of them in Section 17.3.2. The starting question is the following. We understand the local behaviour of smooth maps at regular points by the Local Submersion Theorem 4.7. But what about the local behaviour at *critical* points? In fact, it is often at critical points that something interesting happens. For example, let X be a compact smooth manifold and $f: X \to \mathbb{R}$ be a smooth function. Then we know that f must have a maximum and a minimum. At a point $x \in X$ where $f(x)$ is either a maximal or a minimal value, f does not change in any direction in X. In other words, the derivative df_x must vanish, as $df_x(h)$ is a measure for the change of f in direction h. Hence x is a critical point in our terminology. In the familiar situation in calculus, we know that at the critical points where f has a local maximum or minimum, the second derivative does not vanish. In Morse theory, we are also interested in those critical points where the second derivative is invertible. We will now explain this idea in more detail.

5.4.1 Morse Functions on Euclidean Space

We begin with the study of critical points of a smooth function $f\colon \mathbb{R}^k \to \mathbb{R}$. Locally around a point $c \in \mathbb{R}^k$, by Taylor's Theorem, we can describe f by

$$\begin{aligned} f(x) = f(c) &+ \sum_{i=1}^{k} \frac{\partial f}{\partial x_i}(c) \cdot (x_i - c_i) \\ &+ \frac{1}{2} \sum_{i,j=1}^{k} \frac{\partial^2 f}{\partial x_i \partial x_j}(c) \cdot (x_i - c_i)(x_j - c_j) + o(|x|^3). \end{aligned}$$

If c is a critical point, then $df_c = \left(\frac{\partial f}{\partial x_1}(c), \ldots, \frac{\partial f}{\partial x_k}(c)\right) = 0$ since otherwise df_c was surjective as a linear map $\mathbb{R}^k \to \mathbb{R}$. We may then look at the Hessian matrix of the second partial derivatives of f at c.

Definition 5.28 (Non-degenerate critical points and Morse functions) Let $f\colon \mathbb{R}^k \to \mathbb{R}$ be a smooth function. A point $c \in \mathbb{R}^k$ where df_c vanishes and the Hessian matrix $H(f)_c = \left(\frac{\partial^2 f}{\partial x_i \partial x_j}(c)\right)$ is invertible is called a *non-degenerate critical point*. A smooth function $f\colon \mathbb{R}^k \to \mathbb{R}$ for which all critical points are non-degenerate is called a *Morse function*.

Non-degenerate critical points are easier to study than arbitrary critical points, since they are

Lemma 5.29 *A non-degenerate critical point is isolated, i.e., it has an open neighbourhood which does not contain any other critical points.*

Proof: We define map $g\colon \mathbb{R}^k \to \mathbb{R}^k$ by $g := \left(\frac{\partial f}{\partial x_1}, \ldots, \frac{\partial f}{\partial x_k}\right)$. Then $df_x = 0$ if and only if $g(x) = 0$, and the matrix representing the derivative dg_x is the Hessian of f at x. So, if x is non-degenerate, then not only is $g(x) = 0$, but g also maps a neighbourhood of x diffeomorphically onto a neighbourhood of 0 by the Inverse Function Theorem 3.4. In particular, g is injective in that neighbourhood of x. Thus g can be zero at no other points in this neighbourhood, and f has no other critical point in this neighbourhood. □

The lemma indicates that Morse functions are easier to understand than arbitrary smooth functions. As an application of Sard's Theorem, we can show that every smooth function is arbitrarily close to a Morse function.

Proposition 5.30 (Morse functions on $\mathbb{R}^k$ are generic) *Let $U \subset \mathbb{R}^k$ be an open subset, and $a \in \mathbb{R}^k$ be a point. Let $f : U \to \mathbb{R}$ be a smooth function. We define the smooth function $f_a(x) := f(x) + a^\top \cdot x$. Then the set of $a \in \mathbb{R}^k$ such that f_a is a Morse function is dense in $\mathbb{R}^k$.*

Proof: Let $g : \mathbb{R}^k \to \mathbb{R}^k$ again be the map $g = \left(\frac{\partial f}{\partial x_1}, \ldots, \frac{\partial f}{\partial x_k}\right)$. The derivative of f_a at a point $p \in U$ then satisfies $(df_a)_p = \left(\frac{\partial f_a}{\partial x_1}(p), \ldots, \frac{\partial f_a}{\partial x_k}(p)\right) = g(p) + a$. Hence the critical points of f_a are the points $p \in U$ with $g(p) + a = 0$. Moreover, the Hessian of f_a at p is the matrix dg_p, i.e., $H(f_a)_p = H(f)_p = dg_p$. Hence f_a is a Morse function if and only if $\det(H(f_a)_p) \neq 0$ at all critical points p, and this is the case if and only if $\det(dg_p) \neq 0$ at all p with $g(p)+a = 0$. Thus, f_a is a Morse function if and only if $-a$ is a regular value of g. By Sard's Theorem 5.17, the set of $a \in \mathbb{R}^k$ such that $-a$ is a regular value of g is dense in $\mathbb{R}^k$. Therefore, the set of $a \in \mathbb{R}^k$ such that f_a is a Morse function is dense in $\mathbb{R}^k$. □

5.4.2 Morse Functions on Manifolds

Now we would like to transport the concept of non-degenerate critical points to manifolds. Let X be a smooth manifold of dimension k, and let $f : X \to \mathbb{R}$ be a smooth function. Suppose that a is a critical point for f. Let $\phi : U \to X$ be a local parametrisation with $\phi(0) = a$ and $U \subset \mathbb{R}^k$ open. Then $d(f \circ \phi)_0 = df_a \circ d\phi_0$, and hence 0 is a critical point for the function $f \circ \phi$.

Definition 5.31 We call $a \in X$ a *non-degenerate critical point for f* if $0 \in \mathbb{R}^k$ is a non-degenerate critical point for $f \circ \phi$.

Lemma 5.32 *Being a critical point for f is independent of the choice of local parametrisation.*

Proof: So let $\psi : V \to X$ be another local parametrisation with $\psi(0) = a$. We define $\theta := \psi^{-1} \circ \phi : U \to V$. Since θ is a diffeomorphism, the critical points of $f \circ \phi$ and $f \circ \psi \circ \theta$ are the same. Assuming that c is a critical point of f, i.e., $df_a = 0$, the chain rule implies $H(f \circ \phi)_0 = (d\theta_0)^\top \cdot H(f \circ \psi)_0 \cdot d\theta_0$ for the two Hessian matrices at 0. Since $d\theta_0$ is invertible, we see that $H(f \circ \phi)_0$ is invertible if and only if $H(f \circ \psi)_0$ is invertible. □

An important fact on Morse functions is that they can locally be described in a canonical form. The following result extends our understanding of the local behaviour of smooth maps and is a key result in this section.

Lemma 5.33 (Morse Lemma) *Let X be a smooth manifold and $f\colon X \to \mathbb{R}$ be a smooth function. Suppose that $a \in X$ is a non-degenerate critical point for f. Then there is a local parametrisation $\phi\colon U \to X$ with $\phi(0) = a$ such that $f\circ\phi\colon U \to \mathbb{R}$ has the form $(f\circ\phi)(u) = f(a) - u_1^2 - \ldots - u_s^2 + u_{s+1}^2 + \ldots + u_k^2$ where s is the number of negative eigenvalues of the Hessian matrix of f at a.*

For the proof of Lemma 5.33, we follow Milnor in [34, part I, §2.1]. We are going to use the following general observation.

Lemma 5.34 *Let $\mathbb{B}^k \subset \mathbb{R}^k$ be an open ball around the origin and $f\colon \mathbb{B}^k \to \mathbb{R}$ be a smooth function with $f(0) = 0$. Then $f(x_1, \ldots, x_k) = \sum_{i=1}^k x_i g_i(x_1, \ldots x_k)$ for some suitable smooth functions $g_i\colon \mathbb{B}^k \to \mathbb{R}$ with $g_i(0) = \frac{\partial f}{\partial x_i}(0)$.*

Proof: By the Fundamental Theorem of Calculus, we have

$$f(x_1, \ldots, x_k) = \int_0^1 \frac{df(tx_1, \ldots, tx_k)}{dt} dt = \int_0^1 \sum_{i=1}^k \frac{\partial f}{\partial x_i}(tx_1, \ldots, tx_k) \cdot x_i dt.$$

Hence we may set $g_i(x_1, \ldots x_k) := \int_0^1 \frac{\partial f}{\partial x_i}(tx_1, \ldots, tx_k) \cdot x_i dt.$ □

Proof of Lemma 5.33: After possibly translating, we can assume that a is the origin and that $f(a) = f(0) = 0$. Let $\mathbb{B}^k \subset \mathbb{R}^k$ be a sufficiently small open ball such that there is a local parametrisation $\phi\colon \mathbb{B}^k \to X$ with $\phi(0) = 0$. By Lemma 5.34, there are smooth functions $g_i\colon \mathbb{B}^k \to \mathbb{R}$ such that $f \circ \phi(u) = \sum_{j=1}^k u_j g_j(u_1, \ldots u_k)$ for all $u \in \mathbb{B}^k$. Since 0 is a critical point, we then have $g_j(0) = \frac{\partial(f\circ\phi)}{\partial u_j}(0) = 0$, and hence we can apply Lemma 5.34 to each g_j and find smooth functions $h_{ij}\colon \mathbb{B}^k \to \mathbb{R}$ such that $g_j(x_1, \ldots, x_k) = \sum_{j=1}^k x_j h_{ij}(x_1, \ldots x_k)$.

Thus, we obtain

$$(f \circ \phi)(u_1, \ldots, u_k) = \sum_{i,j=1}^k u_i u_j h_{ij}(u_1, \ldots u_k) \tag{5.2}$$

for all $u \in \mathbb{B}^k$. Moreover, we can assume that $h_{ij} = h_{ji}$ since we can replace h_{ij} with $\bar{h}_{ij} = \frac{1}{2}(h_{ij} + h_{ji})$ and then get $\bar{h}_{ij} = \bar{h}_{ji}$ and $f = \sum u_i u_j \bar{h}_{ij}$. By construction, we have $h_{ij}(0) = \frac{\partial^2(f\circ\phi)}{\partial u_i \partial u_j}(0)$. Thus the matrix H with entry $h_{ij}(0)$ in position (i, j) is the Hessian matrix of the smooth map $f \circ \phi \colon \mathbb{B}^k \to \mathbb{R}$ at the origin. Since 0 is a non-degenerate critical point, the Hessian matrix is invertible. Since H is a real symmetric matrix, there is an orthonormal matrix P such that $P^\top H P$ is a diagonal matrix. Let $\lambda_1, \ldots, \lambda_k$ be the eigenvalues of H, where we order them, and hence the columns in P, such that $\lambda_1, \ldots, \lambda_s$ are negative and the remaining eigenvalues are positive. Note that H does not have eigenvalue 0, since it is invertible by the assumption on a being a non-degenerate critical point. By (5.2), we then get $(f \circ \phi)(u) = u^\top P^\top H P u = \sum_{i=1}^k \lambda_i u_i^2$. Finally, after possibly rescaling each u_i by multiplying with $1/\sqrt{|\lambda_i|}$ and letting U be the resulting open subset in $\mathbb{R}^k$, we get the desired form $(f \circ \phi)(u) = -u_1^2 - \ldots - u_s^2 + u_{s+1}^2 + \ldots + u_k^2$. □

Now we can generalise Proposition 5.30 to smooth manifolds. We will see an application of functions and the following theorem in Section 17.3.2 when we discuss the Poincaré–Hopf Index Theorem and the Euler characteristic.

Theorem 5.35 (Morse functions on manifolds are generic) *Let $X \subset \mathbb{R}^N$ be a smooth manifold and $f \colon X \to \mathbb{R}$ be a smooth function. For $a \in \mathbb{R}^N$, we define the smooth function $f_a \colon X \to \mathbb{R}$ by $f_a(x) := f(x) + a^\top \cdot x$, where we consider x with its coordinates in $\mathbb{R}^N$. Then the set of $a \in \mathbb{R}^N$ such that f_a is a Morse function is dense in $\mathbb{R}^N$.*

Proof: We would like to use Proposition 5.30 for $U \subset \mathbb{R}^k$ open. However, $X \subset \mathbb{R}^N$ is, in general, not open; in fact, it is never open if $\dim X < N$. The strategy is therefore to cover X by open subsets and then to lift the k-dimensional result to $\mathbb{R}^N$. Let $x \in X$ be a point, and let $v_1, \ldots, v_k \in \mathbb{R}^N$ be a basis of $T_x(X)$. The matrix which has the v_i as columns has rank k, and hence it has k linearly independent rows, say $i_1, \ldots, i_k$. Let $\pi \colon \mathbb{R}^N \to \mathbb{R}^k$ be projection defined by $(x_1, \ldots, x_N) \mapsto (x_{i_1}, \ldots, x_{i_k})$. Then the derivative $(d\pi_x)_{|T_x(X)} \colon T_x(X) \to \mathbb{R}^k$ is an isomorphism by construction. Hence, by the Inverse Function Theorem 3.4, the restriction $\pi_{|X} \colon X \to \mathbb{R}^k$ is a local diffeomorphism at x. Hence there are open subsets $U \subset \mathbb{R}^k$, $V \subset \mathbb{R}^N$ and $V \cap X \subset X$, and a diffeomorphism $\phi \colon U \to W$ such that its inverse is the restriction $\pi_{|V \cap X} \colon V \cap X \to U$ of the projection $\pi \colon \mathbb{R}^N \to \mathbb{R}^k$. By repeating this process for all points in X, we can cover X with open subsets $V_\alpha \subset \mathbb{R}^N$ and local parametrisations $\phi_\alpha \colon U_\alpha \to V_\alpha \cap X$ such that their inverses are given by the projections onto a k-tuple $(i_1(\alpha), \ldots, i_k(\alpha))$ of the coordinates. Moreover, it suffices to consider points in X with rational coordinates, and we can therefore assume that countably many $V_\alpha \subset \mathbb{R}^N$ suffice to cover X. We observe that f has a degenerate critical point on X if and only if it has a degenerate critical point on some $V_\alpha \cap X$. Moreover, since the countable intersection of dense subsets in Euclidean space is dense,[1]

[1] This is the Baire Category Theorem for Euclidean space.

it therefore suffices to prove the assertion for each V_α. Hence we fix $V := V_\alpha$ and let $\phi\colon U \to V \cap X$ be the associated local parametrisation with $V \cap X$ and $U \subset \mathbb{R}^k$ open such that the inverse is the restriction $\pi_{|V\cap X}\colon V \cap X \to U$ of the projection $\pi : \mathbb{R}^N \to \mathbb{R}^k$. To simplify the notation, we assume that π is the projection of the first k coordinates, i.e., $(i_1(\alpha), \dots, i_k(\alpha)) = (1, \dots, k)$. Now we write $a \in \mathbb{R}^N$ as $a = (b, c)$ where b denotes the first k coordinates and c denotes the last $N-k$ coordinates. We then write the smooth map $f_a\colon X \cap V \to \mathbb{R}$ as $f_a(x) = f(x) + b^\top \cdot (x_1, \dots, x_k) + c^\top \cdot (x_{k+1}, \dots, x_N)$. Now we can apply Proposition 5.30 to the smooth function $(f \circ \phi)_{(0,c)}\colon U \to \mathbb{R}$ defined by $(f \circ \phi)_{(0,c)}(x) = f(x) + c^\top \cdot (x_{k+1}, \dots, x_N)$. Thus, for every $c \in \mathbb{R}^{N-k}$, we know that the set of $b \in \mathbb{R}^k$ such that $(f \circ \phi)_{(b,c)}\colon U \to \mathbb{R}$ is a Morse function is dense in $\mathbb{R}^k$. Now let W be an open subset of $\mathbb{R}^N$. Then there is a $c \in \mathbb{R}^{N-k}$ such that $W \cap (\mathbb{R}^k \times \{c\})$ is non-empty and open in $\mathbb{R}^k \times \{c\}$. Then we have shown that there is an $a = (b, c)$ in $W \cap (\mathbb{R}^k \times \{c\})$ such that $f_a\colon V_\alpha \to \mathbb{R}$ is a Morse function. Thus, the set of a such that f_a is a Morse function is dense in V_α. □

5.5 Exercises

5.1 Let $z = (a, b) \in \mathbb{S}^1 \subseteq \mathbb{R}^2$ and let $N_z = \{(a, y)\colon y \in \mathbb{R}\}$ be the vertical line intersecting the circle at z. When is $\mathbb{S}^1 \subseteq \mathbb{R}^2$ transverse to $N_z \subseteq \mathbb{R}^2$?

5.2 Are $\mathrm{SL}(n)$ and $\mathrm{O}(n)$ transverse in $\mathrm{M}(n)$?

5.3 Recall the maps $f\colon \mathbb{R} \to \mathbb{R}^2$, $f(t) = \left(\frac{e^t+e^{-t}}{2}, \frac{e^t-e^{-t}}{2}\right)$ and $g\colon \mathbb{R}^2 \to \mathbb{R}$, $g(x, y) = x^2 - y^2$ from Exercise 3.5. Is the set $\mathrm{Im}\,(f)$, the image of f in $\mathbb{R}^2$, a manifold? Is the set $(g \circ f)^{-1}(1)$ a manifold?

5.4 Consider the submanifolds $X = \{(x, y, z) \in \mathbb{R}^3\colon x^2 + y^2 - z^2 = 1\}$ and $Z_a = \{(x, y, z) \in \mathbb{R}^3\colon x^2 + y^2 + z^2 = a\}$ with $a > 0$ in $\mathbb{R}^3$. Show that the intersection of X and Z_a is transverse in $Y = \mathbb{R}^3$ if and only if $a \neq 1$.

5.5 Let $X \xrightarrow{f} Y \xrightarrow{g} Z$ be a sequence of smooth maps between manifolds, and let $W \subset Z$ be a submanifold. Assume that g is transversal to W. Show that $f \pitchfork g^{-1}(W)$ if and only if $(g \circ f) \pitchfork W$.

5.6 Let V be a finite-dimensional real vector space, and let Δ be the diagonal of $V \times V$. For a linear map $A\colon V \to V$, consider the graph $\Gamma(A) = \{(v, Av)\colon v \in V\}$. Show that $\Gamma(A) \pitchfork \Delta$ if and only if $+1$ is not an eigenvalue of A.

5.7 Let $f\colon X \to X$ be a map, and let x be a fixed point of f, i.e., $f(x) = x$. If $+1$ is not an eigenvalue of $df_x\colon T_x(X) \to T_x(X)$, then x is called a *Lefschetz fixed point* of f. The map f is called a *Lefschetz map* if all its fixed points are Lefschetz. Prove that if X is compact and f is Lefschetz, then f has only finitely many fixed points. Hint: Show that the intersection of the graph of f and the diagonal of X is a zero-dimensional submanifold of $X \times X$, and use Exercise 5.6.

5.8 Consider the following intersection in $\mathbb{C}^5 \setminus \{0\}$:

$$\mathbb{S}_k^7 = \{z_1^2 + z_2^2 + z_3^2 + z_4^3 + z_5^{6k-1} = 0\}$$
$$\cap \{|z_1|^2 + |z_2|^2 + |z_3|^2 + |z_4|^2 + |z_5|^2 = 1\}.$$

Prove that $\mathbb{S}_k^7$ is a seven-dimensional manifold by showing that the intersection is transverse in $\mathbb{C}^5 \setminus \{0\}$. Hint: Use the strategy of Example 5.11.

5.9 Prove Theorem 5.26.

Hint: Use Sard's Theorem 5.17 and the strategy of the proof of Theorem 5.25.

5.10 Let f be a smooth function on an open set $U \subset \mathbb{R}^k$. For each $x \in U$, let $H(f)_x$ be the Hessian matrix of f at x. Show that f is a Morse function if and only if $(\det(H(f)_x))^2 + \sum_{i=1}^k \left(\frac{\partial f}{\partial x_i}(x)\right)^2 > 0$ for all $x \in U$.

6
Abstract Smooth Manifolds

Now we would like to define manifolds without referring to a given embedding into some $\mathbb{R}^N$. The key idea that should be preserved in any new definition is that a manifold is a space which locally looks like Euclidean space. And we have learned that the latter should mean that every point has an open neighbourhood which is homeomorphic to an open subset in some $\mathbb{R}^k$. A little additional care, however, is necessary. In the definition of an abstract manifold, we want to avoid certain pathological spaces, such as spaces which are not Hausdorff or which do not have countable bases. There are several reasons for this choice. First of all, manifolds are characterised by how open neighbourhoods of points look like, and we would like to be able to find separate neighbourhoods for distinct points. That is exactly the Hausdorff property. Another reason is that we would like to use that a compact subset is a closed subset. This conclusion requires, in general, that X is a Hausdorff space. Second, on several occasions, we have used the existence of a countable base for the topology of subspaces of $\mathbb{R}^N$, for example in the proof of Sard's Theorem which is a key result that abstract manifolds should also satisfy. In fact, all the results and theory we have developed so far apply to the forthcoming new definition of a smooth manifold as we will show in Chapter 7.

We will first introduce some more terminology from general topology and will then discuss the new definition of a smooth manifold. Then we will look at some interesting new examples like projective space and Grassmannian manifolds. It would have been much more difficult to discuss these examples without the new definition. We will study further new examples in the exercises. In Section 7.1, we will then show that every abstract smooth manifold can, in fact, be embedded into Euclidean space and is therefore a smooth manifold in the sense of our previous definition. Hence all the machinery we have developed can be applied to abstract manifolds. For other presentations of manifolds, one may also look, for example, at [10], [29], and [46].

6.1 Abstract Manifolds

We will now set out to introduce abstract manifolds. This requires some rather technical preparation.

6.1.1 Charts and Atlases

Definition 6.1 Let (X, τ) be a topological space. A *basis for the topology* τ is a collection $\mathcal{B} \subset \tau$ of open sets such that every open set in τ is the union of a subcollection of open sets in $\mathcal{B}$.

Example 6.2 The collection of all open balls is a basis for the standard topology on Euclidean space $\mathbb{R}^N$. If $X \subset \mathbb{R}^N$ is a subspace, then the collection of all intersections of X with open balls in $\mathbb{R}^N$ is a basis for the subspace topology on X.

Definition 6.3 A topological space (X, τ) is called *second-countable* if there exists a countable basis for the topology τ, i.e., if there is a basis $\mathcal{B}$ which consists of countably many sets.

Example 6.4 Euclidean space $\mathbb{R}^N$ is second-countable since the collection of all open balls which have rational radii and which are centred at points having only rational coordinates is a countable basis for the topology on $\mathbb{R}^N$. All subspaces of $\mathbb{R}^N$ with the subspace topology are then also second-countable.

Recall that a continuous map $f: X \to Y$ between topological spaces is called *open* if it sends open subsets in X to open subsets in Y. The following lemma shows that surjective open maps respect bases for topologies.

Lemma 6.5 *Let (X, τ_X) and (Y, τ_Y) be topological spaces, and let $\mathcal{B}_X$ be a basis for the topology τ_X on X. Let $f: X \to Y$ be a continuous map. Assume that f is open and surjective. Then $\mathcal{B}_Y := \{f(U) : U \in \mathcal{B}_X\}$ is a basis for the topology τ_Y on Y. If we assume moreover that X is second-countable, then Y is second-countable.*

Proof: Let $V \subset Y$ be a non-empty open subset. Since f is surjective, $f^{-1}(V)$ is non-empty. Since f is continuous, $f^{-1}(V)$ is open in X. Since $\mathcal{B}_X$ is a basis for the topology on X, we can write $f^{-1}(V)$ as a union of U_α of open sets in $\mathcal{B}_X$. Since f is open, $f(U_\alpha)$ is open in Y, and we have $V = \cup_\alpha f(U_\alpha)$. This proves the first assertion. The second assertion follows from the first, since if $\mathcal{B}_X$ is countable then $\mathcal{B}_Y$ as constructed above is countable. □

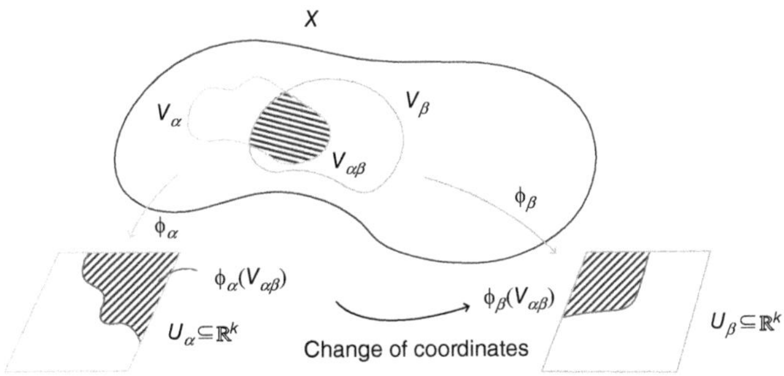

Figure 6.1 If X has an atlas, every point is contained in the domain of a chart. If two charts overlap, we can look at the composition of the chart maps and get a map between open subsets in $\mathbb{R}^k$. We call this the change-of-coordinate map and require this map to be smooth in the sense of calculus.

Next we introduce new terminology for talking about locally Euclidean covers of a space.

Definition 6.6 (Charts) Let X be a topological space. An *n-dimensional chart on X* is a pair (V, ψ) where $V \subset X$ is an open subset and $\psi: V \to U$ is a homeomorphism from V to an open subset $U \subset \mathbb{R}^n$. We say that two n-dimensional charts (V_α, ψ_α) and (V_β, ψ_β) are *compatible* if, for $V_{\alpha\beta} := V_\alpha \cap V_\beta \neq \emptyset$, the change-of-coordinates map

$$(\psi_\beta \circ \psi_\alpha^{-1})_{|\psi_\alpha(V_{\alpha\beta})}: \psi_\alpha(V_{\alpha\beta}) \to \psi_\beta(V_{\alpha\beta})$$

is smooth as a map between open subsets of $\mathbb{R}^n$. See Figure 6.1.

In fact, if two charts are compatible, then the change-of-coordinates maps are diffeomorphisms since they are mutual smooth inverses to each other. We now introduce a name for families of charts which cover the whole space.

Definition 6.7 (Atlas) Let X be a Hausdorff and second-countable topological space. An *n-dimensional atlas on X* is a collection $\mathcal{A} = \{(V_\alpha, \psi_\alpha)\}$ of compatible n-dimensional charts which cover X, i.e., such that $X = \cup_\alpha V_\alpha$. An n-dimensional atlas $\mathcal{A}$ on X is called *maximal* if each chart (V, ψ) on X which is compatible with every chart in $\mathcal{A}$ is already contained in $\mathcal{A}$.

Proposition 6.8 *Let $\mathcal{A}$ be an atlas on a Hausdorff and second-countable topological space. Then $\mathcal{A}$ is contained in a unique maximal atlas.*

Proof: For $\mathcal{A} = \{(V_\alpha, \psi_\alpha)\}$, we define the collection

$$\mathcal{M} := \{(V, \psi): (V, \psi) \text{ is a chart compatible with all charts of } \mathcal{A}\}.$$

We first show that $\mathcal{M}$ is an atlas. Let (V_1, ψ_1) and (V_2, ψ_2) be two charts in $\mathcal{M}$, and let x be a point in $V_1 \cap V_2$. Let (V_α, ψ_α) be a chart in $\mathcal{A}$ with $x \in V_\alpha$. Then both (V_1, ψ_1) and (V_2, ψ_2) are compatible with (V_α, ψ_α) by definition of $\mathcal{M}$. Hence $\psi_\alpha \circ \psi_2^{-1}$ is smooth at $\psi_2(x)$, and $\psi_1 \circ \psi_\alpha^{-1}$ is smooth at $\psi_\alpha(x)$ where we omit from the notation the necessary restrictions of the maps. This implies that $\psi_1 \circ \psi_2^{-1}$ is smooth at $\psi_2(x)$ as well. Similarly, we show that $\psi_2 \circ \psi_1^{-1}$ is smooth at $\psi_1(x)$. Since x was an arbitrary point in $V_1 \cap V_2$, this shows that (V_1, ψ_1) and (V_2, ψ_2) are compatible. Thus, $\mathcal{M}$ is an atlas on X. Moreover, since $\mathcal{A}$ is an atlas, all charts in $\mathcal{A}$ are compatible with charts in $\mathcal{A}$, and we have $\mathcal{A} \subset \mathcal{M}$ by definition of $\mathcal{M}$. Next, we show that $\mathcal{M}$ is maximal. Let (U, ϕ) be a chart which is compatible with all charts of $\mathcal{M}$. Then (U, ϕ) is compatible with all charts of $\mathcal{A}$ as well, and we have $(U, \phi) \in \mathcal{M}$. Now we show that $\mathcal{M}$ is unique. Let $\mathcal{N}$ be an atlas with $\mathcal{A} \subset \mathcal{N}$. Then all charts of $\mathcal{N}$ are compatible with all charts of $\mathcal{A}$, and hence $\mathcal{N} \subset \mathcal{M}$. Moreover, if $\mathcal{N}$ is a maximal atlas with $\mathcal{A} \subset \mathcal{N}$, then, since $\mathcal{N} \subset \mathcal{M}$, every chart (V, ψ) of $\mathcal{M}$ is compatible with every chart of $\mathcal{N}$. Hence, since $\mathcal{N}$ is maximal, we have (V, ψ) is in $\mathcal{N}$, and we get $\mathcal{M} \subset \mathcal{N}$ as well. Thus, $\mathcal{M}$ is the unique maximal atlas containing $\mathcal{A}$. □

6.1.2 Abstract Manifolds

Now we are ready to define smooth manifolds again.

Definition 6.9 (Smooth manifolds) A *smooth manifold* of dimension n is a Hausdorff and second-countable space X together with a maximal n-dimensional atlas.

Remark 6.10 We can think of the maximal atlas on a smooth manifold as its *smooth structure*. Any two compatible atlases then belong to the same smooth structure defined by the unique maximal atlas which contains both. Moreover, to specify a smooth structure, we only need to find some atlas. By Proposition 6.8, there is then a unique maximal atlas which contains the given atlas. This fact will turn out to be very useful when we look at examples soon. This is the reason why we require that the atlas on a smooth manifold should be maximal in Definition 6.9.

Definition 6.11 (Smooth maps between abstract manifolds) Let X be a smooth manifold of dimension n. A continuous map $f: X \to \mathbb{R}^k$ is called *smooth* if, for every chart $\psi_\alpha: V_\alpha \to U_\alpha$, the composition $f \circ \psi_\alpha^{-1}: U_\alpha \to \mathbb{R}^k$ is smooth as a map from an open subset of $\mathbb{R}^n$ to $\mathbb{R}^k$. More generally, let Y a smooth manifold of dimension k and $f\colon X \to Y$ a continuous map. Then f is *smooth at* $x \in X$ if, for every chart $\psi^X\colon V \to U$ on X around x and every chart $\psi^Y\colon V' \to U'$ on Y around $f(x)$, the map

$$\psi^Y \circ f_{|V \cap f^{-1}(V')} \circ (\psi^X)^{-1}_{|U \cap \psi^X(V \cap f^{-1}(V'))} \colon U \cap \psi^X(V \cap f^{-1}(V')) \to U'$$

is a smooth map as a map from an open subset of $\mathbb{R}^n$ to an open subset of $\mathbb{R}^k$. We say that f is *smooth* if it is smooth at every $x \in X$.

Our initial definition of a smooth manifold fits nicely into this picture.

Proposition 6.12 *Let $X \subset \mathbb{R}^N$ be a smooth manifold of dimension n in the sense of Definition 2.11. Then X is a smooth manifold of dimension n in the sense of Definition 6.9.*

Proof: By Example 1.19, X is Hausdorff as every subspace of Euclidean space is Hausdorff. By Example 6.4, X is second-countable. We now show that local parametrisations provide X with an atlas. Let $x \in X$ be a point in X and let $\phi_1\colon U_1 \to V_1$ and $\phi_2\colon U_2 \to V_2$ be local parametrisations around x with open subsets $U_1, U_2 \subset \mathbb{R}^n$ and $V_1, V_2 \subset X$. The overlap of V_1 and V_2 is non-empty since it contains x. Moreover, $V_1 \cap V_2$ is an open subset of X. Since both ϕ_1 and ϕ_2 are homeomorphisms, $\phi_1^{-1}(V_1 \cap V_2) \subset U_1$ and $\phi_2^{-1}(V_1 \cap V_2) \subset U_2$ are both open subsets in $\mathbb{R}^n$. Since both ϕ_1 and ϕ_2 are diffeomorphisms, the transition map $\phi_2^{-1} \circ \phi_1\colon \phi_1^{-1}(V_1 \cap V_2) \to \phi_2^{-1}(V_1 \cap V_2)$ is a diffeomorphism between open subsets in $\mathbb{R}^n$. Thus, local parametrisations, or rather the corresponding local coordinate systems, equip X with an atlas. By Proposition 6.8, there is a unique maximal atlas containing this atlas. This provides X with the structure of a smooth manifold. □

Remark 6.13 Proposition 6.12 shows that all the examples of smooth manifolds contained in Euclidean space are examples of smooth manifolds in the sense of Definition 6.9. In particular, the n-sphere $\mathbb{S}^n \subset \mathbb{R}^{n+1}$ is a smooth manifold, and any of the local parametrisations we studied so far are compatible and determine the same maximal atlas on $\mathbb{S}^n$. We call this the *standard smooth structure* on $\mathbb{S}^n$. It turns out that for $n \geq 7$, there may be different smooth structures on $\mathbb{S}^n$ in general. See [33] and Remark 5.14.

Remark 6.14 One difference between the two ways to define manifolds is that we previously required local parametrisations to be diffeomorphisms. For an abstract manifold X, we use the charts to define what smoothness means for a map on X. A priori, it only makes sense to talk about the smoothness of the change of coordinate maps. A posteriori, we can then check that charts are in fact diffeomorphisms. Similarly, a priori, we only know what smoothness of maps between Euclidean spaces means. We need to use the charts to transfer the concept of smooth maps between manifolds from maps between Euclidean spaces.

6.1.3 Tangent Space of an Abstract Smooth Manifold

There are several different, though equivalent, ways to define the tangent space of an abstract smooth manifold. We will look at just one way which is closest to our intuitive and concrete approach to tangent spaces, and which is similar to the description of tangent vectors as velocity vectors of curves in Exercise 2.16. For other constructions of the tangent space, see for example [29] and [46]. Let X be a smooth manifold of dimension n in the sense of Definition 6.9, and let $x \in X$ be a point.

Definition 6.15 A *smooth curve on* X *through* x is a smooth map $\gamma\colon J \to X$ with $\gamma(0) = x$ and where J is an open interval in $\mathbb{R}$ with $0 \in J$.

We note that by definition of smooth maps, γ is smooth if and only if γ is continuous and there is a chart $\psi\colon V \to U$ with $x \in V$ and $U \subset \mathbb{R}^n$ open such that $\psi \circ \gamma$ as a map from the open subset $\gamma^{-1}(V) \subset \mathbb{R}$ to $\mathbb{R}^n$. We now define an equivalence relation on the set of smooth curves on X through x.

Definition 6.16 We say that two smooth curves $\gamma_1\colon J_1 \to X$ and $\gamma_2\colon J_2 \to X$ on X through x are *equivalent*, denoted $\gamma_1 \sim \gamma_2$, if, for every chart $\psi\colon V \to U$ with $x \in V$ and $U \subset \mathbb{R}^n$ open, we have an equality $d(\psi \circ \gamma_1)_0 = d(\psi \circ \gamma_2)_0$ in $\mathrm{Hom}_{\mathbb{R}}(\mathbb{R}, \mathbb{R}^n)$ of derivatives of smooth maps $(-\varepsilon, \varepsilon) \to \mathbb{R}^n$ where $(-\varepsilon, \varepsilon) \subset (\gamma_1^{-1}(V) \cap \gamma_2^{-1}(V))$ is a sufficiently small open interval in $J_1 \cap J_2$. We denote the set of equivalence classes by $\mathcal{T}_x(X)$.

It is clear from the definition that the set $\mathcal{T}_x(X)$ only depends on the structure of X locally at x.

Lemma 6.17 (One chart is enough) *Two smooth curves* $\gamma_1\colon J_1 \to X$ *and* $\gamma_2\colon J_2 \to X$ *on* X *through* x *are equivalent if and only if there is at least one chart* $\psi\colon V \to U$ *with* $x \in V$ *such that* $d(\psi \circ \gamma_1)_0 = d(\psi \circ \gamma_2)_0$ *in* $\mathrm{Hom}_{\mathbb{R}}(\mathbb{R}, \mathbb{R}^n)$.

Proof: By definition of the equivalence relation, only one direction needs to be proven. So, assume that there is a chart $\psi\colon V \to U$ with $x \in V$ such that $d(\psi \circ \gamma_1)_0 = d(\psi \circ \gamma_2)_0$. Let $\psi'\colon V' \to U'$ be another chart with $x \in V'$. Then we have the following equalities of linear maps in $\mathrm{Hom}_{\mathbb{R}}(\mathbb{R}, \mathbb{R}^n)$

$$
\begin{aligned}
d(\psi' \circ \gamma_1)_0 &= d(\psi' \circ (\psi^{-1} \circ \psi) \circ \gamma_1)_0 = d(\psi' \circ \psi^{-1})_{\psi(x)} \circ d(\psi \circ \gamma_1)_0 \\
&= d(\psi' \circ \psi^{-1})_{\psi(x)} \circ d(\psi \circ \gamma_2)_0 = d(\psi' \circ (\psi^{-1} \circ \psi) \circ \gamma_2)_0 \\
&= d(\psi' \circ \gamma_2)_0.
\end{aligned}
$$

We note that all of the above composed maps are maps between Euclidean spaces, and hence their derivatives are defined and the chain rule from calculus applies. Here we omitted the restriction of the maps to keep the notation simple. This shows $\gamma_1 \sim \gamma_2$ as claimed. □

Definition 6.18 Let X be a smooth manifold. The *tangent space of X at x* is defined as the set $\mathcal{T}_x(X)$ of all equivalence classes of smooth curves on X through x.

Example 6.19 We identify $\mathrm{Hom}_{\mathbb{R}}(\mathbb{R}, \mathbb{R}^n)$ with $\mathbb{R}^n$ via the canonical isomorphism $\mathrm{Hom}_{\mathbb{R}}(\mathbb{R}, \mathbb{R}^n) \to \mathbb{R}^n$ given by $\varphi \mapsto \varphi(1)$. For every $x \in \mathbb{R}^n$, we now construct a canonical isomorphism $\mathcal{T}_x(\mathbb{R}^n) \cong \mathbb{R}^n$. We have the map $\mathcal{T}_x(\mathbb{R}^n) \to \mathbb{R}^n$ given by $[\gamma] \mapsto d\gamma_0$ with inverse defined as follows. For $v \in \mathbb{R}^n$, we define the smooth curve $\gamma_v \colon \mathbb{R} \to \mathbb{R}^n$ through x in $\mathbb{R}^n$ by $\gamma_v(t) := x + vt$. This defines a map $\mathbb{R}^n \to \mathcal{T}_x(\mathbb{R}^n)$ by $v \mapsto [\gamma_v]$. Since the identity is a global chart on $\mathbb{R}^n$, Lemma 6.17 and $d(\gamma_v)_0 = v$, considered as an element of $\mathbb{R}^n \cong \mathrm{Hom}_{\mathbb{R}}(\mathbb{R}, \mathbb{R}^n)$, imply that the mutual compositions are the respective identity maps.

It requires some effort to define a vector space structure on $\mathcal{T}_x(X)$ which we will do later. The derivative of a smooth map, however, is quite easy to define.

Definition 6.20 Let X and Y be smooth manifolds and let $f \colon X \to Y$ be a smooth map. Let $x \in X$ be a point. The *derivative of f at x*, denoted $df_x \colon \mathcal{T}_x(X) \to \mathcal{T}_{f(x)}(Y)$, is defined to be the map which sends the equivalence class $[\gamma]$ in $\mathcal{T}_x(X)$ to the equivalence class $[f \circ \gamma]$ in $\mathcal{T}_{f(x)}(Y)$.

Lemma 6.21 *The derivative df_x is well defined, i.e., if $\gamma_1 \sim \gamma_2$ as smooth curves on X, then $f \circ \gamma_1 \sim f \circ \gamma_2$ as smooth curves on Y.*

Proof: Let $\psi \colon V \to U$ be a chart on X with $x \in V$ such that $d(\psi \circ \gamma_1)_0 = d(\psi \circ \gamma_2)_0$. Let $\psi' \colon V' \to U'$ be a chart on Y with $f(x) \in V'$ such that $\psi' \circ (f \circ \gamma_1)$ and $\psi' \circ (f \circ \gamma_2)$ are smooth maps. To simplify the notation, we let $\bar{f} \colon U \cap \psi(V \cap f^{-1}(V')) \to U'$ denote the composition $\psi' \circ f \circ \psi^{-1}$, where we omit the restrictions to the appropriate open subsets from the notation. We then have the equality of derivatives

$$\begin{aligned} d(\psi' \circ (f \circ \gamma_1))_0 &= d(\bar{f} \circ \psi \circ \gamma_1))_0 = d\bar{f}_{\psi(x)} \circ d(\psi \circ \gamma_1))_0 \\ &= d\bar{f}_{\psi(x)} \circ d(\psi \circ \gamma_2))_0 = d(\bar{f} \circ \psi \circ \gamma_2))_0 \\ &= d(\psi' \circ (f \circ \gamma_2))_0. \end{aligned}$$

We note again that all of the above composed maps are maps between Euclidean spaces, and hence their derivatives are defined and the chain rule from calculus applies. By Lemma 6.17, this shows $f \circ \gamma_1 \sim f \circ \gamma_2$ as smooth curves on Y. □

Lemma 6.22 (Chain rule) *Let $X \xrightarrow{f} Y \xrightarrow{g} Z$ be smooth maps of manifolds. Then $d(g \circ f)_x = dg_{f(x)} \circ df_x$.*

Proof: For all $[\gamma] \in \mathcal{T}_x(X)$, we have

$$\begin{aligned} d(g \circ f)_x([\gamma]) &= [(g \circ f) \circ \gamma] = [g \circ (f \circ \gamma)] = dg_{f(x)}([f \circ \gamma]) \\ &= dg_{f(x)}(df_x([\gamma])). \end{aligned}$$

This proves the lemma. □

In the following, we identify $\mathrm{Hom}_{\mathbb{R}}(\mathbb{R}, \mathbb{R}^n)$ with $\mathbb{R}^n$ via the canonical isomorphism $\mathrm{Hom}_{\mathbb{R}}(\mathbb{R}, \mathbb{R}^n) \to \mathbb{R}^n$ given by $\varphi \mapsto \varphi(1)$.

Proposition 6.23 *Let X be a smooth manifold and let $x \in X$. Let $\psi \colon V \to U$ be a chart on X with $x \in V$. The map $\tau_x \colon \mathcal{T}_x(X) \to \mathbb{R}^n$ defined by $\tau_x([\gamma]) := d(\psi \circ \gamma)_0$ is a bijection with inverse $\tau_x^{-1} \colon \mathbb{R}^n \to \mathcal{T}_x(X)$ given by $v \mapsto [\gamma_v^x]$ where $\gamma_v^x \colon (-\varepsilon, \varepsilon) \to X$ is the smooth curve on X defined by $\gamma_v^x(t) := \psi^{-1}(\psi(x) + tv)$ for a sufficiently small $\varepsilon > 0$.*

Proof: By Lemma 6.17, the maps τ_x and τ_x^{-1} are well defined and independent of the choice of chart ψ. It remains to check that they are mutual inverses. For $v \in \mathbb{R}^n$, we have

$$\begin{aligned} \tau_x(\tau_x^{-1}(v)) &= \tau_x([\gamma_v^x]) = d(\psi \circ \gamma_v^x)_0 = d(\psi(\psi^{-1}(\psi(x) + tv)))_0 \\ &= d(\psi(x) + tv)_0 = v. \end{aligned}$$

For $[\gamma] \in \mathcal{T}_x(X)$, we have $\tau_x^{-1}(\tau_x([\gamma])) = \tau_x^{-1}(d(\psi \circ \gamma)_0) = [\gamma_{d(\psi \circ \gamma)_0}]$, and $[\gamma_{d(\psi \circ \gamma)_0}] = [\gamma]$ since

$$\begin{aligned} d(\psi \circ \gamma_{d(\psi \circ \gamma)_0})_0 &= d(\psi \circ (\psi^{-1}(\psi(x) + td(\psi \circ \gamma)_0)))_0 \\ &= d(\psi(x) + td(\psi \circ \gamma)_0)_0 = d(\psi \circ \gamma)_0. \end{aligned}$$

This proves the assertion. □

We note that the class of a constant curve at x corresponds to 0 under the bijection τ_x. Since the bijections τ_x and τ_x^{-1} are independent of the choice of ψ, we can use them to define a vector space structure on $\mathcal{T}_x(X)$.

Definition 6.24 (Vector space structure) Let X be a smooth manifold and let $x \in X$. We define the zero vector in $\mathcal{T}_x(X)$ to be the class of a constant curve at x, and we define a vector space structure on $\mathcal{T}_x(X)$ by

$$[\gamma_1] + [\gamma_2] := \tau_x^{-1}(\tau_x([\gamma_1]) + \tau_x([\gamma_2]))$$

for all $[\gamma_1], [\gamma_2] \in \mathcal{T}_x(X)$, and $\lambda \cdot [\gamma] := \tau_x^{-1}(\lambda \cdot \tau_x([\gamma]))$ for all $\lambda \in \mathbb{R}$ and $[\gamma] \in \mathcal{T}_x(X)$.

Proposition 6.25 *Let X and Y be smooth manifolds of dimension n and m, respectively. Let $f\colon X \to Y$ be a smooth map and $x \in X$. Let $\psi\colon V \to U$ be a chart on X with $x \in V$ and $\psi'\colon V' \to U'$ be a chart on Y with $f(x) \in V'$. We let $\bar{f}\colon U \cap \psi(V \cap f^{-1}(V')) \to U'$ denote the composition $\psi' \circ f \circ \psi^{-1}$, where we omit the restrictions to the appropriate open subsets from the notation. Then we have a commutative diagram*

$$\begin{array}{ccc} \mathcal{T}_x(X) & \xrightarrow{\quad df_x \quad} & \mathcal{T}_{f(x)}(Y) \\ \Big\downarrow{\scriptstyle \tau_x} & & \Big\downarrow{\scriptstyle \tau_{f(x)}} \\ \mathbb{R}^n & \xrightarrow{\quad d\bar{f}_{\psi(x)} \quad} & \mathbb{R}^m \end{array} . \tag{6.1}$$

In particular, the derivative $df_x\colon \mathcal{T}_x(X) \to \mathcal{T}_{f(x)}(Y)$ is a linear map.

Proof: Once we have shown that diagram (6.1) commutes, it follows from the definition of the vector space structures on $\mathcal{T}_x(X)$ and $\mathcal{T}_{f(x)}(Y)$ that df_x is linear since $d\bar{f}_{\psi(x)}$ is a linear map. To show that the diagram commutes, we compute both compositions. For $[\gamma] \in \mathcal{T}_x(X)$, we have $\tau_{f(x)}(df_x([\gamma])) = \tau_{f(x)}([f \circ \gamma]) = d(\psi' \circ f \circ \gamma)_0$, and, using the chain rule, we have

$$\begin{aligned} d\bar{f}_{\psi(x)}(\tau_x([\gamma])) &= d\bar{f}_{\psi(x)}(d(\psi \circ \gamma)_0) = d(\psi' \circ f \circ \psi^{-1})_{\psi(x)}(d(\psi \circ \gamma)_0) \\ &= d(\psi' \circ f \circ \psi^{-1} \circ \psi \circ \gamma)_0 = d(\psi' \circ f \circ \gamma)_0. \end{aligned}$$

Thus, $\tau_{f(x)}(df_x([\gamma]) = d\bar{f}_{\psi(x)}(\tau_x([\gamma]))$ as claimed. □

Remark 6.26 The commutative diagram (6.1) gives us a concrete way to compute a matrix which represents df_x. By construction of the vector space structures on $\mathcal{T}_x(X)$ and $\mathcal{T}_{f(x)}(Y)$, we have a canonical choice of bases, namely the images of the standard bases of $\mathbb{R}^n$ and $\mathbb{R}^m$ under τ_x and $\tau_{f(x)}$, respectively. Then we get a representing matrix for df_x with respect to these bases as the Jacobian matrix of $d\bar{f}_{\psi(x)}\colon \mathbb{R}^n \to \mathbb{R}^m$.

From the chain rule, we can now deduce that the derivative of a diffeomorphism of smooth manifolds is an isomorphism. A smooth map $f: X \to Y$ is again called a *local diffeomorphism at* x if there is an open subset $U \subset X$ containing x such that $f(U) \subset Y$ is open in Y and $f_{|U}: U \to f(U)$ is a diffeomorphism, and we say that f is a *local diffeomorphism* if it is a local diffeomorphism at every $x \in X$. We can then use the Inverse Function Theorem 3.2 to show that f is a local diffeomorphism at x if df_x is an isomorphism. Moreover, we can define the notions of immersion and submersion in the same way as before, i.e., a smooth map $f: X \to Y$ is called an *immersion* if df_x is injective at all $x \in X$, and f is called a *submersion* if df_x is surjective at all $x \in X$. The notion of a regular value of a smooth map is defined as before, too.

Finally, we show that the two constructions of tangent spaces we now have for smooth manifolds X embedded in Euclidean space agree.

Proposition 6.27 (Comparing tangent spaces) *Let $X \subset \mathbb{R}^N$ be a smooth manifold of dimension n and let $x \in X$ be a point. Let $\phi: U \to V$ be a local parametrisation with $x \in V \subset X$ and $0 \in U \subset \mathbb{R}^n$ open and $\phi(0) = x$. We let $\psi := \phi^{-1}$ be the corresponding chart on X. Then the map defined by $\kappa_x([\gamma]) := d\phi_0(d(\psi \circ \gamma)_0)$ is a natural isomorphism of vector spaces $\kappa_x: \mathcal{T}_x(X) \xrightarrow{\cong} T_x(X)$.*

Proof: By definition, elements of $T_x(X)$ are of the form $d\phi_0(v)$ for some $v \in \mathbb{R}^n$. We can then construct an inverse κ_x^{-1} for κ_x by sending $d\phi_0(v)$ to the class of the curve $\gamma_v: (-\varepsilon, \varepsilon) \to X$ given by $\gamma_v(t) := \phi(tv)$ for a sufficiently small $\varepsilon > 0$. By construction of γ_v, we have $d(\psi \circ \gamma_v)_0 = v$. We then get $\kappa_x(\kappa_x^{-1}(d\phi_0(v))) = \kappa_x([\gamma_v]) = d\phi_0(d(\psi \circ \gamma_v)_0) = v$, and, for $[\gamma] \in \mathcal{T}_x(X)$, we have $\kappa_x^{-1}(\kappa_x([\gamma])) = \kappa_x^{-1}(d\phi_0(d(\psi \circ \gamma)_0)) = [\gamma_{d(\psi\circ\gamma)_0}]$. Since we have $d(\psi \circ \gamma)_0 = d(\psi \circ \gamma_{d(\psi\circ\gamma)_0})_0$, we get $[\gamma] = [\gamma_{d(\psi\circ\gamma)_0}]$. Finally, it follows directly from the construction of the vector space structure on $\mathcal{T}_x(X)$ that κ_x and κ_x^{-1} are linear maps. □

6.1.4 Examples: Circle, Torus, and Klein Bottle

We now look at some examples. We are already familiar with the circle and the torus. We will now learn how to define them as abstract manifolds without placing them in Euclidean space first. The other examples are new.

Example 6.28 (The circle as an abstract manifold) On the real line $\mathbb{R}$, we consider the equivalence relation R given by $x \sim_R x + n$ for all $n \in \mathbb{Z}$. We write $\mathbb{R}/\mathbb{Z}$ for the corresponding quotient space. Let $q: \mathbb{R} \to \mathbb{R}/\mathbb{Z}$ be the quotient map. If $U \subset \mathbb{R}$ is open, then $q^{-1}(q(U)) = \{x + n \in \mathbb{R} : x \in U,\ n \in \mathbb{Z}\}$ is the union of open sets of $\mathbb{R}$ and is therefore open in $\mathbb{R}$ as well. By definition of the quotient topology on $\mathbb{R}/\mathbb{Z}$, this shows that q is an open map. We now show that R is also closed. We observe that $\Gamma(R)$ equals the preimage $f^{-1}(\mathbb{Z}) = \{(x, y) \in \mathbb{R} \times \mathbb{R} : x - y \in \mathbb{Z}\}$ under the continuous map $f : \mathbb{R} \times \mathbb{R} \to \mathbb{R}$ which sends (x, y) to $x - y$. Since the complement of $\mathbb{Z}$ in $\mathbb{R}$ is a union of open subsets, $\mathbb{Z}$ is closed in $\mathbb{R}$. Thus, $\Gamma(R) = f^{-1}(\mathbb{Z})$ is closed in $\mathbb{R}$. By Exercise 1.7, this implies that $\mathbb{R}/\mathbb{Z}$ is a Hausdorff space. By Lemma 6.5, we also know that $\mathbb{R}/\mathbb{Z}$ is second-countable. We now define charts on $\mathbb{R}/\mathbb{Z}$. Let $[x]$ denote the equivalence class of $x \in \mathbb{R}$. Then, for any $[x] \in \mathbb{R}/\mathbb{Z}$ and $-1/2 < \delta < 1/2$, every element of the form $[x + \delta]$ has a unique representative in the open interval $U_x := (-1/2 + x, x + 1/2) \subset \mathbb{R}$. Since q is an open map, this shows that the restriction of the quotient map $q_{|U_x} : U_x \to \mathbb{R}/\mathbb{Z}$ is a homeomorphism onto its image. We let ψ_x denote the inverse of this map. Then the pairs $(V_{[x]} := q(U_x), \psi_{[x]})$ for all $x \in [0, 1)$ define charts. In fact, we can cover $\mathbb{R}/\mathbb{Z}$ with just two of these charts, for example $V_{[0]} = \{[t] \in \mathbb{R}/\mathbb{Z} : t \in (-1/2, 1/2)\} = \mathbb{R}/\mathbb{Z} \setminus \{[1/2]\}$ and $V_{[1/2]} = \{[t] \in \mathbb{R}/\mathbb{Z} : t \in (0, 1)\} = \mathbb{R}/\mathbb{Z} \setminus \{[0]\}$. The transition map from $\psi_{[0]}(V_{[0]} \cap V_{[1/2]}) = (-1/2, 0) \cup (0, 1/2)$ to $(0, 1/2) \cup (1/2, 1) = \psi_{[1/2]}(V_{[0]} \cap V_{[1/2]})$ is the identity $t \mapsto t$ on $(0, 1/2)$ and the translation $(-1/2, 0) \to (1/2, 1)$, $t \mapsto t + 1$, on the other set. Thus the transition map is a diffeomorphism, and the collection $\{(V_{[0]}, \psi_{[0]}), (V_{[1/2]}, \psi_{[1/2]})\}$ is an atlas on $\mathbb{R}/\mathbb{Z}$. The corresponding maximal atlas then defines the structure as a smooth manifold of dimension one on $\mathbb{R}/\mathbb{Z}$. In fact, $\mathbb{R}/\mathbb{Z}$ is diffeomorphic to the circle $\mathbb{S}^1 \subset \mathbb{R}^2$ via the diffeomorphism $\mathbb{R}/\mathbb{Z} \xrightarrow{\cong} \mathbb{S}^1$ defined by $t \mapsto (\cos(2\pi t), \sin(2\pi t))$.

Remark 6.29 Let X and Y be smooth manifolds with atlases $\mathcal{A} = \{(V_\alpha, \psi_\alpha)\}$ and $\mathcal{B} = \{(W_\beta, \phi_\beta)\}$, respectively. Then $X \times Y$ is Hausdorff and second-countable. Moreover, the collection $\{(V_\alpha \times W_\beta, \psi_\alpha \times \phi_\beta)\}$ defines an atlas on $X \times Y$. Hence the product of two smooth manifolds is again a smooth manifold.

Example 6.30 (The torus as an abstract manifold) Now we can define the torus as the product $\mathbb{R}/\mathbb{Z} \times \mathbb{R}/\mathbb{Z}$, i.e., we consider the quotient space of $\mathbb{R} \times \mathbb{R}$ under the equivalence relation $(x, y) \sim (x + m, y + n)$ for all $m, n \in \mathbb{Z}$. Then $\mathbb{R}/\mathbb{Z} \times \mathbb{R}/\mathbb{Z}$ is a manifold since it is the product of manifolds. We can imagine the quotient as follows. We start with a square, and then we glue together opposite edges. See Figure 6.2. Alternatively, we can argue as in Example 6.28. Moreover, we have a diffeomorphism $\mathbb{R}/\mathbb{Z} \times \mathbb{R}/\mathbb{Z} \xrightarrow{\cong} \mathbb{S}^1 \times \mathbb{S}^1 = \mathbb{T}^2$.

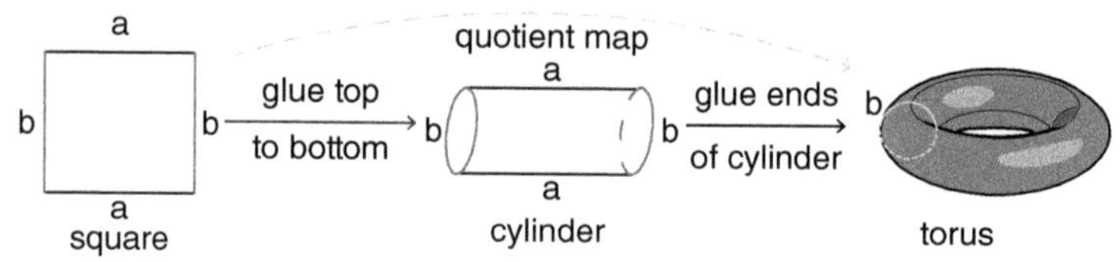

Figure 6.2 The torus $\mathbb{T}^2$ can be constructed by gluing together the opposite edges of a square.

Example 6.31 (The Klein bottle) A slightly more complicated space is the *Klein bottle*, named after the mathematician Felix Klein. This is a two-dimensional space which we cannot embed into $\mathbb{R}^3$. We consider the quotient space of $\mathbb{R} \times \mathbb{R}$ under the equivalence relation $(x, y) \sim (x + m, -y + n)$ for all $m, n \in \mathbb{Z}$. The Klein bottle is defined as the quotient $K := (\mathbb{R} \times \mathbb{R})/\sim$. We can think of the quotient as follows. We start with a rectangle, and then we glue together one pair of opposite edges as we did for the torus. For the other pair of opposite edges, however, we first twist one of them and then glue them together. The additional twist produces an object which is harder to visualise. We can then check as in Example 6.28 that the quotient map $q: \mathbb{R} \times \mathbb{R} \to K$ is open and $\Gamma(R)$ is closed. Hence K is a second-countable Hausdorff space by Lemma 6.5 and Exercise 1.7. It remains to define an atlas. Let $[x, y] \in K$ denote the equivalence class of $(x, y) \in \mathbb{R} \times \mathbb{R}$. Then, for any $[x, y] \in K$ and $-1/2 < \delta, \varepsilon < 1/2$, every element of the form $[x + \delta, y + \varepsilon]$ has a unique representative in the open interval $U_{x,y} := (-1/2 + x, x + 1/2) \times (-1/2 + y, y + 1/2)$. Since q is an open map, this shows that the restriction of the quotient map $q_{|U_{x,y}}: U_{x,y} \to K$ is a homeomorphism onto its image. We let $\psi_{x,y}$ denote the inverse of this map. We can then show that the collection $\{(V_{[x,y]} := q(U_{x,y}), \psi_{[x,y]})\}$, for $[x, y]$ being one of the pairs $\{[0, 0], [0, 1/2], [1/2, 0], [1/2, 1/2]\}$, define an atlas. There are six overlaps we need to check. We only discuss the transition map for one of them. Checking the remaining ones is very similar. We consider $V_{[0,0]} \cap V_{[1/2,1/2]} = K \setminus \{[0, 0], [1/2, 1/2]\}$. We then have $\psi_{[0,0]}(V_{[0,0]} \cap V_{[1/2,1/2]}) = ((-1/2, 0) \cup (0, 1/2)) \times ((-1/2, 0) \cup (0, 1/2))$ and $\psi_{[1/2,1/2]}(V_{[0,0]} \cap V_{[1/2,1/2]}) = ((0, 1/2) \cup (1/2, 1)) \times ((0, 1/2) \cup (1/2, 1))$. The transition map is given by the product of the following two maps: the map on the first coordinate is $x \mapsto x$ on $(0, 1/2)$ and $x \mapsto x + 1$ from $(-1/2, 0)$ to $(1/2, 1)$, and the map on second coordinate is $y \mapsto -y$ from $(-1/2, 0)$ to $(0, 1/2)$ and $y \mapsto -y + 1$ from $(0, 1/2)$ to $(1/2, 1)$. In total, we get that the transition maps are diffeomorphisms, and hence the charts form an atlas. The associated maximal atlas then provides K with the structure of an abstract smooth manifold of dimension two.

Remark 6.32 In Exercise 6.3, we study Hopf manifolds as another important and interesting class of examples. They play a crucial role in complex geometry, since they provide the simplest examples of compact complex manifolds which do not admit a Kähler metric and therefore cannot be embedded into complex projective space. See for example [23] for the theory of complex manifolds.

6.2 Real Projective Space

This is an important example which we can easily describe with the new definition of an abstract manifold, but for which it is not obvious at all how we can embed it into $\mathbb{R}^N$.

Definition 6.33 (Real Projective Space) The *real projective n-space* $\mathbb{RP}^n$ is the set of all straight lines through the origin in $\mathbb{R}^{n+1}$. As a topological space, $\mathbb{RP}^n$ is the quotient space $\mathbb{RP}^n = (\mathbb{R}^{n+1} \setminus \{0\})/\sim$ where the equivalence relation is given by $x \sim y$ if there is a non-zero real number λ such that $x = \lambda y$. A subset V is open in $\mathbb{RP}^n$ if and only if its preimage $U = \{x \in \mathbb{R}^{n+1} \setminus \{0\} : [x] \in V\}$ is open in $\mathbb{R}^{n+1} \setminus \{0\}$.

Note that each line through the origin intersects the unit sphere in two antipodal points. Hence $\mathbb{RP}^n$ can also be described as $\mathbb{S}^n/\sim$ where the equivalence relation is given by $x \sim -x$. As a quotient of $\mathbb{S}^n$, we see that $\mathbb{RP}^n$ is actually compact. As we have learned already, this may turn out to be a useful observation. We will now show that $\mathbb{RP}^n$ is a Hausdorff space. We consider $\mathbb{RP}^n$ as the quotient of $\mathbb{S}^n$ modulo the relation $x \sim -x$. We let $q: \mathbb{S}^n \to \mathbb{RP}^n$ denote the quotient map which sends a point x on the sphere to its equivalence class $[x]$. A subset $V \subset \mathbb{RP}^n$ is open in $\mathbb{RP}^n$ if and only if the preimage $q^{-1}(V) \subset \mathbb{S}^n$ is open in $\mathbb{S}^n$. The following lemma shows that the image $q(U)$ of an open set in $\mathbb{S}^n$ is also open in $\mathbb{RP}^n$.

Lemma 6.34 *The quotient map q is open.*

Proof: Let $U \subset \mathbb{S}^n$ be an open subset. The antipodal map $a: \mathbb{S}^n \to \mathbb{S}^n$, which sends x to $-x$ is a homeomorphism. Hence the subset $-U := \{-x \in \mathbb{S}^n : x \in U\}$ of antipodal points to U is open in $\mathbb{S}^n$. Their union $U \cup -U$ is therefore open in $\mathbb{S}^n$ as well. Since $q^{-1}(q(U)) = U \cup -U$, the subset $q(U)$ is open in $\mathbb{RP}^n$ by the definition of the quotient topology on $\mathbb{RP}^n$ under q. □

Proposition 6.35 *Real projective n-space $\mathbb{RP}^n$ is a Hausdorff space.*

Proof: Let $[x]$ and $[y]$ be two distinct points in $\mathbb{RP}^n$. Let x and $-x$ be the antipodal points in $\mathbb{S}^n$ which correspond to $[x]$, and let y and $-y$ be the antipodal points in $\mathbb{S}^n$ which correspond to $[y]$. Then either y or $-y$ is in the same hemisphere of $\mathbb{S}^n$ as x, and hence $|x-y|/2 \leq 1$ or $|x+y|/2 \leq 1$. We then choose an $\varepsilon > 0$ such that $\varepsilon < \frac{1}{4}\min\{|x-y|, |x+y|\}$. Then we have $\mathbb{B}^{n+1}_\varepsilon(x) \cap \mathbb{B}^{n+1}_\varepsilon(y) = \emptyset$. We define the sets $U_z := \mathbb{B}^{n+1}_\varepsilon(z) \cap \mathbb{S}^n$ for $z = x, y$. Then $q(U_x)$ and $q(U_y)$ are disjoint neighbourhoods of $[x]$ and $[y]$, respectively. By Lemma 6.34, both $q(U_x)$ and $q(U_y)$ are open in $\mathbb{RP}^n$. This shows that $\mathbb{RP}^n$ is Hausdorff. □

Theorem 6.36 *Real projective n-space $\mathbb{RP}^n$ is a compact smooth manifold of dimension n.*

Proof: We already observed that $\mathbb{RP}^n$ is compact as a continuous image of $\mathbb{S}^n$. By Lemmas 6.5 and 6.34, $\mathbb{RP}^n$ is second-countable. By Proposition 6.35, $\mathbb{RP}^n$ is Hausdorff. It remains to construct an atlas. For $x = (x_0, \ldots, x_n) \in \mathbb{R}^{n+1} \setminus \{0\}$, we write $[x]$ for its equivalence class considered as a point in $\mathbb{RP}^n$. We also write $[x] = [x_0 : \ldots : x_n]$. For $0 \leq i \leq n$, let $V_i := \{[x] \in \mathbb{RP}^n : x_i \neq 0\}$. The preimage of V_i in $\mathbb{R}^{n+1}$ is the open subset $\{x \in \mathbb{R}^{n+1} : x_i \neq 0\}$. Hence each V_i is open in $\mathbb{RP}^n$. By varying i, we obtain an open cover of $\mathbb{RP}^n$ since every representative $(x_0, \ldots, x_n)$ of a point $[x] \in \mathbb{RP}^n$ must have at least one coordinate $\neq 0$, otherwise it would be the origin which is excluded. For each i, we have the maps $\phi_i : \mathbb{R}^n \to V_i$ defined by

$$(x_0, \ldots, \widehat{x_i}, \ldots, x_n) \mapsto \left[x_0 : \ldots : x_{i-1} : 1 : x_{i+1} : \ldots : x_n\right]$$

and $\phi_i^{-1} \colon V_i \to \mathbb{R}^n$ defined by

$$[x_0 : \ldots : x_i : \ldots : x_n] \mapsto \frac{1}{x_i}(x_0, \ldots, \widehat{x_i}, \ldots, x_n),$$

where $\widehat{x_i}$ means that x_i is omitted. Note that the quotient $\frac{1}{x_i}$ is well defined on all of V_i. We note that ϕ_i^{-1} is well defined, i.e., it does depend on the chosen representative, since, if $[x_0 : \ldots : x_i : \ldots : x_n] = [\lambda x_0 : \ldots : \lambda x_i : \ldots : \lambda x_n]$ for some $\lambda \neq 0$, then

$$\begin{aligned}\phi_i^{-1}([\lambda x]) &= \frac{1}{\lambda x_i}(\lambda x_0, \ldots, \lambda x_{i-1}, \lambda x_{i+1}, \ldots, \lambda x_n) \\ &= \frac{1}{x_i}(x_0, \ldots, x_{i-1}, x_{i+1}, \ldots, x_n) = \phi_i^{-1}([x]).\end{aligned}$$

It is easy to see that ϕ_i and ϕ_i^{-1} are mutual inverses and are both continuous. Finally, the change-of-coordinate maps are smooth since the composite

$$\phi_i^{-1}(V_i \cap V_j) \xrightarrow{\phi_i} V_i \cap V_j \xrightarrow{\phi_j^{-1}} \phi_j^{-1}(V_i \cap V_j)$$

is just $(x_0, \dots, \widehat{x_i}, \dots, x_n) \mapsto \frac{1}{x_j}(x_0, \dots, x_{i-1}, 1, x_{i+1}, \dots, \widehat{x_j}, \dots, x_n)$ which is smooth whenever $x_j \neq 0$.[1] This shows that we obtain an atlas on $\mathbb{RP}^n$. The corresponding maximal atlas provides $\mathbb{RP}^n$ with the structure of a smooth manifold. □

In the special case of $n = 1$, $\mathbb{RP}^1$ is actually very familiar.

Lemma 6.37 *The circle $\mathbb{S}^1$ and $\mathbb{RP}^1$ are diffeomorphic.*

Proof: Identifying antipodal points on $\mathbb{R}/\mathbb{Z} \cong \mathbb{S}^1$ corresponds to imposing the equivalence relation R_2 given by $x \sim_{R_2} x + n/2$ for all $n \in \mathbb{Z}$ and $x \in \mathbb{R}$. We can then define a continuous map $\mathbb{R}/\mathbb{Z} \to \mathbb{R}/R_2$ by sending x to $x/2$, and a continuous map $f: \mathbb{R}/R_2 \to \mathbb{R}/\mathbb{Z}$ by sending x to $2x$. These maps are mutual inverses to each other and the induced maps on open subsets of $\mathbb{R}$ induced by charts are diffeomorphisms given by multiplying with $1/2$ and 2, respectively. This shows that f is a diffeomorphism. □

Remark 6.38 We can also think of the diffeomorphism $\mathbb{RP}^1 \cong \mathbb{S}^1$ as follows. Consider $\mathbb{RP}^1 = \mathbb{S}^1/\sim$ as the above quotient of $\mathbb{S}^1 = \{(\cos(\pi t), \sin(\pi t)) : t \in [0,2]\}$. Then every point in $\mathbb{RP}^1 \setminus \{[(1,0)]\}$ has a unique representative on the upper hemisphere $\{(\cos(\pi t), \sin(\pi t)) : t \in (0,1)\} \subset \mathbb{S}^1$. Now we get $\mathbb{RP}^1$ from the upper hemisphere by identifying the left and right endpoints. This is exactly how we get a circle from an interval, i.e., we have defined a diffeomorphism $\mathbb{RP}^1 \xrightarrow{\cong} \mathbb{S}^1$ by $[(\cos(\pi t), \sin(\pi t))] \mapsto (\cos(2\pi t), \sin(2\pi t))$.

Remark 6.39 For $n \geq 2$, however, $\mathbb{RP}^n$ and $\mathbb{S}^n$ are not diffeomorphic. In fact, they are not even homotopy equivalent. Proving this fact is not simple, and we will need to develop several useful techniques to show it. See Corollary 13.15.

6.3 Grassmannian

We will now study an important generalisation of real projective space.

Definition 6.40 (Grassmannian) For every $n, k \geq 1$, the set of all k-dimensional linear subspaces in $\mathbb{R}^{n+k}$ is called the *Grassmann manifold*, or just the *Grassmannian*, $\mathrm{Gr}_k(\mathbb{R}^{n+k}) := \{k\text{-dimensional linear subspaces in } \mathbb{R}^{n+k}\}$.

The Grassmannian $\mathrm{Gr}_k(\mathbb{R}^{n+k})$ can be identified with the quotient of the Stiefel manifold $V_k(\mathbb{R}^{n+k})$ of k-frames, i.e., orthonormal sequences $[v_1, \dots, v_k]$ of vectors $v_i \in \mathbb{R}^{n+k}$, modulo the equivalence relation given

[1] Our notation may seem to indicate that j must be bigger than i. But this is not the case. Both cases $i < j$ and $i > j$ work out in the same way.

by $[v_1', \dots, v_k'] \sim [v_1, \dots, v_k]$ if and only if there exists a $T \in \mathrm{O}(k)$, i.e., an invertible orthogonal $(k \times k)$-matrix T, such that $[v_1', \dots, v_k'] = [v_1, \dots, v_k] \cdot T$. In other words, $\mathrm{Gr}_k(\mathbb{R}^{n+k})$ is the quotient of $V_k(\mathbb{R}^{n+k})$ that we get by identifying k-frames which span the same subspace in $\mathbb{R}^{n+k}$. We let $q\colon V_k(\mathbb{R}^{n+k}) \to \mathrm{Gr}_k(\mathbb{R}^{n+k})$ denote the quotient map given by $[v_1, \dots, v_k] \mapsto \mathrm{Span}(v_1, \dots, v_k)$ in $\mathbb{R}^{n+k}$.

Example 6.41 (Projective spaces are Grassmannians) In fact, we have already seen an example of a Grassmannian. For $k = 1$, $\mathrm{Gr}_1(\mathbb{R}^{n+1})$ is the space of one-dimensional vector subspaces in $\mathbb{R}^{n+1}$. Thus, $\mathrm{Gr}_1(\mathbb{R}^{n+1}) = \mathbb{R}\mathrm{P}^n$ is the real projective space of dimension n.

6.3.1 The Canonical Bundle

Before we show that $\mathrm{Gr}_k(\mathbb{R}^{n+k})$ is a smooth manifold, we point out that there are many reasons why Grassmannians are important. We will now sketch one of them. We introduced smooth vector bundles briefly in Section 2.5.2. The tangent bundle on a smooth manifold is an important example. The Grassmannian is equipped with a canonical or tautological vector bundle, which is defined as follows. A point in $\mathrm{Gr}_k(\mathbb{R}^{n+k})$ consists of a k-dimensional vector subspace V in $\mathbb{R}^{n+k}$. We can consider V both as a point in $\mathrm{Gr}_k(\mathbb{R}^{n+k})$ and as a k-dimensional vector space, and we can pick a vector $v \in V$. We let γ_k^{n+k} the space consisting of such pairs, i.e., $\gamma_k^{n+k} = \{(V, v)\colon V \in \mathrm{Gr}_k(\mathbb{R}^{n+k}),\ v \in V\}$. Together with the projection $\pi : \gamma_k^{n+k} \to \mathrm{Gr}_k(\mathbb{R}^{n+k})$ defined by sending (V, v) to V, we obtain a smooth vector bundle on $\mathrm{Gr}_k(\mathbb{R}^{n+k})$ which is called the *canonical bundle*. The canonical bundle is very useful, as we will now explain. Let $X \subset \mathbb{R}^N$ be a smooth manifold of dimension k. Let n be a natural number such that $N = n + k$. Let $x \in X$ be a point and $T_x X$ be the tangent space at x. This is a k-dimensional vector subspace of $\mathbb{R}^{n+k}$. Hence we can consider $T_x X$ as an element in $\mathrm{Gr}_k(\mathbb{R}^{n+k})$. This defines a map $f\colon X \to \mathrm{Gr}_k(\mathbb{R}^{n+k})$ by $x \mapsto T_x(X)$. Now recall that the tangent bundle TX consists of pairs (x, v) with $x \in X$ and $v \in T_x X$. Sending the pair (x, v) to the pair $(T_x X, v)$ defines a map $\bar{f}\colon TX \to \gamma_k^{n+k}$. Assuming the results we will prove in this section, both these maps f and $\bar{f}$ are smooth. Moreover, they fit into a commutative diagram

$$\begin{array}{ccc} TX & \xrightarrow{\ \bar{f}\ } & \gamma_k^{n+k} \\ \downarrow & & \downarrow \\ X & \xrightarrow[\ f\ ]{} & \mathrm{Gr}_k(\mathbb{R}^{n+k}) \end{array} \quad . \tag{6.2}$$

We can show that this actually defines a morphism of vector bundles. This diagram is often called a *generalised Gauss map*. The maps f and $\bar{f}$ are very useful for studying manifolds. Moreover, this picture admits a fascinating generalisation. Let $E \to X$ be any k-dimensional vector bundle over X. Then, for n large enough, there is a continuous map $\kappa \colon X \to \mathrm{Gr}_k(\mathbb{R}^{n+k})$ such that E is the pullback of the canonical vector bundle along this map:

$$\begin{array}{ccc} E \cong \kappa^* \gamma_k^{n+k} & \xrightarrow{\ \bar{\kappa}\ } & \gamma_k^{n+k} \\ \big\downarrow & & \big\downarrow \\ X & \xrightarrow[\ \kappa\ ]{} & \mathrm{Gr}_k(\mathbb{R}^{n+k}) \end{array} \quad . \tag{6.3}$$

Up to homotopy, a relation we will study in Chapter 8, the map κ is in fact uniquely determined by the isomorphism class of $E \to X$. We refer to this phenomenon by saying that $\mathrm{Gr}_k(\mathbb{R}^{n+k})$ is a classifying space for k-dimensional vector bundles. The concept of classifying spaces is very powerful and influential in many areas of mathematics.

Example 6.42 (Canonical line bundle over projective space) Recall from Example 6.41 that, for $k = 1$, we have $\mathrm{Gr}_1(\mathbb{R}^{n+1}) = \mathbb{R}\mathrm{P}^n$. We will now describe the canonical bundle over $\mathbb{R}\mathrm{P}^n$ and will show that it is a non-trivial bundle. Note that since the fibres of γ_1^{n+1} are one-dimensional, we call γ_1^{n+1} the canonical *line bundle* over $\mathbb{R}\mathrm{P}^n$. We consider $\mathbb{R}\mathrm{P}^n$ as the quotient of $\mathbb{S}^n$ under the equivalence relation $x \sim -x$, and write $\{[x] \in \mathbb{R}\mathrm{P}^n$ for the equivalence class of $x \in \mathbb{S}^n$. Then we can write $\gamma_1^{n+1} = \{([x], t \cdot x) \colon x \in \mathbb{S}^n,\ t \in \mathbb{R}\}$ together with the projection map $\pi \colon \gamma_1^{n+1} \to \mathbb{R}\mathrm{P}^n$ which sends $([x], t \cdot x)$ to $[x]$. Now assume γ_1^{n+1} was a trivial bundle. Then, by Lemma 2.42, there would be a nowhere-zero section σ of π. A section σ of π consists of a continuous map $\sigma \colon \mathbb{R}\mathrm{P}^n \to \gamma_1^{n+1}$ with $\sigma([x]) = ([x], s(x) \cdot x)$ for some continuous function $s \colon \mathbb{S}^n \to \mathbb{R}$. The function s has to satisfy $s(-x) = -s(x)$ for all $x \in \mathbb{S}^n$. We consider the following two subsets $A := \{x \in \mathbb{S}^n \colon s(x) > 0\}$ and $B := \{x \in \mathbb{S}^n \colon s(x) < 0\}$. Both A and B are open in $\mathbb{S}^n$ since s is continuous and $\mathbb{R} \setminus \{0\}$ is open in $\mathbb{R}$. If $s(x) \neq 0$ for all $x \in \mathbb{S}^n$, then we would have $\mathbb{S}^n = A \cup B$. This would imply that A and B also are closed in $\mathbb{S}^n$ since they are the complements of open subsets. Now, if A or B is non-empty, then the other one is also non-empty since $s(x) > 0$ if and only if $s(-x) = -s(x) < 0$. Thus, A and B would be non-empty, disjoint open and closed subsets of $\mathbb{S}^n$. However, since $\mathbb{S}^n$ is connected, we cannot have $\mathbb{S}^n = A \cup B$. This implies that there must be an $x_0 \in \mathbb{S}^n$ such that $s(x_0) = 0$. Thus, the section σ must have a zero. Since this argument holds for all sections, we conclude that γ_1^{n+1} is not a trivial bundle by Lemma 2.42.

Example 6.43 (Canonical line bundle over $\mathbb{RP}^1$ is a Möbius band) We now specialise Example 6.42 further and look at the case $n = 1$. Then we can write $\gamma_1^{1+1} = \{([x], t \cdot (\cos(\pi x), \sin(\pi x))\colon x \in [0,1],\ t \in \mathbb{R}\}$, where we use that $\mathbb{RP}^1$ is diffeomorphic to $\mathbb{S}^1$ by Lemma 6.37 and Remark 6.38. For $0 < x < 1$, each point has a unique representative. For $x = 0$ or $x = 1$, however, the points $([0], t \cdot (1,0))$ and $([1], t \cdot (-1,0)) = ([1], -t \cdot (1,0))$ are equal in γ_1^{1+1}. Thus, we obtain γ_1^{1+1} from the product $[0,1] \times \mathbb{R}$ by identifying the outer points $\{0\} \times \mathbb{R}$ and $\{1\} \times \mathbb{R}$ via the equivalence relation $(0,t) \sim (1, -t)$, i.e., we have $\gamma_1^{1+1} \cong ([0,1] \times \mathbb{R})/((0,t) \sim (1, -t))$. This shows that we can think of γ_1^{1+1} as an *open Möbius band* over $\mathbb{RP}^1 \cong \mathbb{S}^1$.

6.3.2 Grassmannians Are Manifolds

Now we show that Grassmannians are manifolds.

Theorem 6.44 *The Grassmannian* $\mathrm{Gr}_k(\mathbb{R}^{n+k})$ *is a compact smooth manifold of dimension* $k \cdot n$.

First, we show that Grassmannians have the required topological properties.

Proposition 6.45 *The Grassmannian* $\mathrm{Gr}_k(\mathbb{R}^{n+k})$ *is compact, second-countable, and Hausdorff.*

Proof: The quotient map $q\colon V_k(\mathbb{R}^{n+k}) \to \mathrm{Gr}_k(\mathbb{R}^{n+k})$ is continuous and surjective. Since $V_k(\mathbb{R}^{n+k})$ is compact by Theorem 4.24, $\mathrm{Gr}_k(\mathbb{R}^{n+k})$ is compact by Lemma 1.13. Moreover, q is an open map since, if $U \subset V_k(\mathbb{R}^{n+k})$ is open, then $q^{-1}(q(U)) = \cup_{T \in \mathrm{O}(k)} U \cdot T$ is open. By Lemma 6.5, this implies that $\mathrm{Gr}_k(\mathbb{R}^{n+k})$ is second-countable since $V_k(\mathbb{R}^{n+k})$ is second-countable. Let $\alpha\colon V_k(\mathbb{R}^{n+k}) \times \mathrm{O}(k) \to V_k(\mathbb{R}^{n+k}) \times V_k(\mathbb{R}^{n+k})$ be the map defined by sending $([v_1, \dots, v_k], T)$ to $([v_1, \dots, v_k], [v_1, \dots, v_k] \cdot T)$. Let $\Gamma := \mathrm{Im}\,(\alpha)$. By Theorem 4.24, $V_k(\mathbb{R}^{n+k})$ is compact and Hausdorff. By Exercise 1.9, $\mathrm{O}(k)$ is compact, and hence $V_k(\mathbb{R}^{n+k}) \times \mathrm{O}(k)$ is compact by Proposition 1.15. Since α is continuous, Γ is compact. Since $V_k(\mathbb{R}^{n+k})$ is Hausdorff, Γ is closed by Proposition 1.21. Now we note that Γ is the graph of the equivalence relation R on $V_k(\mathbb{R}^{n+k})$ defined by $[v_1', \dots, v_k'] \sim [v_1, \dots, v_k]$ if and only if $[v_1', \dots, v_k'] = [v_1, \dots, v_k] \cdot T$ for some $T \in \mathrm{O}(k)$. By Exercise 1.7, this implies that $\mathrm{Gr}_k(\mathbb{R}^{n+k}) = V_k(\mathbb{R}^{n+k})/R$ is Hausdorff. □

Proof of Theorem 6.44: By Proposition 6.45, $\mathrm{Gr}_k(\mathbb{R}^{n+k})$ is Hausdorff, second-countable, and compact. It remains to construct an atlas on $\mathrm{Gr}_k(\mathbb{R}^{n+k})$. We will often refer to a k-dimensional linear subspace as a k-plane. Let $V \subset \mathbb{R}^{n+k}$ be a k-plane in $\mathbb{R}^{n+k}$, and let $V^\perp$ be the orthogonal complement of V in $\mathbb{R}^{n+k}$ with respect to the standard inner product. Define the subspace $\mathcal{U}_V \subset \mathrm{Gr}_k(\mathbb{R}^{n+k})$ consisting of k-planes V' of $\mathbb{R}^{n+k}$ with the property that $V' \cap V^\perp = \{0\}$, i.e.,

$$\mathcal{U}_V = \{V' \in \mathrm{Gr}_k(\mathbb{R}^{n+k}) : V' \cap V^\perp = \{0\}\}.$$

Equivalently, $\mathcal{U}_V$ is the set of all k-planes $V' \subset \mathbb{R}^{n+k}$ which are mapped surjectively onto V by the projection $p\colon \mathbb{R}^{n+k} = V \oplus V^\perp \to V$.

Lemma 6.46 *The subset $\mathcal{U}_V$ is an open neighbourhood of V.*

Proof: To show that $\mathcal{U}_V$ is open in $\mathrm{Gr}_k(\mathbb{R}^{n+k})$, it suffices to show that its preimage $\widetilde{\mathcal{U}}_V$ under q is open in $V_k(\mathbb{R}^{n+k})$ since the topology on the Grassmannian is the quotient topology induced by

$$q\colon V_k(\mathbb{R}^{n+k}) \to \mathrm{Gr}_k(\mathbb{R}^{n+k}),\ [v_1, \dots, v_k] \mapsto \mathrm{Span}(v_1, \dots, v_k) \text{ in } \mathbb{R}^{n+k}.$$

By definition of q and $\mathcal{U}_V$, the set $\widetilde{\mathcal{U}}_V$ consists of all orthonormal k-frames $[v_1', \dots, v_k']$ such that

$$\mathrm{Span}(v_1', \dots, v_k') \cap V^\perp = \{0\}. \tag{6.4}$$

Let $\{w_1, \dots, w_n\}$ be a basis of $W := V^\perp$. Then condition (6.4) is equivalent to saying that the set of vectors $\{v_1', \dots, v_k', w_1, \dots, w_n\}$ forms a basis of $\mathbb{R}^{n+k}$ since they are linearly independent and span an $(n+k)$-dimensional subspace of $\mathbb{R}^{n+k}$. Equivalently, the $(n+k) \times (n+k)$-matrix $M(v_1', \dots, v_k', w_1, \dots, w_n)$ with column vectors $v_1', \dots, v_k', w_1, \dots, w_n$ has non-zero determinant. Keeping the basis of W fixed, we see that $\widetilde{\mathcal{U}}_V$ consists of the preimage of the open subset $\mathbb{R} \setminus \{0\}$ under the map

$$\det\colon V_k(\mathbb{R}^{n+k}) \to \mathbb{R},\ [v_1', \dots, v_k'] \mapsto \det\left(M(v_1', \dots, v_k', w_1, \dots, w_n)\right).$$

Hence $\widetilde{\mathcal{U}}_V$ is an open subset. This proves the lemma. □

Now we are going to define a map from $\mathcal{U}_V$ to Euclidean space. We can think of each $V' \in \mathcal{U}_V$ as the graph of a linear map $V \to W = V^\perp$ as follows. Given $V' \subset \mathcal{U}_V$, let $\mathrm{pr}_V^{V'}$ be the orthogonal projection of V' onto V and $\mathrm{pr}_W^{V'}$ be the orthogonal projection of V' onto W. See Figure 6.3. Note that since $V' \cap V^\perp = \{0\}$ and $\dim V' = \dim V$, we know that $\mathrm{pr}_V^{V'}$ is an isomorphism. Hence we define the map

$$\psi_V : \mathcal{U}_V \to \mathrm{Hom}_{\mathbb{R}}(V, W),\ V' \mapsto \psi_V(V') = \mathrm{pr}_W^{V'} \circ (\mathrm{pr}_V^{V'})^{-1}.$$

The space $\mathrm{Hom}_{\mathbb{R}}(V, W)$ is a real vector space isomorphic to $\mathbb{R}^{nk}$. We define an inverse to ψ_V as follows. Given a linear map $f\colon V \to W$, the graph of f, $\Gamma(f) = \{v + f(v) \in V \oplus W\} \subset V \oplus W \cong \mathbb{R}^{n+k}$, is a linear subspace of dimension k which satisfies $\Gamma(f) \cap W = V \cap W = \{0\}$. Hence $\Gamma(f) \in \mathcal{U}_V$ and we obtain the map $\psi_V^{-1}\colon \mathrm{Hom}_{\mathbb{R}}(V, W) \to \mathcal{U}_V$ defined by $f \mapsto \Gamma(f)$.

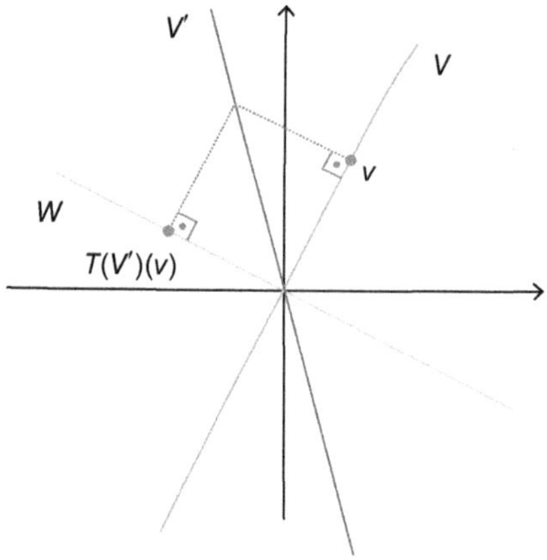

Figure 6.3 A vector v' in V' defines a linear transformation from V to W by taking orthogonal projections.

Lemma 6.47 *The maps ψ_V and ψ_V^{-1} are mutual inverses.*

Proof: First, note that $\Gamma(\psi_V(V')) = V'$ by definition of $\psi_V(V')$ by considering V' as the graph of a map $V \to W$. Now let $f\colon V \to W$ be a linear map. Then

$$\psi_V(\Gamma(f))(v) = \mathrm{pr}_W^{\Gamma(f)} \circ (\mathrm{pr}_V^{\Gamma(f)})^{-1}(v) = \mathrm{pr}_W^{\Gamma(f)}(v + f(v)) = f(v).$$

Hence $\psi_V(\Gamma(f)) = f$. This proves the lemma. □

Lemma 6.48 *The map ψ_V is a homeomorphism.*

Proof: To prove the claim, it remains to show that ψ_V and ψ_V^{-1} are continuous. Note that we consider $\mathrm{Hom}_{\mathbb{R}}(V, W)$ as a topological space by identifying it with $\mathbb{R}^{nk}$. That is, we fix an orthonormal basis $[v_1, \ldots, v_k]$ for V and an orthonormal basis $[w_1, \ldots, w_n]$ for $W = V^{\perp}$. Then we identify linear maps $f\colon V \to W$ with their associated matrix A_f with respect to these bases, which we as usual also consider as an element in $\mathbb{R}^{n\cdot k}$. We show first that ψ_V is continuous. By definition of the topology on $\mathrm{Gr}_k(\mathbb{R}^{n+k})$, ψ_V is continuous if and only if the composite

$$\begin{array}{ccc} \widetilde{\mathcal{U}}_V = q^{-1}(\mathcal{U}_V) & \xrightarrow{\quad q \quad} & \mathcal{U}_V \\ & \searrow^{\psi_V \circ q} & \downarrow{\scriptstyle \psi_V} \\ & & \mathrm{Hom}_{\mathbb{R}}(V, W) \end{array} \tag{6.5}$$

is continuous. Let $[v_1', \ldots, v_k']$ be an orthonormal k-frame for V'. The entries in the matrices of the maps $\mathrm{pr}^W_{V'}$ and $(\mathrm{pr}^{V'}_V)^{-1}$ expressed in the bases of V, W, and V', respectively, depend continuously on the coordinates of the v_i'. Hence entries of the matrix of the map $\psi_V(V') = \mathrm{pr}^W_{V'} \circ (\mathrm{pr}^{V'}_V)^{-1}$ depend continuously on the k-frame $[v_1', \ldots, v_k']$, and $\psi_V \circ q$ is continuous. Now let $f : V \to W$ be a linear map and let A_f be its representing matrix with respect to the chosen bases. Let $v_i'(f) := v_i + f(v_i) = v_i + A_f(v_i) \in \Gamma(f)$ for $i = 1, \ldots, k$. The collection $v_1'(f), \ldots, v_k'(f)$ forms a basis of $\Gamma(f)$ since $v_i \in V$ and $f(v_i) \in W = V^\perp$ and hence any linear combination of the $v_1'(f), \ldots, v_k'(f)$ in $\Gamma(f)$ is the sum of a linear combination in V and W. Let $[\tilde{v}_1(f), \ldots, \tilde{v}_k(f)]$ be the orthonormal k-frame of $\Gamma(f)$ obtained from $v_1'(f), \ldots, v_k'(f)$ by the Gram–Schmidt process. By definition of the $v_i'(f)$, the coordinates of each $v_i'(f)$ depend continuously on the entries in A_f. This implies that the coordinates of the orthonormal k-frame $[\tilde{v}_1(f), \ldots, \tilde{v}_k(f)]$ in $V_k(\mathbb{R}^{n+k})$ depend continuously on the entries in A_f. Since $[\tilde{v}_1(f), \ldots, \tilde{v}_k(f)]$ spans $\Gamma(f) = \psi_V^{-1}(f)$, we get that $\Gamma(f)$ depends continuously on f. This shows that ψ_V^{-1} is also continuous and proves the lemma. □

We have now shown that every point $V \in \mathrm{Gr}_k(\mathbb{R}^{n+k})$ has an open neighbourhood $\mathcal{U}_V$ which is homeomorphic to $\mathbb{R}^{nk}$. It remains to show that the collection of the $(\mathcal{U}_V, \psi_V)$ forms an atlas. Let $V_1, V_2 \in \mathrm{Gr}_k(\mathbb{R}^{n+k})$ be such that $\mathcal{U}_{V_1} \cap \mathcal{U}_{V_2} \neq \emptyset$, where by construction

$$\mathcal{U}_{V_1} \cap \mathcal{U}_{V_2} = \{V' \in \mathrm{Gr}_k(\mathbb{R}^{n+k}) : V' \cap V_1^\perp = \{0\} = V' \cap V_2^\perp\}.$$

We need to show that the transition maps $\psi_{V_1} \circ \psi_{V_2}^{-1}$ are smooth. To do so, we choose orthonormal bases for V_1, $V_1^\perp$, V_2, and $V_2^\perp$. Let $f : V_1 \to V_1^\perp$ be a linear map such that $\Gamma(f) \in \mathcal{U}_{V_1} \cap \mathcal{U}_{V_2}$, and let A_f be the representing matrix with respect to the chosen bases. Let $[\tilde{v}_1(f), \ldots, \tilde{v}_k(f)]$ be the orthonormal k-frame for $\Gamma(f)$ of the proof of Lemma 6.48. The entries in each $\tilde{v}_i(f)$ depend smoothly on the entries of A_f and hence $\Gamma(f)$ depends smoothly on f. Similarly, the entries of the matrix representing the map $\psi_{V_1}(\Gamma(f)) = \mathrm{pr}^{V_1^\perp}_{\Gamma(f)} \circ (\mathrm{pr}^{\Gamma(f)}_V)^{-1}$ depend smoothly on the entries of each $\tilde{v}_i(f)$. This shows that the map $\psi_{V_1} \circ \psi_{V_2}^{-1}$ is smooth at f. Since f was an arbitrary element in the domain, we conclude that $\psi_{V_1} \circ \psi_{V_2}^{-1}$ is smooth and that the collection of the $(\mathcal{U}_V, \psi_V)$ forms an atlas. The associated maximal atlas then provides $\mathrm{Gr}_k(\mathbb{R}^{n+k})$ with the structure of a smooth manifold of dimension $n \cdot k$. □

Remark 6.49 (Tangent space of the Grassmannian) The local coordinate chart ψ_V also gives us a way to understand the tangent space of $\mathrm{Gr}_k(\mathbb{R}^{n+k})$

at V. The map ψ_V behaves like a linear map (if we keep vectors small enough), and the image $\mathrm{Hom}_{\mathbb{R}}(V, V^{\perp})$ of ψ_V is a vector space. Hence the tangent space of $\mathrm{Hom}_{\mathbb{R}}(V, V^{\perp})$ at 0 is $\mathrm{Hom}_{\mathbb{R}}(V, V^{\perp})$ itself. Therefore, we can use $d\psi_V$ to get a linear isomorphism $T_V(\mathrm{Gr}_k(\mathbb{R}^{n+k})) \cong T_0(\mathrm{Hom}_{\mathbb{R}}(V, V^{\perp})) = \mathrm{Hom}_{\mathbb{R}}(V, V^{\perp})$.

6.4 Exercises

6.1 Let X be the set of all lines in $\mathbb{R}^2$, i.e., not just lines through the origin.

(a) Show that X is a smooth manifold of dimension two by showing that we can identify X with an open subset of the real projective plane $\mathbb{R}\mathrm{P}^2$.

(b) Show that there is a bijection between X and the set of equivalence classes $(\mathbb{S}^1 \times \mathbb{R})/\sim$ where $\sim$ is the equivalence relation defined by $(s, x) \sim (y, t)$ if and only if $t = \pm s$ and $y = x$.

6.2 Recall the Hopf map $\hbar\colon \mathbb{S}^3 \to \mathbb{S}^2$, $\hbar(z, w) = (2z\bar{w}, |z|^2 - |w|^2)$, where $\mathbb{S}^3 \subset \mathbb{C}^2$ and $\mathbb{S}^2 \subset \mathbb{C} \times \mathbb{R}$. In this exercise, we study an alternative way to define $\hbar$. Let $\mathbb{C}\mathrm{P}^1$ denote the complex projective space consisting of all complex-one-dimensional linear subspaces in $\mathbb{C}^2$. We can consider $\mathbb{C}\mathrm{P}^1$ as the set of pairs $(z, w) \neq (0, 0)$ of complex numbers modulo the equivalence relation $(z_0, w_0) \sim (z_1, w_1)$ if and only if $z_1 = \lambda z_0$ and $w_1 = \lambda w_0$ for some $\lambda \in \mathbb{C} \setminus \{0\}$. We consider $\mathbb{C}\mathrm{P}^1$ as a topological space as the quotient $(\mathbb{C}^2 \setminus \{(0,0)\})/\sim$ and write $[z : w]$ for the equivalence class of (z, w).

(a) Show that $\mathbb{C}\mathrm{P}^1$ is a two-dimensional abstract smooth manifold. Hint: Follow the outline for $\mathbb{R}\mathrm{P}^n$.

(b) We consider $\mathbb{S}^3$ as a subset in $\mathbb{C}^2$. Show that $\mathbb{C}\mathrm{P}^1$ can also be defined as a quotient $\mathbb{C}\mathrm{P}^1 = \mathbb{S}^3/\sim_s$ by an appropriate equivalence relation $\sim_s$.

(c) We consider $\mathbb{S}^3$ as a subset of $\mathbb{C}^2$ and define the map

$$\varphi\colon \mathbb{S}^3 \to \mathbb{C}\mathrm{P}^1,\ (z, w) \mapsto [z : w].$$

Find a map $h\colon \mathbb{C}\mathrm{P}^1 \to \mathbb{S}^2$ such that the composition

$$\mathbb{S}^3 \xrightarrow{\varphi} \mathbb{C}\mathrm{P}^1 \xrightarrow{h} \mathbb{S}^2$$

equals $\hbar$. Hint: Think of the stereographic projection.

(d) Let $[z_0, w_0]$ be a point in $\mathbb{C}\mathrm{P}^1$. Describe the fibre $\varphi^{-1}([z_0 : w_0])$.

6.3 Let $0 < \lambda < 1$ be a real number. We let $k \in \mathbb{Z}$ act on $\mathbb{C}^n \setminus \{0\}$ by

$$(z_1, \ldots, z_n) \mapsto (\lambda^k z_1, \ldots, \lambda^k z_n).$$

The *Hopf manifold* H_λ^{2n} is defined as the quotient

$$H_\lambda^{2n} := (\mathbb{C}^n \setminus \{0\})/\mathbb{Z},$$

i.e., we identify points z and $\lambda^k z$ for every $k \in \mathbb{Z}$. We can show that H_λ^{2n} is a $2n$-dimensional abstract smooth manifold.[2] To focus on the important features, we will restrict our study to the cases $n = 1$ and $n = 2$. Though the general argument follows the same outline. We are also going to fix a λ and write H^{2n} for H_λ^{2n}.

(a) Show that H^2 is a two-dimensional abstract smooth manifold.[3]

(b) Let $A = \{z \in \mathbb{C} : 1 \leq |z| \leq 2\} \subset \mathbb{C}$. Show that H^2 homeomorphic to the quotient $A/\mathbb{Z}$ and deduce that H^2 is a compact space.

(c) Show that H^4 is a four-dimensional smooth manifold.[4] Hint: It is the same argument as for $n = 1$. We are just practising a bit more and preparing for the next point.

(d) Show that there is a homeomorphism between H^4 and $\mathbb{S}^3 \times \mathbb{S}^1$. Then show that this is actually a diffeomorphism by showing that the compositions with coordinate charts are smooth.[5]

Note: The Hopf surface H_λ^4 is the simplest example of a compact complex manifold which cannot be equipped with a Kähler metric and therefore cannot be holomorphically embedded into complex projective space. This follows from the fact that $H_\lambda^4 \cong \mathbb{S}^3 \times \mathbb{S}^1$, which implies that the first cohomology of a Hopf surface is one-dimensional, whereas every Kähler manifold has an even-dimensional first cohomology.

6.4 For $0 < m \leq n$, the real *Milnor hypersurface* $H(m,n)$ is defined by

$$H(m,n) = \left\{([x_0 : \ldots : x_m], [y_0 : \ldots : y_n]) \in \mathbb{R}\mathrm{P}^m \times \mathbb{R}\mathrm{P}^n : \sum_{i=0}^{m} x_i y_i = 0\right\}.$$

Show that $H(m,n) \subset \mathbb{R}\mathrm{P}^m \times \mathbb{R}\mathrm{P}^n$ is a smooth manifold of dimension $m + n - 1$. Hint: Show that the subsets $V_{ij} = \{([x_0 : \ldots : x_m], [y_0 : \ldots : y_n]) \in H(m,n) : x_i \neq 0, y_j \neq 0\}$ are open and their union for all $i \neq j$ covers $H(m,n)$.

[2] Once we have learned more general theory, we can say that this follows from the fact that the action of $\mathbb{Z}$ on $\mathbb{C}^n \setminus \{0\}$ is *free and discrete*.

[3] Since H^2 is a complex manifold of *complex* dimension one, it is called the *Hopf curve*.

[4] Since H^4 is a complex manifold of *complex* dimension two, it is called the *Hopf surface*.

[5] In general, H^{2n} is diffeomorphic to $\mathbb{S}^{2n-1} \times \mathbb{S}^1$. The previous point for $n = 1$ is a special case of this fact since $A/\mathbb{Z} \cong \mathbb{S}^1 \times \mathbb{S}^1$.

7

Whitney's Embedding Theorems

In the introduction, we claimed that the two approaches to defining a smooth manifold ultimately describe the same class of spaces. We now present a precise formulation of this equivalence, due to Whitney [50, 51], thereby providing justification for our initial, more concrete perspective on smooth manifolds as subsets of Euclidean space. Our first goal is to prove Whitney's embedding theorem for compact, abstract smooth manifolds. We will then refine this result by reducing the dimension of the ambient Euclidean space. Along the way, we introduce two fundamental tools for working with smooth manifolds: *smooth bump functions* and *partitions of unity*, both of which play a crucial role in several proofs later.

7.1 Embedding Abstract Manifolds in Euclidean Space

As before, an *embedding* of smooth manifolds is an immersion which is a diffeomorphism onto its image.

Theorem 7.1 (Whitney) *Let X be a smooth manifold of dimension k in the sense of Definition 6.9. Then there is an embedding $X \hookrightarrow \mathbb{R}^N$ for some N.*

The theorem shows that every smooth manifold is diffeomorphic to a submanifold of $\mathbb{R}^N$. Via such a diffeomorphism, we may consider it then as a subspace of $\mathbb{R}^N$. This is a fundamental result for smooth manifolds. In Section 7.1.2, we provide a proof of Theorem 7.1 for compact manifolds. The non-compact case requires us to first prove further results for abstract manifolds, which would deviate us too much from our main story. We therefore omit the discussion of the non-compact case. See also Remark 7.9. A proof of the general case may be found, for example, in [29, theorem 6.15]. We note that all the examples of abstract manifolds we study in this book are indeed compact. This includes real and complex projective spaces, the torus and the Klein bottle, Grassmannians, Hopf manifolds, and Milnor hypersurfaces.

7.1.1 Smooth Bump Functions

We now introduce a particularly useful class of smooth functions, which will play an important role in many subsequent applications. We first recall some terminology from general topology.

Definition 7.2 (The closure of a subset) Let X be a topological space and A be an arbitrary subset. The *closure of* A in X, denoted $\overline{A}$, is the intersection of all closed subsets in X which contain A.

We are familiar with the closure of subsets in many cases. For example, the closure of an open ball $\mathbb{B}_\varepsilon(0)$ in $\mathbb{R}^n$ is just the closed ball $\overline{\mathbb{B}}_\varepsilon(0) = \{x \in \mathbb{R}^n : |x| \leq \varepsilon\}$. We need the closure of a subset, for example, when we want to talk about the support of a function.

Definition 7.3 (Support of a function) Let X be a smooth manifold and $f\colon X \to \mathbb{R}$ be a smooth function. The *support of* f is defined to be the closed subset $\operatorname{supp}(f) := \overline{\{x \in X \colon f(x) \neq 0\}}$.

For many of the constructions that follow, we require auxiliary functions with specific properties. In particular, they should have bounded support. We begin by giving such functions a name and then show that they exist under increasingly stronger conditions.

Definition 7.4 (Bump function) A *bump function* on $\mathbb{R}^n$ is a smooth function $\mathbb{R}^n \to \mathbb{R}$ which takes the value 1 on some neighbourhood of the origin and the value 0 outside some larger neighbourhood of the origin.

We will now show that bump functions exist. First, we prove a rather simple version to illustrate the argument.

Lemma 7.5 (Bump functions exist) *For every pair of real numbers* $0 < a < b$, *there is a smooth function* $\varphi\colon \mathbb{R}^n \to [0,1] \subset \mathbb{R}$ *such that*

$$\varphi(x) = \begin{cases} 1 & \text{for } |x| \leq a \\ 0 & \text{for } |x| \geq b. \end{cases}$$

Proof: We begin with the function

$$f\colon \mathbb{R} \to \mathbb{R},\ f(t) = \begin{cases} e^{-1/t^2} & \text{for } t > 0 \\ 0 & \text{for } t \leq 0. \end{cases}$$

To show that f is smooth, we only need to think about $t \geq 0$. The ith derivative has the form e^{-1/t^2} times a rational polynomial. Such a product is differentiable and $\lim_{t\to 0} f^{(i)}(t) = 0$ since e^{-1/t^2} goes faster to 0 than any rational polynomial can go to $\pm\infty$. The key feature of the map f is that it provides a smooth transition from 0 to positive values. For given real numbers $a < b$, we then define the function $g(t) := 1 - \frac{f(t-a)}{f(t-a)+f(b-t)}$. The function g is well defined and therefore smooth since $f(t-a) + f(b-t) > 0$ for all t. Moreover, $g(t) = 1$ if and only if $t \leq a$ and $g(t) = 0$ if and only if $t \geq b$. Finally, we define the smooth bump function $\varphi\colon \mathbb{R}^n \to \mathbb{R}$ by $\varphi(x) := g(|x|)$. □

We also need the following more refined version of smooth bump functions.

Proposition 7.6 (Refined smooth bump functions) *Let $U \subset \mathbb{R}^n$ be an open subset and $K \subset U$ be a compact subset. Then there is a smooth function $\varphi\colon \mathbb{R}^n \to \mathbb{R}$ with $\varphi(x) = 1$ for all $x \in K$ and $\varphi(x) = 0$ for all $x \in \mathbb{R}^n \setminus C$ for some closed subset C with $K \subset C \subset U$.*

For the proof of the proposition, we need a lemma in general topology.

Lemma 7.7 (Separating closed subsets) *Let A and C be disjoint closed subsets in $\mathbb{R}^n$. Then there are disjoint open subsets U and V such that $A \subset U$ and $C \subset V$.*

Proof: For each $a \in A$, choose an $\varepsilon_a > 0$ such that $\mathbb{B}_{2\varepsilon_a}(a) \cap C = \emptyset$. This is possible since C is closed and since A and C are disjoint. Similarly, for each $c \in C$, choose an $\varepsilon_c > 0$ such that $\mathbb{B}_{2\varepsilon_c}(c) \cap A = \emptyset$. We define $U := \cup_{a\in A}\mathbb{B}_{\varepsilon_a}(a)$ and $V := \cup_{c\in C}\mathbb{B}_{\varepsilon_c}(c)$. Then U and V are open subsets with $A \subset U$ and $C \subset V$. We claim that U and V are disjoint. For, if $x \in U \cap V$, then $x \in \mathbb{B}_{\varepsilon_a}(a) \cap \mathbb{B}_{\varepsilon_c}(c)$ for some $a \in A$ and $c \in C$. By the triangle inequality, this implies $|a - c| < \varepsilon_a + \varepsilon_c$. But, if $\varepsilon_a \leq \varepsilon_c$, then $|a - c| < 2\varepsilon_c$ and $a \in \mathbb{B}_{2\varepsilon_c}(c)$. And, if $\varepsilon_c \leq \varepsilon_a$, then $|a - c| < 2\varepsilon_a$ and $c \in \mathbb{B}_{2\varepsilon_a}(a)$. Both cases are impossible. □

Proof of Proposition 7.6: Since $\mathbb{R}^n \setminus U$ is closed and K is compact, we can choose by Lemma 7.7 an $\varepsilon > 0$ such that, for each point $x \in K$, we have $\mathbb{B}_{2\varepsilon}(x) \cap U = \emptyset$. Then the $V := \cup_{x\in K}\mathbb{B}_{\varepsilon}(x)$ is an open set containing K with closure $\overline{V} \subset U$ contained in U. Moreover, since K is compact and hence bounded by Theorem 1.16, $\overline{V}$ is also bounded, as we extend K by at most a ball of radius ε. Thus, $\overline{V}$ is closed and bounded and therefore compact by Theorem 1.16. Let χ_V be the characteristic function on V, i.e., the function

$$\chi_V\colon \mathbb{R}^n \to \mathbb{R}, \quad \begin{cases} \chi_V(x) = 1 & \text{for } x \in V \\ \chi_V(x) = 0 & \text{for } x \notin V. \end{cases}$$

The function χ_V is identically 1 on K and has compact support contained in U. But it is not smooth on $\mathbb{R}^n$, not even continuous. Hence we need to modify it to make it smooth. Let ψ be the function

$$\psi \colon \mathbb{R}^n \to \mathbb{R},\ \psi(x) := \begin{cases} \exp\left(\frac{1}{|x|^2-1}\right) & \text{if } |x| < 1 \\ 0 & \text{if } |x| \geq 1. \end{cases}$$

As in the proof of Lemma 7.5, we can show that this is a smooth function. For a given $\varepsilon > 0$, we define $\psi_\varepsilon \colon \mathbb{R}^n \to \mathbb{R}$ by $\psi_\varepsilon(x) := \frac{\psi(x/\varepsilon)}{\int_{\mathbb{R}^n} \psi(y/\varepsilon)dy}$. This is a smooth function with $\int_{\mathbb{R}^N} \psi_\varepsilon dy = 1$, where dy denotes the standard Lebesgue measure on $\mathbb{R}^N$. Then the desired smooth function φ is the convolution $\psi_\varepsilon * \chi_V$ of χ_V and ψ_ε, i.e., the function

$$\varphi \colon \mathbb{R}^n \to \mathbb{R}, x \mapsto \int_{\mathbb{R}^n} \psi_\varepsilon(x-y)\chi_V(y)dy.$$

We note that the integral is well defined and φ is smooth, since the support of χ_V, i.e., the closure $\overline{V}$ of V, is compact. □

7.1.2 Proof of Whitney's Theorem for Compact Manifolds

We will now provide a proof of Whitney's theorem for compact manifolds.

Proof of Theorem 7.1 for compact *manifolds:* Let $\psi_\alpha \colon V_\alpha \to U_\alpha$ be a chart around a point x in the maximal atlas on X with $U_\alpha \subset \mathbb{R}^k$ open where we can assume that U_α contains the origin and $\psi_\alpha(x) = 0$. Let $\varepsilon(\alpha) > 0$ be a real number such that the closed ball $\bar{\mathbb{B}}_\alpha := \bar{\mathbb{B}}^k_{\varepsilon(\alpha)}(0) \subset U_\alpha$ around the origin of radius $\varepsilon(\alpha)$ is contained in U_α. By Proposition 7.6, we can choose a smooth function $\varphi_\alpha \colon \mathbb{R}^k \to \mathbb{R}$ such that $\varphi_\alpha(x) = 1$ for all $x \in \bar{\mathbb{B}}_\alpha$ and $\varphi_\alpha(x) = 0$ for all $x \in \mathbb{R}^n \setminus C_\alpha$ for some closed subset C_α with $\bar{\mathbb{B}}_\alpha \subset C_\alpha \subset U_\alpha$. We let $\tilde{\varphi}_\alpha := \varphi_\alpha \circ \psi_\alpha \colon V_\alpha \to \mathbb{R}$ denote the composition with the homeomorphism ψ_α. We note that $\mathrm{supp}(\tilde{\varphi}_\alpha) \subset V_\alpha$ by construction. The collection of all $\psi_\alpha^{-1}(\mathbb{B}_\alpha)$ then forms an open cover of X. Since X is assumed to be compact, we can choose finitely many charts $\psi_1, \ldots, \psi_n$ such that $X = \cup_i \psi_i^{-1}(\mathbb{B}_i)$. For each $i = 1, \ldots, n$, we define a new map

$$f_i\colon X \to \mathbb{R}^k,\ f_i(x) = \begin{cases} \tilde{\varphi}_i(x)\cdot\psi_i(x) & \text{for } x \in V_i \\ 0 & \text{for } x \in X \setminus \operatorname{supp}(\tilde{\varphi}_i). \end{cases}$$

The map f_i is well defined, since if $x \in V_i \setminus \operatorname{supp}(\tilde{\varphi}_i)$ then both expressions agree to be 0.[1] Moreover, f_i is smooth since its restrictions to the two open subsets V_i and $X \setminus \operatorname{supp}(\tilde{\varphi}_i)$ are smooth. Now we define the map

$$F\colon X \to \mathbb{R}^n \times \mathbb{R}^{nk},\ x \mapsto (\tilde{\varphi}_1(x),\ldots,\tilde{\varphi}_n(x), f_1(x),\ldots,f_n(x)).$$

We note that F is smooth, since the f_i and the $\tilde{\varphi}_i$ are smooth.

Lemma 7.8 *The map F is injective and proper.*

Proof: Since X is compact and the Euclidean space $\mathbb{R}^n \times \mathbb{R}^{nk}$ is Hausdorff by Example 1.19, F is proper by Lemma 1.26. Now we show that F is injective. Assume $F(x) = F(y)$. Then $\tilde{\varphi}_i(x) = \tilde{\varphi}_i(y)$ for all i by construction of F. Since the $\psi_i^{-1}(\mathbb{B}_i)$ cover X, there is at least one i such that $x \in \psi_i^{-1}(\mathbb{B}_i) \subset V_i$. In particular, by construction of $\tilde{\varphi}_i$, this implies $\tilde{\varphi}_i(x) = 1$. Since $\tilde{\varphi}_i(x) = \tilde{\varphi}_i(y)$, we then get $y \in \operatorname{supp}(\tilde{\varphi}_i) \subset V_i$ Thus, x and y lie in the same V_i. Since $f_i(x) = f_i(y)$ and $\tilde{\varphi}_i(x) = \tilde{\varphi}_i(y) = 1$, we must have $\psi_i(x) = \psi_i(y)$. Since ψ_i is a bijection, this shows $x = y$. Thus F is injective. □

Since Euclidean space is a locally compact Hausdorff space, this shows that f is a homeomorphism onto its image by Proposition 1.27. It remains to show that F is an immersion. Then F is a diffeomorphism onto its image and hence an embedding. Let $x \in X$ be a point, and let $\psi_i^{-1}(\mathbb{B}_i)$ be an open set of the above cover such that $x \in \psi_i^{-1}(\mathbb{B}_i)$. Since $\tilde{\varphi}_i = 1$ is constant equal one on $\psi_i^{-1}(\mathbb{B}_i)$, we have $d(f_i)_x = d(\psi_i)_x$. Since ψ_i is a diffeomorphism, $d(\psi_i)_x$ is injective. Thus, $d(f_i)_x$ is injective. This shows that dF_x is always of maximal rank k and hence injective for all $x \in X$. This concludes the proof of Theorem 7.1 for compact manifolds. □

Remark 7.9 One can apply the idea of the above argument also in the non-compact case and use local charts to map pieces of X into $\mathbb{R}^k$. However, when using only finitely many copies of $\mathbb{R}^k$ to accommodate infinitely many neighbourhoods of X, we lose injectivity. The key tool that restores injectivity is a suitable partition of unity, which evens out the troubles caused by overlapping neighbourhoods. Partitions of unity are powerful tools for manifolds, which we will introduce in Definition 7.17 and discuss further in Section 7.3. For the above strategy to work, it is crucial that the topology on

[1] We note that x may be in V_i and V_j for $i \neq j$, but f_i only cares about x being in V_i. The map f_j, however, cares about x being in V_j and so on.

X has a countable basis. We remark that it is possible to construct topological spaces without a countable basis which are locally homeomorphic to Euclidean space, but which cannot be embedded into Euclidean space.

Remark 7.10 We note that there is no analogue of Theorem 7.1 for complex manifolds, i.e., there are many complex manifolds which cannot be holomorphically embedded into $\mathbb{C}^N$. This indicates that Theorem 7.1 is not merely a formal consequence of defining a manifold as a space locally modelled on Euclidean space. It also fundamentally depends on specific properties of the real numbers. Some of these properties are reflected in the existence of smooth bump functions and the existence of partitions of unity. For instance, by Liouville's Theorem in complex analysis, every bounded holomorphic function $\mathbb{C} \to \mathbb{C}$ is constant. Hence there are no holomorphic bump functions. This prevents us from using the strategy of the proof of Theorem 7.1. Alternatively, we could use the maximum principle in complex analysis as an obstruction to the existence of a holomorphic embedding of a compact complex manifold into $\mathbb{C}^N$.

7.2 Whitney's Theorems for Smooth Manifolds

Now that we know that also abstract smooth manifolds can be considered as subspaces of some Euclidean space, we will try to answer the following question.

Question Given a smooth manifold X of dimension k. What is the *minimal* n such that we can be sure that there is an embedding $X \subset \mathbb{R}^n$?

We will approach an answer to this question in two steps. In Section 7.2.1, we show that there always is a one-to-one immersion of X into $\mathbb{R}^{2k+1}$. If X is compact, this immersion will automatically be an embedding as we learned in Section 3.3. In Section 7.2.2, we will then show that also for non-compact smooth k-dimensional manifolds there is always an embedding into $\mathbb{R}^{2k+1}$. In fact, Whitney showed that $N = 2k$ always works. The proof is much harder, and we will not discuss it in this book. See [50] and [51].

7.2.1 Whitney's Immersion Theorem for Smooth Manifolds

Theorem 7.11 (Whitney's Immersion Theorem) *Let $X \subset \mathbb{R}^N$ be a smooth manifold of dimension k. Then there is a one-to-one immersion $X \hookrightarrow \mathbb{R}^{2k+1}$.*

Remark 7.12 Note that Theorem 7.11 does not necessarily give us the minimal N for an individual manifold. For example, we know that $\mathbb{S}^n$ is embedded in $\mathbb{R}^{n+1}$. Theorem 7.16, however, tells us that $N = 2k + 1$ *always works*. The example of real projective space shows that N cannot, in general, be reduced beyond $2k - 1$. For example, if there is an immersion $\mathbb{R}\mathrm{P}^{2^r} \to \mathbb{R}^{2^r+k}$, then k must be at least $2^r - 1$. See for example [36, theorem 4.8].

Proof of Theorem 7.11: We construct the desired one-to-one immersion inductively as follows. Given an injective immersion $f\colon X \hookrightarrow \mathbb{R}^N$ with $N > 2k + 1$, we will show that there exists a vector $a \in \mathbb{R}^N$ such that the composition

$$X \xrightarrow{f} \mathbb{R}^N \xrightarrow{\pi} H := \{b \in \mathbb{R}^N : b \perp a\}$$

of f with the projection map π from $\mathbb{R}^N$ onto the orthogonal complement H of a is still an injective immersion. Since H is an $N - 1$-dimensional vector subspace of $\mathbb{R}^N$, it is isomorphic to $\mathbb{R}^{N-1}$. Thus, after choosing a basis for H, composition with the corresponding diffeomorphism $H \cong \mathbb{R}^{N-1}$ yields an injective immersion into $\mathbb{R}^{N-1}$. By continuing this procedure, we obtain a chain of linear maps $\mathbb{R}^N \to \mathbb{R}^{N-1} \to \cdots \to \mathbb{R}^{2k+1}$ such that the composition $X \hookrightarrow \mathbb{R}^N \to \mathbb{R}^{2k+1}$ is still an injective immersion.

Now we explain the details of the induction step. Assume that we have an injective immersion $f\colon X \hookrightarrow \mathbb{R}^N$ with $N > 2k + 1$. We define two smooth maps h and g

$$\begin{array}{ccc} & & X \times X \times \mathbb{R} \\ & & \downarrow h \\ T(X) & \xrightarrow[g]{} & \mathbb{R}^N \end{array} \tag{7.1}$$

by $h(x, y, t) := t(f(x) - f(y))$ and, using $T_y(\mathbb{R}^N) = \mathbb{R}^N$ at any $y \in \mathbb{R}^N$ and hence the derivative df_x is a smooth map $df_x\colon T_x(X) \to \mathbb{R}^N$, by $g(x, v) := df_x(v)$. By Sard's Theorem, the sets of regular values R_g and R_h for g and h, respectively, are dense subsets in $\mathbb{R}^N$. Thus their intersection is non-empty, and there exists a point in $\mathbb{R}^N$ which is a regular value for both g and h simultaneously. Since $N > 2k + 1$, N is strictly bigger than both $\dim T(X)$ and $\dim(X \times X \times \mathbb{R})$. Thus, the only regular values of g and h are the points in $\mathbb{R}^N$ which are not in the image of g or h. Hence there exists a point $a \in \mathbb{R}^N$ which is neither in the image of g nor in the image of h. Note that, since 0 belongs to both images, we must have $a \neq 0$. Let $H = (\mathrm{Span}\{a\})^{\perp} \subset \mathbb{R}^N$ be the orthogonal complement of a. Let π be the projection of $\mathbb{R}^N$ onto H.

Lemma 7.13 *The map $\pi \circ f\colon X \to H$ is injective.*

Proof: Suppose that $\pi \circ f(x) = \pi \circ f(y)$. Since π is linear, this implies $\pi(f(x) - f(y)) = 0$, i.e., $f(x) - f(y) \in \operatorname{Ker}(\pi) = \operatorname{Span}\{a\}$ in $\mathbb{R}^N$. Thus there is a $t \in \mathbb{R}$ with $f(x) - f(y) = ta$. If $x \neq y$ then $t \neq 0$ since f is injective. But then $a = 1/t(f(x) - f(y)) = h(x, y, 1/t)$, which contradicts the choice of a not being in the image of h. This proves the lemma. □

Lemma 7.14 *The map $\pi \circ f\colon X \to H$ is an immersion.*

Proof: Suppose there was a non-zero vector v in $T_x(X)$ for which $d(\pi \circ f)_x = 0$. Since π is linear, we have $d\pi_{f(x)} = \pi$ and $d(\pi \circ f)_x = \pi \circ df_x$ by the chain rule. Thus $\pi(df_x(v)) = 0$, i.e., $df_x(v)$ lies in the span of a by definition of π and H. That is, $df_x(v) = ta$ for some $t \in \mathbb{R}$. Since f is an immersion, we must have $ta \neq 0$. Since we know $a \neq 0$, this implies $t \neq 0$. Thus, since df_x is linear, $a = df_x(v)/t = df_x(v/t) = g(x, v/t)$, which contradicts the choice of a not being in the image of g. This proves the lemma. □

Lemmas 7.13 and 7.14 show the induction step, which finishes the proof of the theorem. □

Remark 7.15 We note that we could only make the above argument for the existence of a suitable a when $N > 2k + 1$. Hence the induction argument of shrinking N further and further cannot be extended beyond $2k + 1$.

For compact manifolds, one-to-one immersions are embeddings by Corollary 3.17. Together with Theorem 7.1, we have proved Whitney's embedding theorem in the compact case.

Theorem 7.16 (Whitney's Embedding for compact smooth manifolds) *Every compact smooth k-dimensional manifold admits an embedding into $\mathbb{R}^{2k+1}$.*

7.2.2 Whitney's Embedding Theorem

We now show that also every non-compact smooth k-dimensional manifold $X \subset \mathbb{R}^N$ admits an embedding into $\mathbb{R}^{2k+1}$. In order to do so, we would like to modify the immersion of Theorem 7.11 to make it proper. This is a topological, not a differential problem. The key tool will be provided by partitions of unity, which we define next.

Definition 7.17 (Partition of unity) Let X be a smooth manifold and let $\{U_\alpha\}$ be an open cover of X. A sequence of smooth functions $\{\rho_i\colon X \to \mathbb{R}\}$ is called a *partition of unity subordinate to the open cover* $\{U_\alpha\}$ if it satisfies the following properties:

(a) $0 \leq \rho_i(x) \leq 1$ for all $x \in X$ and all i.
(b) Each $x \in X$ has a neighbourhood on which all but finitely many functions ρ_i are identically zero.
(c) For each i, $\mathrm{supp}(\rho_i) \subset U_\alpha$ for some α.
(d) For each $x \in X$, $\sum_i \rho_i(x) = 1$.

We observe that according to (b), the sum in (d) is always finite. We also note that the restriction to manifolds is not essential. In fact, partitions of unity exist on every paracompact topological space, meaning any space in which every open cover admits a locally finite refinement. This class is much broader than smooth manifolds. We postpone the rather technical proof of the existence of partitions of unity on smooth manifolds to Section 7.3. We use the existence of partitions of unity to prove the following lemma.

Lemma 7.18 *On every smooth manifold $X \subset \mathbb{R}^N$, there is a proper smooth function $p\colon X \to \mathbb{R}$.*

Proof: Let $\{U_\alpha\}$ be the collection of open subsets of X that have compact closure and which cover X. Such a collection exists for every subspace in $\mathbb{R}^N$, as we can cover $\mathbb{R}^N$ by open balls with finite radii whose closure is then compact. Let $\{\rho_i\}$ be a subordinate partition of unity. Then

$$p(x) := \sum_{i=1}^{\infty} i\rho_i(x)$$

is a well-defined smooth function since, in an open neighbourhood of every point, it is a finite sum of smooth functions. In order to show that p is proper, we need to show that the preimage of any compact subset of $\mathbb{R}$ is again compact. By Theorem 1.16, every compact subset $K \subset \mathbb{R}$ is contained in a closed interval of the form $[-n,n]$ for some large enough natural number n. Since $[-n,n] \subset \mathbb{R}$ is a Hausdorff space, K is a closed subset of $[-n,n]$ by Proposition 1.21. Hence if we can show that $p^{-1}([-n,n])$ is compact, then $p^{-1}(K)$ is a closed subset of a compact set and therefore also compact by Lemma 1.13. For a given natural number j, assume we have $\rho_1(x) = \cdots = \rho_n(x) = 0$ for any $x \in X$. Then, by definition of a partition of unity, we have $\sum_{i=n+1}^{\infty} \rho_i(x) = 1$ and therefore we get the inequalities $p(x) \geq (n+1)\sum_{i=n+1}^{\infty} \rho_i(x) = n+1 > n$. This shows

$$p^{-1}([-n,n]) \subset \bigcup_{i=1}^{n} \{x \in X : \rho_i(x) \neq 0\}.$$

Since $\{\rho_i\}$ is a partition of unity subordinate to $\{U_\alpha\}$, we have $\mathrm{supp}(\rho_i) \subset U_i$. Since U_i has compact closure and the finite union of compact sets is compact, this shows that $p^{-1}([-n,n])$ is a closed subset in a compact set, and therefore it is also compact by Lemma 1.14. □

Theorem 7.19 (Whitney's Embedding Theorem) *Every smooth k-dimensional manifold $X \subset \mathbb{R}^N$ admits an embedding into $\mathbb{R}^{2k+1}$.*

We note again that the strongest general result is that dimension $2k$ suffices. But this is much harder to prove. See [50] and [51]. Here we follow the argument of [14, chapter 1, §8], which is based on Whitney's ideas.

Proof of Theorem 7.19: By Theorem 7.11, we can find an injective immersion $f: X \hookrightarrow \mathbb{R}^{2k+1}$. The idea is now to replace f with $(f,p): X \hookrightarrow \mathbb{R}^{2k+2}$ for some proper map $p: X \to \mathbb{R}$. Then (f,p) is still an injective immersion, and it is proper, since p is proper. It then remains to reduce the dimension $2k+1$. We now explain the details of the argument. By Lemma 7.18, we can choose a proper smooth function $p: X \to \mathbb{R}$. We define a new map

$$F: X \to \mathbb{R}^{2k+2},\ x \mapsto (f(x), p(x)).$$

The map F is injective since f is injective, and F is an immersion since the rank of dF_x is at least the rank of df_x, which is maximal. Since $2k+2 > 2k+1$, we can apply the argument of the proof of Theorem 7.11 and find a non-zero vector $a \in \mathbb{R}^{2k+2}$ such that $\pi \circ F: X \to H$ is still an injective immersion, where π is the projection onto the orthogonal complement $H = \{v \in \mathbb{R}^{2k+2}: b \perp a\}$ of a in $\mathbb{R}^{2k+2}$. By choosing a basis for H, we obtain a diffeomorphism $H \xrightarrow{\cong} \mathbb{R}^{2k+1}$. Thus, to prove the theorem it suffices to show that $\pi \circ F: X \to H$ is proper. Then the composition $X \xrightarrow{F} \mathbb{R}^{2k+2} \xrightarrow{\pi} H \xrightarrow{\cong} \mathbb{R}^{2k+1}$ is the desired one-to-one immersion. Now we show that $\pi \circ F$ is proper. By rescaling, we can assume $|a| = 1$. In other words, we can assume $a \in \mathbb{S}^{2k+1}$. Moreover, up to rotation which is a diffeomorphism, we can assume that a is neither the north nor the south pole on $\mathbb{S}^{2k+1}$. Now we show that $\pi \circ F$ is proper. We first make the following observation.

Lemma 7.20 *For every $n > 0$, there exists a number $b > 0$ such that*

$$\{x \in X: |(\pi \circ F)(x)| \leq n\} \subset \{x \in X: |p(x)| \leq b\}.$$

Proof of Lemma 7.20: If the assertion were false, then there would be no b bounding $|p(x_i)|$, and hence there would be a sequence of points $\{x_i\}$ in X for which $|(\pi \circ F)(x_i)| \leq n$ and $|p(x_i)| \to \infty$. In particular, after choosing a suitable subsequence, we can assume that $|p(x_i)| > 0$ for all i. By composing

$f\colon X \hookrightarrow \mathbb{R}^{2k+1}$ with the diffeomorphism $\mathbb{R}^{2k+1} \to \mathbb{B}_1^{2k+1}(0)$, $x \mapsto \frac{x}{1+|x|^2}$, we can assume that $|f(x)| < 1$ for all $x \in X$. By definition of the projection onto an orthogonal complement, for every $z \in \mathbb{R}^{2k+2}$, $\pi(z)$ is the unique point in H for which $z - \pi(z)$ is a multiple of a. In particular, $F(x_i) - \pi(F(x_i))$ is a multiple of a for each i, and hence so is the vector $(F(x_i) - \pi(F(x_i)))/p(x_i)$. When i tends to infinity, the sequence $\{F(x_i)/p(x_i) = (f(x_i)/p(x_i), 1)\}$ converges to $(0, \ldots, 0, 1)$ since $|f(x_i)| < 1$ for all i and $p(x_i) \longrightarrow \infty$. We have $|\pi(F(x_i))/p(x_i)| \le n/|p(x_i)|$ by assumption. Thus, when the sequence $\{\pi(F(x_i))/p(x_i)\}$ converges to 0, the sequence of w_i converges to $(0, \ldots, 0, 1)$. However, each $(F(x_i) - \pi(F(x_i)))/p(x_i)$ is a multiple of a. Hence the limit of the sequence $\{(F(x_i) - \pi(F(x_i)))/p(x_i)\}$ must be a multiple of a as well. We conclude that a must be either the north or south pole of $\mathbb{S}^{k+1}$, which contradicts our assumption on a. This proves the lemma. □

Now we can show that $\pi \circ F$ is proper. Let K be a compact subset of H. After choosing a basis for H, we use the resulting linear isomorphism to identify H with $\mathbb{R}^{2k+1}$. Then Theorem 1.16 implies that K is bounded. Thus, there is a real number $n > 0$ such that K is contained in the closed ball $\overline{\mathbb{B}}_n := \bar{\mathbb{B}}_n^{2k+1}(0)$ of radius n around the origin in $\mathbb{R}^{2k+1}$. Moreover, K is a closed subset of $\mathbb{R}^{2k+1}$. Thus, K is a closed subset of $\overline{\mathbb{B}}_n$ by definition of the subspace topology on the latter. This implies $(\pi \circ F)^{-1}(K)$ is a closed subset of $(\pi \circ F)^{-1}(\overline{\mathbb{B}}_n)$. To show that $(\pi \circ F)^{-1}(K)$ is a compact subset of X, it then suffices to show that $(\pi \circ F)^{-1}(\overline{\mathbb{B}}_n)$ is a compact subset of X. Then $(\pi \circ F)^{-1}(K)$ is compact by Lemma 1.14. It remains to show that the set $(\pi \circ F)^{-1}(\overline{\mathbb{B}}_n) = \{x \in X\colon |(\pi \circ F)(x)| \le n\}$ is compact. By Lemma 7.20, there is a real number $b > 0$ such that $(\pi \circ F)^{-1}(\overline{\mathbb{B}}_n) \subset p^{-1}([-b, b])$. Since $\overline{\mathbb{B}}_n$ is closed in $\mathbb{R}^{2k+1}$, $(\pi \circ F)^{-1}(\overline{\mathbb{B}}_n)$ is closed in X, and hence $(\pi \circ F)^{-1}(\overline{\mathbb{B}}_n)$ is closed in $p^{-1}([-b, b])$ by definition of the subspace topology on $p^{-1}([-b, b]) \subset X$. Now we finally use that p is proper. By Exercise 1.4, $[-b, b] \subset \mathbb{R}$ is compact in $\mathbb{R}$. Since p is proper, the set $p^{-1}([-b, b])$ is a compact subset of X. Hence we have shown that $(\pi \circ F)^{-1}(\overline{\mathbb{B}}_n)$ is a closed subset of the compact set $p^{-1}([-b, b])$. By Lemma 1.14, this shows that $(\pi \circ F)^{-1}(\overline{\mathbb{B}}_n)$ is compact as required. This finishes the proof of the theorem. □

7.3 Existence of Partitions of Unity

We now return to the concept of partitions of unity introduced in Definition 7.17. Our goal is to demonstrate, step by step, that partitions of unity exist on all subsets $X \subset \mathbb{R}^N$ of Euclidean space, gradually increasing the level of generality. We begin with a compact subset of Euclidean space and then extend this result to arbitrary subsets in Euclidean space. Finally, we deduce the existence of partitions of unity on compact smooth manifolds in the sense of Definition 6.9. We will not address the proof for non-compact abstract manifolds here.

Theorem 7.21 *Every subspace $X \subset \mathbb{R}^N$ admits partitions of unity.*

Proof of Theorem 7.21: We prove the theorem in a series of lemmas. We begin with the compact case.

Lemma 7.22 *Partitions of unity exist when $X \subset \mathbb{R}^N$ is compact.*

Proof: Let $\{U_\alpha\}$ be an open cover of X. Since X is compact, $\{U_\alpha\}$ has a finite subcover $\{U_1, \dots, U_n\}$. A partition of unity subordinate to the finite subcover is also a partition of unity subordinate to the original cover. We are going to show that we can shrink the cover to an open cover $\{V_1 \dots, V_n\}$ such that $\overline{V}_i \subset U_i$ for each i. Consider the closed subset $A := X \setminus (U_2 \cup \cdots \cup U_n)$ of X. Since $\{U_1, \dots, U_n\}$ covers X, we know $A \subset U_1$. Since A and $X \setminus U_1$ are closed and disjoint subsets of X, we can choose by Lemma 7.7 an open subset V_1 containing A such that V_1 is disjoint from an open subset W which contains $X \setminus U_1$. Since $X \setminus W$ is a closed subset which contains V_1, we know $\overline{V}_1 \subset X \setminus W$ since the closure of V_1 is the intersection of all closed subsets which contain V_1. Since $X \setminus U_1 \subset W$ by the choice of W, we have $X \setminus W \subset X \setminus (X \setminus U_1) = U_1$. Thus we have $\overline{V}_1 \subset U_1$. Since V_1 contains the complement of $U_2 \cup \cdots \cup U_n$ in X, the collection $\{V_1, U_2, \dots, U_n\}$ covers X. Now we proceed by induction as follows. Given open subsets $V_1, \dots, V_{k-1}$ such that $X = \{V_1, \dots, V_{k-1}, U_k, U_{k+1}, \dots, U_n\}$, let A_k be the subset $A_k := X \setminus (V_1 \cup \cdots \cup V_{k-1}) \cup (U_{k+1} \cup \cdots \cup U_n)$. Then A_k is a closed subset of X which is contained in the open set U_k. By Lemma 7.7, we can choose an open subset V_k containing A_k such that $\overline{V}_k \subset U_k$. Then the collection of subspaces $\{V_1, \dots, V_{k-1}, V_k, U_{k+1}, \dots, U_n\}$ covers X. At the nth step of the induction, we are done. Now we repeat the above process to the open cover $\{V_1, \dots, V_n\}$ to find an open cover $\{W_1, \dots, W_n\}$ of X such that $\overline{W}_i \subset V_i$ for each i. For each i, we can choose by Proposition 7.6 a smooth function $\varphi_i : X \to [0, 1]$ such that $\varphi_i(x) = 1$ for all $x \in \overline{W}_i$ and $\varphi_i(x) = 0$ for all $x \in X - V_i$. Since $\varphi_i^{-1}(\mathbb{R} \setminus \{0\}) \subset V_i$, we have $\text{supp}(\varphi_i) \subset \overline{V}_i \subset U_i$. Since $\{W_1, \dots, W_n\}$ covers X, we have $\varphi(x) := \sum_{i=1}^n \varphi_i(x) > 0$ for all $x \in X$. Finally, for each i, we define $\rho_i(x) := \frac{\varphi_i(x)}{\varphi(x)}$. □

Remark 7.23 Note that we needed to modify the open over twice. If we were working with the V_i instead of W_i, then we would have $\operatorname{supp}(\varphi) \subset \overline{U}_i$ instead of $\operatorname{supp}(\varphi) \subset U_i$ as required for a partition subordinate to the cover $\{U_i\}$.

Lemma 7.24 *Partitions of unity exist when $X \subset \mathbb{R}^N$ and $X = X_1 \cup X_2 \cup X_3 \cup \cdots$ where each X_i is compact and $X_i \subset \operatorname{int}(X_{i+1})$.*

Proof: Let $\{U_\alpha\}$ be an open cover of X. For each i, we define the subset $U_\alpha^i := U_\alpha \cap (X_{i+1} \setminus \operatorname{int}(X_{i-2}))$. Then $\{U_\alpha^i\}$ is an open cover of $Y_i := X_i \setminus \operatorname{int}(X_{i-1})$. Since $\operatorname{int}(X_{i-1})$ is an open subset, Y_i is a closed subset of X_i, and therefore Y_i is also compact by Lemma 1.14. Then, for each i, Lemma 7.22 shows that there is a partition of unity φ_α^i on Y_i subordinate to the cover $\{U_\alpha^i\}$. For each $x \in X$, there is an i such that $x \in X_i$ and hence $\varphi_\alpha^j(x) = 0$ for all $j \geq i+2$. Hence, for each $x \in X$, the sum $\varphi(x) := \sum_{\alpha,i} \varphi_\alpha^i(x) > 0$ is a finite sum in some open set containing x. For each α, we define $\rho_\alpha^i(x) := \frac{\varphi_\alpha^i(x)}{\varphi(x)}$. This is a partition of unity subordinate to the open cover $\{U_\alpha\}$. □

Lemma 7.25 *Partitions of unity exist when $X \subset \mathbb{R}^N$ is open.*

Proof: For every $i \geq 1$, we define a subset

$$X_i := \{x \in X\colon |x| \leq i \text{ and the distance to } \mathbb{R}^N \setminus X \text{ is } \geq 1/i\}.$$

Since X_i is the intersection $\overline{\mathbb{B}_i(0)} \cap (X \setminus (\cup_{p \in \mathbb{R}^N \setminus X} \mathbb{B}_{1/i}(p)))$ and therefore closed and bounded in $\mathbb{R}^N$, each X_i is compact by Theorem 1.16. Moreover, for each i, we have $X_i \subset \operatorname{int}(X_{i+1})$, and $X = X_1 \cup X_2 \cup \cdots$. Hence we can apply Lemma 7.24. □

Now we can finish the proof of Theorem 7.21. Let $X \subset \mathbb{R}^N$ be an arbitrary subspace. Let $\{U_\alpha\}$ be an open cover of X. By the definition of the subspace topology on X, for each α, there is a subset $V_\alpha \subset \mathbb{R}^N$ open in $\mathbb{R}^N$ such that $U_\alpha = X \cap V_\alpha$. Let Y be the union of all the V_α in $\mathbb{R}^N$. By Lemma 7.25, there is a partition of unity on Y subordinate to the open cover $\{V_\alpha\}$. By restricting to X, this gives a partition of unity on X subordinate to the open cover $\{U_\alpha\}$. This finishes the proof of Theorem 7.21. □

Corollary 7.26 *Let X be a compact smooth manifold in the sense of Definition 6.9. Then X admits partitions of unity.*

Proof: Let $\{V_\alpha\}$ be an open cover of X. By intersecting with the domains of charts on X, we get a refinement of the cover. Hence we can assume that V_α are the domains of charts on X. Since X is compact, the domains of finitely many charts on X suffice to cover X. We label them $(V_1, \psi_1), \ldots, (V_n, \psi_n)$. Then each $U_i = \psi_i(V_i)$ is an open subset in $\mathbb{R}^k$. Now we can proceed as in the proof of Lemma 7.22 for the finite cover $\{U_1, \ldots, U_n\}$ of the space $Y := U_1 \cup \cdots \cup U_n \subset \mathbb{R}^k$ to obtain a partition of unity $\{\rho_i\}$ subordinate to the cover $\{U_1, \ldots, U_n\}$. Composition of each ρ_i with ψ_i yields a partition of unity $\{\rho_i \circ \psi_i\}$ on X subordinate to the cover $\{V_1, \ldots, V_n\}$ and therefore subordinate to the cover $\{V_\alpha\}$. □

8
Smooth Homotopy

We now continue our study of smooth manifolds as subspaces of Euclidean space. As we saw in Chapter 7, this is not actually a restriction of the theory. In this chapter, we are going to introduce one of the most important concepts in topology: homotopies between maps. The idea of studying maps up to homotopy has turned out to be extremely successful in many areas of mathematics. One motivation for allowing maps to vary up to homotopy is that it is too complicated to try to classify and understand all maps between manifolds. We will soon define invariants, i.e., numbers that we attach to smooth manifolds and the maps between them. These numbers do not change if we modify a map by a homotopy, and, in fact, we will see that this behaviour makes the numbers all the more useful. However, we have also seen a very concrete motivation for introducing homotopies. Let $f : X \to Y$ be a smooth map, and let $y_0 \in Y$ be neighbourhood we are interested in. By Sard's Theorem, we know that in any small open neighbourhood U of y_0 we can find a regular value y for f. So, while y_0 may not be a regular value, it is very close to one. Now we could imagine two ways to remedy the defect of y_0 not being regular. We could replace y_0 with y. However, imagine it is more important to stick to y_0 than to stick to the initial map f. Then we could imagine to smoothly deform f into another smooth map $g : X \to Y$ for which y_0 is indeed a regular value and such that g still has all the good properties of f. In this chapter, we will provide a rigorous framework for making this idea work. This will then open the path to a whole variety of new tools.

8.1 Smooth Homotopies

First, we define continuous homotopies between continuous maps of topological spaces. Then we will focus on smooth homotopies between smooth manifolds.

Definition 8.1 (Homotopy) Let X and Y be two topological spaces and let $I = [0,1]$ denote the unit interval in $\mathbb{R}$. We say that two continuous maps f_0 and f_1 from X to Y are *homotopic*, denoted $f_0 \sim f_1$, if there exists a continuous map $F\colon X \times [0,1] \to Y$ such that $F(x,0) = f_0(x)$ and $F(x,1) = f_1(x)$ for all $x \in X$. The map F is called a *homotopy between* f_0 *and* f_1. We also often write $f_t(x)$ for $F(x,t)$ and write f_t for F.

In other words, a homotopy is a family of continuous maps f_t which continuously interpolates between f_0 and f_1. We can also think of a homotopy f_t as a continuous deformation from f_0 to f_1.

Example 8.2 The maps $f_0\colon \mathbb{R} \to \mathbb{R}^2$, $f_0(x) = (x,0)$ and $f_1\colon \mathbb{R} \to \mathbb{R}^2$, $f_1(x) = (x, \sin x)$ are homotopic with homotopy $F\colon \mathbb{R} \times [0,1] \to \mathbb{R}^2$ given by $F((x,t) = (x, t \sin x)$.

Example 8.3 Let $\gamma\colon [0,1] \to \mathbb{R}^2$ be a loop, i.e., a continuous map with domain such that $\gamma(0) = \gamma(1)$. Then γ is homotopic to the constant map $[0,1] \to \{0\} \subset \mathbb{R}^2$. In fact, this is true when we replace $\mathbb{R}^2$ with any $\mathbb{R}^k$, since $\mathbb{R}^k$ is contractible as we will show in Exercise 8.1.

Example 8.4 In Exercise 8.4, we will show that the antipodal map on the k-sphere $\mathbb{S}^k \to \mathbb{S}^k$, $x \mapsto -x$, which sends a point to the point on the other side of the sphere, is homotopic to the identity on $\mathbb{S}^k$ if k is odd. If k is even, however, the antipodal map is not homotopic to the identity. To prove the latter claim will require more sophisticated tools, which we will develop soon. See for example Exercise 15.1.

For the study of smooth manifolds, it is desirable to strengthen our assumptions on what a homotopy is and to require it to be a smooth family of maps. We first present the necessary definition and will then see that it actually is not such a strong restriction after all. Recall that, for smooth manifolds X and Y embedded in Euclidean space, a map $F\colon X \times [0,1] \to Y$ is smooth in the second variable if we can extend F to a smooth map $X \times (-\varepsilon, 1+\varepsilon) \to Y$ for some $\varepsilon > 0$.

Definition 8.5 (Smooth homotopy) Let X and Y be smooth manifolds. We say that two smooth maps f_0 and f_1 from X to Y are *smoothly homotopic*, denoted $f_0 \sim f_1$, if there exists a smooth map $F\colon X \times [0,1] \to Y$ such that $F(x,0) = f_0(x)$ and $F(x,1) = f_1(x)$ for all $x \in X$. The map F is called a *smooth homotopy between* f_0 *and* f_1. We again often write $f_t(x)$ for $F(x,t)$ and f_t for F.

Thus, a smooth homotopy is a family of smooth functions f_t which smoothly interpolates between f_0 and f_1. To allow only smooth homotopies might seem like a strong restriction. It turns out, however, that we can always approximate a continuous map by a smooth map within its homotopy class. This is the content of the following important theorem.

Theorem 8.6 (Whitney Approximation Theorem) *Let X and Y be smooth manifolds. Let $g_0, g_1 : X \to Y$ be two smooth maps which are continuously homotopic. Then they are* smoothly *homotopic. Let $g : X \to Y$ be a continuous map. Then g is homotopic to a smooth map $f : X \to Y$.*

We will prove this important result in Section 12.2. See Theorem 12.28 in particular. The key tool we will use for the proof is tubular neighbourhoods, which we will introduce and study in Section 12.1.

Now we show that smooth homotopy is an equivalence relation.

Proposition 8.7 (Smooth homotopy is an equivalence relation) *Let X and Y be smooth manifolds. Smooth homotopy defines an equivalence relation on the set of smooth maps from X to Y.*

This result will allow us to consider the equivalence classes of maps under the homotopy relation, i.e., we can form a well-defined quotient set $\mathrm{Hom}(X, Y)/\sim$, where we identify maps if and only if they are homotopic. The equivalence class to which a map belongs is called its *homotopy class*.

Proof of Proposition 8.7: We need to check that $\sim$ is reflexive, symmetric, and transitive. Reflexivity is clear as every map is homotopic to itself via the homotopy $f_t = f$ for all t. For symmetry, suppose $f \sim g$ and let F be a homotopy. Then the map defined by $(x,t) \mapsto F(x, 1-t)$ is a homotopy from g to f. Hence $g \sim f$ as well. For transitivity, suppose $f \sim g$ and $g \sim h$, and let F be a homotopy from f to g and G be a homotopy from g to h. We would like to compose F and G to get a homotopy from f to h. Since we require our homotopies to be smooth, we need to make sure that the transition from F to G is smooth. Hence we need to modify F and G a bit, and here we are fortunate to have smooth bump functions at our disposal. By Lemma 7.5, we can choose a smooth function $\varphi : \mathbb{R} \to \mathbb{R}$ such that

$$\varphi(t) = \begin{cases} 0 & t \leq 1/4 \\ 1 & t \geq 3/4, \end{cases}$$

and we define new homotopies $\tilde{F}(x,t) := F(x, \varphi(t))$ from f to g and $\tilde{G}(x,t) := G(x, \varphi(t))$ from g to h. Now we can define the map

$$H : X \times [0,1] \to Y, \ H(x,t) = \begin{cases} \tilde{F}(x, 2t) & t \in [0, 1/2] \\ \tilde{G}(x, 2t-1) & t \in [1/2, 1]. \end{cases}$$

This map is well–defined and smooth since $\tilde{F}(x,2t) = g(x)$ $t \in (3/8, 1/2]$ and $g(x) = \tilde{G}(x,2t-1)$ for $t \in [1/2, 5/8)$. Thus H is a smooth homotopy from f to h. Hence $\sim$ is also transitive and an equivalence relation. □

Remark 8.8 We emphasise that the role of φ in the proof of Proposition 8.7 is merely to make sure that the transition from F to G is smooth. While $t \mapsto F(x,2t)$ and $t \mapsto G(x,2t-1)$ are smooth in t, their concatenation, i.e., the map given by F for $t \geq 1/2$ and by G for $t \geq 1/2$, may not be smooth at $t = 1/2$. The smooth bump function φ, however, makes both $\tilde{F}(x,2t)$ and $\tilde{G}(x,2t-1)$ constant in t equal to $g(x)$ in an open neighbourhood of $t = 1/2$.

8.2 Simply Connected Spaces

We will now explore some concepts and results related to homotopies. We will formulate parts of this section for topological spaces and not just smooth manifolds, even though we are interested in the latter. Recall that, by Whitney's Approximation Theorem 8.6, we can assume that all homotopies are smooth whenever we consider maps between smooth manifolds, for example maps $\mathbb{S}^k \to \mathbb{S}^n$. Thus, when we study manifolds, we can use the techniques we developed for smooth maps to study homotopies. We begin with the following definition.

Definition 8.9 (Homotopy equivalence) Let X and Y be topological spaces. A continuous map $f: X \to Y$ is called a *homotopy equivalence* if there exists a continuous map $g: Y \to X$ such that $f \circ g \simeq \mathrm{id}_Y$ and $g \circ f \simeq \mathrm{id}_X$. We call g a *homotopy inverse* of f. We say that X and Y are *homotopy equivalent* if there exists a homotopy equivalence $X \to Y$.

Every homeomorphism is also a homotopy equivalence. The converse, however, is not true. For example, $\mathbb{R}^n$ is homotopy equivalent to the one-point space $\{0\}$, consisting of the origin in $\mathbb{R}^n$, since we can contract the identity map on $\mathbb{R}^n$ to the constant map at the origin. But $\mathbb{R}^n$ is not homeomorphic to $\{0\}$ for $n \geq 1$, since any map $\mathbb{R}^n \to \{0\}$ cannot be injective.

Definition 8.10 (Simply connected) A topological space X is called *simply connected* if X is connected and every continuous map $\mathbb{S}^1 \to X$ is homotopic to a constant map.

Proposition 8.11 *Let X and Y be topological spaces. Assume that X and Y are homotopy equivalent. Then X is simply connected if and only if Y is simply connected.*

Proof: Let $f: X \to Y$ be a homotopy equivalence with homotopy inverse $g: Y \to X$. Let $H: Y \times [0,1] \to Y$ be a homotopy between $H(y,1) = f \circ g(y)$ and $H(y,0) = y$ for all $y \in Y$. Assume that X is simply connected, and let $\gamma: \mathbb{S}^1 \to Y$ be a continuous map. The composition $g \circ \gamma: \mathbb{S}^1 \xrightarrow{\gamma} Y \xrightarrow{g} X$ is homotopic to a constant map with value c. Let $h: \mathbb{S}^1 \times [0,1] \to X$ be a homotopy with $h(s,1) = g(\gamma(s))$ and $h(s,0) = c$. Then we can define a homotopy $\Gamma: \mathbb{S}^1 \times [0,1] \to Y$ by

$$\Gamma(s,t) = \begin{cases} H(\gamma(s), 2t) & \text{if } 0 \leq t \leq 1/2 \\ f(h(s, 2(1-t))) & \text{if } 1/2 \leq t \leq 1. \end{cases}$$

This map is well defined, since for $t = 1/2$ we have

$$H(\gamma(s),1) = f \circ g(\gamma(s)) = f(g(\gamma(s)) = f(h(s,1)).$$

Moreover, we have $\Gamma(s,0) = \gamma(s)$ and $\Gamma(s,1) = f(c)$ for all $s \in \mathbb{S}^1$. Thus Γ is a homotopy between γ and the constant map with value $f(c) \in Y$. This shows that Y is simply connected. By applying the argument with the roles of X and Y reversed, we get the assertion. □

Remark 8.12 We note that Proposition 8.11 helps us classify spaces up to homotopy equivalence. Thus, there are spaces that are simply connected and those that are not.

In dimensions two and higher, we are very familiar with examples of simply connected spaces as we will now show.

Theorem 8.13 *For $n \geq 2$, the n-dimensional sphere $\mathbb{S}^n$ is simply connected.*

Remark 8.14 In Exercise 8.2, we use Sard's Theorem 5.17 to show that every smooth map $X \to \mathbb{S}^n$ from a smooth manifold X of dimension k is homotopic to a constant map whenever k is *strictly less* than n. Since every continuous map $\mathbb{S}^1 \to \mathbb{S}^n$ is homotopic to a smooth map $\mathbb{S}^1 \to \mathbb{S}^n$ by Whitney's Approximation Theorem 8.6, this implies the assertion of Theorem 8.13. Since we have not yet proven Theorem 8.6, we give an alternative proof that does not make use of smooth maps. See also for example [40, theorems 59.1 and 59.3].

Proof of Theorem 8.13: Let $p \in \mathbb{S}^n$ be any point. Stereographic projection defines a diffeomorphism $\mathbb{S}^n \setminus \{p\} \cong \mathbb{R}^n$. In $\mathbb{R}^n$ every continuous map is homotopic to a constant map. Hence the assertion follows from the following Lemma 8.15. □

Lemma 8.15 (Paths avoiding a point) *Let $n \geq 2$ and $p \in \mathbb{S}^n$. Every path γ in $\mathbb{S}^n$ with $\gamma(0) \neq p \neq \gamma(1)$ is homotopic to a path in $\mathbb{S}^n \setminus \{p\}$.*

Proof: Let $U = \mathbb{S}^n \setminus \{p\}$ and $V = \mathbb{S}^n \setminus \{-p\}$. Then $\mathbb{S}^n$ is the union of the two open subsets U and V. Both U and V are diffeomorphic to $\mathbb{R}^n$ via stereographic projection. By Exercise 1.4 on Lebesgue numbers, we can find a subdivision $b_0 < \cdots < b_m$ of the unit interval $[0,1]$ such that, for each i, the set $\gamma([b_{i-1}, b_i])$ is contained either in U or in V. If there is an index i such that $\gamma(b_i) \in U \cap V$, i.e., if $\gamma(b_i) \in \{p, -p\}$, then we modify the subdivision as follows. Each of the sets $\gamma([b_{i-1}, b_i])$ and $\gamma([b_i, b_{i+1}])$ is either contained in U or in V. If $\gamma([b_i]) \in U$, then both of these sets must lie in U. Then we may delete b_i from our subdivision to obtain a new subdivision $c_0 < \ldots < c_{m-1}$ that still satisfies the condition that $\gamma([c_{i-1}, c_i])$ is contained either in U or in V for each i. If $\gamma([b_i]) \in V$ we proceed similarly. After a finite number of steps, this leads to a subdivision $a_0 < \cdots < a_n$ of $[0,1]$ such that $\gamma([a_{i-1}, a_i])$ is contained either in U or in V and $\gamma(a_i) \in U \cap V$ for each i. Then we choose a path $\gamma_i : [a_{i-1}, a_i] \to \mathbb{S}^n$ for each i as follows. If $\gamma([a_{i-1}, a_i]) \subset U$, we set $\gamma_i = \gamma|_{[a_{i-1}, a_i]}$. If $\gamma([a_{i-1}, a_i]) \subset V$, then we let γ_i be a path which is homotopic to $\gamma|_{[a_{i-1}, a_i]}$ which avoids p with $\gamma_i(a_{i-1}) = \gamma(a_{i-1})$ and $\gamma_i(a_i) = \gamma(a_i)$. We can find such a path since V is homeomorphic to $\mathbb{R}^n$. Composing the γ_i yields a path $[0,1] \to \mathbb{S}^n$ which is homotopic to γ, has the same start and endpoints, and avoids p. □

However, not all spaces are simply connected. The following discussion shows, for example, that $\mathbb{S}^1$ is *not* simply connected. We will see further examples later. We first introduce yet another property defined via homotopy.

Definition 8.16 A manifold X is called *contractible* if the identity map $\mathrm{id}_X : X \to X$ is homotopic to some constant map $X \to \{x\}$ where x is any point of X.

In Exercise 8.1, we will show that, if X is contractible, then, in fact, all self-maps $X \to X$ are homotopic. In Exercise 8.3, we show that all contractible manifolds are simply connected. The converse, however, is false. As an example, consider $X = \mathbb{S}^2$. By Theorem 8.13, the two-sphere is simply connected, but it is not contractible. For example, the antipodal map is not homotopic to the identity. We will have to develop further techniques and smart invariants to be able to prove this. In Exercise 8.7, we show that $\mathbb{S}^1 \setminus \{x\}$ is contractible for any $x \in \mathbb{S}^1$. The whole circle, however, is not simply connected as we will now show.

Theorem 8.17 *The identity map of the circle $\mathbb{S}^1$ is not homotopic to a constant map. In particular, $\mathbb{S}^1$ is not contractible and not simply connected.*

Remark 8.18 To prove that a homotopy cannot exist is often much harder than to find a homotopy. We will later develop techniques and invariants that will allow us to handle such problems, for example, the Brouwer degree, winding numbers, and intersection numbers. In Exercise 10.2, we will use Brouwer's Fixed Point Theorem, which we will discuss and prove in Section 10.3. Overall, we will see more conceptual reasons for the statement of Theorem 8.17. We now provide a first proof of the theorem.

The proof of Theorem 8.17 is based on the following idea. At first glance, this may seem like a trick coming out of the blue, but there are conceptual reasons why this works. Let $p\colon \mathbb{R} \to \mathbb{S}^1 \subset \mathbb{C}$ be the map defined by $t \mapsto e^{2\pi i t}$. We say that a continuous map $f\colon \mathbb{S}^1 \to \mathbb{S}^1$ *lifts to* $\mathbb{R}$ if there is a continuous map $\tilde{f}\colon \mathbb{S}^1 \to \mathbb{R}$ such that the diagram

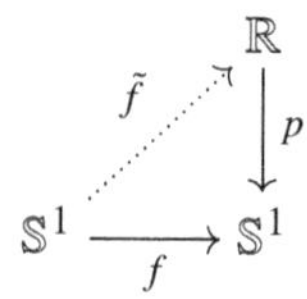

(8.1)

commutes, i.e., such that $f = p \circ \tilde{f}$.

Example 8.19 Let $x \in \mathbb{S}^1$ be a point and let $c_x\colon \mathbb{S}^1 \to \mathbb{S}^1$ be the constant map with value x. Since p is surjective, we can choose a $t \in \mathbb{R}$ such that $p(t) = x$. Then the constant map $\mathbb{S}^1 \to \mathbb{R}$ with value t is a lift of c_x to $\mathbb{R}$.

Example 8.20 The identity map $\mathrm{id}_{\mathbb{S}^1}\colon \mathbb{S}^1 \to \mathbb{S}^1$, however, does not lift to $\mathbb{R}$. To prove this claim, assume there was a continuous lift $\tilde{f}$ of $\mathrm{id}_{\mathbb{S}^1}$ to $\mathbb{R}$. Then there is some $n_0 \in \mathbb{Z}$ such that $\tilde{f}(1) = n_0$ where $1 \in \mathbb{S}^1 \subset \mathbb{C}$. By Example 3.6, p is a local diffeomorphism, and, for example, the restriction of p to the open interval $(n_0, n_0 + 1) \subset \mathbb{R}$ has an inverse $(p_{|(n_0,n_0+1)})^{-1}\colon \mathbb{S}^1 \setminus \{1\} \to (n_0, n_0 + 1)$. We must then have $\tilde{f}(p(t)) = (p_{|(n_0,n_0+1)})^{-1}(p(t)) = t$ for all $t \in (n_0, n_0 + 1)$. Since all the above maps are continuous, we must have $\tilde{f}(p(n_0)) = n_0$. However, we also must have $\tilde{f}(p(n_0 + 1)) = n_0 + 1$. But we know $p(n_0 + 1) = p(n_0) = 1$. This shows that $\tilde{f}$ cannot be defined on $\mathbb{S}^1$.

While Example 8.20 shows that not all maps can be lifted, there are also positive examples.

Lemma 8.21 *Let $f\colon \mathbb{S}^1 \to \mathbb{S}^1$ be continuous maps such that f is not surjective. Then f lifts to $\mathbb{R}$.*

Proof: To simplify the notation, we assume that $1 \in \mathbb{S}^1 \subset \mathbb{C}$ is not in the image. Any other case can be proven similarly. By Example 3.6, p is a local diffeomorphism. Let $(p_{|(0,1)})^{-1}\colon \mathbb{S}^1 \setminus \{1\} \to (0,1)$ be the inverse of the restriction of p to the open interval $(0,1) \subset \mathbb{R}$. We can then define a lift $\tilde{f}$ of f to $\mathbb{R}$ by $\tilde{f}(x) := f((p_{|(0,1)})^{-1}(x))$ for all $x \neq 1$ and $\tilde{f}(1) := f(1)$. Since f is continuous and defined on $\mathbb{S}^1$, $\tilde{f}$ is a well-defined continuous map such that $p \circ \tilde{f} = f$. □

Lemma 8.22 *Let $f, g\colon \mathbb{S}^1 \to \mathbb{S}^1$ be continuous maps such that $f(x) \neq -g(x)$ for all $x \in \mathbb{S}^1$. Assume that g lifts to $\mathbb{R}$. Then f lifts to $\mathbb{R}$.*

Proof: Considering $\mathbb{S}^1 \subset \mathbb{C}$, $f(x) \neq -g(x)$ is equivalent to $f(x)/g(x) \neq -1$ where we use that $g(x) \neq 0$ as $|g(x)| = 1$. Thus, the map $f/g\colon \mathbb{S}^1 \to \mathbb{S}^1$ given by $x \mapsto f(x)/g(x)$ has image contained in $\mathbb{S}^1 \setminus \{-1\}$, where $-1 \in \mathbb{C}$. By Lemma 8.21, there is a lift $\widetilde{f/g}$ of f/g to $\mathbb{R}$. Now let $\tilde{g}$ be the lift of g to $\mathbb{R}$. We then define a map $\tilde{f}\colon \mathbb{S}^1 \to \mathbb{R}$ by $\tilde{f}(x) := \tilde{g}(x) + \widetilde{f/g}(x)$. For all $x \in \mathbb{S}^1$, we can then compute in $\mathbb{C}$ that

$$p(\tilde{f}(x)) = p(\tilde{g}(x) + \widetilde{f/g}(x)) = p(\tilde{g}(x)) \cdot p(\widetilde{f/g}(x)) = g(x) \cdot f(x)/g(x) = f(x).$$

Thus, $\tilde{f}$ is a lift of f. □

Proof of Theorem 8.17: Let $x_0 \in \mathbb{S}^1$ be a point. Let $c_{x_0}\colon \mathbb{S}^1 \to \mathbb{S}^1$ be the constant map with value x_0, and let $f\colon \mathbb{S}^1 \to \mathbb{S}^1$ be a continuous map. Assume there is a homotopy $H\colon \mathbb{S}^1 \times [0,1] \to \mathbb{S}^1$ between c_{x_0} and f, i.e., $H(x,0) = c_{x_0}(x)$ and $H(x,1) = f(x)$ for all $x \in \mathbb{S}^1$. Since $\mathbb{S}^1 \times [0,1]$ is compact by Proposition 1.15, H is uniformly continuous. Thus, we can find a $\delta > 0$ such that $|H(x,t) - H(y,s)| < 2$ whenever $|(x,t) - (y,s)| < \delta$, where the norm is taken in $\mathbb{R}^3$. In particular, we have $|H(x,t) - H(x,s)| < 2$, i.e., $H(x,t) \neq H(x,s)$, for all $x \in \mathbb{S}^1$ whenever $|t - s| < \delta$. By Example 8.19, c_{x_0} lifts to $\mathbb{R}$. By Lemma 8.22, this implies that, for a fixed t with $0 \le t \le \delta/2$, $H(-,t)\colon \mathbb{S}^1 \to \mathbb{S}^1$ lifts to $\mathbb{R}$ as well. Since δ only depends on H and not on x or t, we then get, after finitely many applications of Lemma 8.22, that $f = H(-,1)$ lifts to $\mathbb{R}$. However, by Example 8.20, we know that $\mathrm{id}_{\mathbb{S}^1}$ does not lift to $\mathbb{R}$. Hence there cannot be a homotopy between c_{x_0} and $\mathrm{id}_{\mathbb{S}^1}$. This shows that $\mathbb{S}^1$ is not contractible and, in fact, not simply connected. □

For later use, we are going to study the above situation a little more. Example 8.20 shows what can go wrong when we try to lift self-maps of $\mathbb{S}^1$. The following lemma shows that things get easier when we also lift the domain. This will turn out to be very useful later. See for example Theorems 11.17 and 15.19.

Lemma 8.23 (Lifting self-maps of the circle) *Let $f\colon \mathbb{S}^1 \to \mathbb{S}^1$ be a continuous map, and let $p\colon \mathbb{R} \to \mathbb{S}^1 \subset \mathbb{R}^2$ be the map defined by $t \mapsto e^{2\pi i t}$. Then* *there exists a continuous map* $g\colon \mathbb{R} \to \mathbb{R}$ *such that the following diagram commutes*

$$\begin{array}{ccc} \mathbb{R} & \xrightarrow{g} & \mathbb{R} \\ {\scriptstyle p}\downarrow & & \downarrow{\scriptstyle p} \\ \mathbb{S}^1 & \xrightarrow[f]{} & \mathbb{S}^1 \end{array}, \tag{8.2}$$

i.e., $f \circ p = p \circ g$. *There is then an integer* $q \in \mathbb{Z}$ *which only depends on* f *such that* $g(t+1) = g(t) + q$ *for all* $t \in \mathbb{R}$. *Moreover, if* f *is smooth, then there is a smooth map* g *with the above properties.*

There is a deeper reason why the map g of the lemma exists. In fact, $\mathbb{R}$ is a universal covering space of $\mathbb{S}^1$, and continuous paths can always be lifted to a covering space.

Proof of Lemma 8.23: The idea is that, given any map $f\colon \mathbb{S}^1 \to \mathbb{S}^1$, we can lift f locally and then patch the pieces together to get a map $g\colon \mathbb{R} \to \mathbb{R}$ such that $p(g(t)) = f(p(t))$. See Figure 8.1. We only prove the assertion when f is continuous. To get the smooth case, we can compose with appropriate smooth

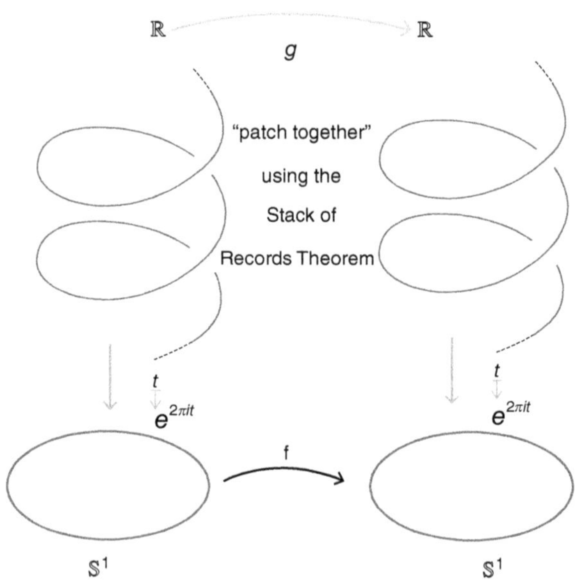

Figure 8.1 We think of $\mathbb{R}$ lying as a spiral above $\mathbb{S}^1$. That makes it easier to imagine how we lift f to a map g, first by local lifts which then are patched together.

bump functions to smoothen out any possible singular points. We first consider the composite $h = f \circ p$:

$$\begin{array}{ccc} [0,1] & & \mathbb{R} \\ {\scriptstyle p}\downarrow & \overset{h}{\searrow} & \downarrow{\scriptstyle p} \\ \mathbb{S}^1 & \xrightarrow{\quad f \quad} & \mathbb{S}^1 \end{array}, \tag{8.3}$$

where p also denotes its restriction to $[0,1] \subset \mathbb{R}$ by a slight abuse of notation. We now make a choice of a preimage point $s_0 \in \mathbb{R}$ with $p(s_0) = h(0)$. The restriction p_{s_0} of the covering map

$$p_{s_0} \colon (s_0 - 1/2, s_0 + 1/2) \to \mathbb{S}^1 \setminus \{p(s_0 + 1/2)\}$$

is a diffeomorphism. Hence we can use the inverse $p_{s_0}^{-1}$ to construct a local lift $p_{s_0}^{-1} \circ h$ in a neighbourhood U_0 of 0 in $\mathbb{R}$ which is small enough such that $h(U_0) \subseteq \mathbb{S}^1 \setminus \{p(s_0 + 1/2)\} =: \mathbb{S}^1_-$:

$$\begin{array}{ccc} U_0 & \xrightarrow{\;p_{s_0}^{-1} \circ h\;} & \mathbb{R} \\ {\scriptstyle p}\downarrow & \overset{h}{\searrow} & {\scriptstyle p}\downarrow \uparrow {\scriptstyle p_{s_0}^{-1}} \\ \mathbb{S}^1 & \xrightarrow{\quad f \quad} & \mathbb{S}^1_- \end{array}. \tag{8.4}$$

Now we continue this procedure to construct enough local lifts to cover all of $[0,1]$. That this works can be shown as follows.[1] We have just constructed a lift of h on the interval $[0,s]$ for some s with $0 < s \leq 1$. Thus the set

$$\begin{aligned} D = \{t \in [0,1] |\ & h \colon [0,t] \to \mathbb{S}^1 \text{ can be lifted to a map} \\ & \tilde{g} \colon [0,t] \to \mathbb{R} \text{ beginning at } s_0\} \end{aligned}$$

is non-empty. Since D is bounded by 1, it has a least upper bound $d \in [0,1]$. We will now show that $d \in D$ and then that $d = 1$. Let U be an open subset of $\mathbb{S}^1$ containing $h(d)$ such that p maps each component of $p^{-1}(U)$ diffeomorphically onto U. Since h is continuous, there is an open subset $W \subset [0,1]$ containing d with $h(W) \subset U$. By definition of D and d, there is a point $s' \in W$ such that $0 < s' < d$ and $s' \in D$. Since we can lift h to $\tilde{g} \colon [0,s'] \to \mathbb{R}$, we have a unique point $\tilde{g}(s') \in \mathbb{R}$. We also have $\tilde{g}(s') \in p^{-1}(h(s'))$. Let $V \subset \mathbb{R}$ denote the component of $p^{-1}(U)$ with $\tilde{g}(s') \in V$. Since $h(s')$ and $h(d)$ are both in U, we can use $(p_{|V})^{-1}$ to define a local lift $(p_{|V})^{-1} \circ h \colon [0,s'] \to \mathbb{R}$ of h on $[s',d]$. Now the previous lift of h

[1] You find a proof of this fact in almost every textbook on algebraic topology in the section on covering spaces and lifting of paths. See for example [47, page 88].

on $[0,s']$ and the new one on $[s',d]$ agree at s' by construction. Hence we can glue them together to get a lift $\tilde{g}\colon [0,d] \to \mathbb{R}$ of h on the entire interval $[0,d]$.[2] Thus $d \in D$ as claimed.

Now we show that $d = 1$. Suppose $d < 1$. For the open set $W \subset [0,1]$ as above, we have $d \in W$. Since W is open, it still contains the open interval $(d-2\varepsilon, d+2\varepsilon)$ for some small $\epsilon > 0$. Then we could use the above argument to extend the lift of h to $[0, d+\varepsilon]$. Thus we would have $d + \varepsilon \in D$, which contradicts that d is the least upper bound of D. Hence we must have $d = 1$ as claimed. Hence we have shown that there is a continuous map $\tilde{g}\colon [0,1] \to \mathbb{R}$ such that the diagram

$$\begin{array}{ccc} [0,1] & \xrightarrow{\tilde{g}} & \mathbb{R} \\ {\scriptstyle p}\downarrow & & \downarrow{\scriptstyle p} \\ \mathbb{S}^1 & \xrightarrow[f]{} & \mathbb{S}^1 \end{array} \tag{8.5}$$

commutes. We then have $p(\tilde{g}(1)) = f(p(1)) = f(p(0)) = p(\tilde{g}(0))$ and hence $g(1) = g(0) + q$ for some fixed $q \in \mathbb{Z}$. Finally, we define $g\colon \mathbb{R} \to \mathbb{R}$ by setting $g(t) := \tilde{g}(t)$ for all $t \in [0,1]$, and $g(t+1) = g(t) + q$ for all $t \in \mathbb{R}$. □

Remark 8.24 It may seem that in the proof Lemma 8.23 we force g to satisfy $g(t+1) = g(t) + q$. However, if there is a map g such that $f \circ p = p \circ g$, then we have $p(g(t+1)) = f(p(t+1)) = f(p(t)) = p(g(t))$ for all t. Since $p(t_1) = p(t_2)$ if and only if $t_1 - t_2 \in \mathbb{Z}$, we get $g(t+1) - g(t) \in \mathbb{Z}$. Since the function $t \mapsto g(t+1) - g(t)$ is assumed to be continuous and takes only values in the discrete space $\mathbb{Z}$, it is locally constant. Since $\mathbb{R}$ is connected, the function must be constant by Lemma 1.30. In other words, for all $t \in \mathbb{R}$, we have $g(t+1) = g(t) + q$ for some fixed $q \in \mathbb{Z}$. This q is a fixed integer depending only on f. We can think of q as measuring how fast f wraps $\mathbb{S}^1$ around itself. We will see in Theorem 15.19 that the number q actually determines f up to homotopy. For example, for $f = \mathrm{id}_{\mathbb{S}^1}$ we have $q = 1$, which agrees with the argument in Example 8.20.

8.3 Homotopy Groups

An important tool to study a space up to homotopy is provided by homotopy groups that we will now define.

[2] This gives a continuous lift, but we can compose with appropriate smooth bump functions to smoothen out any possible singular points.

Definition 8.25 (Homotopy groups) Let X be a topological space, and let $x \in X$ be a point. For $n \geq 1$, the set $\pi_n(X,x)$ is defined as the set of homotopy classes of continuous maps $\mathbb{S}^n \to X$ which send a fixed point in $\mathbb{S}^n$ to x. One can show that $\pi_n(X,x)$ is, in fact, a group which is even abelian when $n \geq 2$ and does not depend on the choice of the point x. We note that the neutral, or null element, of $\pi_n(X,x)$ is the class of the constant map which sends all of $\mathbb{S}^n$ to $\{x\}$. Even though we will not discuss the construction of the group operation in this book, we will refer to $\pi_n(X,x)$ as the nth *homotopy group* of X. For $n = 1$, $\pi_1(X,x)$ is called the *fundamental group* of X.

One may think of $\pi_n(X)$ as an attempt to understand X by studying how we can map a given familiar test space, here the n-sphere, into X. Computing $\pi_n(X)$, however, may be extremely difficult. Many of the tools in differential topology have been invented with the goal to compute $\pi_k(\mathbb{S}^n)$, and for many pairs (k,n), the groups $\pi_k(\mathbb{S}^n)$ have been computed. The larger k is compared to n, however, the less we know about the structure of $\pi_k(\mathbb{S}^n)$.

Example 8.26 The fundamental group of a simply connected space is the trivial group. In particular, Theorem 8.13 implies that the fundamental group of $\mathbb{S}^n$, for all $n \geq 2$, is trivial, i.e., $\pi_1(\mathbb{S}^n) = \{0\}$. On the other hand, Theorem 8.17 shows that $\pi_1(\mathbb{S}^1,x)$ is not trivial. We will say more about $\pi_1(\mathbb{S}^1,x)$ in Sections 15.2 and 16.3. In fact, Hopf's Theorem 16.26 shows that $\pi_n(\mathbb{S}^n)$ is isomorphic to the integers for all $n \geq 1$. To prove this result will require a considerable amount of effort and motivates many of the forthcoming constructions.

Assuming Whitney's Approximation Theorem 8.6 for the moment, it follows from Exercise 8.2 that $\pi_k(X)$ is trivial for all smooth manifolds X of dimension strictly larger than k. In particular, we know $\pi_k(\mathbb{S}^n) = 0$ for $k < n$. A much more interesting question is whether $\pi_k(\mathbb{S}^n)$ is trivial or not for $k > n$. For $n = 1$, one can show that $\pi_k(\mathbb{S}^1)$ is the trivial group for all $k \geq 2$. For $n = 2$, we may ask the following question.

Question Is $\pi_3(\mathbb{S}^2)$ a non-trivial group?

In fact, we know of an interesting map $f\colon \mathbb{S}^3 \to \mathbb{S}^2$, namely the Hopf map $\hbar$. Hence we may ask the concrete question whether the Hopf map is homotopic to a constant map. In Section 11.5.3, we will show that f is not homotopic to a constant map. This implies that $\pi_3(\mathbb{S}^2)$ is non-trivial. This is a famous result due to Heinz Hopf [22] and is one of the highlights of the early stages of differential topology. Showing this result will require the development of important invariants, in particular the Brouwer degree, linking numbers, and the Hopf invariant. We should therefore keep this question in mind as a motivation for the forthcoming chapters.

8.4 Exercises

8.1 Let X be a smooth manifold.

(a) Show that if X is contractible, then all smooth maps $Y \to X$ from any smooth manifold Y to X are homotopic.

(b) Conversely, show that if, for every smooth manifold Y, all maps $Y \to X$ are homotopic, then X is contractible.

(c) Show that $\mathbb{R}^k$ is contractible.

8.2 Let X be a smooth manifold of dimension k. If $k < n$, show that every smooth map $f\colon X \to \mathbb{S}^n$ is homotopic to a constant map. Hint: Use Sard's Theorem 5.17.

8.3 Show that all contractible manifolds are simply connected.

8.4 Show that the antipodal map $\mathbb{S}^k \to \mathbb{S}^k$, $x \mapsto -x$, is homotopic to the identity if k is odd. (We will see later that this is not true when k is even.) Hint: Start with $k = 1$ by using the linear maps defined by $[0,1] \to \mathrm{M}(2)$, $t \mapsto \begin{pmatrix} \cos(\pi t) & -\sin(\pi t) \\ \sin(\pi t) & \cos(\pi t) \end{pmatrix}$.

8.5 Show that every connected smooth manifold X is *path-connected*, i.e., given any two points $x_0, x_1 \in X$, there exists a smooth map $\gamma\colon [0,1] \to X$ with $\gamma(0) = x_0$ and $\gamma(1) = x_1$. Hint: Use that homotopy is an equivalence relation to show that the relation x_0 *and* x_1 *can be joined by a smooth path* is an equivalence relation on X. Then show that the equivalence classes are both open and closed subsets of X. You may also want to have a look at the proof of Proposition 1.36.

8.6 Let X be a connected smooth manifold of dimension $\dim X > 1$, and let $x_0 \in X$ be a point. Show that $X \setminus \{x_0\}$ is still path-connected. Hint: Begin with the strategy sketched for Exercise 8.5. Then you need to use the assumption $\dim X > 1$ at some point.

8.7 Show that $\mathbb{S}^1 \setminus \{x\}$ is contractible for every $x \in \mathbb{S}^1$.

8.8 Let $f, g\colon \mathbb{S}^1 \to \mathbb{S}^1$ be smooth maps such that $f(x) \neq -g(x)$ for all $x \in \mathbb{S}^1$. Show that f and g are homotopic. Hint: Use Exercises 8.2 and 8.7.

8.9 Let X be a smooth manifold. Let $f, g\colon X \to \mathbb{S}^k$ be two continuous maps such that $|f(x) - g(x)| < 2$ for all $x \in X$, where the norm is taken as elements in $\mathbb{R}^{k+1}$. Show that f and g are homotopic. By composition with a bump function, we then get that f and g then they are smoothly homotopic when both are smooth maps. Hint: What does the assumption say geometrically about $f(x)$ and $g(x)$?

8.10 Let $X \subset \mathbb{R}^N$ be a smooth manifold, and let $\pi\colon T(X) \to X$ denote its tangent bundle. A *vector field* on X is a smooth section of π, i.e., a smooth map $\mathbf{v}: X \to T(X)$ such that $\pi \circ \mathbf{v} = \mathrm{Id}_X$. An equivalent way to describe such a section is to give a map $s: X \to \mathbb{R}^N$ such that $s(x) \in T_x(X) \subset \mathbb{R}^N$ for all x, with corresponding $\mathbf{v}(x) = (x, s(x))$. A point $x \in X$ is a *zero of the vector field* $\mathbf{v}$ if $\mathbf{v}(x) = (x, 0)$, or equivalently, if $s(x) = 0$.

(a) Show that if k is odd, there exists a vector field on $\mathbb{S}^k$ having no zeros. Hint: For $k = 1$, use $(x_1, x_2) \mapsto (-x_2, x_1)$.

(b) Prove that if $\mathbb{S}^k$ admits a vector field which has no zeros, then its antipodal map $x \mapsto -x$ is homotopic to the identity. Hint: Show that you may assume $|s(x)| = 1$ everywhere. Now contemplate about $(\cos(\pi t))x + (\sin(\pi t))s(x)$ for $t \in [0,1]$.

8.11 Show that if k is even, then the antipodal map on $\mathbb{S}^k$ is homotopic to the reflection map $r: \mathbb{S}^k \to \mathbb{S}^k$ defined by $r(x_1, \ldots, x_{k+1}) = (-x_1, x_2, \ldots, x_{k+1})$. Hint: Consider the reflections $r_i(x_1, \ldots, x_{k+1}) = (x_1, \ldots, -x_i, \ldots, x_{k+1})$, and show that $r_i \circ r_{i+1}$ is homotopic to the identity on $\mathbb{S}^k$.

8.12 Let X and Y be smooth manifolds, and assume that X is compact. Let $f_0, f_1: X \to Y$ be smooth maps, and let f_t be a smooth homotopy between f_0 and f_1. Prove the following two cases of the *Stability Theorem*:

(a) If f_0 is an immersion, then there is an $\varepsilon > 0$ such f_t is an immersion for all $t \in [0, \varepsilon)$.

(b) If f_0 is an embedding, then there is an $\varepsilon > 0$ such f_t is an embedding for all $t \in [0, \varepsilon)$.

8.13 Show that the assertion of Exercise 8.12 fails to be true in general when X is not compact. That is, give an example of a smooth family $f_t: X \to Y$ such that f_0 is an immersion (respectively an embedding), but there is no $\varepsilon > 0$ such that f_t is an immersion (respectively an embedding) for all $t \in [0, \varepsilon)$. Hint: You may want to try $X = Y = \mathbb{R}$ and start with a smooth function $\rho: \mathbb{R} \to \mathbb{R}$ with $\rho(s) = 1$ for $|s| < 1$ and $\rho(s) = 0$ for $|s| > 2$. Then use ρ to find a suitable homotopy f_t of the identity map.

8.14 Suppose that f_t is a smooth homotopy of smooth functions $\mathbb{R}^k \to \mathbb{R}$. Assume that f_0 is a Morse function on some open subset $U \subset \mathbb{R}^k$ containing a compact set $K \subset \mathbb{R}^k$. Show that every f_t for t sufficiently small is a Morse function. Hint: Use Exercise 5.10.

8.15 Let X be a compact smooth manifold, let $f_0: X \to \mathbb{R}$ be a smooth function, and let f_t be a homotopy of f_0. Assume that f_0 is Morse. Show that there is an $\varepsilon > 0$ such that f_t is a Morse function for all $t \in [0, \varepsilon)$. Hint: Use Exercise 8.14.

9
Manifolds with Boundary

In order to be able to analyse a wider class of phenomena, we now enlarge the class of smooth manifolds. A typical example which we would like to include is the domain of a homotopy $X \times [0,1]$ for a smooth k-dimensional manifold X. This should be a manifold of dimension $k+1$. However, the points on $X \times \{0\}$ and $X \times \{1\}$ do not have an open neighbourhood which is diffeomorphic to $\mathbb{R}^{k+1}$. Another example is the closed unit ball in $\mathbb{R}^k$. So far, such spaces do not qualify as a smooth manifold. From now on, we would like to allow such spaces. The good news is that most of the theorems we have proved so far can be transferred to manifolds with boundary.

9.1 Smooth Manifolds with Boundary

The idea of a manifold with boundary is the same as before. It is a space which locally looks like some model space with a boundary which we understand well. We need, however, a suitable new model space.

9.1.1 Manifolds with Boundary

Definition 9.1 (Euclidean model with boundary) The standard model of a *Euclidean space with boundary* is the half-plane

$$\mathbb{H}^k := \left\{(x_1, \ldots, x_k) \in \mathbb{R}^k \colon x_k \geq 0\right\}$$

in $\mathbb{R}^k$. The *boundary* of $\mathbb{H}^k$, denoted $\partial\mathbb{H}^k$, is the subspace

$$\partial\mathbb{H}^k := \left\{(x_1, \ldots, x_k) \in \mathbb{R}^k \colon x_k = 0\right\} = \mathbb{R}^{k-1} \times \{0\} \subset \mathbb{R}^k.$$

Now a manifold with boundary is a space which locally looks like $\mathbb{H}^k$.

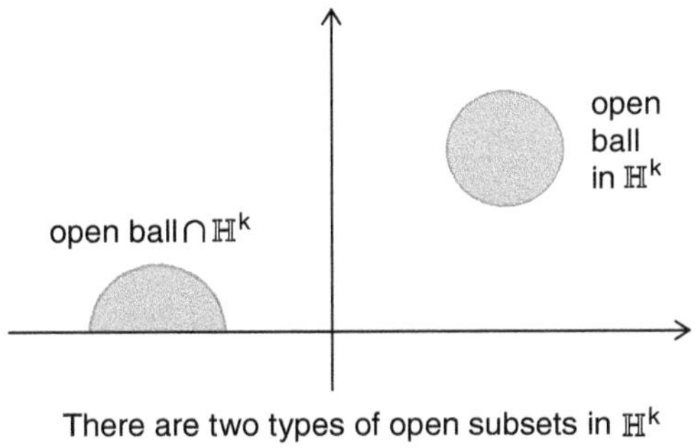

Figure 9.1 There are two types of open balls in $\mathbb{H}^k$. The balls where all points satisfy $x_k > 0$ are well-known. The balls which allow $x_k = 0$ are new.

Definition 9.2 (Manifolds with boundary) A subspace X of $\mathbb{R}^N$ is called a *smooth k-dimensional manifold with boundary* if, for every point x of X, there is an open neighbourhood $V \subset X$ containing x and an open subset $U \subset \mathbb{H}^k$ together with a diffeomorphism $\phi\colon U \to V$. Any such a diffeomorphism is called a *local parametrisation* of X. The *boundary* of X, denoted ∂X, consists of those points that belong to the image of the boundary of $\mathbb{H}^k$ under some local parametrisation. The complement $X \setminus \partial X$ of the boundary is called the *interior* of X, denoted $\mathrm{Int}(X) := X \setminus \partial X$.

We will show below that the boundary and the interior are well defined. A smooth manifold X with $\partial X = \emptyset$ is just a smooth manifold in our previous terminology. In order to make the distinction clear, if necessary, we call them also *boundaryless manifolds* or *manifolds without a boundary*.

Remark 9.3 By mimicking Definition 6.9 using $\mathbb{H}^k$, we obtain the notion of a smooth manifold with boundary independent of an embedding into $\mathbb{R}^N$. We only have to be a little more careful to define smoothness at points on the boundary. We will explain how to do this below when we discuss tangent spaces and derivatives at boundary points. In light of Whitney's Embedding Theorem, however, which also applies to manifolds with boundary, we continue to make the assumption that manifolds are embedded in some Euclidean space. See Figure 9.1.

Example 9.4 (The closed unit interval) The unit interval $[0, 1]$ is a one-dimensional smooth manifold with boundary, and the boundary consists of the two endpoints $\{0, 1\}$. We can choose local parametrisations $\phi_0\colon [0, 1) \to [0, 1]$ given by $x \mapsto x$, and $\phi_1\colon [0, 1) \to [0, 1]$ given by $x \mapsto 1 - x$. We note that

$[0,1)$ is an open subset on $\mathbb{H}^1 = [0,\infty) \subset \mathbb{R}^1$ since $[0,1) = \mathbb{H}^1 \cap (-1,1)$. The interior of $[0,1]$ is the open interval $(0,1)$.

Remark 9.5 (Interior of a manifold vs interior of a subset) Let $X \subset \mathbb{H}^N$ be a smooth manifold with boundary. Then the interior of X as a manifold may differ from the interior of X as a subspace of $\mathbb{H}^N$ as defined in Definition 5.18. For example, let $X = \{(x,y) \in \mathbb{H}^2 : y \leq 1\} \subset \mathbb{H}^2$. Then X is a smooth manifold with boundary, and we have $X \setminus \partial X = \{(x,y) \in \mathbb{H}^2 : 0 < y < 1\}$. The interior of X as a subspace of $\mathbb{H}^2$, however, is the set $\{(x,y) \in \mathbb{H}^2 : 0 \leq y < 1\}$, which contains the x-axis.

Let X be a smooth manifold with boundary. We need to check that our definition of points in the interior and on the boundary is independent of the choice of a local parametrisation. So let $x \in X$ be a point which is in the image of a local parametrisation $\phi\colon U \to V \subset X$ such that $U \subset \mathbb{H}^k$ is an open subset of $\mathbb{H}^k$ which is contained in the interior of $\mathbb{H}^k$. Then U is an open subset of $\mathbb{R}^k$. Now assume x is also in the image of another local parametrisation $\phi'\colon U' \to V' \subset X$. Then $x \in W := V \cap V' \subset X$, and the composition $\phi' \circ \phi^{-1}\colon \phi^{-1}(W) \to (\phi')^{-1}(W)$ is a diffeomorphism. Hence, after possibly shrinking U', we see that U' is also an open subset of $\mathbb{R}^k$. Thus, x being an interior point is well defined. This shows, in particular, that the interior of X, $\mathrm{Int}(X)$, is a boundaryless manifold of the same dimension as X. It remains to show that being a boundary point is also well defined. We show this by proving the following useful result.

Proposition 9.6 (Boundaries are manifolds) *Let X be a smooth k-dimensional manifold with boundary. Then ∂X is a $(k-1)$-dimensional smooth manifold without boundary.*

Proof: Let $x \in X$, and let ϕ and ψ be two local parametrisations around x. After possibly shrinking the domains and codomains, we can assume that $\phi\colon U \to V$ and $\psi\colon W \to V$ are both diffeomorphisms from open sets $U \subset \mathbb{H}^k$, $W \subset \mathbb{H}^k$ to the same open subset $V \subset X$. To show that ∂X is well defined, we need to prove $\phi(\partial U) = \psi(\partial W)$. Moreover, since $\partial U = U \cap \partial\,\mathbb{H}^k$ is an open subset of $\mathbb{R}^{k-1}$, we would get that every point $y \in \partial X$ is contained in a local parametrisation $\phi_{|\partial U}\colon U \cap \partial\,\mathbb{H}^k \to \partial X$. This will show that ∂X is a manifold of dimension $k-1$. By our assumption on ϕ and ψ, it suffices to show $\psi(\partial W) \subset \phi(\partial U)$. The other inclusion will follow by symmetry. Hence we would like to show $\phi^{-1}(\psi(\partial W)) \subset \partial U$. Since both ϕ and ψ are diffeomorphisms, $\phi^{-1} \circ \psi$ is a diffeomorphism. Suppose that the claim is false and there is a point $w \in \partial W$ which is mapped to an interior point

$u = \phi^{-1}(\psi(w))$ of U. Since $u \in \mathrm{Int}(U)$, there is a neighbourhood $\widetilde{U}$ of u which is open in $\mathbb{R}^k$. Thus, the Inverse Function Theorem 3.4, applied to the inverse $\psi^{-1} \circ \phi$ defined on $\widetilde{U}$, implies that the image of $\psi^{-1} \circ \phi$ contains a neighbourhood of w that is open in $\mathbb{R}^k$. This contradicts the assumption $w \in \partial W$. Thus, u cannot have been an interior point. □

Sometimes we have to be careful when we apply our developed concepts to manifolds with boundaries. For example, the product of two manifolds with boundary may not be a smooth manifold anymore.

Example 9.7 (Square is not a manifold) The square $S = [0,1] \times [0,1]$ is not a smooth manifold with boundary. Suppose S was a smooth manifold with boundary. Then the corner $s = (0,0)$ had an open neighbourhood $V \subset S$, and there was a diffeomorphism $f : U \to V$ to an open $U \subset \mathbb{H}^2$ such that $f(\partial V) \subset \partial\mathbb{H}^2$. After shrinking V if necessary, let $F : \tilde{V} \to \mathbb{R}^2$ be a smooth extension of f on a subset $V \subset \tilde{V} \subset \mathbb{R}^2$ open in $\mathbb{R}^2$. Then the derivative dF_s is an isomorphism since f is a diffeomorphism. Writing $F = (F_1, F_2)$, we have $F_2(x,0) = 0 = F_2(0,y)$ for any $(x,0)$ and $(0,y)$ in U since ∂V is mapped to $\partial\mathbb{H}^2 = \{(x,y) \in \mathbb{R}^2 : y = 0\}$. Thus taking partial derivatives yields $\frac{\partial F_2}{\partial x}(s) = 0$ and $\frac{\partial F_2}{\partial y}(s) = 0$. However, this implies that $dF_s(e_1)$ and $dF_s(e_2)$ both lie in $\mathbb{R}\times \{0\} \subset \mathbb{R}^2$, where e_1 and e_2 denote the first and second standard basis vectors in $\mathbb{R}^2$, respectively. In particular, $dF_s(e_1)$ and $dF_s(e_2)$ are linearly dependent. This contradicts that dF_s is an isomorphism, and therefore the diffeomorphism f cannot exist.

However, if only one manifold has a boundary, we do obtain a manifold.

Proposition 9.8 (Products and Boundaries) *The product of a manifold without boundary X and a manifold with boundary Y is a manifold with boundary. Furthermore, $\partial(X \times Y) = X \times \partial Y$, and $\dim(X \times Y) = \dim X + \dim Y$.*

Proof: If $U \subset \mathbb{R}^k$ and $V \subset \mathbb{H}^l$ are open, then $U \times V \subset \mathbb{R}^k \times \mathbb{H}^l = \mathbb{H}^{k+l}$ is open. Moreover, if $\phi : U \to X$ and $\psi : V \to Y$ are local parametrisations, so is $\phi \times \psi : U \times V \to X \times Y$. □

Example 9.9 (Closed cylinder) The closed cylinder $\mathbb{S}^1 \times [0,1]$ is a manifold with boundary, see Figure 9.2, and the boundary consists of $\mathbb{S}^1 \times \{0\} \cup \mathbb{S}^1 \times \{1\}$. More generally, if X is a smooth n-dimensional manifold without boundary, then the domain $X \times [0,1]$ of a homotopy $X \times [0,1] \to Y$ is a smooth $(n+1)$-dimensional manifold with boundary $X \times \{0\} \cup X \times \{1\}$.

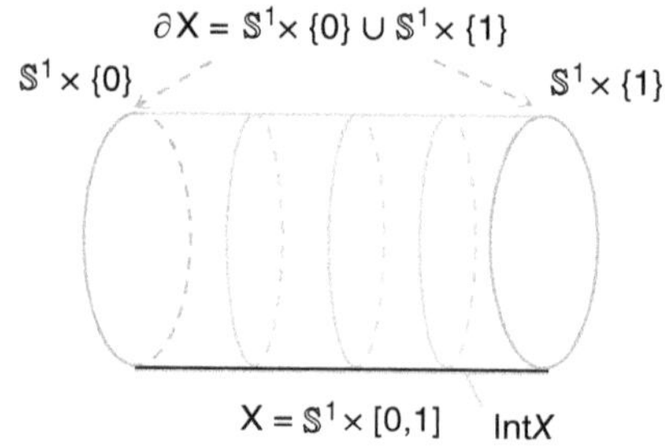

Figure 9.2 A cylinder has the two outer circles as its boundary.

9.1.2 Regular Values of Smooth Functions

We will now show that regular values again provide a natural way to produce smooth manifolds with boundary. First, we consider regular values of smooth functions. We look at other smooth maps in Section 9.2.

Proposition 9.10 (Regular values for real-valued functions) *Let X be a smooth manifold without boundary, and let $f: X \to \mathbb{R}$ be a smooth function with regular value* 0. *Then the subset $\{x \in X: f(x) \geq 0\}$ is a manifold with boundary, and the boundary is $f^{-1}(0)$.*

Proof: The set $\{x \in X: f(x) > 0\}$ is open in X since it is the preimage of the open subset $(0, \infty) \subset \mathbb{R}$ under the continuous map f. It is therefore a submanifold of the same dimension as X. Hence every point in $\{x \in X: f(x) > 0\}$ has an open neighbourhood which is diffeomorphic to an open subset of $\mathbb{R}^k$, $k = \dim X$. It remains to study the preimage of 0. So suppose that x is a point with $f(x) = 0$. By assumption, 0 is a regular value, which means that x is a regular point of f. Hence, locally near x, f looks like the canonical submersion by the Local Submersion Theorem 4.7. For the canonical submersion $f: \mathbb{H}^k \to \mathbb{R}$, $(x_1, \ldots, x_k) \to x_k$, the proposition just states the definition of the boundary of $\mathbb{H}^k$, i.e., $\partial\mathbb{H}^k = f^{-1}(0)$. □

Example 9.11 (Spheres are boundaries) Let $f: \mathbb{R}^{n+1} \to \mathbb{R}$ be the smooth function defined by $f(x_1, \ldots, x_{n+1}) = 1 - \sum_i x_i^2$. Then 0 is a regular value of f, and the closed unit ball $\mathbb{D}^{n+1}$ in $\mathbb{R}^{n+1}$ is the subset $\mathbb{D}^{n+1} = \{x \in \mathbb{R}^{n+1} : f(x) \geq 0\}$, and the boundary $f^{-1}(0)$ of $\mathbb{D}^{n+1}$ is the n-sphere $\mathbb{S}^n$.

9.1.3 Tangent Spaces and Derivatives

We now show that tangent spaces and derivatives are still defined for smooth manifolds with boundary. Since smoothness at a point requires a function to be defined on an open neighbourhood around that point, we need to be a bit more careful at boundary points.

Definition 9.12 Let $U \subset \mathbb{H}^n$ be an open subset and let $x \in U$ be a point. Let $f: U \to \mathbb{R}^k$ be a map which is smooth at x, i.e., there is an open subset $V \subset \mathbb{R}^n$ which contains U and a smooth map $F: V \to \mathbb{R}^k$ such that $f = F_{|U}$. We define the *derivative* df_x of f at x to be the derivative dF_x of F at x.

By construction, df_x is a linear map from all of $\mathbb{R}^n$ to $\mathbb{R}^k$. Note that, if x is an interior point of U, then df_x is just the derivative of f as defined previously. For points $x \in \partial U$ on the boundary, however, we get a new definition. In fact, before we can move on, we have to show that df_x is well defined at points on the boundary.

Lemma 9.13 *The derivative df_x is independent of the choice of F.*

Proof: Let $\tilde{F}$ be another local extension of f around x. After possibly taking the intersection of the domains, we can assume that both F and $\tilde{F}$ are defined on the same open subset $V \subset \mathbb{R}^n$. We need to show $dF_x = d\tilde{F}_x$. This is clear when $x \in \text{Int}(U)$, since then we can choose $V \subset \text{Int}(U)$. Now let $x \in \partial U$ be a point on the boundary. We can then choose a sequence of points $\{x_i\}_{i\in\mathbb{N}}$ which converges to x, such that $x_i \in \text{Int}(U) \cap V$. Since the derivative of a smooth map at a point depends continuously on the point, this implies that the sequence of maps dF_{x_i} converges dF_x and the sequence of maps dG_{x_i} converges dG_x. Here we use that the space $\text{Hom}(\mathbb{R}^n, \mathbb{R}^k)$ of linear maps from $\mathbb{R}^n$ to $\mathbb{R}^k$ is isomorphic to $\mathbb{R}^{nk}$, and hence it makes sense to talk about convergent sequences. Moreover, since F and $\tilde{F}$ agree with f on $\text{Int}(U) \cap V$, we have $dF_{x_i} = d\tilde{F}_{x_i}$ for all i. This implies that the limits of the sequences $\{dF_{x_i}\}$ and $\{d\tilde{F}_{x_i}\}$ agree as well. Thus, we have $dF_x = d\tilde{F}_x$ as required. □

Let $X \subset \mathbb{R}^N$ be a smooth manifold with boundary, and $x \in X$. Let $\phi: U \to X$ be a local parametrisation with $U \subset \mathbb{H}^n$ open. Let $u \in U$ be the point with $\phi(u) = x$. Note that we cannot assume $u = 0$ when x is an interior point. Then we form the derivative $d\phi_u: \mathbb{R}^n \to \mathbb{R}^N$, no matter what kind of point x is.

Definition 9.14 Let X be a smooth n-dimensional manifold with boundary. We define the *tangent space* to X at x, denoted $T_x(X)$, to be the image of $\mathbb{R}^n$ in $\mathbb{R}^N$ under the linear map $d\phi_u$.

It follows again from the chain rule that $T_x(X)$ does not depend as a subspace of $\mathbb{R}^N$ on the choice of ϕ.

Definition 9.15 Let $X \subset \mathbb{R}^N$ and $Y \subset \mathbb{R}^M$ be smooth manifolds with boundary of dimensions n and k, respectively. Let $f: X \to Y$ be a smooth map, and let $x \in X$ be a point. Let $\phi \colon U \to X$ with $\phi(u) = x$ and $\psi \colon V \to Y$ with $\psi(v) = f(x)$ be local parametrisations and let $\theta := \psi^{-1} \circ f_{|\phi(U)} \circ \phi$. Note that $v = \theta(u)$). We define the *derivative* $df_x \colon T_x(X) \to T_{f(x)}(Y)$ of f at x to be the unique linear map which makes the following diagram commutative

$$\begin{array}{ccc} T_x(X) & \xrightarrow{\;df_x\;} & T_y(Y) \\ {\scriptstyle d\phi_u}\uparrow & & \uparrow{\scriptstyle d\psi_v} \\ \mathbb{R}^n & \xrightarrow[d\theta_u]{} & \mathbb{R}^k \end{array} \quad . \tag{9.1}$$

9.2 Regular Values and Transversality

We now transfer the important concept of transversality to smooth manifolds with boundary. This requires again some care.

9.2.1 Fibres over Regular Values with Boundary

Recall that transversality is formulated as a criterion on tangent spaces and derivatives. We would like to formulate a similar criterion for maps between smooth manifolds with boundary. As we learned above, the boundary ∂X of a smooth k-dimensional manifold with boundary X is a manifold of dimension $k - 1$ without boundary. Let $x \in \partial X$ be a point on the boundary. We have $\dim T_x(\partial X) = k - 1$ and $\dim T_x(X) = k$. Moreover, since ∂X is a submanifold of X, we know that $T_x(\partial X) \subset T_x(X)$ is a vector subspace of codimension 1 in $T_x(X)$, i.e., the difference of dimensions equals 1.

Notation 9.16 (Restriction to boundary) For a smooth map $f \colon X \to Y$, we introduce the notation $\partial f := f_{|\partial X}$ for the restriction of f to ∂X. The derivative of ∂f at x is just the restriction of df_x to the subspace $T_x(\partial X)$, i.e., $d(\partial f)_x = (df_x)_{|T_x(\partial X)}$ as linear maps $T_x(\partial X) \to T_{f(x)}(Y)$.

Now let $f \colon X \to Y$ be a smooth map from a smooth manifold with boundary X to a boundaryless manifold Y, and let $Z \subset Y$ be a submanifold. We would like to know under which circumstances is $f^{-1}(Z)$ a submanifold with boundary in X, i.e., a subset of X which is itself a smooth manifold with boundary. Moreover, we would like to have control over the boundary in the sense that we would like to have

$$\partial f^{-1}(Z) = f^{-1}(Z) \cap \partial X. \tag{9.2}$$

We will now look at an example which shows that it is not enough to require that f is transverse to Z in the previous sense, i.e., $\operatorname{Im}(df_x) + T_{f(x)}(Z) = T_{f(x)}(Y)$. See also Exercise 9.1.

Example 9.17 Let $\sigma : \mathbb{H}^2 \to \mathbb{R}$ be the restriction of the canonical submersion defined by $(x_1, x_2) \mapsto x_2$. Then our previous condition for transversality is not sufficient to guarantee that (9.2) is satisfied, as we will now explain. The derivative $d\sigma_{(x_1,x_2)} : \mathbb{R}^2 \to \mathbb{R}$ is just the projection onto the second factor. Hence it is surjective at every point (x_1, x_2). In particular, 0 is a regular value for σ. Let $Z := \{0\}$. Then we have $\sigma^{-1}(Z) = \{(x_1,x_2) \in \mathbb{R}^2 : x_2 = 0\} = \partial\mathbb{H}^2$. Since 0 is a regular value, we know that $\sigma^{-1}(Z)$ is a submanifold of dimension one. However, the boundary does not satisfy the condition of (9.2), since $\partial\left(\sigma^{-1}(Z)\right) = \emptyset$ whereas $\sigma^{-1}(Z) \cap \partial X = \partial\mathbb{H}^2 \neq \emptyset$.

The example shows that we need to impose an additional transversality condition on ∂f. We begin with regular values.

Theorem 9.18 (With boundary: fibres of regular values) *Let g be a smooth map of a smooth k-dimensional manifold X with boundary to a boundaryless n-manifold Y. Let $y \in Y$ be a regular value for* both $f : X \to Y$ *and $\partial f : \partial X \to Y$. Assume $f^{-1}(y) \neq \emptyset$. Then the preimage $f^{-1}(y)$ is a $(k - n)$-dimensional smooth manifold with boundary such that $\partial(f^{-1}(y)) = f^{-1}(y) \cap \partial X$.*

Remark 9.19 Recall that $d(\partial f)_x$ is the restriction of df_x to the subspace $T_x(\partial X) \subset T_x(X)$. Hence, if $x \in \partial X$ is a regular point for ∂f, then it is also a regular point for df_x. However, not every point in $f^{-1}(y)$ is in ∂X, and we also need the points $x \in \operatorname{Int}(X) \cap f^{-1}(y)$ to be regular points.

Proof of Theorem 9.18: Whether $f^{-1}(y)$ is a manifold with boundary is a local question. This means that it suffices that each point in $f^{-1}(y)$ has an open neighbourhood which is a manifold with boundary. So let $x \in X$ be a point with $f(x) = y$. After choosing suitable local parametrisations, we can assume that f is a map of the form $f : \mathbb{H}^k \to \mathbb{R}^n$. If x is an interior point in X, we can apply the Preimage Theorem 4.12 for boundaryless manifolds to show that $f^{-1}(y)$ is a smooth manifold without boundary in an open neighbourhood around x. Suppose now that $x \in \partial X$. Since f is smooth at x, there is an open subset $U \subset \mathbb{R}^k$ and a smooth map $F : U \to \mathbb{R}^n$ such that $F_{U\cap\mathbb{H}^k} = f_{U\cap\mathbb{H}^k}$. After possibly replacing U with a smaller open subset, we can assume that F has no critical points in U. Then $F^{-1}(y)$ is a smooth manifold by the Preimage Theorem 4.12 for boundaryless manifolds. We need to show that $f^{-1}(y) = F^{-1}(y) \cap \mathbb{H}^k$ is a manifold with boundary. In order to show this, we set $Z := F^{-1}(y)$ and let $\pi : Z \to \mathbb{R}$, $(x_1, \ldots, x_k) \mapsto x_k$ be the projection to the last coordinate. Then π is a smooth map, and Z is a smooth manifold with $Z \cap \mathbb{H}^k = \{x \in Z : \pi(x) \geq 0\}$. By Proposition 9.10, it now suffices to show that 0 is a regular value for π. So, assume there was an $x \in Z$ with both $\pi(x) = 0$, i.e., $x \in Z \cap \partial\mathbb{H}^k$, and $d\pi_x = 0$. We want to show that the assumption $d\pi_x = 0$ leads to a contradiction. To do so, first note that π is a linear map, and therefore $d\pi_x = \pi$. Thus, $d\pi_x = \pi : T_x(Z) \to \mathbb{R}$ being trivial, just means that the last coordinate of every vector in $T_x(X)$ is 0, i.e., if $d\pi_x = 0$ then $T_x(Z) \subset T_x(\partial\mathbb{H}^k) = \mathbb{R}^{k-1}$. Hence we want to show $T_x(Z) \not\subset \mathbb{R}^{k-1}$. By Proposition 4.13, the tangent space to $Z = F^{-1}(y)$ at x is the kernel of dF_s, i.e., $T_x(Z) = \text{Ker}\,(dF_s = df_s)$ where $df_x = dF_x$ by definition of df_x. We know that $d(\partial f)_x$ is the restriction of $df_x : \mathbb{R}^k \to \mathbb{R}$ to $\mathbb{R}^{k-1}$, i.e., $d(\partial f)_x = (df_x)_{|\mathbb{R}^{k-1}}$. Thus, if $T_x(Z) = \text{Ker}\,(df_x) \subseteq \mathbb{R}^{k-1}$, then $\text{Ker}\,(df_x) = \text{Ker}\,(d(\partial f)_x)$. Now, finally, we apply the assumption of regularity of y. Since y is a regular value of both f and ∂f, we know that both df_x and $d(\partial f)_x$ are surjective. This implies $\dim \text{Ker}\,(df_x) = k - n$ and $\dim \text{Ker}\,(d(\partial f)_x) = k - 1 - n$. This contradicts $\text{Ker}\,(df_x) = \text{Ker}\,(d(\partial f)_x)$. Thus, we must have $T_x(Z) = \text{Ker}\,(df_x) \not\subset \mathbb{R}^{k-1}$. Hence we know $d\pi_x \neq 0$ and therefore $d\pi_x$ is surjective, and 0 is a regular value, which proves the theorem. □

9.2.2 Transverse Preimages of Manifolds with Boundary

We can now extend Theorem 5.2 to smooth manifolds with boundary.

Theorem 9.20 (Preimages of manifolds with boundary) *Let X be a smooth manifold with boundary and let Y be a smooth manifold without boundary. Let $Z \subseteq Y$ be a submanifold without boundary. Let $f : X \to Y$ be a smooth map. Suppose that* both *$f : X \to Y$ and $\partial f : \partial X \to Y$ are transverse to Z in Y. Then the preimage $f^{-1}(Z)$ is a smooth manifold with boundary such that $\partial(f^{-1}(Z)) = f^{-1}(Z) \cap \partial X$. If $f^{-1}(Z) \neq \emptyset$, then the codimension of $f^{-1}(Z)$ in X equals the codimension of Z in Y.*

Proof: The restriction of f to the boundaryless manifold $\text{Int}(X)$ is transverse to Z. Hence, by Theorem 5.2 boundaryless manifolds, the intersection $f^{-1}(Z) \cap \text{Int}(X)$ with the interior of X is a manifold without boundary of correct codimension. Thus, it remains to examine $f^{-1}(Z)$ in a neighbourhood of a point $x \in f^{-1}(Z) \cap \partial X$. Let k denote the codimension of Z in Y. As in the boundaryless case, we can choose a submersion $h : W \to \mathbb{R}^k$ defined on an open neighbourhood W of $f(x)$ in Y to $\mathbb{R}^k$ such that $Z \cap W = h^{-1}(0)$. Then $h \circ f$ is defined in a neighbourhood V of x in X, and $f^{-1}(Z) \cap V = (h \circ f)^{-1}(0)$. Now let $\phi : U \to X$ be a local parametrisation around x

where U is an open subset of $\mathbb{H}^n$. We define $g := h \circ f \circ \phi\colon U \to \mathbb{R}^k$. Since $\phi\colon V \to \phi(V)$ is a diffeomorphism, the set $f^{-1}(Z)$ is a manifold with boundary in a neighbourhood of x if and only if $(f \circ \phi)^{-1}(Z) = g^{-1}(0)$ is a smooth manifold with boundary near $u = \phi^{-1}(x) \in \partial U$. Since both f and ∂f are transverse to Z by assumption, we know that 0 is a regular value of g. Hence we can apply Theorem 9.18, which finishes the proof. □

9.2.3 Sard's Theorem with Boundary

Finally, also Sard's Theorem has a version with boundary.

Theorem 9.21 (Sard's Theorem with boundary) *Let X be a smooth manifold with boundary and let Y be a smooth manifold without boundary. Let $f\colon X \to Y$ be a smooth map. Then the set of regular values for both f and ∂f is a dense subset in Y.*

Proof: For every point $x \in \partial X$, we have $d(\partial f)_x = (df_x)_{|T_x(\partial X)}$ as linear maps $T_x(\partial X) \to T_{f(x)}(Y)$. This implies that, if $d(\partial f)_x$ is surjective, then df_x is surjective. Hence if ∂f is regular at x, so is f. Thus, the set of $y \in Y$ which are regular value of both f and ∂f is the intersection of the set of regular values for $f\colon \mathrm{Int}(X) \to Y$ and the set of regular values for $\partial f\colon \partial X \to Y$. Since $\mathrm{Int}(X)$ and ∂X are manifolds without boundary, both sets of regular values are dense subsets of Y by Sard's Theorem 5.17. Since the intersection of two dense subsets is again a dense subset, this proves the theorem. □

9.3 Exercises

9.1 Let $X = \{(x, y) \in \mathbb{R}^2 : x \geq -1\}$, $Y = \mathbb{R}$, and $f\colon X \to Y$, $f(x, y) = x^2 + y^2$.
   (a) What is the boundary of X? Show that 1 is a regular value of f. Is 1 a regular value of ∂f?
   (b) Determine $f^{-1}(1)$, $\partial(f^{-1}(1))$, and $f^{-1}(1) \cap \partial X$. Why does the answer not contradict the Preimage Theorem for manifolds with boundary?

9.2 Let $U \subset \mathbb{R}^k$ and $V \subset \mathbb{H}^k$ be open neighbourhoods of 0. Show that there exists no diffeomorphism between V and U. Hint: Use the Inverse Function Theorem 3.4.

9.3 Let $f\colon X \to Y$ be a diffeomorphism of smooth manifolds with boundary. Show that ∂f maps ∂X diffeomorphically onto ∂Y. Hint: Use the Inverse Function Theorem 3.4.

9.4 Consider the smooth maps $F\colon \mathbb{R}\times[-1/2,1/2]\to\mathbb{R}^3$, given by $(t,s)\mapsto(\cos t,\, \sin t, s)$, and $G\colon \mathbb{R}\times[-1/2,1/2]\to\mathbb{R}^3$, given by

$$(t,s)\mapsto((1+s\cos(t/2))\cos t,(1+s\cos(t/2))\sin t, s\sin(t/2)).$$

We define X to be the image of F in $\mathbb{R}^3$, and $Y = G$ to be the image of G in $\mathbb{R}^3$.

(a) Show that X is a two-dimensional manifold with boundary whose boundary is diffeomorphic to the disjoint union of two copies of the unit circle. Convince yourself that X is a cylinder obtained by starting with a rectangular surface and then gluing two opposite edges together.

(b) Show that Y is a smooth two-dimensional manifold with boundary whose boundary is diffeomorphic to just one copy of the unit circle. Convince yourself that Y is a closed *Möbius band* obtained by starting with a rectangular surface and then gluing two opposite edges after twisting one edge once.

9.5 Suppose that X is a smooth manifold with boundary and $x\in\partial X$. Let $\phi\colon U\to X$ be a local parametrisation with $\phi(0)=x$, where U is an open subset of $\mathbb{H}^k$. Then $d\phi_0\colon\mathbb{R}^k\to T_x(X)$ is an isomorphism. Define the upper half-space $H_x(X)$ in $T_x(X)$ to be the image of $\mathbb{H}^k$ under $d\phi_0$, i.e., $H_x(X) := d\phi_0(\mathbb{H}^k)$.

(a) Prove that $H_x(X)$ does not depend on the choice of local parametrisation.

(b) Show that there are precisely two unit vectors in $T_x(X)$ that are orthogonal to $T_x(\partial X)$ and that one lies inside $H_x(X)$, the other outside. The one in $H_x(X)$ is called the inward unit normal vector to the boundary, and the other is the outward unit normal vector to the boundary. Denote the *outward unit normal vector* by $n(x)$.

(c) If $X\subset\mathbb{R}^N$, we consider $n(x)$ as an element in $\mathbb{R}^N$ and get a map $n\colon\partial X\to\mathbb{R}^N$. Show that the map n is smooth.

9.6 Recall that the graph of a map $f\colon X\to Y$ is the subset of $X\times Y$ defined by $\Gamma(f)=\{(x,f(x))\in X\times Y\colon x\in X\}$. We define $F\colon X\to\Gamma(f)$ by $F(x)=(x,f(x))$. We now assume that X and Y are smooth manifolds and f is a smooth map. We assume that X has a boundary and Y is without boundary.

(a) Show that F is a diffeomorphism and conclude that $\Gamma(f)$ is a smooth manifold with boundary.

(b) Show that $\partial\Gamma(F)=\{(x,f(x))\in\Gamma(f)\colon x\in\partial X\}$.

10

Brouwer Fixed Point Theorem

We are now equipped with the basic theory of smooth manifolds. Next, we set out to study some more sophisticated properties and tools. In this chapter, we will first classify all smooth one-dimensional manifolds. This classification will be useful for many arguments later in the book. As a first consequence, we prove a famous and important theorem on the existence of fixed points of certain self-maps due to Brouwer [9].

10.1 One-Manifolds

The following theorem gives us a complete list of smooth one-dimensional manifolds. Note that, since every manifold is the disjoint union of its connected components, it suffices to classify connected manifolds.

Theorem 10.1 (Classification of one-manifolds) *Let X be a connected smooth one-dimensional manifold.*

(a) If X is compact and without boundary, then X is diffeomorphic to $\mathbb{S}^1$.
(b) If X is compact with boundary, then X is diffeomorphic to $[0, 1]$.
(c) If X is non-compact without boundary, then X is diffeomorphic $(0, 1)$.
(d) If X is non-compact with boundary, then X is diffeomorphic to either $[0, 1)$ *or* $(0, 1]$.

Before we prove Theorem 10.1, we outline the idea of why the theorem should be true. We assume that X is non-empty. Assume that X is as in (a). Then each point has an open neighbourhood diffeomorphic to $(-1, 1)$. By compactness, finitely many such neighbourhoods $U_1, \ldots, U_n$ cover X. If n was equal to 1, then $X \cong (-1, 1)$. However, an open interval is not compact. Thus, there must be at least two neighbourhoods. Since X is connected, these

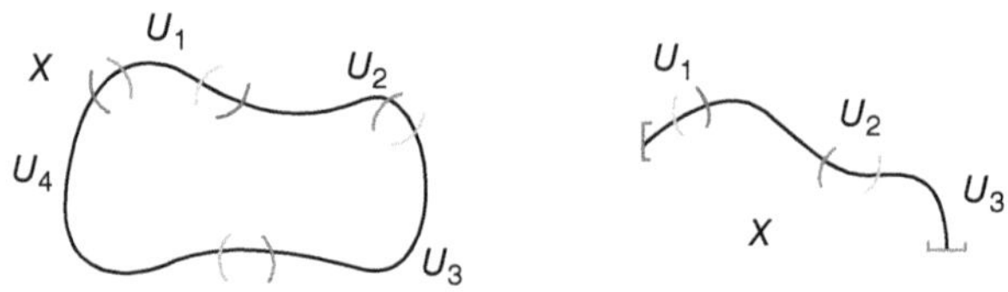

Figure 10.1 We can cover a compact one-manifold by finitely many open subsets which look like an open interval. Either we get back to the beginning and get a circle or we stop with another boundary point. This is the heuristic idea for the classification. There are some details to be straightened out, though.

two neighbourhoods must overlap. The union of these two intervals has to be either an open interval if they overlap on one side or a circle if they overlap on both sides. But if their union is an open interval, there has to be another chart, by the compactness of X. Since there are only finitely many U_i, we must eventually arrive at the situation where the neighbourhoods intersect on both sides and form a circle. See Figure 10.1. Then one can use this to construct a diffeomorphism to $\mathbb{S}^1$.

Now let X be as in (b). Since X has at least one boundary point, there must be an open neighbourhood of the boundary point. This neighbourhood must be diffeomorphic to $[a, b)$ for some $a < b$ in $\mathbb{R}$. Since this interval is not compact, there must be another open neighbourhood in X. This neighbourhood either intersects another boundary point which would yield us $X \cong [a, c]$ for some c, or it does not contain a boundary point. In the latter case, the union of the neighbourhoods is diffeomorphic to a half-open interval $[a, d)$ which is not compact. Hence there has to be another neighbourhood. Since X is compact, this process will end after finitely many steps when we eventually get that X is the union of neighbourhoods which is diffeomorphic to a closed interval. See again Figure 10.1. If X is as in (c) or (d), then the above processes may not terminate, and we obtain diffeomorphisms with open or half-open intervals. As we will see soon, the details of the proof are a little more involved.

We will now begin the proof of Theorem 10.1. We will follow Milnor's proof using arc-lengths given in the appendix of [35]. In this section, X will always denote a connected smooth one-dimensional manifold. First, we introduce some new terminology.

Definition 10.2 (Parametrisation by arc-length) Let I be an interval, i.e., a connected subset of $\mathbb{R}$. A smooth map $f : I \to X$ is called a *parametrisation by arc-length* if f maps I diffeomorphically onto an open subset of X and if the velocity vector $df_t(1) \in T_{f(t)}X \cong \mathbb{R}$ has length one for each $t \in I$, where $df_t : \mathbb{R} = T_t I \to T_{f(t)}X$.

Note that the definition implies that I can have a boundary only if X has a boundary. The proof of the classification theorem relies on the following lemma.

Lemma 10.3 (Number of components) *Let $f: I \to X$ and $g: J \to X$ be two parametrisations by arc-length. Then $f(I) \cap g(J)$ has at most two components. If it has only one component, then f can be extended to a parametrisation by arc-length of the union $f(I) \cup g(J)$. If it has two components, then X must be diffeomorphic to $\mathbb{S}^1$.*

Proof: We write $U := f(I) \cap g(J) \subset X$. Note that U is open in X since both $f(I)$ and $g(J)$ are open. In order to determine the number of components of U, we define the map $\kappa: f^{-1}(U) \xrightarrow{f_|} U \xrightarrow{(g_|)^{-1}} g^{-1}(U)$ where $f_|$ and $g_|$ denote the restrictions of f and g, respectively, such that the maps are defined. By definition of a parametrisation by arc-length, κ is a local diffeomorphism with derivative equal to ± 1 and maps open subsets of I diffeomorphically onto open subsets of J. In particular, it is an open map. We consider the graph of κ, denoted by $\Gamma \subset I \times J$, given by $\Gamma = \{(s,t) \in I \times J: f(s) = g(t)\}$. Since Γ is the preimage of the diagonal Δ in $X \times X$ under the continuous map $(f,g): I \times J \to X \times X$ and since Δ is closed in $X \times X$, Γ is a closed subset of $I \times J$. Moreover, $d\kappa_{(s,t)}$ takes values in the discrete set $\{\pm 1\}$. Hence $d\kappa_{(s,t)}$ is locally constant and is therefore constant on each connected component of Γ by Lemma 1.30. Thus, thinking of $I \times J$ as a rectangle with sides I and J, Γ consists of line segments of slope either $+1$ or -1. We need to understand how many such line segments there are. Note that the image of κ equals the projection of the components of Γ to J. Let K be a connected component of Γ.

- Claim: The endpoints of K lie on the boundary of $I \times J$.

Proof of the claim: Since Γ is closed, K is closed in $I \times J$ as well. Assume that K ends in the interior of $I \times J$ with endpoint a. Then the projection of K onto J has endpoint $\kappa(a)$. But since κ is a local diffeomorphism, the image of κ contains an open neighbourhood around $\kappa(a)$. Hence $\kappa(a)$ cannot be an endpoint of a subinterval of J. This proves the claim. □

Moreover, since κ is one-to-one, each of the four edges of the rectangle $I \times J$ can contain at most one endpoint of a component of Γ. Since each component has two endpoints, Γ can have at most two components. This proves the first assertion in the lemma. Now we show the second assertion by considering the two cases.

First, we assume Γ has one component. Then κ has a constant slope and Γ is just a line segment in the rectangle $I \times J$. Hence we can extend κ to a linear

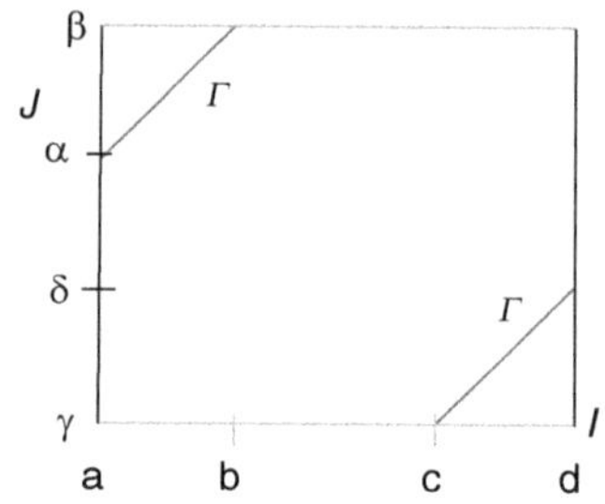

Figure 10.2 The case when Γ has two components. The slope is the same for both components, here $+1$. All four edges of the rectangle $I \times J$ are hit by Γ.

map $L\colon \mathbb{R} \to \mathbb{R}$. Since both f, g, and L are smooth and have equal derivatives, we can glue f and $g \circ L_{L^{-1}(J)}$ together to get an extension

$$F\colon I \cup L^{-1}(J) \to f(I) \cup g(J),$$

which is a parametrisation by arc-length of the union $f(I) \cup g(J)$. This proves the second assertion of the lemma in the first case.

Second, we assume Γ has two components. Then all four edges are hit by the components. This implies that the slope of both components must be the same. Let us assume the slope is $+1$. The case with slope -1 is analogous. Let $a < d$ be the endpoints of I and $\gamma < \beta$ be the endpoints of J. Let $\alpha \in J$ be the starting point and $b \in I$ be the endpoint of the first component of Γ, and let $c \in I$ be the starting point and $\delta \in J$ be the endpoint of the second component. See Figure 10.2.

After possibly translating the interval J in $\mathbb{R}$, we can assume that $\gamma = c$ and $\delta = d$. By the previous arguments, we then have $a < b \le c = \gamma < d = \delta \le \alpha < \beta$. Now we can define a map $h\colon \mathbb{S}^1 \to X$ as follows. We write $\theta := 2\pi t/(\alpha - a)$ and define

$$h(\cos\theta, \sin\theta) := \begin{cases} f(t) & \text{for } a < t < d \\ g(t) & \text{for } c < t < \beta. \end{cases}$$

We need to check that h is well defined. First, if t varies between a and β, θ takes all values in $[0, 2\pi]$, since $\alpha < \beta$. Thus h is defined on all points of $\mathbb{S}^1$. Second, we need to check what happens on the overlaps. We have two intervals on which h is defined by two a priori different terms: for t between $c = \gamma$ and $d = \delta$, and for t between α and β. However, by the definition of Γ, we have $f(c+r) = g(\gamma+r)$ for $0 \le r \le d-c = \delta-\gamma$. Similarly, we have $f(a+r) = g(\alpha+r)$ for $0 \le r \le b-a = \beta-\alpha$. Hence h is well defined and, since f and g are smooth, h is smooth. Moreover, h is an open map, since f and g are. Thus

$h(\mathbb{S}^1)$ is an open subset of X. Since h is continuous and $\mathbb{S}^1$ is compact, $h(\mathbb{S}^1)$ is compact by Lemma 1.13. Thus, since X is Hausdorff, $h(\mathbb{S}^1)$ is a closed subset of X by Proposition 1.21. Since X is connected, this implies that $h(\mathbb{S}^1) = X$. Since f and g are local diffeomorphisms, so is h. Thus, h is a diffeomorphism. This finishes the proof of the second assertion of the lemma. □

Proof of the Classification Theorem 10.1: Let $f: I \to X$ be a parametrisation by arc-length. If there exists a parametrisation by arc-length $g: J \to X$ with $I \cap J \neq \emptyset$, then we can glue these parametrisations together to get a parametrisation by arc-length $I \cup J \to X$ by Lemma 10.3. Hence, after possibly extending f as far as possible to the left and then as far as possible to the right, we can assume that f is maximal in the sense that it cannot be extended over any larger interval as a parametrisation by arc-length. If X is not diffeomorphic to $\mathbb{S}^1$, we will prove that f is surjective. This will imply that f is a diffeomorphism, since we already know it is injective and a local diffeomorphism. If the open set $f(I) \subset X$ was not all of X, then there would be a limit point x of $f(I)$ in $X \setminus f(I)$. Then we can parametrise a neighbourhood of x by arc-length. Lemma 10.3 then implies that f can be extended over a larger interval. This contradicts the assumption that f is maximal. The assertion of the theorem now follows from the comparison with the different types of intervals in $\mathbb{R}$. □

10.2 Boundaries and Retractions

As a first consequence of the classification of one-manifolds we can deduce the following fact about the boundary of a one-manifold.

Proposition 10.4 (Boundary of one-manifolds) *The boundary of a compact one-dimensional manifold with boundary consists of an even number of points.*

Proof: A compact one-manifold with boundary X is the disjoint union of finitely many connected components. By Theorem 10.1, each component is diffeomorphic to a copy of $[0, 1]$. Hence the boundary of each component consists of two points. The boundary of X consists of these finitely many pairs of points. □

This result has an important consequence for the possible boundaries of one-manifolds. First, we introduce some new terminology.

Definition 10.5 (Retraction) Let X be a smooth manifold and $Z \subset X$ be a submanifold. A *retraction* onto Z is a smooth map $g: X \to Z$ such that $g_{|Z}$ is the identity on Z.

There is an important restriction on the existence of such retractions for manifolds with boundary.

Proposition 10.6 (No retraction onto boundary) *Let X be a compact smooth manifold with boundary. Then there is no retraction of X onto its boundary.*

Proof: Suppose there is such a smooth map $g\colon X \to \partial X$ such that the restriction $\partial g\colon \partial X \to \partial X$ is the identity. By Sard's Theorem 9.21, we can choose a regular value $z \in \partial X$ of g. Since ∂g is the identity, all values in ∂X are regular for ∂g. Hence z is regular for both g and ∂g. By the Preimage Theorem 9.20, we know that $g^{-1}(z)$ is a submanifold of X with boundary

$$\partial(g^{-1}(z)) = g^{-1}(z) \cap \partial X. \tag{10.1}$$

Moreover, the codimension of $g^{-1}(z)$ in X equals the codimension of $\{z\}$ in ∂X, namely $\dim X - 1$ as $\{z\}$ has dimension 0. Hence $g^{-1}(z)$ is one-dimensional. Since it is a closed subset in the compact manifold X, it is also compact. Thus, if g is a retraction onto ∂X, then, by Proposition 10.4, $\#\partial\left(g^{-1}(z)\right)$ must be even. Let $i\colon \partial X \hookrightarrow X$ denote the inclusion of ∂X into X. By definition of ∂g as the restriction of g to ∂X, we have for just set-theoretic reasons

$$(\partial g)^{-1}(z) = (g_{|\partial X})^{-1}(z) = (g \circ i)^{-1}(z) = i^{-1}(g(z)) = g^{-1}(z) \cap \partial X.$$

Hence, using (10.1), we have $(\partial g)^{-1}(z) = \partial(g^{-1}(z))$. Since $\partial g = g_{|\partial X} = \mathrm{Id}_{\partial X}$, this implies $\partial(g^{-1}(z)) = (\partial g)^{-1}(z) = \{z\}$. In particular, this shows that $\partial(g^{-1}(z))$ consists of an odd number of points. However, this contradicts the conclusion in (10.1). Hence the retraction g cannot exist. □

10.3 Brouwer Fixed Point Theorem

We will now show that Proposition 10.6 has the following important consequence.

Theorem 10.7 (Brouwer Fixed Point Theorem) *Let $f\colon \mathbb{D}^n \to \mathbb{D}^n$ be a continuous map of the closed unit ball $\mathbb{D}^n = \{x \in \mathbb{R}^n \colon |x| \leq 1\} \subset \mathbb{R}^n$ into itself. Then f has a* fixed point, *i.e., there is an $x \in \mathbb{D}^n$ with $f(x) = x$.*

Before we prove the theorem, we consider the assertion in dimension one.

Remark 10.8 (Dimension one) We have seen Theorem 10.7 for $n = 1$ in calculus, namely, every continuous map $f\colon [0,1] \to [0,1]$ has a fixed point. We can prove this fact as follows. We define the function $g(x) = f(x) - x$,

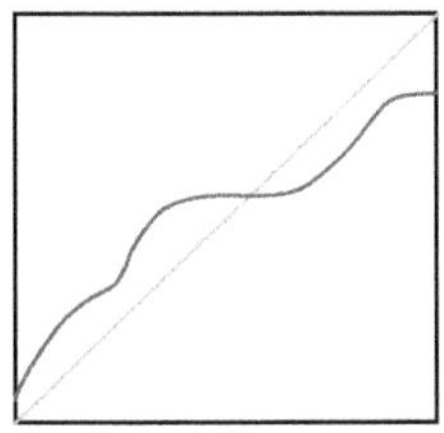

Figure 10.3 In dimension one, this is the Intermediate Value Theorem. The graph has to cross the diagonal which consists of fixed points of f.

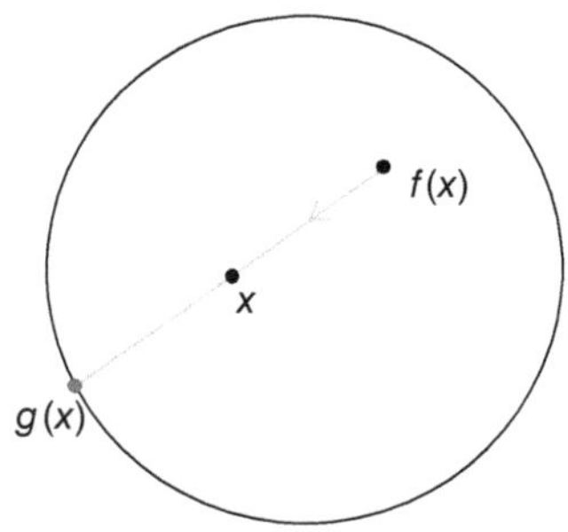

Figure 10.4 We construct g by extending the line from $f(x)$ to x until we hit the boundary. If this map existed for all x, then we would have found a retraction onto the boundary. This would be a contradiction to Proposition 10.6.

which is a continuous map from $[0, 1]$ to itself. We have $g(0) \geq 0$ and $g(1) \leq 0$ since $f(0) \geq 0$ and $f(1) \leq 1$. If $g(0) = 0$ or $g(1) = 1$, we are done. And if $g(0) > 0$ and $g(1) < 1$, then the Intermediate Value Theorem implies that there is an $x_0 \in (0, 1)$ with $g(x_0) = 0$, i.e., $f(x_0) = x_0$. See Figure 10.3.

We first prove the theorem for smooth maps and will then deduce Theorem 10.7 from the smooth case.

Theorem 10.9 (Brouwer Fixed Point Theorem for smooth maps) *Let $f\colon \mathbb{D}^n \to \mathbb{D}^n$ be a smooth map. Then f has a fixed point.*

Proof of Theorem 10.9: Suppose that there exists an f without fixed points. We will show that such an f would allow us to construct a retraction $g\colon \mathbb{D}^n \to \partial\mathbb{D}^n$. However, since $\mathbb{D}^n$ is compact, we have proved in Proposition 10.6 that such a retraction cannot exist. So suppose $f(x) \neq x$ for all $x \in \mathbb{D}^n$. Then, for every $x \in \mathbb{D}^n$, the two distinct points x and $f(x)$ determine a line. Let $g(x)$ be the point where the line segment starting at $f(x)$ and passing through x hits the boundary $\partial\mathbb{D}^n$. This defines a map $g\colon \mathbb{D}^n \to \partial\mathbb{D}^n$. See Figure 10.4.

However, if $x \in \partial\mathbb{D}^n$, then $g(x) = x$ by construction of g. Hence $g\colon \mathbb{D}^n \to \partial\mathbb{D}^n$ is the identity on $\partial\mathbb{D}^n$. To show that g is a retraction onto the boundary, it remains to show that g is smooth. So let us describe $g(x)$ more explicitly. As a point on the line from $f(x)$ to x, $g(x)$ can be written as $g(x) = x + t(x) \cdot \mathbf{v}$ with $\mathbf{v} := \frac{x-f(x)}{|x-f(x)|}$ for some real number $t = t(x)$. Note that, since we assume $x \neq f(x)$, the vector $\mathbf{v}$ is always defined. In fact, it is the unit vector pointing from $f(x)$ to x. Moreover, since f is smooth, $\mathbf{v}$ depends smoothly on x. We now calculate $t(x)$ and show that $t(x)$ depends smoothly on x. Since $g(x)$ is a point on the boundary of $\mathbb{D}^n$, we know $|g(x)| = 1$, and $t = t(x)$ is determined by the equation $1 = |g(x)|^2 = (x + t\mathbf{v}) \cdot (x + t\mathbf{v}) = x \cdot x + 2tx \cdot \mathbf{v} + t^2\mathbf{v} \cdot \mathbf{v}$ or, equivalently,

$$0 = (\mathbf{v} \cdot \mathbf{v})t^2 + (2x \cdot \mathbf{v})t + x \cdot x - 1. \tag{10.2}$$

By definition of $\mathbf{v}$, we know $\mathbf{v} \cdot \mathbf{v} = |\mathbf{v}|^2 = 1$. Since $\mathbf{v}$ points from $f(x)$ to x, we know that t must be positive. Now we just need to find the positive solution of the quadratic equation (10.2) in t and get

$$t = -x \cdot \mathbf{v} + \sqrt{(x \cdot \mathbf{v})^2 - x \cdot x + 1},$$

where $(x \cdot \mathbf{v})^2 - x \cdot x + 1$ is positive, since $x \cdot x = |x|^2 \leq 1$ and $(x \cdot \mathbf{v})^2 > 0$. Since the scalar product and square root involved depend smoothly on x, we see that $t = t(x)$ depends smoothly on x. Hence g is smooth. This finishes the proof of Theorem 10.9. □

Remark 10.10 (Dimension one) Note that, for $n = 1$, in the above proof, we would have constructed a map $g\colon [-1, 1] \to \{-1, 1\}$ which would send -1 to -1 and 1 to 1. Such a map cannot be continuous, since $[-1, 1]$ is connected.

Next, we will prove Brouwer's theorem for continuous maps. The idea is to deduce Theorem 10.7 from Theorem 10.9 by approximating f by a smooth map.

Lemma 10.11 (Approximation by a smooth map) *Let $f\colon \mathbb{D}^n \to \mathbb{D}^n$ be a continuous map. For every $\varepsilon > 0$, there exists a smooth map $P\colon \mathbb{D}^n \to \mathbb{D}^n$ such that $|P(x) - f(x)| < 2\varepsilon$ for all $x \in \mathbb{D}^n$.*

Proof: Since $\mathbb{D}^n$ is compact, we can apply Weierstrass' Approximation Theorem, which says the following: given $\varepsilon > 0$, there is a polynomial function $Q\colon \mathbb{D}^n \to \mathbb{R}^n$ with $|Q(x) - f(x)| < \varepsilon$ for all $x \in \mathbb{D}^n$. However, Q may send points in $\mathbb{D}^n$ to points outside of $\mathbb{D}^n$. In order to remedy this defect, we replace Q with $P(x) := \frac{Q(x)}{1+\varepsilon}$. The map P is still a polynomial and hence smooth. Since $|f(x)| \leq 1$, P satisfies $(1+\varepsilon)|P(x)| \leq |Q(x) - f(x)| + |f(x)| < \varepsilon + 1$

where we apply the triangle inequality. Hence $|P(x)| \leq 1$ and P is a map $\mathbb{D}^n \to \mathbb{D}^n$. Moreover, we have

$$\begin{aligned}(1+\varepsilon)|P(x)-f(x)| &= |Q(x)-(1+\varepsilon)f(x)| \\ &= |Q(x)-f(x)+\varepsilon f(x)| \\ &\leq |Q(x)-f(x)|+\varepsilon|f(x)| < 2\varepsilon,\end{aligned}$$

where we use that $|f(x)| \leq 1$. Since $1+\varepsilon > 1$, this shows $|P(x)-f(x)| < 2\varepsilon$ for all $x \in \mathbb{D}^n$. This proves the lemma. □

Proof of Theorem 10.7: We suppose that $f(x) \neq x$ for all $x \in \mathbb{D}^n$. Since $\mathbb{D}^n$ is compact, then the continuous function $\mathbb{D}^n \to \mathbb{R}$, $x \mapsto |f(x)-x|$, has a minimum μ. Since $f(x) \neq x$ for all x, we must have $\mu > 0$. Now, for $\varepsilon = \mu/2$, we can find, by Lemma 10.11, a smooth map $P\colon \mathbb{D}^n \to \mathbb{D}^n$ such that $|P(x)-f(x)| < \mu$ for all $x \in \mathbb{D}^n$. However, since $|f(x)-x| \geq \mu$ for all $x \in \mathbb{D}^n$, the triangle inequality yields

$$\begin{aligned}\mu \leq |f(x)-x| &= |f(x)-P(x)+P(x)-x| \\ &\leq |f(x)-P(x)|+|P(x)-x|.\end{aligned}$$

This implies that $|P(x)-x| > 0$, and therefore $P(x) \neq x$ for all $x \in \mathbb{D}^n$. Hence P is a smooth map from $\mathbb{D}^n$ to itself without a fixed point. This contradicts Theorem 10.9 and completes the proof. □

Example 10.12 (Counter-example on an open ball) The assertion of Theorem 10.9 is not true for an open ball. Let $\mathbb{B}^n_1(0) = \{x \in \mathbb{R}^n : |x| < 1\}$ be the open ball of radius one around the origin in $\mathbb{R}^n$. Recall the diffeomorphism $\varphi\colon \mathbb{B}^n_1(0) \to \mathbb{R}^n$, $x \mapsto \frac{x}{\sqrt{1-|x|^2}}$, from Exercise 2.3. Let $T\colon \mathbb{R}^n \to \mathbb{R}^n$ denote the translation defined by $x \mapsto x+1$. The composition $f := \varphi^{-1} \circ T \circ \varphi\colon \mathbb{B}^n_1(0) \to \mathbb{B}^n_1(0)$ is a smooth map. We then have $\varphi^{-1}(T(\varphi(x))) = x$, i.e., $T(\varphi(x)) = \varphi(x)$. Thus, if f had a fixed point, T would have a fixed point. However, a non-trivial translation does not have a fixed point. Hence, f cannot have a fixed point.

10.4 Exercises

10.1 Prove the Theorem of *Perron–Frobenius*: let A be an $(n \times n)$-matrix A with only non-negative entries. Then A has a real non-negative eigenvalue. Hint: We can assume that A is invertible, otherwise 0 is an eigenvalue. Let A also denote the associated linear map of $\mathbb{R}^n$, and consider the map $v \to Av/|Av|$ restricted to $\mathbb{S}^{n-1} \to \mathbb{S}^{n-1}$. Show that this maps the first quadrant $Q = \{(x_l, \ldots, x_n) \in \mathbb{S}^{n-1} : \text{ all } x_i \geq 0\}$ into itself. Now use the fact that there is a homeomorphism $\mathbb{D}^{n-1} \to Q$ to get a continuous map $\mathbb{D}^{n-1} \to \mathbb{D}^{n-1}$.

10.2 In this exercise, we use the following fact: a space X is simply connected if it is connected and every continuous map $f: \mathbb{S}^1 \to X$ can be extended to a continuous map $F: \mathbb{D}^2 \to X$ such that $F_{|\mathbb{S}^1} = f$. We are now going to show that $X := \mathbb{R}^2 \setminus \{(0,0)\}$ is not simply connected using Brouwer's Fixed Point Theorem 10.7. This implies that $\mathbb{S}^1$ is not simply connected.
   (a) Consider the antipodal map $a: \mathbb{S}^1 \to \mathbb{S}^1$ which sends p to $-p$. Composing with the inclusion $\mathbb{S}^1 \subset X = \mathbb{R}^2 \setminus \{(0,0)\}$, we consider a as a map $f: \mathbb{S}^1 \to X$. Show that, if X was simply connected, then this would induce a map $g: \mathbb{D}^2 \to X \to \mathbb{S}^1 \hookrightarrow \mathbb{D}^2$.
   (b) Still assuming $X = \mathbb{R}^2 \setminus \{(0,0)\}$ was simply connected, show that g does not have a fixed point.
   (c) Conclude that $X = \mathbb{R}^2 \setminus \{(0,0)\}$ is not simply connected.

10.3 Deduce from Exercise 10.2 that the composed map

$$\iota: \mathbb{S}^1 \xrightarrow{\text{id}} \mathbb{S}^1 \subset \mathbb{R}^2 \setminus \{(0,0)\}, \; p \mapsto p,$$

is not homotopic to the constant map $c: \mathbb{S}^1 \to \mathbb{R}^2 \setminus \{(0,0)\}$ sending all points to $(1,0)$. Hint: Use that $\mathbb{R}^2 \setminus \{(0,0)\}$ is not contractible, since it is not simply connected.

11
The Brouwer Degree Modulo 2

We will now define an important invariant for smooth maps, the Brouwer degree. Despite its rather simple definition, the degree is a surprisingly powerful tool. In this chapter, we define the degree with values in $\mathbb{Z}/2$, the integers modulo 2. This has the advantage that the degree is defined for a large class of maps and the definition only requires what we have learned so far. We then study some interesting applications of the degree modulo 2. In particular, we introduce winding numbers modulo 2 and prove the Borsuk–Ulam Theorem in Section 11.4, and we introduce the Hopf invariant modulo 2 and compute it for the Hopf map in Section 11.5. We will define an integer-valued version of the Brouwer degree in Section 15.

11.1 The Brouwer Degree of Maps Modulo 2

First, we outline the idea. Let $f\colon X \to Y$ be a smooth map. We will always assume in this chapter that X is compact and without boundary. Moreover, we will assume that X and Y have the same dimension. The idea for the degree is rather simple: pick a regular value y for f. By the Stack of Records Theorem 4.28, the fibre $f^{-1}(y)$ is a finite set. Hence we may count how many points there are in $f^{-1}(y)$. Since we also know that $\#f^{-1}(y)$ is locally constant, we may want to use the number $\#f^{-1}(y)$ to characterise f. However, there are two caveats. First, the number $\#f^{-1}(y)$ may not be constant for all regular values. For example, when f is not surjective all $y \notin \operatorname{Im} f$ are regular, with $\#f^{-1}(y) = 0$, but there may also be regular values in $\operatorname{Im} f$ with $\#f^{-1}(y) > 0$. Second, since describing maps and manifolds up to homotopy seems to be a much more promising task, we would like to have a number that only depends on the homotopy class of f. Luckily, we can remedy both defects of our initial idea and construct an invariant which only depends on the homotopy class of

f and not on the choice of regular value. The price we pay is that we only get a number modulo 2, i.e., an invariant which takes only the value 0 or 1. This may seem to be quite a coarse invariant. We will see, however, that the mod 2 degree is rather powerful after all.

11.1.1 The Homotopy Lemma Mod 2

The first step for the construction is to understand how the number of points in the fibre over a regular value depends on the homotopy class of a map.

Lemma 11.1 (Homotopy Lemma mod 2) *Let X and Y be smooth manifolds. We assume that* $\dim X = \dim Y$, *Y is connected and that X is compact and without boundary. Let $f, g\colon X \to Y$ be two smooth maps. Assume f and g are smoothly homotopic. If $y \in Y$ is a regular value for both f and g, then* $\#f^{-1}(y) \equiv \#g^{-1}(y)$ *modulo 2.*

Proof: Let $F\colon X \times [0,1] \to Y$ be a smooth homotopy between $F_0 = f$ and $F_1 = g$. First, we suppose that y is a regular value for F with $F^{-1}(y) \neq \emptyset$. By the Preimage Theorem 9.18 with boundary, this implies $F^{-1}(y)$ is a submanifold with boundary of $X \times [0,1]$ of dimension $\dim(X \times [0,1]) - \dim Y = 1$. Moreover, the boundary of $F^{-1}(y)$ is

$$\begin{aligned}\partial F^{-1}(y) &= F^{-1}(y) \cap \partial(X \times [0,1]) = (F_0^{-1}(y) \times \{0\}) \cup (F_1^{-1}(y) \times \{1\}) \\ &= (f^{-1}(y) \times \{0\}) \cup (g^{-1}(y) \times \{1\}).\end{aligned}$$

Since X is compact, $X \times [0,1]$ is compact by Proposition 1.15. Since $\{y\}$ is a closed subset of Y, $F^{-1}(y)$ is closed in $X \times [0,1]$. This implies that $F^{-1}(y)$ is compact by Lemma 1.14. By Proposition 10.4, this implies that $\partial F^{-1}(y)$ consists of an even number of points. Thus, computing modulo 2, we get $\#f^{-1}(y) + \#g^{-1}(y) \equiv 0$ and hence $\#f^{-1}(y) \equiv \#g^{-1}(y)$. Now suppose that y is not a regular value of F but still a regular value for f and g. By Proposition 4.30, the functions $y' \mapsto \#f^{-1}(y')$ and $y' \mapsto \#g^{-1}(y')$ are both locally constant on the set of regular values $y' \in Y$ of f and g, respectively. Thus there is an open neighbourhood $U_1 \subset Y$ of y consisting only of regular values of f so that $\#f^{-1}(y') = \#f^{-1}(y)$ for all $y' \in U_1$. Similarly, there is an open neighbourhood $U_2 \subset Y$ of y consisting only of regular values of g so that $\#g^{-1}(y') = \#g^{-1}(y)$ for all $y' \in U_2$. The intersection $U_1 \cap U_2$ is an open subset of Y. Hence, by Sard's Theorem 9.21, there is a regular value z of F in $U_1 \cap U_2$. By the first case, we have $\#f^{-1}(z) = \#g^{-1}(z)$ modulo 2 and we get

$$\#f^{-1}(y) = \#f^{-1}(z) \equiv \#g^{-1}(z) = g^{-1}(y) \mod 2. \qquad \square$$

11.1.2 The Isotopy Lemma

In order to make our construction independent of the choice of a regular value, we will need a special type of homotopy which preserves more information than homotopies in general.

Definition 11.2 (Isotopy) An *isotopy* is a smooth homotopy h_t in which each map h_t is a diffeomorphism. We say that two diffeomorphisms are *isotopic* if they can be joined by an isotopy. An isotopy is called *compactly supported* if the maps h_t are all equal to the identity map when restricted to the complement of some fixed compact set.

The following result will allow us to move points on connected manifolds via a family of diffeomorphisms. The fact that every map in the homotopy family is a diffeomorphism makes it much easier to keep track of the fibres of maps.

Theorem 11.3 (Isotopy Theorem) *Let Y be a connected smooth manifold. Given any two points y and z in Y, there exists a diffeomorphism $h\colon Y \to Y$ such that $h(y) = z$ and h is isotopic to the identity. Moreover, the isotopy can be chosen to be compactly supported.*

We provide a proof of Theorem 11.3 and an extension to finitely many points in Section 11.3. Using the Homotopy Lemma 11.1 and the Isotopy Theorem 11.3, we can now prove the following key result.

Theorem 11.4 *Let $f\colon X \to Y$ be a smooth map from a compact manifold X to a connected manifold Y and $\dim X = \dim Y$. Assume y and z are regular values of f. Then $\#f^{-1}(y) \equiv \#f^{-1}(z)$ modulo 2. This common number $\#f^{-1}(y)$ modulo 2 only depends on the homotopy class of f.*

Proof: Let y and z be two regular values of f. Since Y is connected, we can apply the Isotopy Theorem 11.3 and find a diffeomorphism $h\colon Y \to Y$ which is isotopic to the identity and with $h(y) = z$. Since h is a diffeomorphism, z is a regular value of the composite $h \circ f$. Since h is homotopic to the identity, the composite $h \circ f$ is homotopic to f. Hence the Homotopy Lemma 11.1 implies that $\#(h \circ f)^{-1}(z) \equiv \#f^{-1}(z)$ modulo 2. The left-hand fibre is given by

$$(h \circ f)^{-1}(z) = f^{-1}(h^{-1}(z)) = f^{-1}(y),$$

where we use that h is injective, i.e., $h^{-1}(z) = \{y\}$. In particular, we have $\#(h \circ f)^{-1}(z) = \#f^{-1}(y)$. Thus we can conclude $\#f^{-1}(y) \equiv \#f^{-1}(z)$ modulo 2. To prove the second assertion, assume that $g\colon X \to Y$ is a smooth map which is smoothly homotopic to f. By Sard's Theorem 9.21, there exists an element $y \in Y$ which is a regular value for both f and g. By the Homotopy Lemma 11.1, this implies $\#f^{-1}(y) \equiv \#g^{-1}(y)$ modulo 2. □

Remark 11.5 We emphasise that Theorem 11.4 only shows that the residue class modulo 2 of $\#f^{-1}(y)$ is constant for all regular values and only depends on the smooth homotopy class of f. The number $\#f^{-1}(y)$ itself, however, may not be constant. For example, if $\deg_2(f) = 0$, the fibre $f^{-1}(y)$ can be empty for some regular values, while for others it can be non-empty and consist of an even number of points. All we know is that the parity of $\#f^{-1}(y)$ does not change, but the number $\#f^{-1}(y)$ itself may vary.

11.1.3 A Well-Defined Invariant

As a consequence of the above observations, we conclude that the number $\#f^{-1}(y)$ is independent of the choice of y and only depends on the smooth homotopy class of f. Hence we make the following definition.

Definition 11.6 (Degree modulo 2) Let X be a compact smooth manifold X without boundary and Y a connected smooth manifold with $\dim X = \dim Y$. Then the *mod* 2 *degree of* f, denoted $\deg_2(f)$, is the mod 2 residue class of $\#f^{-1}(y)$ for any regular value y of f.

While the values of $\deg_2$ are rather restricted, $\deg_2$ is a powerful invariant nevertheless. Moreover, it is easy to calculate, as we will now see in some first examples. We keep the assumptions on X and Y of Definition 11.6.

Example 11.7 (Constant maps have degree zero) Let $f\colon X \to Y$ be a constant map with value y_0. If $\dim Y \geq 1$, then there is a point $y \in Y$ with $y \neq y_0$. Since this y is a regular value for f with $f^{-1}(y) = \emptyset$, we have $\deg(f) = 0$.

Example 11.8 (Mod 2 degree of a diffeomorphism) Let $f\colon X \to Y$ be a diffeomorphism. Then we have $\#f^{-1}(y) = 1$ for every $y \in Y$, and $\deg_2(f) = 1$. For example, the reflection $r_i\colon \mathbb{S}^n \to \mathbb{S}^n$, $r_i(x_1, \ldots, x_{n+1}) = (x_1, \ldots, -x_i, \ldots, x_{n+1})$ and the antipodal map $a\colon \mathbb{S}^n \to \mathbb{S}^n$ have $\deg_2(r_i) = \deg_2(a) = 1$.

Example 11.9 (Degree of self-maps of $\mathbb{S}^1$) For $q \in \mathbb{Z}$, the restriction of complex multiplication $z \to z^q$ defines a smooth map $f_q\colon \mathbb{S}^1 \to \mathbb{S}^1$. For $q \neq 0$, we have $\#f_q^{-1}(1) = q$ and 1 is a regular value. We have $\deg_2(f_q) = q$ modulo 2 for all $q \in \mathbb{Z}$. We learn from this example that $\deg_2$ cannot distinguish between, for example, $z \mapsto z^{-3}$ and $z \mapsto z^5$, even though these maps are not homotopic.

We now prove a useful property of the degree that we will take advantage of in the following sections.

Lemma 11.10 ($\deg_2$ is additive) *Let X and Z be compact smooth manifolds without boundary, and let Y be a connected smooth manifold with* $\dim X = \dim Z = \dim Y$. *Let $f: X \sqcup Z \to Y$ be a smooth map. Then* $\deg_2$ *is additive, i.e., we have* $\deg_2(f) = \deg_2(f_{|X}) + \deg_2(f_{|Z})$.

Proof: Let y be a regular value for f. Then $\deg_2(f) = \#f^{-1}(y)$, and the number of elements in $\#f^{-1}(y)$ is the sum of the number of elements lying in X and Z, respectively. These two numbers are $\deg_2(f_{|X})$ and $\deg_2(f_{|Z})$, respectively. □

11.2 First Applications of the Mod 2 Degree

There are many applications of the invariant $\deg_2$. In this section, we have a glimpse at a few of them. We will see further applications later in this chapter and in the exercises.

11.2.1 Obstructions via the Degree

First, we look at some cases where the degree works as an *obstruction*, i.e., it allows us to show that certain structures do not exist.

Theorem 11.11 *A compact manifold without boundary of dimension at least one is not smoothly contractible.*

Proof: The identity map on X has degree one, since $\#(\mathrm{id}^{-1}(y)) = 1$ for every $y \in X$. Thus $\deg_2(\mathrm{id}) = 1$. Now let $c: X \to X$ be a constant map with value $x_0 \in X$. Then every $x \neq x_0$ is a regular value for c. Since $\dim X \geq 1$, there are points $x \neq x_0$. Thus $\deg_2(c) = 0$. Hence we can conclude $\deg_2(\mathrm{id}) \neq \deg_2(c)$. Since $\deg_2$ is invariant under smooth homotopy, the identity is not smoothly homotopic to a constant map, which proves the theorem. □

As a consequence of this theorem, we deduce in Exercise 11.5 a special case of the assertion of Proposition 10.6.

Remark 11.12 We note that an empty boundary is crucial for the assertion of Theorem 11.11. The closed interval $[0, 1]$ provides a simple example of a smooth manifold of dimension one with boundary which is smoothly contractible. The map $H\colon [0,1] \times [0,1] \to [0,1]$ given by $H(s,t) = s \cdot t$ is a smooth homotopy between the constant map with value zero and the identity map of $[0,1]$.

Remark 11.13 In Section 12.2, we will strengthen the assertion of Theorem 11.11. See Theorem 12.29. There, we will show that smooth maps are smoothly homotopic if and only if they are homotopic, i.e., via a continuous homotopy. This will imply that a compact manifold without boundary of dimension at least one is not contractible.

The degree also provides a computable obstruction to the existence of an extension of a given map.

Theorem 11.14 (Boundary Theorem for $\deg_2$) *Let Y be a connected smooth manifold without boundary. Let $X = \partial W$ be the boundary for some compact manifold W and $\dim X = \dim Y$. Assume that $f\colon X \to Y$ is a smooth map which can be extended to all of W, i.e., there is a smooth map $F\colon W \to Y$ with $F_{|\partial W} = f$. Then $\deg_2(f) = 0$.*

Note that when W is compact, then the closed subset $X = \partial W$ is also compact. Hence $\deg_2(f)$ is defined.

Proof: Let $F\colon W \to Y$ be an extension of f. By Sard's Theorem 9.21, there is an element $y \in Y$ which is a regular value for both f and F. If $f^{-1}(y) = \emptyset$, we have $\deg_2(f) = 0$. So, we assume that f is surjective. By the Preimage Theorem 9.18, this implies $F^{-1}(y)$ is a submanifold of W of dimension $\dim W - \dim Y = 1$ with boundary $\partial F^{-1}(y) = F^{-1}(y) \cap \partial W = F^{-1}(y) \cap X = (F_{|X})^{-1}(y) = f^{-1}(y)$. Since $\{y\}$ is a closed subset of Y, $F^{-1}(y)$ is closed in W. By Lemma 1.14, this implies that $F^{-1}(y)$ is compact. Proposition 10.4 then shows that $\#\partial(F^{-1}(y))$ is even. Thus, we get $\deg_2(f) \equiv \#f^{-1}(y) = \#\partial(F^{-1}(y)) \equiv 0$ modulo 2. □

Corollary 11.15 (Obstruction for extending maps) *Let Y be a connected smooth manifold. Let W be a compact manifold with $\dim W = \dim Y + 1$ and $f\colon \partial W \to Y$ a smooth map. If $\deg_2(f) \neq 0$, then f cannot be extended to a smooth map $W \to Y$ on all of W.*

Example 11.16 (Existence of zeros for complex-valued functions) Suppose that $p\colon \mathbb{R}^2 \to \mathbb{R}^2$ is a smooth map and $W \subset \mathbb{R}^2$ is a two-dimensional, compact, smooth manifold with boundary. Question: how can we decide whether there is a $z \in W$ with $p(z) = 0$? Assume that p has no zeros on the boundary ∂W. Then $p/|p|\colon \partial W \to \mathbb{S}^1$ is defined and smooth as a map of compact

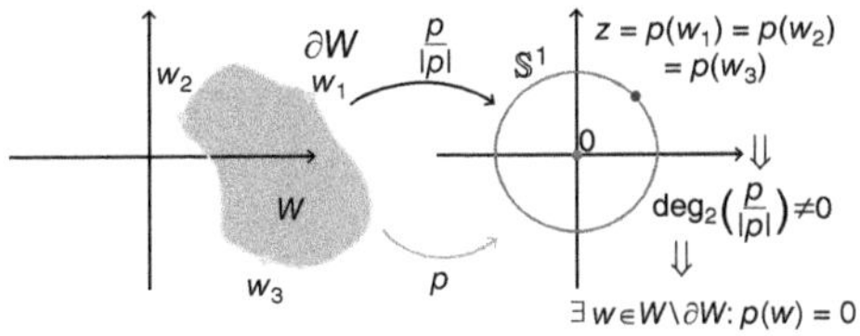

Figure 11.1 The degree is able to detect a zero of the map p in the region W.

one-manifolds. Now, if p has no zeros inside W, then $p/|p|$ is defined on all of W, i.e., the map $p/|p|: \partial W \to \mathbb{S}^1$ can be extended to a smooth map $W \to \mathbb{S}^1$ on all of W. If this is the case, we just learned in Theorem 11.14 that we must have $\deg_2(p/|p|) = 0$. Thus, if the mod 2 degree of $p/|p|: \partial W \to \mathbb{S}^1$ is non-zero, then the function p does have a zero inside W. Note that calculating $\deg_2(p/|p|)$ simply consists of picking a point $z \in \mathbb{S}^1$, which we could think of as a direction, and counting the number of times we find a $w \in \partial W$ with $p(w) = z$, i.e., how often $p(w)$ points in the chosen direction. See Figure 11.1. Our observation shows that this simple procedure tells us whether p has a zero inside W. If you learned about complex analysis, then this should remind you of the Residue Theorem and Cauchy's formula. We look at an application of this discussion in Exercise 11.6.

11.2.2 The Mod 2 Degree of Self-Maps of the Circle

We now look at an important special type of maps for which the degree is defined: smooth maps $\mathbb{S}^1 \to \mathbb{S}^1$. The following result gives us a method to compute their degree based on what we learned in calculus.

Theorem 11.17 (Mod 2 degree of self-maps of the circle) *Let $f: \mathbb{S}^1 \to \mathbb{S}^1$ be a smooth map, and let $p: \mathbb{R} \to \mathbb{S}^1 \subset \mathbb{C}$ be the map defined by $t \mapsto e^{2\pi i t}$. Let $g: \mathbb{R} \to \mathbb{R}$ be a smooth map such that $f \circ p = p \circ g$. By Lemma 8.23, we know that such a map g exists, and there is a unique $q \in \mathbb{Z}$ such that $g(t+1) = g(t) + q$ for all $t \in \mathbb{R}$. We then have* $\deg_2(f) = q$ *modulo* 2.

Proof: First, note that if f is not surjective, then we can pick a point $y \notin f(\mathbb{S}^1)$. This y is a regular value for f. Since $\#f^{-1}(y) = 0$, we must have $\deg_2(f) = 0$. Alternatively, we could use that the stereographic projection map $\mathbb{S}^1 \setminus \{y\} \to \mathbb{R}$ is a diffeomorphism. Since $\mathbb{R}$ is contractible, this shows that $\mathbb{S}^1 \setminus \{y\}$ is contractible. Hence f is a map to a contractible space and therefore homotopic to a constant map and has degree 0. To compute q, we note that

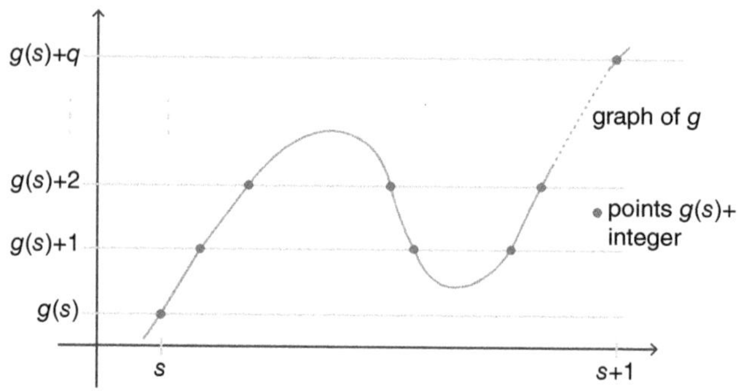

Figure 11.2 An application of the Intermediate Value Theorem in calculus. We count how many times the graph of g hits integer values, marked by the horizontal lines. We know that if the graph drops below a given value, it has to pass it again to reach the value $g(s)+q$ eventually. Counting modulo 2, we get q crossings of the horizontal lines.

$p \circ g$ is not surjective, since otherwise f would be surjective. Hence we must have $g(t+1) = g(t)$, i.e., $q = 0$.

Now we assume that f is surjective. Let y be a regular value of f. We have $\deg_2(f) = \#f^{-1}(y)$ modulo 2. Let $x \in f^{-1}(y)$ and $s \in \mathbb{R}$ be such that $p(s) = x$. We then have $y = f(x) = f(p(s)) = p(g(s)) = p(g(s)+k)$ for every integer k. Moreover, since $p(u) = p(u')$ if and only if $u - u' \in \mathbb{Z}$, the points of the form $g(s)+k$ are the only points with $p(u) = y$. Hence we need to count how often we have $g(t) - g(s) \in \mathbb{Z}$ for $t \in [s, s+1]$. See Figure 11.2. We assume that $q \geq 0$. If $q < 0$, then we use a similar argument with $-q$. Since y is a regular value, we have $\det(df_x) \neq 0$ and hence $dg(t_k) \neq 0$ for all t_k with $g(t_k) = g(s)+k$. By the Intermediate Value Theorem in calculus, we then know that the smooth function $g_{|[s,s+1]}\colon [s, s+1] \to \mathbb{R}$ takes the value $g(s)+k$ for each integer $k \in \{0, 1, \ldots, q-1\}$ an odd number of times and the values $g(s)+k$ for any other integer k an even number of times. Since $p(g(s+1)) = p(g(s))$, we only need to take one of s and $s+1$ into account. Hence, modulo 2, there are q many points t in the interval $[s, s+1)$ such that $p(g(t)) = p(g(s)) = f(y)$. Thus, modulo 2, there are q many points in $f^{-1}(y)$, and we have shown $\deg_2(f) = q$. □

11.3 Proof of the Isotopy Theorem

We now provide the proof of the Isotopy Theorem, an important result for the construction of the Brouwer degree. We will then show a generalisation.

Proof of Theorem 11.3: We call two points y and z *isotopic* if the statement of Theorem 11.3 holds. This defines an equivalence relation on the set of points in Y. The proof of the theorem will consist of showing that each equivalence class is open. Since equivalence classes are disjoint, this will show that Y is the disjoint union of open subsets. Since Y is connected, this implies there can only be one equivalence class, i.e., the assertion of the theorem holds for all points in Y. To prove that the equivalence classes are open, we will first construct an isotopy h_t on Euclidean space $\mathbb{R}^k$ such that h_0 is the identity, each h_t is the identity outside some specified small ball around 0, and $h_1(0)$ is any desired point sufficiently close to 0. Then we transport the isotopy to Y as follows. Given $y \in Y$, choose a local parametrisation $\phi\colon \mathbb{R}^k \to V \subset Y$ with $\phi(0) = y$. Let $\mathbb{B}$ be an open ball around 0 such that h_t on $\mathbb{R}^k$ is the identity outside $\mathbb{B}$. Since ϕ is a diffeomorphism onto V, $\phi(\mathbb{B})$ is an open subset in Y contained in V. Then the map $H_t := \phi \circ h_t \circ \phi^{-1}$ extends to an isotopy on all of Y by defining H_t to be the identity outside $\phi(\mathbb{B})$. The existence of the isotopy H_t then shows that y is isotopic to all points in the open subset $\phi(\mathbb{B})$ of Y. Hence the isotopy-equivalence class of y is open.

Now we construct the isotopy h_t on $\mathbb{R}^k$. We begin with the case $k = 1$. Given $\varepsilon > 0$, by Lemma 7.5 we can find a smooth function $\varphi\colon \mathbb{R} \to \mathbb{R}$ that vanishes outside $(-\varepsilon, \varepsilon)$ and equals 1 at 0. For $z \in \mathbb{R}^1$ and $t \in [0,1]$, we define the smooth map $h_t\colon \mathbb{R}^1 \to \mathbb{R}^1$ by $h_t(x) := x + t \cdot \varphi(x) \cdot z$. Then h_0 is the identity, and $h_t(x) = x$ if $x \notin (-\varepsilon, \varepsilon)$ for all t, and $h_1(0) = z$. To show that h_t is an isotopy, we compute its derivative $h_t'(x) = 1 + t \cdot \varphi'(x) \cdot z$ with respect to x. Since $|\varphi'(x)|$ vanishes outside a compact set, it must be bounded. Hence, there is an $\varepsilon' > 0$ such that $|t \cdot \varphi'(x) \cdot z| < 1$ for all $z \in (-\varepsilon', \varepsilon')$, $t \in [0,1]$ and $x \in \mathbb{R}$. Thus $h_t'(x) > 0$ for all t and x, and h_t is strictly increasing. By the Inverse Function Theorem 3.2, this implies that the inverse of h_t is also smooth. Hence h_t is a diffeomorphism of $\mathbb{R}$ for all $z \in (-\varepsilon', \varepsilon')$.

Now let $k > 2$. Given any point p in $\mathbb{R}^k$ near the origin, we may rotate the coordinate axes so that the point lies on the first axis. Hence, writing $\mathbb{R}^k = \mathbb{R} \times \mathbb{R}^{k-1}$, we can assume that p has coordinates of the form $p = (z, 0)$ with $z \in \mathbb{R}$. Given $\varepsilon > 0$, let again $\varphi\colon \mathbb{R} \to \mathbb{R}$ be a smooth function that vanishes outside $(-\varepsilon, \varepsilon)$ and equals 1 at 0. By Lemma 7.5, we can choose a function $\sigma\colon \mathbb{R}^{k-1} \to \mathbb{R}$ such that $\sigma(0) = 1$ and σ is zero outside some small ball $\mathbb{B}_\delta^{k-1}(0)$ of radius δ. For $z \in \mathbb{R}$ and $t \in [0,1]$, we define

$$h_t\colon \mathbb{R}^k \to \mathbb{R}^k,\ h_t(x,y) := (x + t \cdot \sigma(y) \cdot \varphi(x) \cdot z, y)$$

for $(x,y) \in \mathbb{R} \times \mathbb{R}^{k-1}$. We need to check that h_t has the desired properties. We have $h_t(x,y) = (x,y)$ unless $|x| < \varepsilon$, $|y| < \delta$, and $t > 0$, and we have $h_1(0,0) = (z,0)$ by the choice of σ and ρ. It remains to show that h_t

is a diffeomorphism of $\mathbb{R}^k$ when $|z|$ is small. Since $|\varphi'(x)|$ vanishes outside a compact set, it must be bounded. Since σ is also bounded, there is an $\varepsilon' > 0$ such that $|t \cdot \sigma(y) \cdot \varphi'(x) \cdot z| < 1$ for all t, x, y, and $|z| < \varepsilon'$. Hence, for each fixed $y \in \mathbb{R}^{k-1}$, h_t restricts to a diffeomorphism of $\mathbb{R} \times \{y\}$. Since h_t is the identity on $\mathbb{R}^{k-1}$, h_t is a bijection on all of $\mathbb{R}^k$. Since h_t is smooth, it remains to check that its inverse is also smooth. The derivative of h_t at $(x, y) \in \mathbb{R} \times \mathbb{R}^{k-1}$ is represented by the following matrix in the standard bases

$$\begin{pmatrix} 1 + t \cdot \sigma(y) \cdot \varphi'(x) \cdot z & * \cdots * \\ 0 & \\ \vdots & I_{k-1} \\ 0 & \end{pmatrix},$$

where I_{k-1} is the $(k-1) \times (k-1)$-identity matrix. Again, since $|\varphi'(x)|$ and σ are bounded, there exists $\varepsilon' > 0$ such that $1 + t \cdot \sigma(y) \cdot \varphi'(x) \cdot z$ is positive for all t, x, y, and $|z| < \varepsilon'$, and hence the matrix has strictly positive determinant. By the Inverse Function Theorem 3.2, this implies that the inverse of h_t is smooth. □

We can extend the assertion of Theorem 11.3 to any finite number of points. We will use this result later.

Theorem 11.18 *Let Y be a connected manifold of dimension strictly bigger than 1, and let $\{y_1, \ldots, y_n\}$ and $\{z_1, \ldots, z_n\}$ be two sets of distinct points in Y. Then there exists a diffeomorphism $h \colon Y \to Y$ which is isotopic to the identity with $h(y_1) = z_1, \ldots, h(y_n) = z_n$. Moreover, the isotopy may be chosen to be compactly supported.*

Proof: The proof is by induction. Theorem 11.3 is the case $n = 1$. Now we assume $n \geq 2$ and that the assertion is true for $n-1$. Then we have a compactly supported isotopy $h'_t \colon Y \setminus \{y_n, z_n\} \to Y \setminus \{y_n, z_n\}$ such that $h'_1(y_i) = z_i$ for all $i < n$ and $h'_0 = \mathrm{id}$. Since $\dim Y > 1$, the punctured manifold $Y \setminus \{y_n, z_n\}$ is path-connected by Exercise 8.6. Since the isotopy h'_t has compact support, there are open neighbourhoods around y_n and z_n in Y on which the h'_t are all equal to the identity. Hence we can extend the family h'_t to a family of diffeomorphisms of Y that fixes those two points. Now we apply the induction hypothesis again to the punctured manifold $Y \setminus \{y_1, \ldots, y_{n-1}, z_1, \ldots, z_{n-1}\}$ and the points y_n, z_n. Then we get a compactly supported isotopy h''_t with $h''_1(y_n) = z_n$ and $h''_0 = \mathrm{id}$. By the same argument as for h'_t, we can extend h''_t to an isotopy on all of Y such that all h''_t satisfy $h''_t(y_i) = z_i$ for all $i < n$. Then $h_t := h''_t \circ h'_t$ is the desired isotopy. □

11.4 Winding Numbers and the Borsuk–Ulam Theorem

Now we study an important application of the degree modulo 2. First, we introduce a new invariant based on $\deg_2$.

11.4.1 Winding Numbers Modulo 2

Let X be a compact, connected smooth manifold of dimension $n-1$, and let $f: X \to \mathbb{R}^n$ be a smooth map. Let z be a point of $\mathbb{R}^n$ which is not in the image of f. We would like to understand how the points $f(x)$ wind around z. To make this more precise, we define the smooth map $u: X \to \mathbb{S}^{n-1}$ by

$$u(x) := \frac{f(x) - z}{|f(x) - z|}.$$

We would like to know how often the vector $f(x) - z$ points in a given direction, i.e., how often $u(x)$ has a given value. We learned in Section 11.1.3 that the degree of u is an invariant that encodes this information. For a regular value y of u, we have $\deg_2(u) = \#u^{-1}(y)$ modulo 2. That y is regular for u means that the image of f and the line through z and y are transverse. See Figure 11.3.

Definition 11.19 (Winding number) We call $\deg_2(u)$ the *winding number of f around z*. We denote it by $W_2(f,z) := \deg_2(u)$.

The goal of this section is to prove the following famous result.

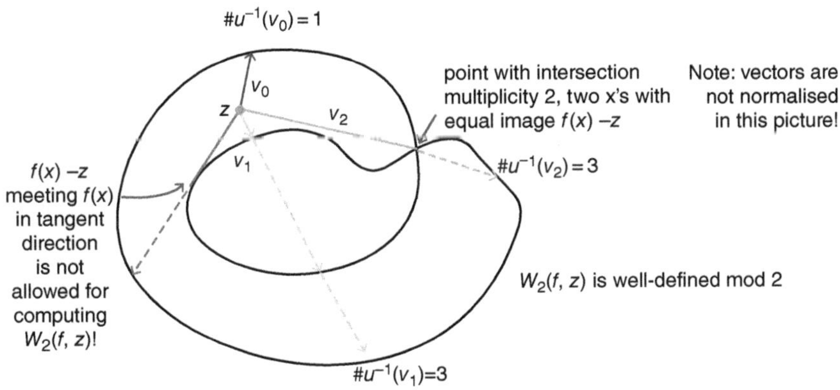

Figure 11.3 The winding number of f around z: we count the number of times the vector $f(x) - z$ points in a given direction, where we neglect tangential points. Note also that some points on the graph may contribute multiple times. In the end, we take our count modulo 2.

Theorem 11.20 (Borsuk–Ulam Theorem) *Let $f\colon \mathbb{S}^n \to \mathbb{R}^{n+1} \setminus \{0\}$ be a smooth map. Suppose that f is odd, i.e., f satisfies the anti-symmetry condition*

$$f(-x) = -f(x) \text{ for all } x \in \mathbb{S}^n. \tag{11.1}$$

Then the winding number of f equals one, i.e., $W_2(f,0) = 1$.

In other words, every map that is anti-symmetric around the origin winds around the origin an odd number of times. Before we approach the proof, we observe that Theorem 11.20 is equivalent to the following assertion.

Theorem 11.21 *Let $f\colon \mathbb{S}^n \to \mathbb{S}^n$ be a smooth map which sends antipodal points to antipodal points, i.e., $f(-x) = -f(x)$. Then $\deg_2(f) = 1$.*

Lemma 11.22 *Theorems 11.20 and 11.21 are equivalent.*

Proof: First, assume that Theorem 11.20 is true. Given a smooth map $f\colon \mathbb{S}^n \to \mathbb{S}^n$ with $f(-x) = -f(x)$, we can consider it as a map $f\colon \mathbb{S}^n \to \mathbb{S}^n \subset \mathbb{R}^{n+1}$. Then we have $f = f/|f|$ and therefore $1 = W_2(f,0) = \deg_2(f/|f|) = \deg_2(f)$. Now assume Theorem 11.21 is true. Given a smooth map $f\colon \mathbb{S}^n \to \mathbb{R}^{n+1} \setminus \{0\}$ with $f(-x) = -f(x)$, then $f/|f|$ is a well-defined smooth map $f/|f|\colon \mathbb{S}^n \to \mathbb{S}^n$. Hence we have $1 = \deg_2(f/|f|) = W_2(f,0)$ by definition of the winding number of f around 0. □

11.4.2 Winding Numbers and Boundaries

In order to prove Theorems 11.20 and 11.21, we follow the strategy in [14, chapter 2, §6] and first investigate the relationship between winding numbers and boundaries.

Theorem 11.23 (Winding numbers and boundaries) *Suppose that $X = \partial D$ is the boundary of a compact n-dimensional manifold D with boundary. Let $f\colon X \to \mathbb{R}^n$ be a smooth map. Assume that $F\colon D \to \mathbb{R}^n$ is a smooth map which extends f, i.e., $\partial F = f$. Suppose that z is a regular value of F that does not belong to the image of f. Then $F^{-1}(z)$ is a finite set, and $W_2(f,z) = \#F^{-1}(z)$ modulo 2. In other words, f winds X around z as often as F hits z, at least modulo 2.*

Proof: First, we assume $F^{-1}(z) = \emptyset$, i.e., $\#F^{-1}(z) = 0$. Since $F(x) \neq z$ for all x, the map $u\colon X = \partial D \to \mathbb{S}^{n-1}$, $u(x) = \frac{f(x)-z}{|f(x)-z|}$ can be extended to a smooth map $D \to \mathbb{S}^{n-1}$ by $x \mapsto \frac{F(x)-z}{|F(x)-z|}$. Hence, by the Boundary Theorem 11.14 for $\deg_2$, we have $W_2(f,z) = \deg_2(u) = 0$ modulo 2. Now we assume

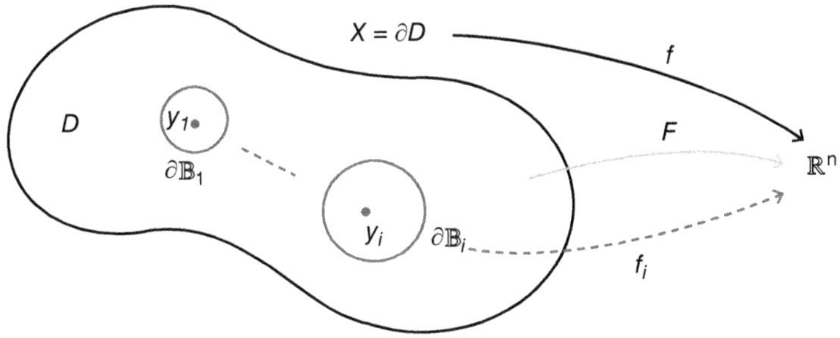

Figure 11.4 We split the contributions of the f_i and f by considering them as restrictions of F.

$F^{-1}(z) \neq \emptyset$. Since D is compact and of dimension n, $F^{-1}(z)$ is a zero-dimensional closed submanifold of D, and hence compact by Lemma 1.14. Since z is a regular value for F, $F^{-1}(z)$ is discrete by the Stack of Records Theorem 4.28. By Lemma 1.17, this implies that $F^{-1}(z)$ is a finite set. Say $F^{-1}(z) = \{y_1, \ldots, y_m\}$. Let $\mathbb{B}_i$ be the image of a closed ball in $\mathbb{R}^n$ around y_i. By the Stack of Records Theorem 4.28, we can choose the radii of the $\mathbb{B}_i$ small enough such that $\mathbb{B}_i \cap \mathbb{B}_j = \emptyset$ and $\mathbb{B}_i \cap \partial D = \emptyset$ for all $i \neq j$. We define $f_i := F_{|\partial\mathbb{B}_i} : \partial\mathbb{B}_i \to \mathbb{R}^n$ to be the restriction of F to $\partial\mathbb{B}_i$. See Figure 11.4. Now we observe that the subset $\widetilde{D} := D \setminus (\cup_i \mathrm{Int}(\mathbb{B}_i))$ is a closed submanifold of D with boundary the disjoint union of the boundaries of D and the $\mathbb{B}_i$, i.e., $\partial\widetilde{D} = \partial D \sqcup \partial\mathbb{B}_1 \sqcup \cdots \sqcup \partial\mathbb{B}_m$. By the choice of the $\mathbb{B}_i$'s, we have $F^{-1}(z) \cap \widetilde{D} = \emptyset$. Hence $F^{-1}(z) \cap \widetilde{D} = (F_{|\widetilde{D}})^{-1}(z) = \emptyset$. Thus the winding number of $\partial F_{|\widetilde{D}}$ at z is zero. Since degrees and hence winding numbers are additive with respect to connected components by Lemma 11.10, this implies

$$0 = W_2(\partial F_{|\widetilde{D}}, z) = W_2(f, z) + W_2(f_1, z) + \cdots + W_2(f_m, z).$$

Thus, modulo 2, we have $W_2(f,z) = W_2(f_1,z) + \cdots + W_2(f_m,z)$. Now it remains to show $W_2(f_i,z) = 1$ for each $i = 1, \ldots, m$. For then we get the desired equality $W(f,z) = m = \#F^{-1}(z)$ modulo 2. Since z is a regular value and $\dim D = n$, dF_{y_i} is an isomorphism. Thus, by the Inverse Function Theorem 3.4, we can choose the radius of $\mathbb{B}_i$ small enough such that $F_{|\mathbb{B}_i}$ is a diffeomorphism onto its image, which contains z. By continuity, this implies also that $f_i = \partial F_{|\mathbb{B}_i}$ is one-to-one onto the boundary of $F(\mathbb{B}_i)$. After possibly rescaling and translating, we are reduced to showing the following. Let $\mathbb{B}$ be the closed unit ball in $\mathbb{R}^n$ and $F : \mathbb{B} \to \mathbb{B}$ be a diffeomorphism. Let $f = \partial F : \mathbb{S}^{n-1} \to \mathbb{S}^{n-1}$. Then $\#F^{-1}(0) = W_2(f,0) = 1$ modulo 2. The latter follows from $W_2(f,0) = \deg_2(f) = \#f^{-1}(v) = 1$ for any $v \in \mathbb{S}^{n-1}$. □

Proof of Theorem 11.21: The proof will proceed by induction. We begin with the case $n = 1$. By Lemma 8.23, we can find a smooth map $g\colon \mathbb{R} \to \mathbb{R}$ with $f \circ p = p \circ g$ and $g(t+1) = g(t) + q$ for some $q \in \mathbb{Z}$. The assumption on f implies $f(p(t+1/2)) = f(-p(t)) = -f(p(t))$, where we use $p(t+1/2) = \exp(2\pi i(t+1/2)) = -\exp(2\pi i t) = -p(t)$. Hence

$$p(g(t+1/2)) = f(p(t+1/2)) = -f(p(t)) = -p(g(t)).$$

We have $p(s_1) = -p(s_2)$ if and only if $\exp(2\pi i s_1) = \exp(2\pi i s_2 + \pi i)$, and hence $p(s_1) = -p(s_2)$ if and only if $s_1 = s_2 + m/2$ for some odd $m \in \mathbb{Z}$. This implies $g(t+1/2) = g(t) + m/2$ for some odd $m \in \mathbb{Z}$. Applied to $t = 1/2$, this yields $g(1) = g(1/2+1/2) = g(1/2)+m/2 = g(0)+m/2+m/2 = g(0)+m$. Thus $q = m$ is odd. Hence, by Theorem 11.17, we have $\deg_2(f) = q = 1 \bmod 2$. This finishes the case $n = 1$.

Now we show the induction step. Let $n \geq 2$ and assume that the theorem is true for $n-1$. Let $f\colon \mathbb{S}^n \to \mathbb{R}^{n+1} \setminus \{0\}$ be a smooth map which satisfies the symmetry condition $f(-x) = -f(x)$ for all $x \in \mathbb{S}^n$. We consider $\mathbb{S}^{n-1}$ to be the equator of $\mathbb{S}^n$, embedded by $(x_1, \ldots, x_n) \mapsto (x_1, \ldots, x_n, 0)$. The idea is to compute $W_2(f, 0)$ by counting how often f intersects a line L in $\mathbb{R}^{n+1}$. By choosing L disjoint from the image of the equator, we can use the induction hypothesis to show that the equator winds around L an odd number of times. Then it is easy to calculate the intersection of f with L once we know the behaviour of f on the equator.

Let $g\colon \mathbb{S}^{n-1} \to \mathbb{R}^{n+1} \setminus \{0\}$ be the restriction of f to the equator. By Sard's Theorem 5.17, we can choose $y \in \mathbb{S}^n$ which is regular for both smooth maps $g/|g|\colon \mathbb{S}^{n-1} \to \mathbb{S}^n$ and $f/|f|\colon \mathbb{S}^n \to \mathbb{S}^n$. The symmetry condition implies that y is regular for both these maps if and only if $-y$ is regular for both maps, since the derivatives at preimages of y and $-y$ just differ by multiplying with (-1). Since $\dim \mathbb{S}^{n-1} < \dim \mathbb{S}^n$, the only way y can be a regular value of $g/|g|$ is when y is not in the image. Hence neither y nor $-y$ is in the image of $g/|g|$. Thus, for the line $L := \mathbb{R} \cdot y = \operatorname{Span}(y)$, we have y is a regular value of g if and only if $\operatorname{Im}(g) \cap L = \emptyset$. That y is regular for $f/|f|$ means by definition $\operatorname{Im}(d(f/|f|)_x) = T_y(\mathbb{S}^n)$. The tangent space to $\mathbb{S}^n$ at y is the orthogonal complement of the line pointing from the origin in direction of y. The map $x \mapsto f(x)/|f(x)|$ is the composite of f and the map $x \mapsto x/|x|$, which is smooth in dimensions $n \geq 2$. The derivative of the map $x \mapsto x/|x|$ satisfies $\operatorname{Im}(d(x/|x|)_x) = (\operatorname{Span}(x))^{\perp} \subset \mathbb{R}^{n+1}$, i.e., $\operatorname{Ker}(d(x/|x|)_x) = \operatorname{Span}(x)$. This implies $\operatorname{Ker}(d(f/|f|)_x) = \operatorname{Span}(f(x)) \cap \operatorname{Im}(df_x)$. Thus, we have $\operatorname{Im}(d(f/|f|)_x) = T_y(\mathbb{S}^n)$ if and only if $\operatorname{Ker}(d(f/|f|)_x) = \{0\}$, which is the case if and only if $\operatorname{Span}(f(x)) \cap \operatorname{Im}(df_x) = \{0\}$. Moreover, since $L = \operatorname{Span}(f(x))$ is just a line, the latter intersection is empty if and only if $\mathrm{Span}(f(x)) \not\subset \mathrm{Im}\,(df_x)$, which is equivalent to $L + \mathrm{Im}\,(df_x) = \mathbb{R}^{n+1}$. Summarising the argument, we have obtained

$$y \text{ is a regular value of } \frac{f}{|f|} \iff f \pitchfork L = \mathrm{Span}(y). \tag{11.2}$$

Now we are going to exploit these observations to calculate $W_2(f,0)$. By definition, we have

$$W_2(f,0) = \deg_2\left(\frac{f-0}{|f-0|}\right) = \deg_2(f/|f|) = \#(f/|f|)^{-1}(y) \mod 2.$$

By symmetry, we have $\#(f/|f|)^{-1}(y) = \#(f/|f|)^{-1}(-y)$. From (11.2) we know $f^{-1}(L) = \{x \in \mathbb{S}^n : f(x) \in L\} = \{x \in \mathbb{S}^n : f(x)/|f(x)| = \pm y\}$ and hence $f^{-1}(L) = (f/|f|)^{-1}(y) \cup (f/|f|)^{-1}(-y)$. Thus $\#(f/|f|)^{-1}(y) = \frac{1}{2}\#f^{-1}(L)$. Hence we need to calculate the number $\frac{1}{2}\#f^{-1}(L)$, at least modulo 2. By symmetry, we can do this on the upper hemisphere $\mathbb{S}^n_+$ of $\mathbb{S}^n$, i.e., the points on $\mathbb{S}^n$ with $x_{n+1} \geq 0$. Let f_+ be the restriction of f to $\mathbb{S}^n_+$. By the choice of y, L does not meet the equator, and hence no point on the equator is in $f^{-1}(L)$. This implies $\frac{1}{2}\#f^{-1}(L) = \#f_+^{-1}(L)$ where f_+ denotes the restriction of f to $\mathbb{S}^n_+$. The upper hemisphere is a manifold with boundary

$$\partial\mathbb{S}^n_+ = \{x = (x_1, \ldots, x_{n+1}) : \sum_i x_i^2 = 1 \text{ and } x_{n+1} = 0\} = \mathbb{S}^{n-1}$$

being the equator. Now we would like to apply Theorem 11.23 to f_+ and $g = \partial f_+$ and use the induction hypothesis. However, the target of f_+ has dimension $n+1$, whereas both the theorem and the induction hypothesis require the target to be a Euclidean space of dimension n. So we need to fix this. The key is that the orthogonal complement $L^\perp$ of L in $\mathbb{R}^{n+1}$ is a vector space of dimension n. By choosing a basis of $L^\perp$, we can identify it with $\mathbb{R}^n$.

To complete the argument, let $\pi : \mathbb{R}^{n+1} \to L^\perp$ be the orthogonal projection onto $L^\perp$. Since g is symmetric and π is linear, $\pi \circ g \colon \mathbb{S}^{n-1} \to L^\perp$ is symmetric, i.e., $\pi(g(-x)) = \pi(-g(x)) = -\pi(g(x))$. Moreover, we have $\pi(g(x)) = 0$ if and only if $g(x) \in L$, and hence $\pi(g(x)) \neq 0$ for all $x \in \mathbb{S}^{n-1}$ by the definition of π and the choice of L. Thus, after choosing a basis for $L^\perp$, we can consider $\pi \circ g$ as a map $\pi \circ g \colon \mathbb{S}^{n-1} \to \mathbb{R}^n \setminus \{0\}$. Now we apply the induction hypothesis to $\pi \circ g$ and get $W_2(\pi \circ g, 0) = 1$. To finish, recall $f_+ \pitchfork L$ and hence $(\pi \circ f_+) \pitchfork \{0\}$ for $\pi \circ f_+ \colon \mathbb{S}^k \to L^\perp$. In other words, 0 is a regular value of $\pi \circ f_+$. Hence we can apply Theorem 11.23 to $\pi \circ f_+$ and its boundary map $\partial(\pi \circ f_+) = \pi \circ g$ to get $W_2(\pi \circ g, 0) = \#(\pi \circ f_+)^{-1}(0)$. By the choice of L, we have $\pi(f_+(x)) = 0$ if and only if $f_+(x) \in L$, and hence $(\pi \circ f_+)^{-1}(0) = f_+^{-1}(L)$. This shows

$$W_2(f,0) = \#f_+^{-1}(L) = W_2(\pi \circ g, 0) = 1. \qquad \square$$

Corollary 11.24 *Let $f: \mathbb{S}^n \to \mathbb{R}^{n+1} \backslash \{0\}$ be a smooth map such that $f(-x) = -f(x)$ for all x. Then f intersects every line through 0 at least once.*

Proof: Let L be a line in $\mathbb{R}^{n+1}$ through the origin. If f never hits L, then $\# f^{-1}(L) = 0$ and $f \pitchfork L$. By repeating the above proof, we would get $W_2(f,0) = \# f^{-1}(L) = 0$ which would contradict Theorem 11.20. □

11.5 Linking Numbers and the Hopf Invariant Modulo 2

As another application of $\deg_2$, we now introduce and study the Hopf invariant modulo 2. We will discuss the $\mathbb{Z}$-valued Hopf invariant in Section 15.3. There are many different ways to define this influential invariant. Here we follow Milnor's outline [35, problems 13–15] of Hopf's original approach using linking numbers.

11.5.1 Mod 2 Linking Number

We begin with the following definition.

Definition 11.25 (Linking number mod 2) For $k \geq 1$, let $X, Y \subset \mathbb{R}^{k+1}$ be two disjoint smooth manifolds. We assume that X and Y are compact and without boundary of dimensions $\dim X = m$ and $\dim Y = n$ such that $m + n = k$. We define the map $\lambda: X \times Y \to \mathbb{S}^k$ by $\lambda(x,y) := \frac{x-y}{|x-y|}$. The *mod 2 linking number* $L_2(X,Y)$ of X and Y is defined to be the mod 2 degree of λ, i.e., $L_2(X,Y) := \deg_2(\lambda)$. See Figure 11.5.

Note that, since X and Y are disjoint, the map λ is well defined and smooth. The order of X and Y does not matter modulo 2, i.e., $L_2(X,Y) = L_2(Y,X)$.

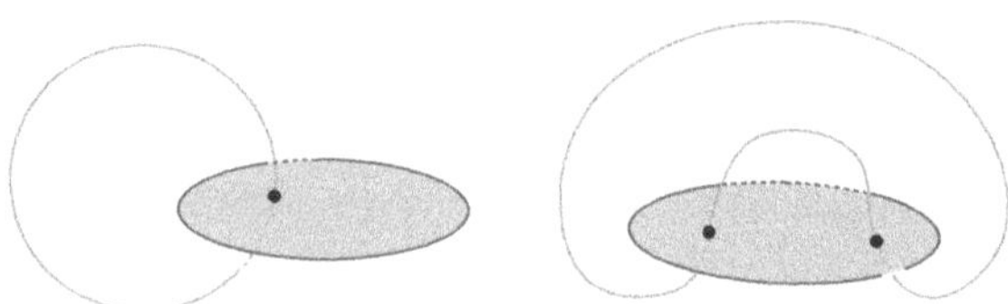

Figure 11.5 The shaded disk is a compact manifold D with boundary. In both cases, the curve intersects the manifold D. The linking number detects the intersection. However, multiple intersections may occur. The circles on the left-hand side are linked and cannot be moved apart. The curves on the right-hand side are not linked and can be moved.

For the oriented linking number which we will introduce in Section 15.3, this will be different.

Example 11.26 (Linking number of two circles in $\mathbb{R}^3$) Suppose $X, Y \subset \mathbb{R}^3$ are two circles embedded in $\mathbb{R}^3$. Let us try to understand why λ detects how the two circles are linked. First, let us assume that X and Y are not linked. In this case, we can move them completely apart in $\mathbb{R}^3$ using an isotopy. Then $\lambda\colon X \times Y \to \mathbb{S}^2$ is not surjective, since the vectors $x - y$ cannot cover all directions of lines through the origin in $\mathbb{R}^3$. Hence every point p in $\mathbb{S}^2$ which is not in the image of λ is a regular value. Since $\lambda^{-1}(p) = \emptyset$ and hence $\#\lambda^{-1}(p) = 0$, we have $\deg_2(\lambda) = 0$. For example, if X consists of points with strictly positive first coordinate and Y consists of points with strictly negative first coordinate, then $x - y$ will always have strictly positive first coordinate. Hence, the subset of $\mathbb{S}^2$ consisting of points with negative first coordinate is not contained in the image of λ. See also Lemma 11.27. Second, assume that X and Y are linked once as on the left-hand side in Figure 11.5. In this case, we can express every direction of lines through the origin in $\mathbb{R}^3$. This implies that $\#\lambda^{-1}(p) = 1$ for every $p \in \mathbb{S}^2$.

Lemma 11.27 (Linking number and boundaries) *Assume that X is the boundary of a smooth $(m+1)$-manifold W which is disjoint from Y. Then we have $L_2(X, Y) = 0$.*

Proof: Since Y does not have a boundary, the product $W \times Y$ is a smooth manifold with boundary $\partial(W \times Y) = \partial W \times Y = X \times Y$. Since W and Y are disjoint, λ extends to a smooth map on $W \times Y$. Then the Boundary Theorem 11.14 implies that we have $\deg_2(\lambda) = 0$. □

Now we extend the definition of the linking number to submanifolds X, Y in $\mathbb{S}^{k+1}$ as follows.

Definition 11.28 (Linking number on spheres) Let X and Y be compact submanifolds of $\mathbb{S}^{k+1}$ without boundary with $\dim X + \dim Y = k$. Since $\mathbb{S}^{k+1}$ is connected and since X and Y are closed subsets, there must be a point p which is not contained in either X or Y. We identify $\mathbb{S}^{k+1} \setminus \{p\}$ with $\mathbb{R}^{k+1}$ via the diffeomorphism defined by stereographic projection ϕ_p from p. Then we consider the images $\phi_p(X)$ and $\phi_p(Y)$ of X and Y, respectively, under this stereographic projection as submanifolds of $\mathbb{R}^{k+1}$. Now we define the *mod 2-linking number* $L_2(X, Y)$ of X and Y to be $L_2(\phi_p(X), \phi_p(Y))$ of $\phi_p(X)$ and $\phi_p(Y)$ as in Definition 11.25. The linking number of two empty submanifolds of $\mathbb{S}^{k+1}$ is defined to be zero.

We can show that $L_2(X, Y)$ does not depend on the choice of p using the fact that the stereographic projection is a diffeomorphism.

11.5.2 Mod 2 Hopf Invariant

The main situation we are interested in is the following.

Definition 11.29 (Hopf invariant modulo 2) For $n \geq 1$, let $f\colon \mathbb{S}^{2n-1} \to \mathbb{S}^n$ be a smooth map. Let $w \neq z \in \mathbb{S}^n$ be two regular values for f. Then $f^{-1}(w)$ and $f^{-1}(z)$ are compact submanifolds of $\mathbb{S}^n$ without boundary, and their linking number $L_2(f^{-1}(w), f^{-1}(z))$ is defined. We define the *mod 2 Hopf invariant of* f, denoted $H_2(f)$, to be the mod 2 linking number of $^{-1}(w)$ and $f^{-1}(z)$, i.e.,

$$H_2(f) := L_2(f^{-1}(w), f^{-1}(z)).$$

Our next goal is to show that this number does not depend on the choice of w and z and only depends on the homotopy class of f.

Lemma 11.30 (Linking number is homotopy invariant) *Let* $F\colon \mathbb{S}^{2n-1} \times [0,1] \to \mathbb{S}^n$ *be a smooth homotopy between* $f_0, f_1\colon \mathbb{S}^{2n-1} \to \mathbb{S}^n$. *Let* $w, z \in \mathbb{S}^n$ *be regular values for* F, *and both* f_0 *and* f_1. *Then we have*

$$L_2(f_0^{-1}(w), f_0^{-1}(z)) = L_2(f_1^{-1}(w), f_0^{-1}(z)) = L_2(f_1^{-1}(w), f_1^{-1}(z)). \tag{11.3}$$

Proof: Since w is a regular value for F and both f_0 and f_1, the Preimage Theorem 9.18 implies that the subset $F^{-1}(w) \subset \mathbb{S}^{2n-1} \times [0,1] \subset \mathbb{R}^{2n+1}$ is a submanifold with boundary $\partial F^{-1}(w) = F^{-1}(w) \cap (\mathbb{S}^{2n-1} \times \{1\} \sqcup \mathbb{S}^{2n-1} \times \{0\})$, i.e., $\partial F^{-1}(w) = f_1^{-1}(w) \times \{1\} \sqcup f_0^{-1}(w) \times \{0\}$. Moreover, since $\mathbb{S}^{2n-1} \times [0,1]$ is compact, the closed subsets $F^{-1}(w)$ and $\partial F^{-1}(w)$ are compact by Lemma 1.14. Since $f_0^{-1}(z)$ is a compact manifold without boundary, $F^{-1}(w) \times f_0^{-1}(z)$ is a compact manifold with boundary given by

$$\partial(F^{-1}(w) \times f_0^{-1}(z)) = (f_1^{-1}(w) \times \{1\} \times f_0^{-1}(z)) \sqcup (f_0^{-1}(w) \times \{0\} \times f_0^{-1}(z)).$$

Thus the linking maps maps $\lambda_0\colon f_0^{-1}(w) \times f_0^{-1}(z) \to \mathbb{S}^{2n-2}$ and $\lambda_1\colon f_1^{-1}(w) \times f_0^{-1}(z) \to \mathbb{S}^{2n-2}$ are the restrictions of $\lambda_F\colon F^{-1}(w) \times f_0^{-1}(z) \to \mathbb{S}^{2n-2}$, $\lambda_F(x,y) = \frac{x-y}{|x-y|}$, to the two boundary components. By the Boundary Theorem 11.14 for $\deg_2$, this implies $\deg_2(\lambda_0 \sqcup \lambda_1) = 0$ where $\lambda_0 \sqcup \lambda_1$ denotes the induced map $(f_0^{-1}(w) \times f_0^{-1}(z)) \sqcup (f_1^{-1}(w) \times f_0^{-1}(z)) \to \mathbb{S}^{2n-2}$. Since $\deg_2$ is by construction additive on connected components by Lemma 11.10, this implies $L_2(f_0^{-1}(w), f_0^{-1}(z)) + L_2(f_1^{-1}(w), f_0^{-1}(z)) \equiv 0$ modulo 2. Thus we have $L_2(f_0^{-1}(w), f_0^{-1}(z)) \equiv L_2(f_1^{-1}(w), f_0^{-1}(z))$ modulo 2. This proves the first equality in (11.3). Applying the same argument for the second factor proves the second equality. □

Lemma 11.31 *The linking number $L_2(f^{-1}(w), f^{-1}(z))$ is locally constant as a function of w and of z.*

Proof: Because of the symmetry, we only prove the assertion for w. By Proposition 4.9, the set of regular values is an open subset of $\mathbb{S}^n$. Hence we can choose an $\varepsilon > 0$ such that every $v \in \mathbb{S}^n$ with $|v - w| < \varepsilon$ is a regular value for the linking map. Let $v \in \mathbb{S}^n$ with $|v - w| < \varepsilon$. Then we can choose a family of smooth rotations $r_t : \mathbb{S}^n \to \mathbb{S}^n$ so that $r_1(w) = v$, and so that

- r_t is the identity for $0 \le t < \delta$, for some $0 < \delta < 1$,
- r_t equals r_1 for $1 - \delta < t \le 1$, and
- each $r_t^{-1}(v)$ lies on the great circle from w to v on $\mathbb{S}^n$ which implies that it is a regular value for f.

Now we can define a smooth homotopy $F : \mathbb{S}^{2n-1} \times [0, 1] \to \mathbb{S}^n$ by setting $F(x,t) := r_t(f(x))$. By our choice of r_t and v, note that, for each fixed t, v is a regular value for the composition $r_t \circ f : \mathbb{S}^{2n-1} \to \mathbb{S}^n$. This implies that v is also a regular value for the map F. Now we can apply Lemma 11.30 to deduce the assertion. □

Proposition 11.32 (Hopf invariant is well defined) *For $n \ge 1$, let $f : \mathbb{S}^{2n-1} \to \mathbb{S}^n$ be a smooth map. The number $L_2(f^{-1}(w), f^{-1}(z))$ does not depend on the choice of regular values w and z and only depends on the smooth homotopy class of f.*

Proof: By Lemma 11.31, the map $(w,z) \mapsto L_2\left(f^{-1}(w), f^{-1}(z)\right)$ is a locally constant function. Since Y is connected, this map is constant by Lemma 1.30. Now let $g : \mathbb{S}^{2n-1} \to \mathbb{S}^n$ be a smooth map such that there is a smooth homotopy F between f and g. By Sard's Theorem 5.17, the sets of regular values for f, g, and F are dense in Y. Hence we can find elements w and z in Y which are regular values for f, g, and F simultaneously. Now we can apply Lemma 11.30 to deduce that $H_2(f) = H_2(g)$. □

Remark 11.33 Denoting by $\pi_{2n-1}(\mathbb{S}^n)$ the $(2n-1)$-homotopy group of $\mathbb{S}^n$, we can view the mod 2 Hopf invariant as a map $H_2 : \pi_{2n-1}(\mathbb{S}^n) \to \mathbb{Z}/2$.

11.5.3 The Mod 2 Hopf Invariant of the Hopf Map

Now we are going to compute the Hopf invariant for the Hopf map $\hbar$. Recall that we may consider $\mathbb{S}^3$ as a subset of $\mathbb{C}^2$, i.e., $\mathbb{S}^3 = \{(z_0, z_1) \in \mathbb{C}^2 : |z_0|^2 + |z_1|^2 = 1\}$, and $\mathbb{S}^2$ as a subset of $\mathbb{C}\times\mathbb{R}$, i.e., $\mathbb{S}^2 = \{(z, x) \in \mathbb{C}\times\mathbb{R} : |z|^2+x^2 = 1\}$.

Then $\hbar$ is the map $\mathbb{S}^3 \to \mathbb{S}^2$ given by $\hbar(z_0, z_1) = (2z_0\bar{z}_1, |z_0|^2 - |z_1|^2)$. We will now prove the following famous result due to Hopf [22].

Theorem 11.34 (Hopf invariant of the Hopf map is non-trivial) *We have $H_2(\hbar) = 1$ in $\mathbb{Z}/2$. In particular, $\hbar$ is not homotopic to a constant map.*

Remark 11.35 The theorem shows that $\hbar$ is a non-contractible map between spheres where the dimension of the domain is bigger than the codomain. Moreover, we have shown in Exercise 2.10 that the Hopf map realises $\mathbb{S}^3$ as a disjoint union of fibres which each look like $\mathbb{S}^1$. Since the Hopf invariant is the linking number of any two distinct circles on $\mathbb{S}^3$, this shows that all these disjoint circles are linked to each other and cannot be pulled apart.

There are different ways to prove Theorem 11.34. The proof we will now provide is not the most elegant one. However, it is concrete and only uses the techniques we have developed so far.

Proof of Theorem 11.34: By definition of $H_2(\hbar)$, we need to choose two distinct regular values w and z of $\hbar$ and calculate the mod 2 linking number of $\hbar^{-1}(w)$ and $\hbar^{-1}(z)$. We showed in Exercise 4.12 that $\hbar$ is a submersion. Thus, every point in $\mathbb{S}^2$ is a regular value. Therefore, we can for example choose $w = a = (0,0,1)$ and $z = b = (0,1,0)$ on $\mathbb{S}^2 \subset \mathbb{R}^3 \cong \mathbb{C} \times \mathbb{R}$. The fibre of $\hbar$ over a is $\hbar^{-1}(a) = \{(z_0, 0) \in \mathbb{S}^3 \subset \mathbb{C}^2 \colon |z_0|^2 = 1\}$. In Exercise 4.12, we determined the fibre over b as $\hbar^{-1}(b) = \{(x_0, y_0, x_1, y_1) \in \mathbb{S}^3 \colon y_0 = x_1, y_1 = -x_0\}$. To calculate the linking number of $\hbar^{-1}(a)$ and $\hbar^{-1}(b)$, we need to choose a point on $\mathbb{S}^3$ disjoint from these two subsets and stereographically project $\mathbb{S}^3$ from this point onto $\mathbb{R}^3$. By our choice of a and b, we get that the north pole $N = (0,0,0,1)$ is neither in $\hbar^{-1}(a)$ nor in $\hbar^{-1}(b)$. Recall that the formula for the stereographic projection $\phi_N^{-1} \colon \mathbb{S}^3 \setminus \{N\} \to \mathbb{R}^3$ is, with the notation we use here, given by $(x_0, y_0, x_1, y_1) \mapsto \frac{1}{1-y_1}(x_0, y_0, x_1)$. Hence we get

$$\mathbb{S}_a := \phi_N^{-1}(\hbar^{-1}(a)) = \{\mathbf{x} = (x_0, x_1, x_2) \in \mathbb{R}^3 \colon x_0^2 + x_1^2 = 1 \text{ and } x_2 = 0\}$$

and

$$\mathbb{S}_b := \phi_N^{-1}(\hbar^{-1}(b)) = \{\mathbf{y} = (y_0, y_1, y_2) \in \mathbb{R}^3 \colon y_1 = y_2 \text{ and } y_0^2 + y_1^2 = (1 - y_0)^2/2\}.$$

Now we can calculate $H_2(\hbar)$ as $\deg_2(\lambda)$ with $\lambda \colon \mathbb{S}_a \times \mathbb{S}_b \to \mathbb{S}^2$, $\lambda(\mathbf{x}, \mathbf{y}) = (\mathbf{x} - \mathbf{y})/(|\mathbf{x} - \mathbf{y}|)$. To compute the degree of λ, we pick a convenient point of $\mathbb{S}^2$ and determine the fibre over this point. Then we check that we actually picked a regular value. So let us look at the point $p = (1,0,0) \in \mathbb{S}^2$. The equation $\lambda(\mathbf{x}, \mathbf{y}) = p$ implies $x_1 = y_1 = 0$ and $x_0 - y_0 = |x_0 - y_0|$. The latter condition implies that $x_0 - y_0$ is positive. This does not look very helpful at first glance, but we also know $1 = x_0^2 + x_1^2 = x_0^2$, i.e., $x_0 = \pm 1$, and $y_0^2 = \frac{(1-y_0)^2}{2}$, i.e., $y_0 = \pm\sqrt{2} - 1$. Hence we can check for the four possible permutations of the signs whether they yield $x_0 - y_0 \geq 0$ and get three points $(\mathbf{x}, \mathbf{y})$ in $\mathbb{S}_a \times \mathbb{S}_b$ with $\lambda(\mathbf{x}, \mathbf{y}) = p$: one point with $x_0 = 1$, $y_0 = \sqrt{2} - 1$, one with $x_0 = 1$, $y_0 = -\sqrt{2} - 1$, and one with $x_0 = -1$, $y_0 = -\sqrt{2} - 1$. Hence, once we have shown that p is a regular value, we will have proved $H_2(\hbar) = \deg_2(\lambda) \equiv 1$ modulo 2.

It remains to show that p is a regular value, i.e., we need to show that the derivatives of λ at these three points are isomorphisms. To do so, we show that the determinant of the derivative of λ at each point is non-zero. Since $\mathbb{S}_a$ is the unit circle in the xy-plane, the tangent space of $\mathbb{S}_a$ at a point $\mathbf{x}$ is given by the space $T_{\mathbf{x}}\mathbb{S}_a = \{\mathbf{u} = (u_0, u_1, u_2) \in \mathbb{R}^3 : u_2 = 0 \text{ and } u_0 x_0 + u_1 x_1 = 0\}$. Similarly, $\mathbb{S}_b$ lies in the plane $\mathbb{P}$ in $\mathbb{R}^3$ of points $\mathbf{y} = (y_0, y_1, y_2)$ with $y_1 = y_2$. Let $\mathbb{S}_b$ be the map $g_b \colon \mathbb{P} \to \mathbb{R}$ defined by $g_b(y_0, y_1, y_1) = y_0^2 + y_1^2 = (1 - y_0)^2/2$. After a simple computation, we get

$$\mathbb{S}_b = g_b^{-1}(1) = \{\mathbf{y} = (y_0, y_1, y_1) \in \mathbb{P} : 2y_1^2 + y_0^2 + 2y_0 = 1\},$$

and 1 is a regular value for g_b. The derivative of g_b as a map from $\mathbb{P} \to \mathbb{R}$ is given by the (2×1)-matrix $d(g_b)_{\mathbf{y}} = (2y_0 + 2, 2y_1)$. Here we could also consider g_b as a map $\mathbb{R}^3 \to \mathbb{R}^2$. Hence, by Proposition 4.13, we get

$$T_{\mathbf{w}}\mathbb{S}_b = \{\mathbf{u} = (u_0, u_1, u_2) \in \mathbb{R}^3 : u_2 = u_1 \text{ and } u_0(y_0 + 1) + u_1 y_1 = 0\}.$$

Now we calculate the derivative of λ. First, we do this for the map $\tilde{\lambda} \colon \mathbb{R}^3 \times \mathbb{R}^3 \to \mathbb{R}^3$ given by $\tilde{\lambda}(\mathbf{x}, \mathbf{y}) = (\mathbf{x} - \mathbf{y})/|\mathbf{x} - \mathbf{y}|$. This will help us since $\tilde{\lambda}$ is constant on $\mathbb{S}_a \times \mathbb{S}_b$. Hence $d\tilde{\lambda}_{(\mathbf{x},\mathbf{y})}$ sends the subspace $T_{\mathbf{x}}\mathbb{S}_a \times T_{\mathbf{y}}\mathbb{S}_b \in \mathbb{R}^3 \times \mathbb{R}^3$ to $T_p\mathbb{S}^2 \subset \mathbb{R}^3$. We determine $d\tilde{\lambda}_{(\mathbf{x},\mathbf{y})}$ by computing its partial derivatives $\frac{\partial \tilde{\lambda}_i}{\partial x_j}(\mathbf{x}, \mathbf{y})$ and $\frac{\partial \tilde{\lambda}_i}{\partial y_j}(\mathbf{x}, \mathbf{y})$ with respect to the variables x_0, x_1, x_2 and y_0, y_1, y_2. For $i \neq j$, we have

$$\frac{\partial \tilde{\lambda}_i}{\partial x_j}(\mathbf{x}, \mathbf{y}) = \frac{(x_i - y_i)(x_j - y_j)}{|\mathbf{x} - \mathbf{y}|^3} = \frac{\partial \tilde{\lambda}_i}{\partial y_j}(\mathbf{x}, \mathbf{y}).$$

For $i = j$, we get

$$\frac{\partial \tilde{\lambda}_i}{\partial x_i}(\mathbf{x}, \mathbf{y}) = \frac{1}{|\mathbf{x} - \mathbf{y}|^3} \cdot \begin{cases} (x_1 - y_1)^2 + (x_2 - y_2)^2 & \text{if } i = 0 \\ (x_0 - y_0)^2 + (x_2 - y_2)^2 & \text{if } i = 1 \\ (x_0 - y_0)^2 + (x_1 - y_1)^2 & \text{if } i = 2. \end{cases}$$

Now we evaluate these formulae at the points $(\mathbf{x},\mathbf{y})$ with $\lambda(\mathbf{x},\mathbf{y}) = p$. For each such point, we found $x_1 = x_2 = y_1 = y_2 = 0$. Hence we get

$$\frac{\partial \tilde{\lambda}_i}{\partial x_j}(\mathbf{x},\mathbf{y}) = 0 = \frac{\partial \tilde{\lambda}_i}{\partial y_j}(\mathbf{x},\mathbf{y}) \text{ for } i \neq j,$$

$$\frac{\partial \tilde{\lambda}_i}{\partial x_i}(\mathbf{x},\mathbf{y}) = \frac{1}{|x_0 - y_0|} \cdot \begin{cases} 0 & \text{if } i = 0 \\ 1 & \text{if } i = 1,2, \end{cases}$$

and

$$\frac{\partial \tilde{\lambda}_i}{\partial y_i}(\mathbf{x},\mathbf{y}) = \frac{1}{|x_0 - y_0|} \cdot \begin{cases} 0 & \text{if } i = 0 \\ -1 & \text{if } i = 1,2. \end{cases}$$

Now we are equipped to study the linear map $d\lambda_{(\mathbf{x},\mathbf{y})} \colon T_{\mathbf{x}}\mathbb{S}_a \times T_{\mathbf{y}}\mathbb{S}_b \to T_p\mathbb{S}^2$. A basis of $T_p\mathbb{S}^2$ is given by the vectors $\mathbf{e}_2 = (0,1,0)$ and $\mathbf{e}_3 = (0,0,1)$. At each of the points $(\mathbf{x},\mathbf{y})$ with $\lambda(\mathbf{x},\mathbf{y}) = \mathbf{p}$, the vector $\mathbf{a} = (0,1,0)$ is a basis of $T_{\mathbf{x}}\mathbb{S}_a$, and the vector $\mathbf{b} = (0,1,1)$ provides a basis of $T_{\mathbf{y}}\mathbb{S}_b$. The map $d\lambda_{(\mathbf{x},\mathbf{y})}$ sends $\mathbf{a}$ to $\frac{1}{|x_0 - y_0|} \cdot (0,1,0)$ and $\mathbf{b}$ to $\frac{1}{|x_0 - y_0|} \cdot (0,-1,-1)$. Hence we have $d\lambda_{(\mathbf{x},\mathbf{y})}(\mathbf{a}) = \frac{1}{|x_0 - y_0|} \cdot \mathbf{e}_2$ and $d\lambda_{(\mathbf{x},\mathbf{y})}(\mathbf{b}) = -\frac{1}{|x_0 - y_0|} \cdot \mathbf{e}_2 - \frac{1}{|x_0 - y_0|} \cdot \mathbf{e}_3$. These two vectors form a basis of $T_p\mathbb{S}^2$, and we see that $(\mathbf{x},\mathbf{y})$ is a regular point. Since this is true for all three points in the fibre of $p \in \mathbb{S}^2$, we conclude that p actually is a regular value. □

11.6 Exercises

11.1 Let $\mathbb{T} = \mathbb{S}^1 \times \mathbb{S}^1 \subset \mathbb{C} \times \mathbb{C}$ denote the torus, where we consider the unit circle $\mathbb{S}^1$ as a subspace of the field of complex numbers $\mathbb{C}$. We define the map $f \colon \mathbb{T} \to \mathbb{C}$ by $f(z,w) = z^2 \cdot w$ using multiplication in $\mathbb{C}$. Compute $\deg_2(f)$.

11.2 Let $X \xrightarrow{f} Y \xrightarrow{g} Z$ be a sequence of smooth maps between manifolds, with X and Y compact, Y and Z connected. Assume that all three manifolds are boundaryless and $\dim X = \dim Y = \dim Z$. Show that the degree is multiplicative, i.e., $\deg_2(g \circ f) = \deg_2(g) \cdot \deg_2(f)$.

11.3 Let $f \colon X \to Y$ be a smooth map between smooth manifolds, with X compact, Y connected, and $\dim X = \dim Y$.
(a) Show that if $\deg_2(f) \neq 0$, then f is surjective.
(b) Show that if Y is not compact, then $\deg_2(f) = 0$.

11.4 Let $X = Y = \mathbb{S}^1 \subset \mathbb{R}^2$ be the unit circle and assume that $f \colon \mathbb{S}^1 \to \mathbb{S}^1$ is a smooth map without fixed points. Show that f is surjective. Hint:

Show that f is homotopic to the antipodal map $\alpha\colon \mathbb{S}^1 \to \mathbb{S}^1$, $x \mapsto -x$. What is $\deg_2(\alpha)$? Use Exercise 11.3.

11.5 Use the degree to prove the following slightly less general version of Proposition 10.6: let X be a compact smooth manifold with boundary ∂X. Assume that ∂X is connected. Then there is no smooth map $f\colon X \to \partial X$, which restricts to the identity on the boundary, i.e., $f_{|\partial X} = \mathrm{id}_{\partial X}$.

11.6 Let m be an odd number, and let $p(z) = z^m + a_1 z^{m-1} + \cdots + a_m$ be a complex polynomial. Show that there exists a $w \in \mathbb{C}$ such that $p(w) = 0$.

11.7 Let $f_1, \ldots, f_k$ be k smooth real-valued odd functions, i.e., $f_i : \mathbb{S}^k \to \mathbb{R}$ with $f(-x) = -f(x)$. Show that $f_1, \ldots, f_k$ must have a common zero, i.e., there is an $x \in \mathbb{S}^k$ such that $f_1(x) = \cdots = f_k(x) = 0$. Hint: Use Theorem 11.20 and Corollary 11.24.

11.8 Let $g_1, \ldots, g_k$ on $\mathbb{S}^k$ be smooth real-valued functions. Show that there exists a point $p \in \mathbb{S}^k$ such that $g_1(p) = g_1(-p), \ldots, g_k(p) = g_k(-p)$. Hint: Use Exercise 11.7.

11.9 Let $p_1, \ldots, p_n$ be real polynomials in $n+1$ variables. Assume each p_i is homogeneous of odd order, i.e., there is an odd number m_i such that $p_i(\lambda x) = \lambda^{m_i} p_i(x)$ for all $\lambda \in \mathbb{R}$. We consider each p_i as a smooth function $\mathbb{R}^{n+1} \to \mathbb{R}$ by sending x to $p_i(x)$. Show that there is a line through the origin in $\mathbb{R}^{n+1}$ on which all the p_i simultaneously vanish. Hint: Use Theorem 11.20 and the previous exercises.

11.10 Show that there is no smooth map $f\colon \mathbb{S}^2 \to \mathbb{S}^1$ with $f(-p) = -f(p)$ for all $p \in \mathbb{S}^2$. Hint: Assume such a map f exists. Then we can define the map $g\colon \mathbb{B}^2 = \{(x,y) \in \mathbb{R}^2 \colon x^2 + y^2 \leq 1\} \to \mathbb{S}^1$ by $g(x,y) := f\left(x, y, \sqrt{1 - x^2 - y^2}\right)$. Show that g satisfies $g(-q) = -g(q)$ for all $q \in \mathbb{S}^1 = \partial\mathbb{B}^2$. What is the degree modulo 2 of g? Conclude that f cannot exist.

11.11 Use Exercise 11.10 to prove the following special case of our previous observations: for every smooth map $\mathbb{S}^2 \to \mathbb{R}^2$, there is a point p on $\mathbb{S}^2$ such that $f(p) = f(-p)$.

11.12 Assume the following continuous version of the assertion of Exercise 11.11: for every continuous map $\mathbb{S}^2 \to \mathbb{R}^2$, there is a point p on $\mathbb{S}^2$ such that $f(p) = f(-p)$. Deduce from this fact the following version of Invariance of Dimension: an open subset in $\mathbb{R}^2$ cannot be homeomorphic to an open subset in $\mathbb{R}^n$ for $n \geq 3$.

12

Tubular Neighbourhoods and Transversality

In this chapter, we will study several advanced tools in differential topology. We begin with a generalisation of the Inverse Function Theorem. Next, we introduce the normal bundle of a submanifold, a smooth vector bundle that is complementary to the tangent bundle. Like the tangent bundle, the linear structure of its fibres provides additional geometric control. We then prove that every smooth manifold embedded in Euclidean space admits an open neighbourhood, called a tubular neighbourhood, which closely resembles its normal bundle. Both the normal bundle and tubular neighbourhoods are powerful tools for performing geometric constructions, and we study the following applications in this chapter. First, we prove Whitney's Approximation Theorem, which enables us to replace continuous maps between manifolds with smooth ones, which greatly expands the scope of differential topology. Then we prove Ehresmann's Fibration Theorem on fibre bundles, a generalisation of the concept of vector bundles. Finally, we discuss Thom's Transversality Theorems, which play a central role in intersection theory and numerous other applications. We refer to [14] for a shorter exposition of the material in this chapter.

12.1 Normal Bundles and Tubular Neighbourhoods

In this section, we introduce two key tools in differential topology, of which we will see several important applications in the following sections and chapters.

12.1.1 A Generalisation of the Inverse Function Theorem

For several constructions in this chapter, we need the following generalisation of the Inverse Function Theorem.

Theorem 12.1 (Generalisation of the Inverse Function Theorem) *Let X and Y be smooth manifolds, and let Z be a submanifold of X. Let $f\colon X \to Y$ be a smooth map. We assume that f maps Z diffeomorphically onto $f(Z)$, and that, for all $x \in Z$, $df_x\colon T_x(X) \to T_{f(x)}(Y)$ is an isomorphism. Then f maps an open neighbourhood of Z in X diffeomorphically onto an open neighbourhood of $f(Z)$ in Y.*

We note that if Z is a single point, then the theorem reduces to the usual Inverse Function Theorem 3.4. For the proof of Theorem 12.1, we recall the following notion from general topology.

Definition 12.2 (Locally finite subcovers) An open cover $\{U_\alpha\}$ of a smooth manifold X is called *locally finite* if each point of X possesses an open neighbourhood that intersects only finitely many of the sets U_α.

Lemma 12.3 *Every open cover $\{V_\alpha\}$ on a manifold admits a locally finite refinement $\{U_\alpha\}$.*

Proof: For $X \subset \mathbb{R}^N$, this follows from the fact that all metric spaces satisfy the assertion of the lemma. This is a fact due to Stone. See for example [42, theorem 9.1.7] or [6]. □

Proof of Theorem 12.1: Since df_x is an isomorphism for all $x \in Z$, the Inverse Function Theorem 3.4 implies that, for each $x \in Z$, there exists an open neighbourhood V_x in X around x on which $f_{|V_x}$ is a diffeomorphism. Then $f(V_x)$ is an open subspace in Y. The union of all V_x is an open neighbourhood of Z. The collection of the $f(V_x)$ is an open cover of $f(Z)$, and the union is an open neighbourhood of $f(Z)$. Now we need to show that we can define an inverse of f. By Lemma 12.3, we can choose a locally finite subcover $\{U_i\}$ of $f(Z)$ in Y. For each U_i, there is a local inverse $g_i\colon U_i \to V_i \subset X$ of $f_{|V_i}$. We define $W := \{y \in U_l : g_l(y) = g_j(y) \text{ when } y \in U_i \cap U_j\}$. On the subset W, we can define an inverse $g\colon W \to X$ of f by $g(y) = g_i(y)$ for any i. This is well defined by the construction of W. Since the g_i are local inverses of f, we have $f(Z) \subset W$. To show that we get a smooth inverse, we need to show that W contains an open subset which still contains $f(Z)$. Let $x \in Z$. Since $f(x) \in f(Z)$, we can find a k such that $f(x) \in U_k$. Assume U_k is not contained in W. Since the cover $\{U_i\}$ is locally finite, only finitely many of its members intersect U_k, say $U_1, \dots, U_n$. Then, for $i = 1, \dots, n$, we set C_{ik} to be the closure of the set $\{y \in U_i \cap U_k\colon g_i(y) \neq g_k(y)\}$. Since the union of a finite collection of closed subsets is closed, $C_k := C_{1k} \cup \dots \cup C_{nk}$ is closed in Y. Hence $U_x := U_k \setminus C_k$ is open in Y. By definition of W and the C_k, we know $U \subset W$. It remains to show that $f(x)$ is still in U_x, i.e., that it does not belong to one of the closures C_{ik}. By definition of the g_i and since $f_{|V_x}$ is a diffeomorphism, we have $g_i(f(z)) = z = g_k(f(z))$ for all $z \in V_x \cap g_i(U_i) \cap g_k(U_k)$ for each $i = 1, \ldots, n$. Hence the finite intersection $f(V_x) \cap U_k \cap U_1 \cap \cdots \cap U_n$ is an open subset which is not contained in any of the sets $\{y \in U_i \cap U_k : g_i(y) \neq g_k(y)\}$. Thus $f(x)$ is not contained in C_k. Hence we have shown that $U_x \subset W$ is an open subset containing $f(x)$. Then the union U of the U_x for all x in Z is an open subset of W containing $f(Z)$. Then $V := f^{-1}(U)$ is an open subset of X containing Z such that $f_{|V} : V \to U$ has the smooth inverse $g_{|U}$. This proves the theorem. □

12.1.2 Normal Bundle

Let $X \subset \mathbb{R}^m$ be a smooth manifold without boundary. We now seek a deeper understanding of how X is embedded in its ambient space. To this end, we introduce an important geometric construction.

Definition 12.4 (The Normal Bundle) For each $x \in X$, we define $N_x(X)$, the *normal space of X at x*, to be the orthogonal complement of $T_x(X)$ in $\mathbb{R}^m$. The *normal bundle $N(X)$* is then defined to be the space

$$N(X) := \{(x, v) \in X \times \mathbb{R}^m : v \in N_x(X)\}$$

together with the projection map $\sigma : N(X) \to X$ defined by $\sigma(x, v) = x$.

Remark 12.5 Unlike the tangent bundle, $N(X)$ is not intrinsic to the manifold X, i.e., $N(X)$ depends on the relationship between X and the surrounding space $\mathbb{R}^m$.

The normal bundle $N(X)$ is actually a smooth manifold itself. In order to show this, we must recall a fact from linear algebra.

Lemma 12.6 *Let $A : \mathbb{R}^m \to \mathbb{R}^k$ be a surjective linear map. Then the transpose $A^\top : \mathbb{R}^k \to \mathbb{R}^m$ maps $\mathbb{R}^k$ isomorphically onto the orthogonal complement of the kernel of A in $\mathbb{R}^m$, i.e., $A^\top : \mathbb{R}^k \xrightarrow{\cong} (\mathrm{Ker}\, A)^\perp$.*

Proof: First, we note that $A^\top$ is injective. We recall that $A^\top$ is characterised by the dot product equation $Av \cdot w = v \cdot A^\top w$ for all $v \in \mathbb{R}^m, w \in \mathbb{R}^k$. If $A^\top w = 0$, then $Av \cdot w = v \cdot A^\top w = 0$, so that $w \perp \mathrm{Im}\, A$. Since A is surjective, w must be zero. Now, if $v \in \mathrm{Ker}\, A$, i.e., $Av = 0$, then $0 = Av \cdot w = v \cdot A^\top w$. Thus $\mathrm{Im}\, A^\top \perp \mathrm{Ker}\, A$. Hence $A^\top$ maps $\mathbb{R}^k$ injectively into the orthogonal complement of $\mathrm{Ker}\, A$. Since $\mathrm{Ker}\, A$ has dimension $m - k$, its complement has dimension k, so $A^\top$ is surjective, too. □

Proposition 12.7 (Normal bundles are manifolds) *Let $X \subset \mathbb{R}^m$ be a smooth n-dimensional manifold without boundary. Then $N(X)$ is a smooth manifold of dimension m, and the projection $\sigma\colon N(X) \to X$ is a submersion.*

Proof: Let x be any point in X. The inclusion $X \subset \mathbb{R}^m$ is an immersion, and hence, by the Local Immersion Theorem 3.11, we can find an open subset $V \subset \mathbb{R}^m$ containing x and local coordinate functions $(u_1, \ldots, u_m)\colon V \to \mathbb{R}^m$ such that $X \cap V = \{v \in V : u_{n+1}(v) = \cdots = u_m(v) = 0\}$. We define the smooth map $\varphi\colon V \to \mathbb{R}^{m-n}$ by $\varphi(v) := (u_{n+1}(v), \ldots, u_m(v))$. We then set $U := \varphi^{-1}(0) = X \cap V$. Note that, since V is open in $\mathbb{R}^m$, U is open in X. Then U is a smooth manifold, and we can consider the normal bundle $N(U)$ of $U \subset \mathbb{R}^m$. We observe that, by construction, $N(U)$ equals $N(X) \cap (U \times \mathbb{R}^m)$. Thus $N(U)$ is open in $N(X)$. For each $x \in U$, $d\varphi_x\colon \mathbb{R}^m \to \mathbb{R}^{m-n}$ is surjective and has kernel $T_x(U) = T_x(X)$ by Proposition 5.5. Therefore, by Lemma 12.6, we get an isomorphism $(d\varphi_x)^\top\colon \mathbb{R}^{m-n} \xrightarrow{\cong} (T_x(X))^\perp = N_x(X)$. This shows that the map $\psi\colon U \times \mathbb{R}^{m-n} \to N(U)$ given by $\psi(x, v) := (x, d\varphi_x^\top(v))$ is bijective. Moreover, ψ defines an embedding of $U \times \mathbb{R}^{m-n}$ into $U \times \mathbb{R}^m$ since it is the identity on the first factor and an injective linear map on the second factor. Hence ψ is a diffeomorphism. Thus we have shown that $N(U)$ is a smooth manifold with local parametrisation ψ of dimension $\dim N(U) = \dim U + m - n = n + m - n = m$. Since every point of $N(X)$ has such an open neighbourhood $N(U)$, $N(X)$ is a smooth manifold. This proves the first assertion. For the second assertion, we note that $\sigma \circ \psi\colon U \times \mathbb{R}^k \to U$ is the projection onto the first factor, which is a submersion. Thus $d(\sigma \circ \psi)_{(u,w)}$ is surjective at every point (u, w). Hence $d\sigma_u$ is surjective at every u, and σ is a submersion. □

Remark 12.8 Proposition 12.7 implies that $\sigma\colon N(X) \to X$ is a smooth vector bundle in the sense of Section 2.5.2, which justifies to call it the normal bundle.

12.1.3 The Tubular Neighbourhood Theorem

Now we are going to study an important application of normal bundles. We begin with a lemma that provides the existence of an open neighbourhood of a smooth manifold in its ambient Euclidean space. We note that the existence of partitions of unity is crucial for the proof of the following result.

Lemma 12.9 (ε-Neighbourhood Lemma) *Let $X \subset \mathbb{R}^n$ be a smooth manifold without boundary. Then there is a smooth function $\varepsilon\colon X \to \mathbb{R}^{>0}$ such that every open subset U of $\mathbb{R}^n$ with $X \subset U$ contains the subspace*

$$X^\varepsilon = \{y \in \mathbb{R}^n : |y - x| < \varepsilon(x) \textit{ for some } x \in X\},$$

which is open in $\mathbb{R}^n$. *If* X *is compact,* ε *can be chosen to be constant.*

Proof: Let U be an open subset of $\mathbb{R}^n$ containing X. For each point $x \in X$, we can find a small radius ε_x such that the open ball $\mathbb{B}_{2\varepsilon_x}(x)$ is contained in U. We set $U_x^{\varepsilon_x} := X \cap \mathbb{B}_{\varepsilon_x}(x)$. We then define the open subset $U_x := \{y \in \mathbb{R}^n : |y - x'| < \varepsilon_x \text{ for some } x' \in U_x\}$. We now show that $U_x^{\varepsilon_x}$ is contained in U. Let $y \in U_x$. Then there is an $x' \in U_x$ with $|y - x'| < \varepsilon_x$. Since $x' \in U_x$ means $|x' - x| < \varepsilon_x$, the triangle inequality implies $|y - x| \leq |y - x'| + |x' - x| < 2\varepsilon_x$. Thus $y \in \mathbb{B}_{2\varepsilon_x}(x) \subset U$ by the choice of ε_x. This proves $U_x \subset U$. The collection of all U_x forms an open cover $\{U_x\}$ of $X \subset \mathbb{R}^n$. By the existence of partitions of unity for subsets in $\mathbb{R}^n$, we can choose a partition of unity $\{\rho_i\}$ subordinate to the cover $\{U_x\}$. Now we define the function $\varepsilon\colon X \to \mathbb{R}^{>0}$ by setting $\varepsilon(x) := \sum_i \rho_i(x)\varepsilon_x$. Note that ε is a smooth function, since all the ρ_i are smooth. Now we define X^ε as in the lemma. It remains to show $X^\varepsilon \subset U$. Let $y \in X^\varepsilon$. Then there is a $x \in X$ such that $|y - x| < \varepsilon(x)$. For this x, only finitely many of the numbers $\rho_i(x)$ are non-zero. This shows that there are finitely many points $x_1, \ldots, x_m$ such that $y \in U_{x_1} \cap \cdots \cap U_{x_m}$. Let ε_{x_m} be the maximum of the finitely many numbers $\varepsilon_{x_1}, \ldots, \varepsilon_{x_n}$. Then, since $\sum_i \rho_i(x) = 1$, we have $\varepsilon(x) \leq \varepsilon_{x_m}$. Hence $|y - x| < \varepsilon(x) \leq \varepsilon_{x_m}$ which implies $y \in U_{x_m}^{\varepsilon_{x_m}} \subset U$. Thus $X^\varepsilon \subset U$. If X is compact, we can reduce $\{U_x\}$ to a finite cover $U_{x_1}, \ldots, U_{x_n}$, and let ε be the maximum of the ε_{x_j}. □

Now we would like to realise such neighbourhoods X^ε of X using the normal bundle $N(X)$ of X in $\mathbb{R}^n$. Since $N(X)$ is a smooth manifold in $\mathbb{R}^n \times \mathbb{R}^n$, we can define the smooth map

$$h\colon N(X) \to \mathbb{R}^n, \ (x, v) \mapsto x + v.$$

Geometrically, h maps each normal space $N_x(X)$ to the affine subspace containing x, which is orthogonal to T_xX. The map h connects the open neighbourhood X^ε with the normal bundle in the following way.

Definition 12.10 (Tubular neighbourhoods) Let $X \subset \mathbb{R}^n$ be a smooth manifold without boundary. A *tubular neighbourhood of* X is an open subset X^ε of $\mathbb{R}^n$ containing X such that h maps an open subspace $N^\varepsilon(X) \subset N(X)$ diffeomorphically onto X^ε where the space $N^\varepsilon(X)$ is given by

$$N^\varepsilon(X) := \big\{(x, v) \in \mathbb{R}^n : |v| < \varepsilon(x)\big\}$$

for a smooth function $\varepsilon\colon X \to \mathbb{R}^{>0}$.

A key feature of smooth manifolds embedded in Euclidean space is that they always have a tubular neighbourhood. We now prove this important fact and then study some of its consequences.

Theorem 12.11 (Tubular Neighbourhood Theorem) *Let $X \subset \mathbb{R}^n$ be a smooth manifold without boundary. Then X has a tubular neighbourhood X^ε together with a submersion $\pi\colon X^\varepsilon \to X$ such that the restriction of π to X is the identity.*

Proof: Let $X_0 = N^0(X) = \{(x,0)\colon x \in X\} \subset N(X)$. The map $h\colon N(X) \to \mathbb{R}^n$ defined by $h(x,v) = x + v$ maps X_0 diffeomorphically onto X. Our goal is to show that the diffeomorphism on X_0 extends to an open neighbourhood around X_0. To do this, we show that $dh_{(x,0)}$ is an isomorphism for every point $(x,0)$ in X_0 as follows. First, we observe that, since $h_{|X_0}\colon X_0 \to X$ is a diffeomorphism, $dh_{(x,0)}$ maps $T_{(x,0)}(X_0) \subset T_{(x,0)}N(X)$ isomorphically onto $T_x(X) \subset \mathbb{R}^n$. Second, since the restriction of h to the normal space $N_x(X) \subset N(X)$ is given by $v \mapsto x + v$, we see that $dh_{(x,0)}$ maps $T_{(x,0)}(\{x\} \times N_x(X)) \subset T_{(x,0)}(N(X))$ isomorphically onto $N_x(X) \subset \mathbb{R}^n$. Since we have $T_x(\mathbb{R}^n) = T_x(X) \oplus N_x(X)$, this shows that $dh_{(x,0)}$ is surjective for every x. Since $\dim T_{(x,0)}(N(X)) = n$, this implies that $dh_{(x,0)}$ is an isomorphism for every x.

Hence the assumptions of the generalised Inverse Function Theorem 12.1 are satisfied, and we can conclude that h maps an open neighbourhood of X_0 in $N(X)$ diffeomorphically onto an open neighbourhood of X in $\mathbb{R}^n$. By the ε-Neighbourhood Lemma 12.9, every open neighbourhood of X contains an X^ε for a smooth function $\varepsilon\colon X \to \mathbb{R}^{>0}$. It follows directly from the definition of h that X^ε is the image under h of the open neighbourhood $N^\varepsilon(X)$ of X_0 in $N(X)$ for the same function ε. Thus, we have shown that there is a smooth map $h^{-1}\colon X^\varepsilon \to N^\varepsilon(X) \subset N(X)$. We define the map π by

$$\pi = \sigma_{|N^\varepsilon(X)} \circ h^{-1}\colon X^\varepsilon \xrightarrow{h^{-1}} N^\varepsilon(X) \xrightarrow{\sigma} X.$$

Since h^{-1} is a diffeomorphism and σ is a submersion, we conclude that π a is submersion. It follows directly from the definition of h and σ that the restriction of π to $X \subset X^\varepsilon$ is the identity. □

Remark 12.12 We emphasise that a key point of the theorem is that we do not only have X^ε, but that we also know that h restricts to a diffeomorphism between X^ε and $N^\varepsilon(X)$. Moreover, for many applications, it is crucial that we have in addition the submersion π.

Remark 12.13 It is crucial that we can find an open neighbourhood of X_0 in $N(X)$ which is diffeomorphic to an open neighbourhood of Z in Y. For it is clear that X is diffeomorphic to X_0 which is closed in $N(X)$. The difference may become apparent when we look at the dimensions, i.e., $\dim X^{\varepsilon} = n$, whereas $\dim X < n$ in general.

Remark 12.14 Recall from Lemma 12.9 that, if X is compact, then $\varepsilon > 0$ can be chosen constant, and X^{ε} is the open set of points in $\mathbb{R}^n$ with distance less than ε from X. If ε is sufficiently small, then each point $w \in X^{\varepsilon}$ possesses a unique closest point in X. Writing $\pi(w)$ for this unique point, defines the map $\pi\colon X^{\varepsilon} \to X$ in this case. See also Exercise 12.2.

12.1.4 The Relative Normal Bundle

We can now consider the normal bundle of a submanifold $Z \subset X$.

Definition 12.15 (Normal Bundle to a submanifold) Let $X \subset \mathbb{R}^m$ be a smooth manifold without boundary, and let Z be a submanifold of X. We define the *normal bundle to Z in X* to be the space

$$N(Z,X) := \{(z,v)\colon z \in Z, v \in T_z(X) \text{ and } v \perp T_z(Z)\}$$

together with the projection map $\sigma\colon N(Z,X) \to Z$ defined by $\sigma(z,v) = z$.

We think of $N(Z,X)$ as the *relative normal bundle*, since we take the normal space within the tangent space of X, not in the ambient Euclidean space. Note that for any $z \in Z$, the preimage $\sigma^{-1}(z) = N_z(Z,X)$ is the space of normal vectors to Z at z in $T_z(X)$.

Proposition 12.16 (Normal bundles are manifolds revisited) *The normal bundle $N(Z,X)$ is a smooth manifold of dimension equal to* $\dim X$. *The projection map $\sigma\colon N(Z,X) \to Z$, $\sigma(z,v) = z$, is a submersion.*

Proof: Let z be a point in Z. Let n be the codimension of Z in $\mathbb{R}^m$, i.e., $\dim Z = m - n$, and k be the codimension of Z in X. The inclusion $Z \subset X$ is an immersion, and hence, by the Local Immersion Theorem 3.11, we can find an open subset $V \subset \mathbb{R}^m$ containing z and local coordinate functions $(u_1,\ldots,u_m)\colon V \to \mathbb{R}^m$ such that $Z \cap V = \{v \in Z\colon u_1(v) = \cdots = u_n(v) = 0\}$ and such that we have $X \cap V = \{v \in X\colon u_{k+1}(v) = \cdots = u_n(v) = 0\}$. Let φ be the smooth map given by $\varphi = (u_1,\ldots,u_n)\colon V \to \mathbb{R}^n$. We set $U := Z \cap V$. We showed in the proof of Proposition 12.7 that the map $\psi\colon U \times \mathbb{R}^n \to N_U(Z,\mathbb{R}^m) := (U \times \mathbb{R}^m) \cap N(Z,\mathbb{R}^m)$ defined by $\psi(u,v) := \big(u, d\varphi_u^{\top}(v)\big)$

is a local parametrisation of $N(Z,\mathbb{R}^m) = N(Z)$. By restricting ψ to elements in $U \times \mathbb{R}^k \subset U \times \mathbb{R}^n$, we get a smooth map ϕ defined as the composite

$$\begin{array}{ccc} U \times \mathbb{R}^k & \xrightarrow{\psi} & N_U(Z,\mathbb{R}^m) \\ & \searrow^{\phi} & \downarrow{\scriptstyle \mathrm{id}\times p} \\ & & N_U(Z,X) \end{array}, \tag{12.1}$$

where $N_U(Z,X) := (U \times \mathbb{R}^m) \cap N(Z,X)$ and p is the map induced by the orthogonal projection $p_z \colon \mathbb{R}^m \to T_z(X)$ at each z. Note that, for a vector $w \in \mathbb{R}^m$ which satisfies $w \perp T_z(Z)$, we have $p(w) \in T_z(X)$ and $p(w) \perp T_z(Z)$. Let $\tilde{\varphi} = (u_{k+1}, \ldots, u_n)\colon V \to \mathbb{R}^{n-k}$. We observe that, by our choice of φ and $\tilde{\varphi}$, we know $T_z(Z) = (\operatorname{Ker} d\varphi_z) \subset \operatorname{Ker}(d\tilde{\varphi}_z) = T_z(X)$ and the orthogonal projection p_z varies smoothly with z. At each $z \in U$, the dimension of the kernel of the composite

$$\mathbb{R}^n \xrightarrow{d\varphi_z^\top} N_z(Z,\mathbb{R}^m) \xrightarrow{p_z} N_z(Z,X)$$

is $\dim \operatorname{Ker}(p_z) = \dim N_z(Z,\mathbb{R}^m) - \dim N_z(Z,X)$, since $d\varphi_z^\top$ is an isomorphism. This dimension is $\dim N_z(Z,\mathbb{R}^m) - \dim N_z(Z,X) = n-k$. Thus, the map $\phi\colon U \times \mathbb{R}^k \to N_U(Z,Y)$ is a diffeomorphism, being the identity on the first factor and a linear isomorphism on the second factor. This shows ϕ is a local parametrisation of $N(Z,X)$. Since $N_U(Z,X)$ is open in $N(Z,X)$ and every point in $N(Z,X)$ lies in such an $N_U(Z,X)$, we conclude that $N(Z,X)$ is a smooth manifold of dimension $\dim N(Z,X) = \dim U + \dim \mathbb{R}^k = \dim X$. The composition $\sigma \circ \phi\colon U \times \mathbb{R}^k \to U$ is the projection onto the first factor, and hence a submersion. Thus $d(\sigma \circ \phi)_{(u,v)}$ is surjective at every point (u,v). Hence $d\sigma_u$ is surjective at every u, and σ is a submersion. □

Remark 12.17 Proposition 12.16 implies that $\sigma : N(Z,X) \to Z$ is a smooth vector bundle, again justifying the terminology.

Example 12.18 (Normal bundle to sphere) Consider $\mathbb{S}^{k-1}$ as a submanifold of $\mathbb{S}^k$ via the embedding $(x_1,\ldots,x_k) \mapsto (x_1,\ldots,x_k,0)$. The tangent space $T_p(\mathbb{S}^{k-1})$ is embedded in $T_p(\mathbb{S}^k)$ as the subspace consisting of vectors with the last coordinate being 0. Let $v_k \in T_p(\mathbb{S}^k)$ be a vector which spans the orthogonal complement of $T_p(\mathbb{S}^{k-1})$ in $T_p(\mathbb{S}^k)$. Then we can define a map $\mathbb{S}^{k-1} \times \mathbb{R} \to N(\mathbb{S}^{k-1},\mathbb{S}^k)$ by $(p,\lambda) \mapsto (p,\lambda v_k)$. This map is a diffeomorphism with inverse given by $(p,\lambda v_k) \mapsto (p,\lambda)$.

Remark 12.19 Recall that an n-dimensional vector bundle which is diffeomorphic to the product of the base space with $\mathbb{R}^n$ is called *trivial*. Hence we just showed that $N(\mathbb{S}^{k-1}, \mathbb{S}^k)$ is a trivial one-dimensional bundle. We get a similar result when we consider $\mathbb{S}^{k-1} \subset \mathbb{R}^k$ for $k \geq 2$. Then, at any $p \in \mathbb{S}^{k-1}$, the unit vector $p/|p|$ spans the normal complement to $T_p(\mathbb{S}^{k-1})$ in $\mathbb{R}^k$. Hence there is a diffeomorphism $\mathbb{S}^{k-1} \times \mathbb{R} \to N(\mathbb{S}^{k-1}, \mathbb{R}^k)$ given by $(p, \lambda) \mapsto (p, \lambda p/|p|)$. This shows that $N(\mathbb{S}^{k-1}, \mathbb{R}^k)$ is a trivial one-dimensional bundle over $\mathbb{S}^{k-1}$.

12.1.5 Tubular Neighbourhood: The Relative Case

Now we are going to extend the tubular neighbourhood theorem to submanifolds. We begin with the neighbourhood lemma.

Lemma 12.20 (ε-Neighbourhood Lemma – relative version) *Let $X \subset \mathbb{R}^n$ be a smooth manifold without boundary, and let $Z \subset X$ be a submanifold without boundary. There is a smooth function $\varepsilon\colon Z \to \mathbb{R}^{>0}$ such that every open subset U of X with $Z \subset U$ contains the subspace*

$$Z^\varepsilon = \{x \in X\colon |x - z| < \varepsilon(z) \text{ for some } z \in Z\},$$

which is open in X. If Z is compact, ε can be chosen to be constant.

Proof: First, we consider Z as a manifold in $\mathbb{R}^n$. By Lemma 12.9, there is an ε-neighbourhood $\widetilde{Z^\varepsilon} \subset \mathbb{R}^n$, which is contained in every open subset $V \subset \mathbb{R}^n$ with $Z \subset V$. Let $U \subset X$ be an open subset in X with $Z \subset U$. Then there is an open subset $V \subset \mathbb{R}^n$ such that $U = V \cap X$ by definition of the subspace topology. Since $Z \subset V$, we have $Z \subset \widetilde{Z^\varepsilon} \subset V$. We can then set $Z^\varepsilon := (\widetilde{Z^\varepsilon} \cap X) \subset U$, and Z^ε is open in X with $Z \subset Z^\varepsilon$. □

Theorem 12.21 (Tubular Neighbourhoods – relative version) *Let $X \subset \mathbb{R}^n$ be a smooth manifold without boundary, and let Z be a submanifold of X without boundary. Then there is a smooth function $\varepsilon\colon Z \to \mathbb{R}^{>0}$ such that there is a diffeomorphism of the open neighbourhood Z^ε of Z in X to an open neighbourhood $N^\varepsilon(Z, X)$ of $Z_0 := Z \times \{0\}$ in $N(Z, X)$. Moreover, there is a submersion $Z^\varepsilon \to Z$, which restricts to the identity on Z.*

Proof: By the Tubular Neighbourhood Theorem 12.11 applied to $X \subset \mathbb{R}^n$, we have the open neighbourhood $X^{\varepsilon_X} \subset \mathbb{R}^n$ for some positive smooth function ε_X and a submersion $\tilde{\pi}\colon X^{\varepsilon_X} \to X$. We define the smooth map $h\colon N(Z, X) \to \mathbb{R}^n$ by $h(z, v) = z + v$. The inverse image $W := h^{-1}(X^{\varepsilon_X}) \subset N(Z, X)$ is an open neighbourhood of Z_0 in $N(Z, X)$. Since $h(z, 0) = z$ for all $z \in Z$, the composition $f = \tilde{\pi} \circ h\colon W \xrightarrow{h} X^{\varepsilon_X} \xrightarrow{\tilde{\pi}} X$ is the identity when we restrict it to Z_0. By the same argument as in the proof of Theorem 12.11, we can show that $dh_{(z,0)}$ is an isomorphism at every point of Z_0 in $N(Z,X)$. Since $d\tilde{\pi}_z$ is the identity for all $z \in Z \subset X^{\varepsilon X}$, $df_{(z,0)}$ is an isomorphism for every $(z,0) \in Z_0 \subset N(Z,X)$. Hence the assumptions of the generalised Inverse Function Theorem 12.1 are satisfied for the map f and the submanifold Z, and we can conclude that there is an open neighbourhood V of Z_0 in $N(Z,X)$ which is mapped diffeomorphically onto an open neighbourhood U of Z in X by $f = \tilde{\pi} \circ h$. By Lemma 12.20, we can then find a positive smooth function $\varepsilon\colon Z \to \mathbb{R}^{>0}$ such that $Z \subset Z^{\varepsilon} \subset U$, and Z^{ε} is diffeomorphic to an open neighbourhood $N^{\varepsilon}(Z,X)$ of Z_0 in $N(Z,X)$. The composition

$$\pi : Z^{\varepsilon} \xrightarrow{\cong} N^{\varepsilon}(Z,X) \xrightarrow{\sigma_{N^{\varepsilon}(Z,X)}} Z$$

is the desired submersion. □

Finally, we make the following observation that we will use later.

Corollary 12.22 *Let $X \subset \mathbb{R}^n$ be a smooth manifold without boundary, and let Z be a submanifold of X. Then $\sigma : N(Z,X) \to Z$ is a trivial vector bundle over Z if and only if there is a submersion $g\colon U \to \mathbb{R}^k$ defined on an open subset $U \subset X$ such that $Z = g^{-1}(0)$.*

Proof: First, we assume that $\sigma : N(Z,X) \to Z$ is trivial, i.e., there is a diffeomorphism $\varphi\colon Z \times \mathbb{R}^k \to N(Z,X)$ which commutes with σ and the projection to Z. By Theorem 12.21, there are open neighbourhoods $Z \subset Z^{\varepsilon} \subset X$ and $N^{\varepsilon}(Z,X) \subset N(Z,X)$ with a diffeomorphism $Z^{\varepsilon} \to N^{\varepsilon}(Z,X)$. Composition and restriction yield the desired submersion $g\colon Z^{\varepsilon} \to N^{\varepsilon}(Z,X) \to \mathbb{R}^k$ such that $Z = g^{-1}(0)$. Now we assume that there is a submersion $g\colon U \to \mathbb{R}^k$ with $Z = g^{-1}(0)$. Hence, for each $z \in Z \subset U \subset X$, $dg_z\colon T_zU = T_zX \to \mathbb{R}^k$ is a surjective linear map. This implies by Lemma 12.6 that the transpose $(dg_z)^{\top}$ maps $\mathbb{R}^k$ isomorphically onto the orthogonal complement of T_zZ in T_zX for every $z \in Z$. Thus, by definition of $N(Z,X)$, the map $\varphi\colon Z \times \mathbb{R}^k \to N(Z,X)$ defined by $\varphi(z,v) = (z,(dg_z)^{\top}(v))$ is a diffeomorphism which commutes with σ and the projection to Z. This shows that $\sigma : N(Z,X) \to Z$ is trivial. □

12.2 Whitney Approximation Theorem

In this section, we establish that every continuous map between smooth manifolds can be approximated by a smooth map. More precisely, we prove that any continuous map is homotopic to a smooth one. The central tools enabling this result are partitions of unity and tubular neighbourhoods. See for example also [29, chapter 6].

12.2.1 Whitney Approximation Theorem for Functions

Let X and Y be smooth manifolds embedded in Euclidean space $\mathbb{R}^N$, and let $A \subset X$ be a closed subset. Recall that we say that a map $f\colon A \to Y$ is smooth at $x \in A$ if there is an open neighbourhood U of a and a smooth map $F\colon U \to Y$ such that F and f agree on $U \cap A$. Using partitions of unity, we now show that we can choose a smooth map F which works for all $x \in A$ when Y is Euclidean space.

Proposition 12.23 (Extending smooth maps) *Let X be a smooth manifold with or without boundary, and let $A \subset X$ be a closed subset. Let $f\colon A \to \mathbb{R}^k$ be a smooth map. Then, for every open subset $U \subset X$ containing A, there exists a smooth map $F\colon X \to \mathbb{R}^k$ such that $F_A = f$ and* $\mathrm{supp}(F) \subset U$.

Proof: Let $U \subset X$ be an open subset containing A. For each $x \in A$, we choose an open subset U_x containing x and a smooth map $F_x\colon U_x \to \mathbb{R}^k$ such that F_x and f agree on $U_x \cap A$. After replacing U_x with $U_x \cap U$, we can assume that $U_x \subset U$. The collection of open sets $\{X \setminus A, U_x\}_{x\in A}$ is an open cover of X. Since X has a countable basis for its topology, we can choose countably many $x_i \in A$ such that $\{X \setminus A, U_{x_i}\}_{i=1}^{\infty}$ is still an open cover of X. By Theorem 7.21, we can choose a partition of unity $\{\rho_0, \rho_i\}$ subordinate to this cover such that $\mathrm{supp}(\rho_0) \subset X \setminus A$ and $\mathrm{supp}(\rho_i) \subset U_{x_i}$. For x_i, the product $\rho_i \cdot F_{x_i}\colon x \mapsto \rho_i(x) \cdot F_{x_i}(x)$ is smooth on U_{x_i}, and, since $\rho_i(x) \cdot F_{x_i}(x) = 0$ for $x \in U_{x_i} \setminus \mathrm{supp}(\rho_i)$, we can extend it to a smooth map $X \to \mathbb{R}^k$ by defining it to be zero on $X \setminus \mathrm{supp}(\rho_i)$. Then we define the map

$$F\colon X \to \mathbb{R}^k,\ F(x) := \sum_{x_i} \rho_i(x) \cdot F_{x_i}(x).$$

We note that, by definition of a partition of unity, for every x, there is an open neighbourhood of x such that only finitely many terms of the sum defining F are non-zero. This implies that F is a smooth map. Since $\{\rho_0, \rho_i\}$ is a partition of unity, we have $\rho_0(x) + \sum_i \rho_i(x) = 1$ for all $x \in X$. For $x \in A$, we have $\rho_0(x) = 0$ and $F_{x_i}(x) = f(x)$ for every x_i. Thus, for every $x \in A$, we have

$$F(x) = \sum_{x_i} \rho_i(x) \cdot F_{x_i}(x) = (\rho_0(x) + \sum_i \rho_i(x)) \cdot f(x) = f(x).$$

This shows that F and f agree on A. Finally, by for example [42, proposition 2.1.9], the union of a locally finite collection of closed subsets is closed.

Since the collection of the supports of a partition of unity is locally finite, we conclude that $\text{supp}(F) \subset \overline{\bigcup_{x_i} \text{supp}(\rho_i)} = \bigcup_{x_i} \text{supp}(\rho_i) \subset U$. This finishes the proof of the proposition. □

Now we show that maps into Euclidean space can be approximated in the following sense.

Definition 12.24 (ε-close functions) Let $\varepsilon\colon X \to \mathbb{R}^{>0}$ be a positive continuous function. We say that two maps $f, g\colon X \to \mathbb{R}^n$ are *ε-close* if $|f(x) - g(x)| < \varepsilon(x)$ for all $x \in X$.

Theorem 12.25 (Approximating maps into Euclidean space) *Let X be a smooth manifold with or without boundary, and let $f\colon X \to \mathbb{R}^n$ be a continuous map. For every positive continuous function $\varepsilon\colon X \to \mathbb{R}^{>0}$, there exists a smooth map $g\colon X \to \mathbb{R}^n$ which is ε-close to f. Moreover, if f is already smooth on a closed subset $A \subset X$, then g can be chosen equal to f on A.*

Proof: Let $A \subset X$ be a closed subset and assume that $f_{|A}$ is smooth. By Proposition 12.23, there is an open subset $U \subset X$ with $A \subset U$ and a smooth map $F\colon U \to \mathbb{R}^n$ such that $f_{|A} = F_{|A}$. We define the set U_0 as

$$U_0 := \{x \in X : |F(x) - f(x)| < \varepsilon(x)\}.$$

This is an open subset in X, since the function $x \mapsto \varepsilon(x) - |F(x) - f(x)|$ is continuous and U_0 is the inverse image of the open subset $\mathbb{R}^{>0}$ under this function. Note that $X \setminus A \subset U_0$, since $F(x) - f(x) = 0$ for all $x \in A$. If $A = \emptyset$, we set $U_0 := \emptyset$. Now let $x \in X \setminus A$ and U_x be an open neighbourhood of x contained in $X \setminus A$ which is small enough such that $|F(y) - f(x)| < \frac{1}{2}\varepsilon(x) < \varepsilon(y)$ for all $y \in U_x$. We can find such a neighbourhood since x is fixed and ε and F are continuous maps. The collection $\{U_x : x \in X \setminus A\}$ of all such open neighbourhoods for all x is an open cover of $X \setminus A$. Since X has a countable basis for its topology, we can choose a countable subcover $\{U_{x_i}\}_{i=1}^{\infty}$ of $X \setminus A$. Then $\{U_0, U_{x_i}\}$ is an open cover of X. Let $\{\rho_0, \rho_i\}$ be a smooth partition of unity subordinate to the open cover $\{U_0, U_{x_i}\}$. We define the map

$$g\colon X \to \mathbb{R}^n,\ g(x) := \rho_0(x)F(x) + \sum_{i=1}^{\infty} \rho_i(x) f(x_i).$$

Since F, ρ, and ρ_i are smooth and the $f(x_i)$ are fixed values, g is smooth. Moreover, we have $g_{|A} = F_{|A} = f_{|A}$. It remains to show that g and f are ε-close. Since $\sum_{i \geq 0} \rho_i(x) = 1$ and $|F(x) - f(x_i)| < \frac{1}{2}\varepsilon(x)$ for all $x \in X$, we get

$$
\begin{aligned}
|g(x) - f(x)| &= \left| \rho_0(x)F(x) + \sum_{i=1}^{\infty} \rho_i(x) f(x_i) - \left(\rho_0(x) + \sum_{i=1}^{\infty} \rho_i(x) \right) f(x) \right| \\
&\leq \rho_0(x)|F(x) - f(x)| + \sum_{i=1}^{\infty} \rho_i(x)|f(x_i) - f(x)| \\
&< \rho_0(x)\varepsilon(x) + \sum_{i=1}^{\infty} \rho_i(x)\varepsilon(x) = \varepsilon(x).
\end{aligned}
$$

This finishes the proof. □

12.2.2 Whitney Approximation Theorem between Manifolds

Now we can extend the approximation to smooth manifolds embedded in $\mathbb{R}^n$. Recall that two maps $f, g\colon X \to Y$ are said to be *homotopic relative to a subset* $A \subset X$ if there is a homotopy $H\colon X \times [0,1] \to Y$ between f and g such that $f(x) = H(x,t) = g(x)$ for all $x \in A$ and all $t \in [0,1]$.

Theorem 12.26 (Whitney Approximation Theorem) *Let X be a smooth manifold with or without boundary and let $Y \subset \mathbb{R}^n$ be a smooth manifold without boundary. Let $f\colon X \to Y$ be a continuous map. Then f is homotopic to a smooth map $X \to Y$. Moreover, if f is already smooth on a closed subset $A \subset X$, then the homotopy can be chosen relative to A.*

Proof: By Theorem 12.11, there is a tubular neighbourhood $Y^\varepsilon \subset \mathbb{R}^n$ of Y in $\mathbb{R}^n$ with associated positive smooth function $\varepsilon\colon Y \to \mathbb{R}^{>0}$ together with a submersion $\pi\colon Y^\varepsilon \to Y$ such that the restriction of π to X is the identity. We define $\tilde{\varepsilon}$ to be the positive continuous function given as the composition $\tilde{\varepsilon} = \varepsilon \circ f\colon X \to \mathbb{R}^{>0}$. By Theorem 12.25, there is a smooth map $g\colon X \to \mathbb{R}^n$ with $g_{|A} = f_{|A}$ and which is $\tilde{\varepsilon}$-close to f. Now we define the map $H\colon X \times [0,1] \to Y$ as the composition of π with a straight-line homotopy between f and g, i.e.,

$$H(x,t) = \pi((1-t)f(x) + tg(x)).$$

We will now show that H is well defined. We know that, for all $x \in X$, $|g(x) - f(x)| < \tilde{\varepsilon}(x) = \varepsilon(f(x))$. Hence $g(x)$ is contained in the open ball around $f(x)$ with radius $\varepsilon(f(x))$. Thus, for all x, $g(x)$ and the whole line segment between $f(x)$ and $g(x)$ are contained in Y^ε by definition of Y^ε. This shows that H is well defined. Hence we conclude that H is a homotopy between $H(x,0) = \pi(f(x)) = f(x)$ and $H(x,1) = \pi(g(x))$. Moreover, we have $H(x,t) = f(x)$ for all $x \in A$. The map $\pi \circ g\colon X \xrightarrow{g} Y^\varepsilon \xrightarrow{\pi} Y$ is smooth and the desired smooth map homotopic to f. □

Remark 12.27 (No boundary is important) The assumption on Y being without boundary is crucial. For a simple example, let $X = \mathbb{R}$ and $Y = [0, \infty)$. Consider the continuous map $f: X \to Y$ given by $f(x) = |x|$ and let $A = [0, \infty)$. Then there does not exist a smooth map $g: X \to Y$ with $g_{|A} = f_{|A}$. For f is smooth on the open interval $(0, \infty)$ with $\frac{d}{dx} f(x) = 1$ for all $x \in (0, \infty)$. However, g would have to satisfy $g(x) \in Y$, i.e., $g(x) \geq 0$ for all $x \in X$. From what we learned in calculus, we can deduce that such a map g cannot be smooth. As a consequence, we get that there is no smooth homotopy $h: X \times [0,1] \to Y$ with $h(x,t) = f(x)$ for all $x \in A$ and all $t \in [0,1]$.

Theorem 12.28 (Homotopic maps are smoothly homotopic) *Let X be a smooth manifold with or without boundary, and let Y be a smooth manifold without boundary. Let $f, g: X \to Y$ be smooth maps. Then, if f and g are homotopic, then they are smoothly homotopic. Moreover, if f and g are homotopic are relative to some closed subset $A \subset X$, then they are smoothly homotopic relative to A.*

Proof: Let $h: X \times [0,1] \to Y$ be a continuous homotopy between f and g relative to A. We would like to apply Whitney's Theorem 12.26 to h. However, $X \times [0,1]$ may not be a manifold with boundary in our sense. Hence, we first extend h to a continuous map $\tilde{h}: X \times \mathbb{R} \to Y$ by setting

$$\tilde{h}(x,t) := \begin{cases} h(x,t), & t \in [0,1] \\ h(x,0), & t \leq 0 \\ h(x,1), & t \geq 1. \end{cases}$$

The restriction of $\tilde{h}$ to $X \times \{0\} \cup X \times \{1\}$ is smooth, since it is equal to $f \circ \pi_1$ on $X \times \{1\}$ and $g \circ \pi_1$ where $\pi_1: X \times \mathbb{R} \to X$ denotes the projection. If h is a homotopy relative to A, then $\tilde{h}$ is also smooth on $A \times [0,1]$. Now, since $X \times \mathbb{R}$ is a smooth manifold, with or without boundary, Whitney's Theorem 12.26 implies that there is a smooth map $H: X \times \mathbb{R} \to Y$ whose restriction to $X \times \{0\} \cup X \times \{1\} \cup A \times [0,1]$ agrees with $\tilde{h}$. By restricting H to $X \times [0,1]$, we get the desired smooth homotopy between f and g relative to A. □

We can now prove the following stronger form of Theorem 11.11.

Theorem 12.29 (Compact manifolds are not contractible) *Let X be a connected compact smooth manifold of dimension at least one without boundary. Then X is not contractible, i.e., the identity map on X is not homotopic to a constant map.*

Proof: By Whitney's Theorem 12.26, we know that if there were a continuous homotopy between the identity and a constant map, then there would be a smooth homotopy between them. Since such a smooth homotopy does not exist by Theorem 11.11. □

12.3 Brouwer Invariance of Domain

We now use Whitney's theorem, the Brouwer degree, and the Borsuk–Ulam Theorem to prove the following fundamental result in topology.

Theorem 12.30 (Brouwer Invariance of Domain) *For $n \geq 1$, let $U \subset \mathbb{R}^n$ be an open subset, and let $f: U \to \mathbb{R}^n$ be an injective smooth map. Then $f(U)$ is open in $\mathbb{R}^n$.*

Remark 12.31 We note that the general form of Brouwer's Invariance of Domain Theorem in [8] only requires the map to be continuous. Since the proof we present below uses the Brouwer degree and the Borsuk–Ulam Theorem in their form developed in this book, we make the additional assumption that the map is smooth. This suffices for the application in the proof of Theorem 13.17.

For the proof of Theorem 12.30, we are going to use the following terminology from general topology. See also [42, section 3.4].

Definition 12.32 (Locally path-connected spaces) A topological space X is called *locally path-connected at a point $x \in X$* if, for each neighbourhood U of x, there is a path-connected neighbourhood V of x such that $V \subseteq U$. The space X is called *locally path-connected* if it is locally connected at each point.

Example 12.33 Euclidean space $\mathbb{R}^n$ is locally path-connected since the collection of all open balls forms a basis for the topology on $\mathbb{R}^n$, and open balls are path-connected. Every smooth manifold of dimension n is then locally path-connected since it is locally homeomorphic to $\mathbb{R}^n$.

Let X be a topological space and $x \in X$. Recall that the *path-component of x in X* is the maximal path-connected subset of X containing x. This is the set of all points $y \in X$ for which there is a path between x and y.

Proposition 12.34 *A topological space is locally path-connected if and only if the path-components of each open subset of X are open.*

Proof: If the path-components of each open subset of X are open, then X is locally path-connected by definition. Now we assume that X is locally path-connected. Let $U \subset X$ be an open subset. Let $P \subset U$ be a path-component of U, and let $x \in P$. By the assumption, there is a path-connected neighbourhood V of x such that $V \subseteq U$. Since $x \in P \cap V$, $P \cup V$ is path-connected and contained in U. Since P is a maximal path-connected subset of U, we have $P \cup V = P$ and hence $V \subseteq P$. Thus, P is a neighbourhood of x, and P is open. □

Now we are ready to prove Brouwer's theorem on the invariance of domain.

Proof of Theorem 12.30: For $n = 1$, the open subsets of $\mathbb{R}$ are exactly the countable disjoint unions of open intervals. An injective smooth map f then sends such a disjoint union to another such disjoint union. Thus, f is an open map. Now we assume $n \geq 2$. To show that $f(U)$ is open in $\mathbb{R}^n$, it suffices to show that every point in $f(U)$ has a neighbourhood which is open in $\mathbb{R}^n$ and which is contained in $f(U)$. Let $x \in U$. Since U is open, we can find an $\varepsilon > 0$ such that the open ball $\mathbb{B}^n_{2\varepsilon}(x)$ is contained in U. Then the closed ball $\bar{\mathbb{B}}^n_\varepsilon(x)$ is still contained in U. We are going to show that $f(\mathbb{B}^n_\varepsilon(x))$ is open in $\mathbb{R}^n$. After rescaling and translating, we can assume that $x = 0$ and $\varepsilon = 1$, i.e., we can assume $\mathbb{B}^n_\varepsilon(x) = \mathbb{B}^n$, $f(\mathbb{B}^n) \subset f(U)$, $\partial\bar{\mathbb{B}}^n_\varepsilon(x) = \partial\bar{\mathbb{B}}^n = \mathbb{S}^{n-1}$, and $f(\mathbb{S}^{n-1}) \subset f(U)$ as well. We then want to show that $f(\mathbb{B}^n)$ is an open subset of $\mathbb{R}^n$. Since $\mathbb{S}^{n-1}$ is compact and f is continuous, $f(\mathbb{S}^{n-1})$ is compact by Lemma 1.13. Since $\mathbb{R}^n$ is Hausdorff, this implies that $f(\mathbb{S}^{n-1})$ is closed in $\mathbb{R}^n$ by Proposition 1.21. Hence $\mathbb{R}^n \setminus f(\mathbb{S}^{n-1})$ is an open neighbourhood of $f(0)$. This implies that $\mathbb{R}^n \setminus f(\mathbb{S}^{n-1})$ is locally path-connected since it is locally homeomorphic to $\mathbb{R}^n$. By Proposition 12.34, this implies that the path-components of $\mathbb{R}^n \setminus f(\mathbb{S}^{n-1})$ are open subsets. Since $\mathbb{R}^n \setminus f(\mathbb{S}^{n-1})$ is open in $\mathbb{R}^n$, this implies that the path-components of $\mathbb{R}^n \setminus f(\mathbb{S}^{n-1})$ are open in $\mathbb{R}^n$, too. Let P denote the path-component of $f(0)$ in $\mathbb{R}^n \setminus f(\mathbb{S}^{n-1})$. Since $\mathbb{B}^n$ is path-connected and f is continuous, $f(\mathbb{B}^n)$ is path-connected and hence $f(\mathbb{B}^n) \subseteq P$. We will now show that $P \subseteq f(\mathbb{B}^n)$ and hence $f(\mathbb{B}^n) = P$. Let $y \in P$. Assume that $y \notin f(\mathbb{B}^n)$. In particular, we then have $f(x) - y \neq 0$, and we can define the constant map $c_y\colon \mathbb{S}^{n-1} \to \mathbb{S}^{n-1}$ by $c_y(z) := (f(0) - y)/|f(0) - y|$. Since $y \in \mathbb{R}^n \setminus f(\mathbb{S}^{n-1})$, we can define the smooth map $g_y\colon \mathbb{S}^{n-1} \to \mathbb{S}^{n-1}$ by $g_y(z) := (f(z) - y)/|f(z) - y|$. The map $F\colon \mathbb{S}^{n-1} \times [0,1] \to \mathbb{S}^{n-1}$ defined by $F(z,t) := (f(tz) - y)/|f(tz) - y|$ is then a homotopy from c_y to g_y. Now let $g\colon \mathbb{S}^{n-1} \to \mathbb{S}^{n-1}$ be the smooth map defined by $g(z) := (f(z) - f(0))/|f(z) - f(0)|$. Since $y \in P$, there is a continuous path γ from y to $f(0)$. Hence we can define the map $G\colon \mathbb{S}^{n-1} \times [0,1] \to \mathbb{S}^{n-1}$ by $G(z,t) := (f(z) - \gamma(t))/|f(z) - \gamma(t)|$. Then G is a homotopy from g_y to g. Since f is smooth and injective, the map $\varphi\colon \mathbb{S}^{n-1} \to \mathbb{S}^{n-1}$ given by $\varphi(z) := (f(z) - f(-z))/|f(z) - f(-z)|$ is well defined and smooth. The map $H\colon \mathbb{S}^{n-1} \times [0,1] \to \mathbb{S}^{n-1}$ defined by

$H(z,t) := (f(z) - f(-tz))/|f(z) - f(-tz)|$ is then a homotopy from g to φ. Since being homotopic is an equivalence relation, the homotopies F, G, and H show that there is a continuous homotopy between φ and the constant map c_y. By Whitney's Theorem 12.28, this implies that there is a smooth homotopy between φ and the constant map c_y. Since the degree is invariant under smooth homotopy, this would imply $\deg_2(\varphi) = 0$ by Example 11.7 as $n - 1 \geq 1$. However, since φ is a smooth map satisfying $\varphi(-z) = -\varphi(z)$, the Borsuk–Ulam Theorem 11.21 implies that $\deg_2(\varphi) \neq 0$. Hence the claimed homotopy between φ and the constant map c_y cannot exist, and the assumption that $y \notin f(\mathbb{B}^n)$ must have been wrong. This proves $f(\mathbb{B}^n) = P$ and hence finishes the proof of the theorem. □

Corollary 12.35 *Let X and Y be smooth n-dimensional manifolds, and let $f: X \to Y$ be a smooth injective map. Then f is an open map, i.e., for every open subset $V \subset X$, $f(V)$ is open in Y.*

Proof: Let $V \subset X$ be open. Let $x \in V$ and hence $f(x) \in f(V)$. It suffices to show that $f(x)$ has an open neighbourhood in Y which is contained in $f(V)$. Then $f(V)$ is the union of these open neighbourhoods for all $x \in V$, and hence $f(V)$ is open in Y. We can assume that V is the domain of a chart (V, ψ_X) with homeomorphism $\psi_X: V \to U$ with $U \subset \mathbb{R}^n$ open. After shrinking V if necessary, we can assume that there is a chart (W, ψ_Y) on Y with $f(V) \subset W$. Moreover, we can assume that $\psi_Y: W \to \mathbb{R}^n$ is a homeomorphism between W and $\mathbb{R}^n$. Since f is smooth and injective, the map $\bar{f} := \psi_Y \circ f \circ (\psi_X)^{-1}: U \to \mathbb{R}^n$ is smooth and injective. By Theorem 12.30, $\bar{f}(U)$ is open in $\mathbb{R}^n$. Since ψ_Y is a homeomorphism, this implies that $(\psi_Y)^{-1}(\bar{f}(U))$ is open in Y. Since $\psi_X: V \to U$ is a homeomorphism, we have $(\psi_Y)^{-1}(\bar{f}(U)) = f(V)$. This proves the assertion. □

Corollary 12.36 (Brouwer Invariance of domain for smooth manifolds) *Let X be a compact smooth n-dimensional manifold, and let Y be a connected smooth n-dimensional manifold. Let $f: X \to Y$ be a smooth map which is a homeomorphism onto its image. Then f is surjective and hence a homeomorphism.*

Proof: Since X is compact and f is continuous, $f(X)$ is compact by Lemma 1.13. Since Y is Hausdorff, this implies that $f(X)$ is closed in Y by Proposition 1.21. Since f is smooth and injective, Corollary 12.35 implies that $f(X)$ is open in Y, too. Since Y is connected, this implies $f(X) = Y$. □

Remark 12.37 Note that if the map f in Corollary 12.36 is also an immersion, then the Inverse Function Theorem implies that f is open and hence a diffeomorphism. Theorem 12.30, however, makes it possible to conclude that f is a homeomorphism without additional local assumptions.

12.4 Ehresmann Fibration Theorem

We now study another important consequence of the existence of tubular neighbourhoods. To begin, we introduce the definition of a particularly interesting class of maps.

Definition 12.38 (Locally trivial fibrations) Let X and Y be smooth manifolds. A smooth map $f : X \to Y$ is called a *locally trivial fibration* if for each $y \in Y$ there is an open neighbourhood $U \subset Y$ of y and a diffeomorphism $\varphi : f^{-1}(U) \to U \times f^{-1}(y)$ such that the diagram

$$\begin{array}{ccc} f^{-1}(U) & \xrightarrow{\quad\varphi\quad} & U \times f^{-1}(y) \\ & {\scriptstyle f_{|f^{-1}(U)}} \searrow \quad \swarrow {\scriptstyle p_U} & \\ & U & \end{array} \tag{12.2}$$

commutes where p_U denotes the projection onto the first factor.

We note that, if Y is connected, then the fibres of a locally trivial fibration $f : X \to Y$ are all diffeomorphic.

Examples 12.39 Every projection $p_X : X \times Z \to X$ is a locally trivial fibration. In fact, it is globally trivial. Every smooth vector bundle $\pi : E \to Y$ is a locally trivial fibration. Moreover, for a smooth vector bundle $\pi : E \to Y$, we can form the *sphere bundle* $f : \mathbb{S}(E) \to Y$ as follows. We set $\mathbb{S}(E) := \{v \in E : |v| = 1 \text{ if } \pi(v) = y\} \subset E$, where the norm is taken in the Euclidean space in which E is embedded. We let f be the restriction of π to $\mathbb{S}(E)$. Then f is a locally trivial fibration. Finally, we remark that the Hopf map $\hbar : \mathbb{S}^3 \to \mathbb{S}^2$ is a locally trivial fibration. This will become clear after we have proven Theorem 12.40.

The following famous theorem due to Ehresmann [11] provides a sufficient criterion for a map to be a locally trivial fibration based on the concepts we have studied so far.

Theorem 12.40 (Ehresmann Fibration Theorem) *Let $f: X \to Y$ be a proper submersion. Then f is a locally trivial fibration. In particular, if X is compact, every submersion is a locally trivial fibration.*

Remark 12.41 Let $y_0 \in Y$ be a given basepoint in Y. Theorem 12.40 tells us that we can think of a proper submersion $f: X \to Y$ as a family of diffeomorphic copies of the fibre $f^{-1}(y_0)$ varying over the points in Y. We note that the second assertion follows from the first since every continuous map with a compact domain is proper. Ehresmann's theorem is highly influential in many areas of geometry and topology. For example, in complex and algebraic geometry, it implies that the higher direct images of a constant sheaf along f form a local system on Y. See for example [48].

Proof of Theorem 12.40: We assume that $f^{-1}(y)$ is non-empty for $y \in Y$. Since f is a submersion, y is a regular value for f. Hence, by the Preimage Theorem 4.12, $Z := f^{-1}(y)$ is a submanifold of X of dimension $\dim Z = \dim X - \dim Y$. By Theorem 12.21, there exist a tubular neighbourhood Z^ε of Z which is open in X and a submersion $\pi : Z^\varepsilon \to Z$ such that the restriction to Z is the identity. Then we can define the map $q := (\pi, f_{|Z^\varepsilon}): Z^\varepsilon \to Z \times Y$. Since f is proper, $Z = f^{-1}(y)$ is compact. Since $\pi_{|Z} = \mathrm{id}_Z$, q is a diffeomorphism onto its image. Moreover, by Proposition 4.13, we have $T_z(Z) = \mathrm{Ker}\,(df_z)$. Since df_z is surjective onto T_yY and $d\pi_z$ is surjective onto T_zZ, dq_z is a surjective linear map. Since the dimension of T_zZ^ε is equal to $\dim X = \dim Z + \dim Y$, dq_z is an isomorphism. Hence we can apply the Inverse Function Theorem 12.1 to $Z \subset Z^\varepsilon$ and q. Thus there is a neighbourhood $Z \subset W \subset Z^\varepsilon$ of Z which is open in Z^ε such that $q_{|W}$ is a diffeomorphism onto an open neighbourhood of $q(Z)$ in $Z \times Y$. Finally, since f is proper and Y is Hausdorff, f is a closed map by Proposition 1.27. Hence, by Exercise 1.2, there is an open neighbourhood $U \subset Y$ around y such that $Z \subset f^{-1}(U) \subset W$. Then we have $q(f^{-1}(U)) = U \times Z$. This shows that $\varphi := q_{f^{-1}(U)}: f^{-1}(U) \to U \times f^{-1}(y)$ is a diffeomorphism such that $f_{|f^{-1}(U)} = p_U \circ \varphi$. This finishes the proof of the theorem. □

12.5 Thom Transversality Theorems

We now use normal bundles and the existence of tubular neighbourhoods to show that submanifolds can always be slightly deformed, and smooth maps can be adjusted up to homotopy to achieve transversality. One of the principal applications of the results developed in this section will be the construction of an intersection theory for smooth manifolds, which we will present in Chapter 13. The results in this section are due to René Thom, see [44, 45], in particular. Our exposition follows [14, chapter 2, §3].

12.5.1 Thom Transversality Theorem

We begin with the following extension of Sard's Theorem 9.21.

Theorem 12.42 (Thom Transversality Theorem) *Let X, Y, and S be smooth manifolds. We assume that S and Y do not have a boundary. Let Z be a submanifold of Y without boundary. Let $F\colon X \times S \to Y$ be a smooth map. For $s \in S$, we let f_s denote the map $x \mapsto f_s(x) = F(x,s)$. We assume that both F and ∂F are transverse to Z. Then the set of $s \in S$ such that both f_s and ∂f_s are transverse to Z is a dense subset of S.*

We observe that the image of $T_{(x,s)}(X \times S)$ under $d(f_s)_x$ is potentially smaller than the image under $dF_{(x,s)}$. This is why we have to work to prove the theorem.

Proof of Theorem 12.42: Since both F and ∂F are transverse to Z, the Preimage Theorem 9.20 implies that $F^{-1}(Z)$ is a submanifold of $X \times S$ with boundary $\partial F^{-1}(Z) = F^{-1}(Z) \cap (\partial X \times S)$. Let $\pi\colon X \times S \to S$ be the projection map. We will show that whenever $s \in S$ is a regular value for the restriction $\pi\colon F^{-1}(Z) \to S$, then $f_s \pitchfork Z$, and whenever s is a regular value for $\partial\pi\colon \partial F^{-1}(Z) \to S$, then $\partial f_s \pitchfork Z$. By Sard's Theorem 9.21, the set of regular values for $\pi_{|F^{-1}(Z)}$ is a dense subset in S, and the assertion of the theorem follows. Now let $s \in S$ be a regular value for the restriction $\pi\colon F^{-1}(Z) \to S$. Suppose $f_s(x) = z \in Z$. We need to show $d(f_s)_x(T_x(X)) + T_z(Z) = T_z(Y)$. Let $w \in T_z(Y)$. We need to find a vector $v \in T_x(X)$ such that $d(f_s)_x(v) - w \in T_z(Z)$. Since $F(x,s) = z$ and F is transverse to Z, we have $dF_{(x,s)}(T_{(x,s)}(X \times S)) + T_z(Z) = T_z(Y)$. Hence there exists a vector $a \in T_{(x,s)}(X \times S)$ such that $dF_{(x,s)}(a) - w \in T_z(Z)$. Since $T_{(x,s)}(X \times S) = T_x(X) \times T_s(S)$, we can write a as a pair (b,c) for vectors $b \in T_x(X)$ and $c \in T_s(S)$. The map $d\pi_{(x,s)}\colon T_x(X) \times T_s(S) \to T_s(S)$ is the projection onto the second factor. Since s is a regular value for $\pi_{|F^{-1}(Z)}$, the restriction of $d\pi_{(x,s)}$ to $T_{(x,s)}(F^{-1}(Z))$ is surjective. In particular, the fibre of $(d\pi_{(x,s)})_{|T_{(x,s)}(F^{-1}(Z))}$ over $c \in T_s(S)$ is non-empty. Thus, there is some vector of the form (b',c) in $T_{(x,s)}(F^{-1}(Z))$. Since the restriction $F_{|F^{-1}(Z)}$ is a map to Z, we have $dF_{(x,s)}(b',c) \in T_z(Z)$. Since $d(f_s)_x$ is the restriction of $dF_{(x,s)}$ to $T_x(X) \times \{0\} \subset T_x(X) \times T_s(S)$, the vector $v := b - b' \in T_x(X)$ satisfies

$$\begin{aligned} d(f_s)_x(v) - w &= dF_{(x,s)}(b - b',0) - w \\ &= dF_{(x,s)}((b,c) - (b',c)) - w \\ &= (dF_{(x,s)}(b,c) - w) - dF_{(x,s)}(b',c). \end{aligned}$$

Since both $dF_{(x,s)}(b,c) - w$ and $dF_{(x,s)}(b',c)$ belong to $T_z(Z)$, we see that the vector $d(f_s)_x(v) - w$ belongs to $T_z(Z)$ as we set out to prove. As a special instance of what we just proved, for the case of the boundaryless manifold ∂X and the map $\partial F\colon (\partial X) \times S \to Y$, we then also get $\partial f_s \pitchfork Z$. □

12.5.2 Transversality Is Generic

Theorem 12.42 shows that transversality for smooth maps $X \to \mathbb{R}^N$ to Euclidean space is generic in the following sense. Let $f\colon X \to \mathbb{R}^N$ be a smooth map, and let $Z \subset \mathbb{R}^N$ be a smooth manifold. Let $\mathbb{B}$ be an open ball in $\mathbb{R}^N$. We define the smooth map $F\colon X \times \mathbb{B} \to \mathbb{R}^N$ by $F(x,s) := f(x) + s$. The derivative of F at (x,s) is $dF_{(x,s)} = (df_x, \mathrm{id}_{\mathbb{R}^N})\colon T_x(X) \times \mathbb{R}^N \to \mathbb{R}^N$. Thus $dF_{(x,s)}$ is surjective at every (x,s), and F is a submersion. Similarly, $d(\partial F)_{(x,s)} = (d(\partial f)_x, \mathrm{id}_{\mathbb{R}^N})$, and hence ∂F is a submersion. This implies that F and ∂F are transverse to Z. Hence, by Theorem 12.42, the set of $s \in \mathbb{B}$ such that the map $x \mapsto f_s(x) = f(x) + s$ is transverse to Z is dense in $\mathbb{B}$. Moreover, the map

$$X \times [0,1] \to \mathbb{R}^N,\ (x,t) \mapsto f_{ts}(x) = f(x) + t \cdot s$$

provides a smooth homotopy between f and f_s. This shows that we can always find a small perturbation f_s of f such that f_s is transverse to Z. Now we would like to generalise this result to an arbitrary smooth manifold $Y \subset \mathbb{R}^N$ without boundary and smooth map $f\colon X \to Y$. Given a submanifold $Z \subset Y$, we have just learned how to vary a smooth map $f\colon X \to Y \subset \mathbb{R}^N$ as a family of maps $X \to \mathbb{R}^N$ such that $f_s \pitchfork Z$ for small s, where we consider Z as a submanifold in $\mathbb{R}^N$. It remains to project these maps down to Y such that a small perturbation f_s of f remains transverse to the given submanifold $Z \subset Y$. The key tool to do this is a suitable tubular neighbourhood. We show that this works in two steps. First, we show that we can vary f in a family of smooth maps to obtain a submersion. Then we use Theorem 12.42 to show transversality.

Proposition 12.43 (Creating families of submersions) *Let X be a smooth manifold, and let $Y \subset \mathbb{R}^N$ be a smooth manifold without boundary. Let $\mathbb{B}$ be the open unit ball in $\mathbb{R}^N$. Let $f\colon X \to Y$ be a smooth map. Then there is a smooth map $F\colon X \times \mathbb{B} \to Y$ such that $F(x,0) = f(x)$, and, for every fixed*

$x \in X$, the map $F(x, -)\colon \mathbb{B} \to Y$, defined by $s \mapsto F(x, s)$ is a submersion. In particular, both F and ∂F are submersions.

Proof: By Theorem 12.11, there is a tubular neighbourhood Y^ε of Y in $\mathbb{R}^N$ together with a submersion $\pi : Y^\varepsilon \to Y$. We define the map $F\colon X \times \mathbb{B} \to Y$ by $F(x, s) := \pi(f(x) + \varepsilon(f(x))s)$. Since π restricts to the identity on Y, we have $F(x, 0) = \pi(f(x) + 0) = f(x)$. For fixed x, the map $\varphi\colon \mathbb{B} \to Y^\varepsilon$, $s \mapsto f(x) + \varepsilon(f(x))s$ is the translation of a linear map. Thus $d\varphi_s$ is just given by multiplying a vector in $T_s(\mathbb{B}) = \mathbb{R}^N$ by the real number $\varepsilon(f(x)) > 0$ to get a vector in $T_{\varphi(s)}(Y^\varepsilon) \subset \mathbb{R}^N$. This implies that $d\varphi_s$ is surjective, and φ is a submersion. Since the diagram

$$\begin{array}{ccc} & & Y^\varepsilon \\ & \overset{\varphi}{\nearrow} & \downarrow \pi \\ \mathbb{B} & \xrightarrow[F(x,-)]{} & Y \end{array} \tag{12.3}$$

commutes and the composition of two submersions is a submersion, we get that the map $F(x, -) = \pi \circ \varphi\colon \mathbb{B} \overset{\varphi}{\to} Y^\varepsilon \overset{\pi}{\to} Y$, $s \mapsto F(x, s)$, is a submersion. Hence the restriction $F_{|\{x\}\times\mathbb{B}}\colon \{x\} \times \mathbb{B} \to Y$ of F to the submanifold $\{x\} \times \mathbb{B}$ is a submersion for every $x \in X$. Since every point of $X \times \mathbb{B}$ lies in one of these submanifolds, F must be a submersion as well. The same argument applied to ∂F and ∂X shows that ∂F is a submersion. □

Theorem 12.44 (Transversality Homotopy Theorem) *Let X be a smooth manifold, and let $Y \subset \mathbb{R}^N$ be a smooth manifold without boundary. Let Z be a submanifold of Y without boundary. Let $f\colon X \to Y$ be a smooth map. Then there exists a smooth map $g\colon X \to Y$ which is homotopic to f such that g and ∂g are transverse to Z.*

Proof: Let $F\colon X \times \mathbb{B} \to Y$ be the submersion of Proposition 12.43. By the Transversality Theorem 12.42, the set of $s \in \mathbb{B}$ such that $f_s \pitchfork Z$ and $\partial f_s \pitchfork Z$ is a dense and therefore a non-empty subset of $\mathbb{B}$. Moreover, each f_s is homotopic to f, the homotopy being $(x, t) \mapsto F(x, ts)$. Thus, we can pick an $s \in \mathbb{B}$ such that $f_s \pitchfork Z$ and $\partial f_s \pitchfork Z$ and set $g := f_s$. □

As a special case, we consider the following application.

Corollary 12.45 (Removing intersections) *Let Y be a smooth manifold without boundary and let X be a compact submanifold of Y. Assume that X intersects another submanifold $Z \subset Y$ with $\dim X + \dim Z < \dim Y$. Then there is a homotopy $\iota_t\colon X \times [0, 1] \to Y$ such that ι_t is an embedding for each t and $\iota_1(X) \cap Z = \emptyset$.*

Proof: Let $\iota\colon X \hookrightarrow Y$ denote the inclusion of X in Y. Let $F\colon X \times \mathbb{B} \to Y$ be the submersion of Proposition 12.43 associated to ι. Then there is an $s \in \mathbb{B}$ such that $F_s\colon X \to Y$ is transverse to Z. However, since $\dim X + \dim Z < \dim Y$, F_s can only be transverse to Z if $F_s^{-1}(Z) = \emptyset$. Moreover, by Theorem 12.42, we can find such an $s \in \mathbb{B}$ with $|s|$ arbitrarily small. Since ι is an embedding, we can apply the Stability Theorem of Exercise 8.12 and find an s small enough so that F_{ts} is an embedding for every $t \in [0,1]$. We then set $\iota_t = F_{ts}\colon X \times [0,1] \to Y$ and get $\iota_1(X) \cap Z = \emptyset$. □

12.5.3 Thom Extension Theorem

The Tubular Neighbourhood Theorem 12.11 actually allows us to prove a stronger form of Theorem 12.44. We first need some terminology.

Definition 12.46 (Transversality on subsets) Let $f\colon X \to Y$ be a smooth map, $Z \subset Y$ a submanifold, and $C \subset X$ be a subset. We will say f is *transverse to Z on C* if the transversality condition

$$\mathrm{Im}\,(df_x) + T_{f(x)}(Z) = T_{f(x)}(Y) \tag{12.4}$$

is satisfied for all $x \in C \cap f^{-1}(Z)$.

Remark 12.47 Note that, even if C is a submanifold, the condition of Definition 12.46 is different from requiring that the restriction of f to C is transverse to Z since (12.4) involves $\mathrm{Im}\,(df_x) = df_x(T_x(X))$ and not, in general, smaller space $\mathrm{Im}\,(d(f_{|C})_x) = df_x(T_x(C))$.

Theorem 12.48 (Thom Extension Theorem) *Let X be a smooth manifold, and let Y be a smooth manifold without boundary. Let Z be a submanifold of Y without boundary. Let C be a closed subset of X. Let $f\colon X \to Y$ be a smooth map. Assume that $f \pitchfork Z$ on C and $\partial f \pitchfork Z$ on $C \cap \partial X$. Then there exists a smooth map $g\colon X \to Y$ homotopic to f such that $g \pitchfork Z$ and $\partial g \pitchfork Z$, and an open subset $U \subset X$ with $C \subset U$ such that $g_{|U} = f_{|U}$.*

For the proof of Theorem 12.48, we need the following lemma.

Lemma 12.49 *Let $C \subset X$ be a closed subset, and let $U \subset X$ be an open subset which contains C. Then there exists a smooth function $\varphi\colon X \to [0,1]$ that is equal to 1 outside U and equal to 0 on an open neighbourhood of C.*

Proof: Since U is open and $C \subset U$, we can choose, for each $c \in C$, an $\varepsilon_c > 0$ such that $\mathbb{B}_{2\varepsilon_c}(c) \subset U$. We set $V := \cup_{c\in C}\mathbb{B}_{\varepsilon_c}(c)$, and let $\overline{V}$ denote the closure. By construction, C is contained in the interior of $\overline{V}$. This implies that $X \setminus \overline{V}$

is an open neighbourhood of C. By Theorem 7.21, there is a partition of unity $\{\rho_i\}$ subordinate to the open cover $\{U, X \setminus \overline{V}\}$ of X. We note that we proved the existence of partitions of unity for arbitrary subsets in $\mathbb{R}^N$, and hence, in particular, also for smooth manifolds with boundary in $\mathbb{R}^N$. Then we define φ as the sum of those ρ_i that vanish outside of $X \setminus \overline{V}$. Then φ is a smooth function with $\varphi = 1$ on U by definition of a partition of unity subordinate to the cover $\{U, X \setminus \overline{V}\}$. □

Proof of Theorem 12.48: First, we show $f \pitchfork Z$ on an open subset containing C. Let $x \in C$. If $x \notin f^{-1}(Z)$, then, since Z is closed, $X \setminus f^{-1}(Z)$ is a neighbourhood of x on which $f \pitchfork Z$. If $x \in f^{-1}(Z)$, then there is an open neighbourhood W of $f(x)$ in Y and a submersion $\pi\colon W \to \mathbb{R}^k$ such that $f \pitchfork Z$ at a point $x' \in f^{-1}(Z \cap W)$ precisely when $\pi \circ f$ is regular at x'. If $\pi \circ f$ is regular at x, then it is regular in an open neighbourhood of x by Proposition 4.9. Thus, we have $f \pitchfork Z$ on an open neighbourhood of every point $x \in C$, and so $f \pitchfork Z$ on an open neighbourhood U of C in X. Let φ be a smooth function as in Lemma 12.49 for the closed subset C and the open subset U in X. We set $\tau := \varphi^2\colon X \to [0,1]$, i.e., $\tau(x) = \varphi(x) \cdot \varphi(x)$ in $[0,1]$. We have $d\tau_x = 2\varphi(x)d\varphi_x$, and hence $\varphi(x) = 0 = \tau(x)$ implies $d\tau_x = 0$. Let $\mathbb{B}$ be the unit ball in $\mathbb{R}^N$. Let $F\colon X \times \mathbb{B} \to Y$ be the submersion of Proposition 12.43.

- Claim: The map $G\colon X \times \mathbb{B} \to Y$, $G(x,s) := F(x, \tau(x)s)$, is transverse to Z.

Proof of the claim: We first assume $\tau(x) \neq 0$. Suppose $(x,s) \in G^{-1}(Z)$. Then the map $\mathbb{B} \to Y$, $r \mapsto G(x,r)$, is a submersion since it is the composition of the diffeomorphism $r \mapsto \tau(x)r$ and the submersion $r \mapsto F(x,r)$. Hence G is regular at (x,s), so certainly $G \pitchfork Z$ at (x,s). Now we assume $\tau(x) = 0$. We need to check that the image of the derivative $dG_{(x,s)}$ is still large enough. We define the map $\mu\colon X \times \mathbb{B} \to X \times \mathbb{B}$ by setting $\mu(x,s) := (x, \tau(x)s)$. To calculate the derivative $d\mu_{(x,s)}\colon T_x(X) \times \mathbb{R}^N \to T_x(X) \times \mathbb{R}^N$, we apply the product rule to the second coordinate and remember that $\tau(x)$ and $d\tau_x(v)$ are both elements in $\mathbb{R}$ for any $v \in T_x(X)$. Then we get

$$d\mu_{(x,s)}(v,w) = (v, \tau(x) \cdot w + d\tau_x(v) \cdot w).$$

We observe that $G = F \circ \mu$. Hence, in order to calculate the derivative of G, we can apply the chain rule. With $\tau(x) = 0$ and $d\tau_x = 0$ and the formula for $d\mu_{(x,s)}$, we then get $dG_{(x,s)}(v,w) = dF_{(x,s)}(v,0)$. Moreover, since $F(x,0) = f(x)$ for all x by construction of F, we know $F_{|X \times \{0\}} = f$. This implies $dF_{(x,s)}(v,0) = dF_{(x,0)}(v,0) = df_x(v)$. Hence we get $dG_{(x,s)}(v,w) = df_x(v)$ and therefore

$$\mathrm{Im}\,(dG_{(x,s)}) = \mathrm{Im}\,(df_x(v)) \subset T_{f(x)}(Y). \tag{12.5}$$

The assumption $\tau(x) = 0$ implies $x \in U$ by the definition of φ and τ. By the choice of U above, this implies $f \pitchfork Z$ at x. Hence (12.5) implies $G \pitchfork Z$ at (x,s). This proves the claim. □

The same argument shows $\partial G \pitchfork Z$. Now we can apply the Transversality Theorem 12.42 to $G\colon X \times S \to Y$ and conclude that we can pick and fix an s for which the map $g(x) := G(x,s)$ satisfies $g \pitchfork Z$ and $\partial g \pitchfork Z$. The map G is then a homotopy between the maps $f = F_{|X\times\{0\}} = G_{|X\times\{0\}}$ and $G_{|X\times\{s\}} = g$. Finally, if x is in the open neighbourhood of C on which $\tau = 0$, then $g(x) = G(x,s) = F(x,0) = f(x)$. This proves the theorem. □

Corollary 12.50 (Extension of maps on boundaries) *Let X be a smooth manifold, and let Y be a smooth manifold without boundary. Let Z be a submanifold of Y without boundary. Let $f\colon X \to Y$ be a smooth map such that $\partial f \pitchfork Z$. Then there exists a map $g\colon X \to Y$ homotopic to f such that $\partial g = \partial f$ and $g \pitchfork Z$. In particular, suppose there is a smooth map $h\colon \partial X \to Y$ transverse to Z. Then, if h extends to any smooth map $X \to Y$, then it extends to a smooth map $X \to Y$ that is transverse to Z.*

Proof: The boundary ∂X is closed in X, and the assertion follows from Theorem 12.48 with $C = \partial X$. □

As a consequence, we can prove the following strengthening of Corollary 12.45, which we will use in Section 16.3.

Proposition 12.51 (Removing intersections – revisited) *Let Y be a smooth manifold without boundary and let X be a compact submanifold of Y. Let $Z \subset Y$ be a closed submanifold with* $\dim X + \dim Z < \dim Y$. *We assume that X intersects Z, but not in ∂X, i.e., $\partial X \cap Z = \emptyset$. Then there is an open subset $U \subset Y$, which contains $X \cap Z$, and there is a homotopy $\iota_t\colon X \times [0,1] \to Y$ such that ι_t is an embedding for each t and $\iota_1(X) \cap Z = \emptyset$ and such that $\iota_1(X)$ and X are equal outside U.*

Proof: Let $\iota\colon X \hookrightarrow Y$ denote the inclusion of X in Y, and let $G\colon X \times S \to Y$ be the corresponding family of maps we constructed in the proof of Theorem 12.48. Since both X and Z are closed in Y, the intersection $X \cap Z$ is closed in Y and we set $C := X \cap Z$. Then there is an $s \in S$ such that $G_s := G(-,s)\colon X \to Y$ is transverse to Z. However, since $\dim X + \dim Z < \dim Y$, G_s can only be transverse to Z if $G_s^{-1}(Z) = \emptyset$. Moreover, by Theorem 12.42, we can find such an $s \in S$ with $|s|$ arbitrarily small. By the Stability Theorem of Exercise 8.12, since ι is an embedding, we can find an s which is small enough so that G_{ts} is an embedding for every $t \in [0,1]$. We set $\iota_t = G_{ts} : X \times [0,1] \to Y$ and get $\iota_1(X) \cap Z = \emptyset$. By the construction of G, there is an open neighbourhood U of $X \cap Z$ such that $\iota_1(X) \setminus U = X \setminus U$. □

12.6 Exercises

12.1 Prove the *General Position Lemma*: Let X and Y be submanifolds of $\mathbb{R}^N$. Show that the set of $a \in \mathbb{R}^N$ such that the translate $X + a$ is transverse to Y is dense in $\mathbb{R}^N$.

12.2 Let X be a compact submanifold of $\mathbb{R}^n$, and $w \in \mathbb{R}^n$. Let $\sigma : N(X) \to X$ denote the normal bundle.

(a) Show that there exists a (not necessarily unique) point $x \in X$ closest to w, and prove that $w - x \in N_x(X)$. Hint: If $c(t)$ is a curve on X with $c(0) = x$, then the smooth function $|w - c(t)|^2$ has a minimum at 0. Now use the fact that we have shown in Exercise 2.16 that there is a unique correspondence between tangent vectors at x and velocity vectors at 0 of curves $c\colon (-a,a) \to X$ with $c(0) = x$.

(b) Let $h\colon N(X) \to \mathbb{R}^n$, $h(x,v) = x+v$, be the map used in the proof of the Tubular Neighbourhood Theorem 12.11. We know that h maps a neighbourhood of X in $N(X)$ diffeomorphically onto $X^\varepsilon \subset \mathbb{R}^n$ for some small constant $\varepsilon > 0$. We define the smooth map $\pi = \sigma \circ h^{-1}$. Use the previous point to show that, if $w \in X^\varepsilon$, then $\pi(w)$ is the unique point of X closest to w.

12.3 Let $X \subset \mathbb{R}^N$ be a smooth manifold, and let $k \geq 1$ be an integer. Show that almost every vector space V of dimension k in $\mathbb{R}^N$ is transverse to X, i.e., $V + T_x(X) = \mathbb{R}^N$ for every $x \in X$. Hint: Use that the set $S \subset \left(\mathbb{R}^N\right)^k$ consisting of all linearly independent k-tuples of vectors in $\mathbb{R}^N$ is open in $\mathbb{R}^{Nk}$. Show that the map $\mathbb{R}^k \times S \to \mathbb{R}^N$ defined by $((t_1,\dots,t_k), v_1,\dots,v_k) \mapsto t_1v_1 + \cdots + t_kv_k$ is a submersion, and use the Transversality Theorem 12.42.

12.4 Let $f\colon \mathbb{R}^n \to \mathbb{R}^n$ be a smooth map with $n > 1$, and let $K \subset \mathbb{R}^n$ be compact and $\varepsilon > 0$. Show that there exists a map $g\colon \mathbb{R}^n \to \mathbb{R}^n$ such that dg_x is never 0, and $|f(x) - g(x)| < \varepsilon$ for all $x \in K$. Hint: Let $\mathrm{M}(n)$ be the space of $n \times n$-matrices. First show that the map $F\colon \mathbb{R}^n \times \mathrm{M}(n) \to \mathrm{M}(n)$, defined by $F(x, A) = df_x + A$, is a submersion. Pick A

so that $F_A \pitchfork \{0\}$ for $F_A \colon x \mapsto (x, A)$ as in the main text. Now use this knowledge to construct g. At some point along this way, you will have used $n > 1$. Make sure you see where and how it has been used.

12.5 Show that the assertion of Exercise 12.4 is false for $n = 1$, i.e., find f, ε, $K \subset \mathbb{R}$ such that we cannot find such a g. Hint: You could contemplate the Mean Value Theorem and the Intermediate Value Theorem from calculus.

13
Intersection Theory Modulo 2

In this chapter, we generalise the construction of the mod 2 Brouwer degree. Let $f: X \to Y$ be a smooth map, and let $Z \subset Y$ be a closed submanifold. We consider the preimage $f^{-1}(Z)$. Under suitable conditions, this preimage is a finite set, allowing us to count its points. Analogous to $\deg_2$, this leads to an interesting invariant known as the mod 2 intersection number. After defining intersection numbers, we will explore several applications. Additional examples can be found in the exercises. The techniques and results developed in Chapter 12 will play a crucial role in constructing intersection numbers.

13.1 Intersection Numbers Modulo 2

We begin with a simple case that will provide the intuition needed for the technical details to come in this chapter.

13.1.1 Intersecting Manifolds

We will make the following assumptions in this section. We will say that two submanifolds X and Z inside Y have *complementary dimension* if $\dim X + \dim Z = \dim Y$. We assume that all manifolds are without boundary for the moment. When $X \pitchfork Z$, then their intersection $X \cap Z$ is a smooth manifold with $\operatorname{codim}(X \cap Z)$ in X being equal to $\operatorname{codim} Z$ in Y by Theorem 5.9. Since $\operatorname{codim} Z = \dim X$, $X \cap Z$ is a zero-dimensional manifold. If we further assume that both X and Z are closed and that at least one of them, say X, is compact, then $X \cap Z$ must be a finite set of points. We are going to think of this finite number of points in $X \cap Z$ as the *intersection number* of X and Z. When X and Z are not transverse, however, it makes in general little sense to count the points in $X \cap Z$. But we have learned that we can deform smooth manifolds

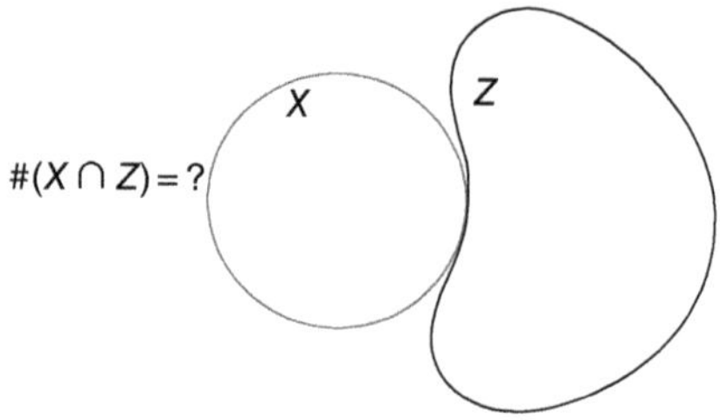

Figure 13.1 A tangential or non-transverse intersection is not stable. Any slight perturbation will change the number of intersection points.

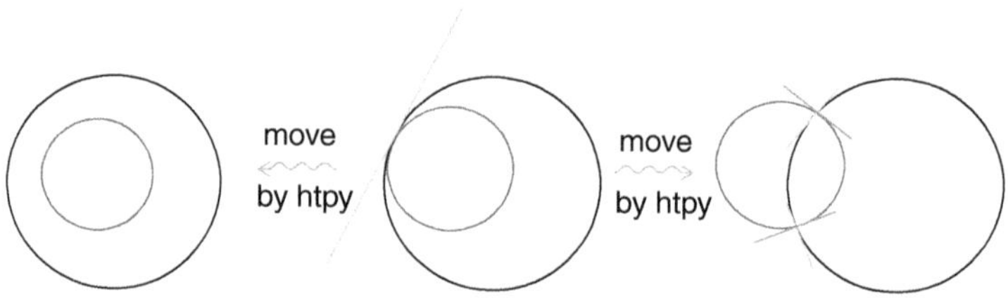

Figure 13.2 We can move the two circles in the plane and either move them apart or make them intersect transversely. In both cases, we get a stable number of intersection points. However, 0 is not equal to 2, unless we work modulo 2. That is what we are going to do.

to make intersections transverse, namely, we can alter them in homotopic families. See Figure 13.1. Moreover, by Exercise 8.12, embeddings form a stable class of maps, i.e., for any homotopy i_t of an embedding i_0, there is an $\varepsilon > 0$ such that i_t is still an embedding for all $t < \varepsilon$. Thus, given an embedding $\iota\colon X \hookrightarrow Y$, any slight deformation of i gives us another embedding $\iota\colon X \hookrightarrow Y$ and thus produces a smooth manifold that is a diffeomorphic copy of X. If $\iota' \pitchfork Y$, then we get an intersection number $\#(\iota'(X) \cap Y)$. However, the resulting intersection number may depend on how we move or deform X. For example, take two circles in $\mathbb{R}^2$. See Figure 13.2. Assume that their intersection is not transverse, i.e., they touch each other at a point such that both tangent spaces agree and together just span a line. Then we can move the circles by a simple translation $x \mapsto x + a$ in direction a such that they intersect either in two points or in no points. In both cases, the intersection is transverse, but the intersection numbers do not agree. We observe, however, that the parity of the intersection numbers is preserved, i.e., up to a multiple of 2 the intersection numbers agree.

13.1.2 Intersection Number Modulo 2

The above observation is the starting point for the following definition.

Definition 13.1 (Mod 2 intersection numbers) Let X and Y be smooth manifolds without boundary, and let $Z \subset Y$ be a closed submanifold. We assume that X is compact and $\dim X + \dim Z = \dim Y$. Let $f\colon X \to Y$ be a smooth map which is transverse to Z in Y. Then $f^{-1}(Z)$ is a closed submanifold of X of codimension equal to $\dim X$. Hence $f^{-1}(Z)$ is of dimension zero, and therefore a finite set. We define the *mod* 2 *intersection number* of the map f with Z, denoted $I_2(f, Z)$, to be the number of points in $f^{-1}(Z)$ modulo 2, i.e.,

$$I_2(f, Z) := \# f^{-1}(Z) \mod 2.$$

For an arbitrary smooth map $g\colon X \to Y$, we can choose, by the Transversality Homotopy Theorem 12.44, a smooth map $f\colon X \to Y$ that is homotopic to g and transverse to Z. Then we define $I_2(g, Z) := I_2(f, Z)$.

We have made choices in the definition, and we need to check that the intersection number does not depend on these choices. The key technical result that allows us to show independence is the Extension Theorem 12.48. For the following lemmas, we assume that X, Y, and $Z \subset Y$ are as in Definition 13.1.

Lemma 13.2 *Assume that $f_0\colon X \to Y$ and $f_1\colon X \to Y$ are homotopic and both transverse to Z. Then there is a smooth homotopy $F\colon X \times [0,1] \to Y$ such that $F \pitchfork Z$ and $\partial F \pitchfork Z$.*

Proof: Let H be a smooth homotopy between f_0 and f_1. The product $X \times [0,1]$ is a compact smooth manifold with boundary. We let C be the boundary of $X \times [0,1]$, i.e., C is the closed subset $C := \partial(X \times [0,1]) = X \times \{0\} \cup X \times \{1\}$. The restriction ∂H of H to the boundary of $X \times [0,1]$ is given by the two maps $f_0 = H(-,0)\colon X \to Y$ and $f_1 = H(-,1)\colon X \to Y$. Thus, the assumption $f_0 \pitchfork Z$ and $f_1 \pitchfork Z$ implies $H \pitchfork Z$ on C and $\partial H \pitchfork C$ on $C \cap \partial X$. By the Extension Theorem 12.48, there is a smooth map $F\colon X \times [0,1] \to Y$ with $F \pitchfork Z$ and $\partial F \pitchfork Z$, and $F = H$ on an open neighbourhood of $C = \partial(X \times [0,1])$. The latter means that F is still a smooth homotopy between $f_0 = H(-,0)$ and $f_1 = H(-,1)$. □

Now we are ready to prove the following important observations.

Lemma 13.3 *If $f_0\colon X \to Y$ and $f_1\colon X \to Y$ are homotopic and both transverse to Z, then $I_2(f_0, Z) = I_2(f_1, Z)$.*

Proof: Let $F: X \times I \to Y$ be a homotopy of f_0 and f_1. By Lemma 13.2, we may assume that $F \pitchfork Z$. By the Preimage Theorem 9.18, this implies $F^{-1}(Z)$ is a submanifold of $X \times [0,1]$ such that $\operatorname{codim} F^{-1}(Z)$ in $X \times [0,1]$ equals $\operatorname{codim} Z$ in Y. Hence $\dim F^{-1}(Z) = 1$, since we assume that $\dim X + \dim Z = \dim Y$. Moreover, the boundary $\partial F^{-1}(Z)$ is $F^{-1}(Z) \cap \partial(X \times [0,1])$ which equals $f_0^{-1}(Z) \times \{0\} \cup f_1^{-1}(Z) \times \{1\}$. Since X is compact and Z is closed in Y, $F^{-1}(Z)$ is compact by Lemma 1.14. Hence, by Proposition 10.4, $\partial F^{-1}(Z)$ consists of an even number of points. Thus, $I_2(f_0, Z) = \# f_0^{-1}(Z)$ and $\# f_1^{-1}(Z) = I_2(f_1, Z)$ are equal modulo 2. □

Lemma 13.4 *If $g_0, g_1 : X \to Y$ are homotopic, then $I_2(g_0, Z) = I_2(g_1, Z)$.*

Proof: By the Transversality Homotopy Theorem 12.44, we can choose maps $f_0 \pitchfork Z$ and $f_1 \pitchfork Z$ such that $g_0 \sim f_0$, $I_2(g_0, Z) = I_2(f_0, Z)$, and $g_1 \sim f_1$, $I_2(g_1, Z) = I_2(f_1, Z)$. Since homotopy is a transitive relation, we have $f_0 \sim g_0 \sim g_1 \sim f_1$ and hence $f_0 \sim f_1$. By Lemma 13.3, this implies $I_2(g_0, Z) = I_2(f_0, Z) = I_2(f_1, Z) = I_2(g_1, Z)$. □

Remark 13.5 (Intersection number and Brouwer degree mod 2) Assume that X is compact, Y is connected, and $\dim X = \dim Y$. Let $f: X \to Y$ be a smooth map. Then we recover the mod 2 Brouwer degree as a mod 2 intersection number as $I_2(f; Z) = \deg_2(f)$ with $Z = \{y\}$ for any $y \in Y$. The difference in the two approaches to the degree is that in Section 11.1 we used the Isotopy Theorem 11.3 to show that, for the given smooth map f, we can use any regular value to calculate the degree. For $I_2(f; Z)$, however, we use the machinery of Thom Transversality of Section 12.5 to replace f with a map for which any given y is a regular value if necessary. While the construction in Section 11.1 may be more direct and simpler, Thom Transversality provides us with a slightly more general approach.

13.1.3 Intersection of Manifolds and Self-Intersections

Now we can consider intersection of manifolds as a special case.

Definition 13.6 (Mod 2 intersection numbers of submanifolds) Let Y be a smooth manifold without boundary, let $X \subset Y$ be a compact submanifold, and let $Z \subset Y$ be a closed submanifold. Assume $\dim X + \dim Z = \dim Y$. Let $\iota: X \hookrightarrow Y$ be the inclusion map. Then we define the *mod 2 intersection number of X with Z*, denoted by $I_2(X, Z)$, by $I_2(X, Z) := I_2(\iota, Z)$.

Remark 13.7 Note that if $X \pitchfork Z$, then $I_2(X,Z) = \#(X \cap Z)$. In general, however, we have to move or deform X into a transverse position with Z. That is, we choose a homotopy $\iota_t \colon X \times [0,1] \to Y$ such that ι_1 is still an embedding and such that ι_1 is transverse to Z. This means that $\iota_1(X)$ is still a submanifold in Y, and we have $\iota_1(X) \pitchfork Z$. Then we obtain $I_2(X,Z)$ as $I_2(\iota_1, Z) = \#(\iota_1(X) \cap Z)$.

Example 13.8 (Obstruction to pull apart circles) If $I_2(X,Z) \neq 0$, then no matter how X is moved or deformed, X cannot be pulled entirely away from Z. For example, on the torus $Y = \mathbb{S}^1 \times \mathbb{S}^1 \subset \mathbb{C} \times \mathbb{C}$, the two circles $\mathbb{S}^1 \times \{1\}$ and $\{1\} \times \mathbb{S}^1$ have complimentary dimensions and non-zero mod 2 intersection number. Hence we cannot move them continuously such that they do not intersect. See Exercise 13.7.

Remark 13.9 (Self-intersection number) If $\dim Y = 2 \dim X$, then we may consider $I_2(X,X)$ as the *mod 2 self-intersection number* of X. By definition, we calculate $I_2(X,X)$ by first deforming one copy of X inside Y, i.e., we choose a homotopy $\iota_t \colon X \times [0,1] \to Y$ from the inclusion ι_0 of X in Y to an embedding ι_1 such that ι_1 is transverse to X. Then we obtain $I_2(X,X)$ as $I_2(\iota_1, X) = \#(\iota_1(X) \cap X)$.

Example 13.10 (The Möbius band) An interesting example is the central circle X on the open *Möbius band*. See Figure 13.3. In Exercise 13.9, we show that $I_2(X,X) = 1$. Without actually calculating the intersection number though, we can already deduce from the Intermediate Value Theorem in calculus that we cannot continuously deform X such that it does not intersect itself anymore.

13.1.4 Obstruction to Extensions to Boundaries

The next theorem shows that intersection numbers can also be used as obstructions. We will see applications of the following results in the exercises.

Theorem 13.11 (Boundary Theorem for intersection numbers) *Suppose that X is the boundary of some compact manifold W and $g \colon X \to Y$ is a smooth map. If g can be extended to all of W, then $I_2(g,Z) = 0$ for every closed submanifold Z in Y which satisfies* $\dim X + \dim Z = \dim Y$.

Proof: Let $G \colon W \to Y$ be an extension of g, i.e., G is smooth and $\partial G = g$. From the Transversality Homotopy Theorem 12.44, we obtain a map $F \colon W \to Y$ homotopic to G with $F \pitchfork Z$ and $\partial F \pitchfork Z$. We write $f := \partial F$. Then

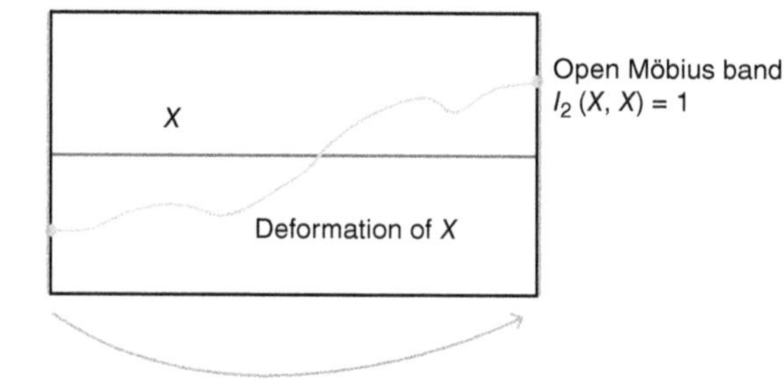

Figure 13.3 We can construct the Möbius band by taking a rectangle and gluing the opposite sides after a twist. The central line then becomes a central circle. If we move it a bit from its initial position, the new curve has to intersect the original central line since the two ends of the curve must be on opposite sides of the central line because of the twist.

$f \simeq g$ and hence $I_2(g, Z) = I_2(f, Z) = \# f^{-1}(Z)$ modulo 2. Now, since Z is closed in Y and X is compact, $F^{-1}(Z)$ is a compact submanifold by Lemma 1.14. Since X is the boundary of W, we have $\dim W = \dim \partial W + 1 = \dim X + 1$, and hence $\dim F^{-1}(Z) = \dim X + 1 - \dim Y + \dim Z = 1$. This shows that $F^{-1}(Z)$ is a compact one-dimensional manifold with boundary. Thus, by Proposition 10.4, $\#\partial(F^{-1}(Z))$ is even and hence $\#\partial(F^{-1}(Z)) = \#(\partial F)^{-1}(Z) = \# f^{-1}(Z)$ vanishes modulo 2. □

Corollary 13.12 (Obstruction for extending maps) *Let W be a compact manifold and $f : \partial W \to Y$ be a smooth map. If there is at least one closed submanifold $Z \subset Y$ such that $I_2(f, Z) \neq 0$, then g cannot be extended to a smooth map $W \to Y$ on all of W.*

13.2 Intersecting Circles on $\mathbb{S}^n$ versus $\mathbb{RP}^n$

In this section, we discuss an example of how intersection theory can be used both to distinguish between smooth manifolds and to show the existence of non-trivial elements in the fundamental group. Recall from Section 6.2 real projective n-space $\mathbb{RP}^n$, which we can consider as a quotient of $\mathbb{S}^n$ under the equivalence relation $x \sim -x$. We showed that $\mathbb{RP}^n$ is an n-dimensional compact smooth manifold, and the quotient map $\mathbb{S}^n \to \mathbb{RP}^n$, which sends a point x in $\mathbb{S}^n$ to its equivalence class $[x]$ in $\mathbb{RP}^n$, is a smooth map. In Lemma 6.37, we showed that $\mathbb{RP}^1$ and $\mathbb{S}^1$ are diffeomorphic. We now show that $n = 1$ is the exception. First, we observe that the intersection theory for circles on $\mathbb{S}^n$ is rather simple.

Example 13.13 (Intersecting circles on $\mathbb{S}^2$) We define a smooth map $j\colon \mathbb{S}^1 \to \mathbb{S}^2$, $(\cos t, \sin t) \mapsto (\cos t, \sin t, 0)$, which maps $\mathbb{S}^1$ onto the intersection of $\mathbb{S}^2$ with the xy-plane in $\mathbb{R}^3$. This map is homotopic to the constant map $\mathbb{S}^1 \to \mathbb{S}^2$ with value $(0,0,1)$. We can show this directly using the homotopy $(\cos t, \sin t, s) \mapsto (s\cos t, s\sin t, 1-s)$. Hence we can smoothly move $j(\mathbb{S}^1)$ away from itself and see that, on $\mathbb{S}^2$, the self-intersection number $I_2\left(j(\mathbb{S}^1), j(\mathbb{S}^1)\right)$ is zero. Note that j is also homotopic to, for example, the map $j'\colon \mathbb{S}^1 \to \mathbb{S}^2$ defined by $j'(t) = (0, \sin t, \cos t)$, and $j(\mathbb{S}^1) \cap j'(\mathbb{S}^1)$ consists of the two points $(0,1,0)$ and $(0,-1,0)$. Hence, modulo 2, we still get that $I_2(j(\mathbb{S}^1), j(\mathbb{S}^1))$ is zero.

Recall from Section 8.2 that a space X is called *simply connected* if every continuous map $\mathbb{S}^1 \to X$ is homotopic to a constant map. We proved in Theorem 8.13 that $\mathbb{S}^n$ is simply connected for $n \geq 2$. We will now use this fact to show that $\mathbb{S}^n$ and $\mathbb{RP}^n$ are not homeomorphic for $n \geq 2$. In fact, we will prove a stronger statement via the following result.

Theorem 13.14 *For every $n \geq 1$, $\mathbb{RP}^n$ is not simply connected, i.e., there is a loop $\mathbb{S}^1 \to \mathbb{RP}^n$ which is not homotopic to a constant map.*

Proof: We recall from Lemma 6.37 that there is a diffeomorphism $\mathbb{RP}^1 \cong \mathbb{S}^1$. The case $n = 1$ then follows from the fact that $\mathbb{S}^1$ is not simply connected, which follows from $\deg_2(\mathrm{id}_{\mathbb{S}^1}) = 1$. So now we assume $n \geq 2$. Our strategy is to find a smooth map $f\colon \mathbb{S}^1 \to \mathbb{RP}^n$ and a suitable closed submanifold $Z \subset \mathbb{RP}^n$ such that $I_2(f, Z)$ is non-zero. Then we know that f is not homotopic to a constant map. For if it were, then f would be homotopic to a constant map whose image is disjoint from Z, since $\mathbb{S}^1$ is path-connected. Then we would have $f \pitchfork Z$, and hence $I_2(f, Z) = 0$. See also Exercise 13.2. This is not possible if our strategy works out. Then we use Whitney's Approximation Theorem 12.28, which implies that if there was a continuous homotopy between f and a constant map, then there also was a smooth homotopy. Since the latter does not exist, the former cannot exist either.

So let us find suitable f and Z. Recall that, given a smooth manifold X, a smooth map $f\colon \mathbb{S}^1 \to X$ is equivalent to a smooth map on the open interval $g\colon (-\varepsilon, 2\pi + \varepsilon) \to X$ such that $g(0) = g(2\pi)$ for some $\varepsilon > 0$. Hence we may define a smooth map $f\colon \mathbb{S}^1 \to \mathbb{RP}^n$ by

$$f(t) := [\cos(t/2) : \sin(t/2) : 0 : \ldots : 0].$$

Now we embed $\mathbb{S}^{n-1}$ as a submanifold of $\mathbb{S}^n$ by requiring the coordinate x_0 to be zero, and let $Z \subset \mathbb{RP}^n$ be its image under the quotient map $\mathbb{S}^n \to \mathbb{RP}^n$, i.e.,

$$Z := \left\{[x] = [x_0 : x_1 : \ldots : x_n] \in \mathbb{RP}^n : x_0 = 0\right\}.$$

Recall from Section 6.2, for $0 \le i \le n$, the subsets $V_i = \{[x] \in \mathbb{RP}^n : x_i \neq 0\}$ and homeomorphisms

$$\phi_i^{-1} \colon V_i \to \mathbb{R}^n,\ [x_0 : \ldots : x_i : \ldots : x_n] \mapsto \frac{1}{x_i}(x_0, \ldots, \widehat{x_i}, \ldots, x_n).$$

Then the subsets $V_i^Z := Z \cap V_i$ and maps $(\phi_i^{-1})_{|V_i^Z} \colon V_i^Z \to \mathbb{R}^{n-1}$ for $1 \le i \le n$ provide charts for Z. This shows that Z is a submanifold of dimension $n-1$ of $\mathbb{RP}^n$. Since $\mathbb{S}^{n-1}$ is a closed subset in $\mathbb{S}^n$, Z is closed in $\mathbb{RP}^n$ by definition of the quotient topology on $\mathbb{RP}^n$. Hence the assumptions for applying intersection theory are satisfied. The only point of Z which is hit by the image of f is the point $[0 : 1 : 0 : \ldots : 0] \in Z$. By describing f as a map $[0, 2\pi] \to \mathbb{RP}^n$ which extends to a smooth map over the boundary points of $[0, 2\pi]$, we get $f^{-1}(Z) = \{\pi\} \subset [0, 2\pi]$. Hence the preimage $f^{-1}(Z)$ consists of exactly one point. It remains to show that f and Z are transverse in $\mathbb{RP}^n$. We can do this by studying the impact of the derivative df_π of f at π and the tangent space of Z locally, i.e., in a chart around the point $[0 : 1 : 0 : \ldots : 0]$. Under the map $\phi_1 \colon \mathbb{R}^n \to V_1$, the point $[0 : 1 : 0 : \ldots : 0] \in V_1$ corresponds to the origin. For $\mathbb{S}^1$, we choose the local parametrisation $\psi \colon (\pi - \varepsilon, \pi + \varepsilon) \to \mathbb{S}^1$, $t \mapsto (\cos(t/2), \sin(t/2))$. Then we consider the diagram

$$\begin{array}{ccc} \mathbb{S}^1 & \xrightarrow{\ f\ } & \mathbb{RP}^n \\ \uparrow{\scriptstyle \psi} & & \downarrow{\scriptstyle \phi_1^{-1}} \\ (\pi - \varepsilon, \pi + \varepsilon) & \xrightarrow[\tilde{f}]{\quad} & \mathbb{R}^n \end{array}, \tag{13.1}$$

where $\tilde{f}$ is defined as the composition of the other maps, such that the diagram commutes. We can compute $\tilde{f}$ as $\tilde{f}(t) = \frac{1}{\sin(t/2)} \cdot (\cos(t/2), 0, \ldots, 0)$. The derivative of $\tilde{f}$ at $t = \pi$ is then the linear map

$$d\tilde{f}_\pi \colon \mathbb{R} \to \mathbb{R}^n,\ x \mapsto (-1/2) \cdot (x, 0, \ldots, 0),$$

given by multiplying x with the number $\tilde{f}'(t) = -(1/2) \cdot \frac{1}{\sin^2(t/2)}$ for $t = \pi$ and embedding as the first coordinate. The key consequence of this computation is that the image of $d\tilde{f}_\pi$ in $\mathbb{R}^n$ is the line $\mathbb{R} \times \{0\} \subset \mathbb{R}^n$. On the other hand, the tangent space at the origin of the image of Z under ϕ_1^{-1} is the subspace $\{0\} \times \mathbb{R}^{n-1} \subset \mathbb{R}^n$. Thus this tangent space and the image of $d\tilde{f}_\pi$ together span all of $\mathbb{R}^n$. By definition of the smooth structure on the manifold $\mathbb{RP}^n$, the map $\phi_1 \colon \mathbb{R}^n \to V_1$ is a diffeomorphism. Thus the induced map on tangent spaces is an isomorphism. Since tangent spaces are determined locally, this shows that f and Z are transverse in $\mathbb{RP}^n$. Thus we have proved $I_2(f, Z) = \#f^{-1}(Z) = 1$ as we set out to show. □

Corollary 13.15 *For $n \geq 2$, there is no homotopy equivalence between $\mathbb{S}^n$ and $\mathbb{RP}^n$, and therefore no homeomorphism between $\mathbb{S}^n$ and $\mathbb{RP}^n$.*

Proof: By Theorem 8.13, $\mathbb{S}^n$ is simply connected for $n \geq 2$, whereas, by Theorem 13.14, $\mathbb{RP}^n$ is not. By Proposition 8.11, this implies that there cannot be a homotopy equivalence between $\mathbb{S}^n$ and $\mathbb{RP}^n$ for $n \geq 2$. Since a homeomorphism is also a homotopy equivalence, we also get the second assertion. □

Theorem 13.14 implies that the fundamental group of $\mathbb{RP}^n$ is non-trivial. Using techniques from algebraic topology, we can say even more.

Remark 13.16 (Fundamental group of $\mathbb{RP}^n$) In Exercise 13.11, we will show that, in a suitable sense, $2[f] = 0$ in the fundamental group $\pi_1(\mathbb{RP}^2, x_0)$ of $\mathbb{RP}^2$, where x_0 is the point $[1 : 0 : \ldots : 0]$. Hence $[f]$ generates a subgroup $\mathbb{Z}/2$ inside $\pi_1(\mathbb{RP}^2)$. One can then show that $\pi_1(\mathbb{RP}^2, x_0) \cong \mathbb{Z}/2$ and extend this computation, for example, using induction and the Seifert–van Kampen Theorem, to show $\pi_1(\mathbb{RP}^n, x_0) \cong \mathbb{Z}/2$ for $n \geq 2$.

13.3 $\mathbb{R}^n$ as a Commutative Division Algebra

We will now discuss an important and at first glance maybe surprising application of Corollary 13.15. An algebra structure on $\mathbb{R}^n$ is a $\mathbb{R}$-bilinear map

$$\mathbb{R}^n \times \mathbb{R}^n \to \mathbb{R}^n, (a,b) \mapsto ab,$$

which we simply write as a multiplication. Note that it is not required that this multiplication has an identity element. Such a map is called a *division algebra* structure if there are no zero-divisors, i.e., $ab = 0$ implies $a = 0$ or $b = 0$. It is called *commutative* if $ab = ba$ for all $a, b \in \mathbb{R}^n$. Examples of finite-dimensional commutative division algebras over $\mathbb{R}$ are given by the fields $\mathbb{R}$ itself and $\mathbb{C} = \mathbb{R}^2$. However, there is not much more. See also [15, section 2.B].

Theorem 13.17 *For $n \geq 3$, there is no commutative division algebra structure on $\mathbb{R}^n$.*

Proof: Suppose there is a commutative division algebra structure on $\mathbb{R}^n$. Then we can define a map $f \colon \mathbb{S}^{n-1} \to \mathbb{S}^{n-1}$ by $f(x) := x^2/|x^2|$. This is well defined, since $x \neq 0$ implies $x^2 \neq 0$ in a division algebra. The map f is smooth, since the multiplication map $\mathbb{R}^n \times \mathbb{R}^n \to \mathbb{R}^n$ is bilinear and hence smooth. Since f satisfies $f(-x) = f(x)$ for all x, f induces a smooth map on the quotient $\bar{f} \colon \mathbb{RP}^{n-1} \to \mathbb{S}^{n-1}$. Now we show that $\bar{f}$ would be a homeomorphism. First, we show that $\bar{f}$ is injective. Assume $f(x) = f(y)$. This implies $x^2 = \alpha^2 y^2$ for $\alpha = \sqrt{|x^2|/|y^2|} > 0$. Thus we have $x^2 - \alpha^2 y^2 = 0$. Using commutativity and the fact that α is a real number and multiplication is $\mathbb{R}$-bilinear, we see that this equation factors into $(x + \alpha y)(x - \alpha y) = 0$. Since we assume that the multiplication has no zero-divisors, this implies $x = \pm\alpha y$. Since $|x| = |y| = 1$ and α is a real number, this implies $|\alpha| = 1$ and hence $x = \pm y$. Thus x and y determine the same point in $\mathbb{R}\mathrm{P}^{n-1}$. This proves that $\bar{f}$ is injective. Since $\mathbb{R}\mathrm{P}^{n-1}$ is compact and $\mathbb{S}^{n-1}$ is Hausdorff, $\bar{f}$ is therefore a homeomorphism onto its image by Corollary 1.23. Since $\mathbb{R}\mathrm{P}^{n-1}$ and $\mathbb{S}^{n-1}$ are smooth manifolds of the same dimension with $\mathbb{R}\mathrm{P}^{n-1}$ compact and $\mathbb{S}^{n-1}$ connected, the invariance of domain of Corollary 12.36 implies that $\bar{f}$ is a homeomorphism. However, by Corollary 13.15, we know that for $n \geq 3$ there cannot be a homeomorphism between $\mathbb{R}\mathrm{P}^{n-1}$ and $\mathbb{S}^{n-1}$. Hence our initial assumption must have been wrong. □

Remark 13.18 There is a famous generalisation of Theorem 13.17. Let $\mathbb{R}^n \times \mathbb{R}^n \to \mathbb{R}^n$ be a division algebra structure on $\mathbb{R}^n$, without assuming it is commutative. For $n = 4$, there are the Hamiltonians, or quaternions, $\mathbb{H} \cong \mathbb{R}^4$ with a multiplication which is associative and almost as good as the one in $\mathbb{C}$ and $\mathbb{R}$, but it is not commutative. For $n = 8$, there are the octonions $\mathbb{O} \cong \mathbb{R}^8$. The multiplication is neither commutative nor associative. By the work of Bott–Milnor [5], and independently by Kervaire [25], we know that this is it, i.e., there is no division algebra structure on $\mathbb{R}^n$ for $n \neq 1, 2, 4, 8$. This is an important result, and surprisingly it is equivalent to a topological problem on the behaviour of tangent spaces on spheres. This problem is also closely related to the Hopf invariant one problem, solved by Adams [1]. The proof goes way beyond the methods we discuss in this book, unfortunately. There is also a very nice argument using complex K-theory by Adams–Atiyah [2]. See also [31, chapter 24].

13.4 Exercises

13.1 Let $X \xrightarrow{f} Y \xrightarrow{g} Z$ be a sequence of smooth maps between manifolds with X compact, and let $W \subset Z$ be a submanifold. Assume that g is transverse to W, so that $g^{-1}(W)$ is a submanifold in Y. Show $I_2(f, g^{-1}(W)) = I_2(g \circ f, W)$. In particular, verify that if one of these two intersection numbers is defined, then the other one is defined as well. Hint: Use Exercise 5.5.

13.2 Let X be a compact manifold without boundary.

(a) Assume $\dim X \geq 1$: Show that if $f\colon X \to Y$ is homotopic to a constant map, then $I_2(f, Z) = 0$ for all complementary dimensional closed submanifolds Z in Y. Hint: Show that if $\dim Z < \dim Y$, then f is homotopic to a constant map $X \to \{y\}$ for some $y \notin Z$. You may assume that Y is connected.

(b) For $\dim X = 0$, show that the corresponding assertion of the previous point is wrong. Hint: If X consists of just one point, for which Z is $I_2(f, Z) \neq 0$?

13.3 Use intersection numbers to show that $\mathbb{S}^1$ is not simply connected. Hint: Consider the identity map, and use Exercise 13.2.

13.4 Show that intersection theory is trivial in contractible smooth manifolds without boundary, i.e., if Y is boundaryless, its identity map is homotopic to a constant map, and $\dim Y > 0$, then $I_2(f, Z) = 0$ for every smooth map $f\colon X \to Y$ such that X compact and Z closed, $\dim X \geq 1$ and $\dim X + \dim Z = \dim Y$. In particular, intersection theory is trivial in Euclidean space.

13.5 Prove that no compact boundaryless manifold – other than the one-point space – is contractible. Hint: Consider the identity map.

13.6 Let $f\colon X \to \mathbb{S}^k$ be a smooth map with X compact and $0 < \dim X < k$. Show that, for all closed submanifolds $Z \subset \mathbb{S}^k$ of dimension complementary to X, $I_2(f, Z) = 0$. Hint: Use Sard's Theorem 5.17 to show that there exists a $p \notin f(X) \cap Z$. Now use a stereographic projection and the previous exercises.

13.7 Show that $\mathbb{S}^2$ and the torus $\mathbb{T} = \mathbb{S}^1 \times \mathbb{S}^1$ are not diffeomorphic. Hint: Use Exercise 13.6.

13.8 Let Y be a smooth manifold. Two compact submanifolds X and Z of the same dimension in Y are called *cobordant* in Y if there exists a compact manifold with boundary $W \subset Y \times [0, 1]$ such that

$$\partial W = X \times \{0\} \cup Z \times \{1\}.$$

The manifold W is called a *cobordism* between X and Z. See Figure 13.4.

(a) Show that if we can deform X into Z, i.e., if there is a smooth homotopy from the inclusion i_0 of X in Y to an embedding $i_1\colon X \hookrightarrow Y$ with $i_1(X) = Z$ such that each i_t is an embedding, then X and Z are cobordant.

Note that the standard image of a cobordism, a pair of pants as in Figure 13.4, illustrates that the converse is false: X and Z are cobordant, but we cannot deform X into Z, since X has one connected component whereas Z has two.

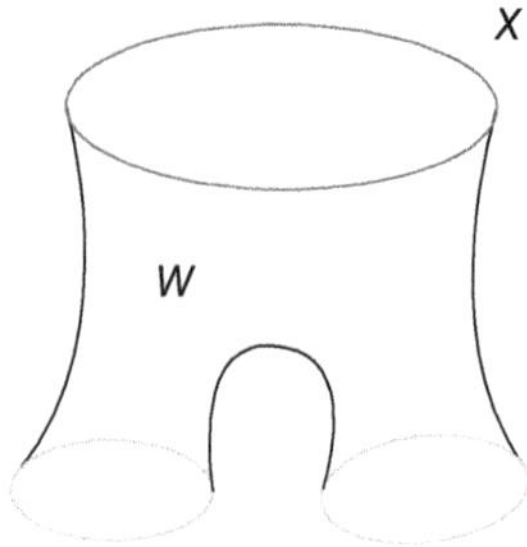

Figure 13.4 A pair of pants is a cobordism between the upper boundary X and the lower boundary Z. While X is connected, Z has two connected components.

(b) Show that if X and Z are cobordant in Y, then for every compact submanifold C in Y with dimension complementary to X and Z, i.e., $\dim X + \dim C = \dim Z + \dim C = \dim Y$ (where $\dim X = \dim Z$ because they are cobordant), we have

$$I_2(C, X) = I_2(C, Z).$$

Hint: Let f be the restriction to W of the projection map $Y \times [0,1] \to Y$, and use the Boundary Theorem 13.11.

13.9 Let $Z \subset X$ be a compact submanifold of X with $\dim Z = \frac{1}{2} \dim X$. Show that if there is a submersion $g\colon U \to \mathbb{R}^k$ defined on an open subset $U \subset X$ such that $Z = g^{-1}(0)$, then $I_2(Z, Z) = 0$. Hint: Use Theorem 12.21 and Corollary 12.22.

13.10 Let X be the open *Möbius band* given as the quotient of $[-1,1] \times (-1,1)$ modulo the equivalence relation $(-1, x) \sim (1, -x)$. Let Z denote the central circle given by $Z = \{(x, 0) : x \in [-1, 1]\} \subset X$.

(a) Show that $I_2(Z, Z) = 1$. See Figure 13.3.

(b) Use Exercise 13.9 to conclude that there cannot be a submersion $g\colon X \to \mathbb{R}$ such that $Z = g^{-1}(0)$.

13.11 Let $f\colon \mathbb{S}^1 \to \mathbb{RP}^2$ be the map defined by sending $t \in [0, 2\pi]$ to the point $[\cos(t/2) : \sin(t/2) : 0]$.

(a) Show that f is smooth, injective, and proper.

(b) We write $2f$ for the map $2f\colon \mathbb{S}^1 \to \mathbb{RP}^2$ defined by sending $t \in [0, 2\pi]$ to $[\cos(t) : \sin(t) : 0]$. Show that $2f$ is homotopic to a constant map by constructing a concrete homotopy. Do you see why the homotopy works for $2f$ but not for f?

14
Orientation

We have not yet made use of a fundamental property of the real numbers, namely that $\mathbb{R}$ is an *ordered* field. In particular, this means that we can specify two directions on the real number line: one pointing in the direction of increasing numbers and one pointing in the direction of decreasing numbers. We will think of such a distinction of directions as an orientation of the number line. Our goal for this chapter is to exploit this observation for smooth manifolds. This will lead us to the concept of an orientation of a manifold which has plenty of interesting applications. First of all, it will help us classify smooth manifolds as we will see that some manifolds can be oriented while others cannot. Secondly, orientations will allow us to improve our definition of the Brouwer degree and remedy the defect that it only has values $+1$ and -1.

14.1 Towards Integer-Valued Invariants

In Section 11.2.2, we studied smooth maps $f\colon \mathbb{S}^1 \to \mathbb{S}^1$ on the circle. Given such an f, we showed that it has a lift $g\colon \mathbb{R} \to \mathbb{R}$ such that $g(t+1) = g(t)+q$ for all t for some fixed $q \in \mathbb{Z}$. The integer q is a number that is attached to f. We then showed that $\deg_2(f) = q$ modulo 2. It would be nice if we had a notion of a degree for f with integer values so that we could write an equation like $\deg(f) = q$ in $\mathbb{Z}$. If we knew in addition that the number $\deg(f)$ is homotopy invariant, i.e., only depends on the class of f under the equivalence relation on the set of maps given by homotopy, then we could use the degree to distinguish between possibly infinitely many different homotopy classes of maps $\mathbb{S}^1 \to \mathbb{S}^1$, or even maps $\mathbb{S}^n \to \mathbb{S}^n$ for all $n \geq 1$.

So let us think about how we came up with the formula $\deg_2(f) = q$ modulo 2. We looked at the graph of g in Figure 11.2. Then we used the Intermediate Value Theorem and observed that the graph crosses the horizontal

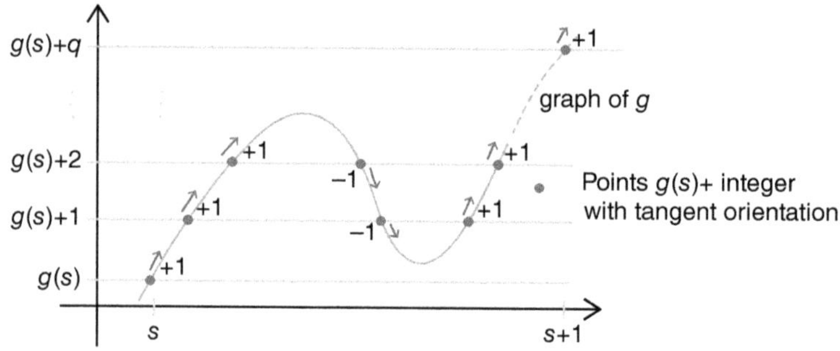

Figure 14.1 The idea is to keep track of how the graph passes the horizontal lines: if g increases while passing the horizontal line we count the intersection as $+1$, and if g decreases we count it as -1.

lines $g(s)+k$ for an integer $k \in \{0, 1, \dots, q-1\}$, an odd number of times, and lines $g(s)+k$ for any other integer k, an even number of times. We would like to count all the values t with $g(t) = g(s) + k$. But there might be many more such values than just q. How can we fix this?

The idea is to remember *how* the graph of g crosses the lines $g(s) + k$: as t increases, $g(t)$ either increases while crossing $g(s) + k$ or decreases. In other words, the derivative of g is either positive or negative at t_k with $g(t_k) = g(s) + k$. Note that we cannot have $g'(t_k) = 0$, since we started with a regular value $y = f(x)$, and hence $\det(df_x) \neq 0$ and therefore also $g'(t) \neq 0$ for all t with $p(g(t)) = y$, where $p(t) = e^{2\pi it}$. Now, if $g'(t_k) > 0$, we may count t_k with value $+1$; if $g'(t_k) < 0$, we may count t_k with value -1. See Figure 14.1. Counting all $t \in [s, s+1)$ with $g(t) = g(s) + k$ for $k \in \mathbb{Z}$, this gives us the desired value q. Hence if we can improve our definition of the degree such that we can use this way of counting points on the graph of g, we would indeed get $\deg(f) = q$.

To generalise this procedure, we need to describe $g'(t_k) > 0$ or $g'(t_k) < 0$ in more general terms. In higher dimensions, we cannot just attach a sign to $g'(t)$ or df_x. However, an alternative way to describe the sign of the derivative of $g'(t)$ is to say that g preserves the orientation of the coordinate axes. More generally, thinking of the axes as the tangent spaces of copies of $\mathbb{R}$, we could say that the derivative of g does or does not preserve the orientation of the tangent spaces. A line has exactly two orientations, depending on whether we walk in one direction or the other. It turns out that this perspective can be generalised to higher dimensions. Since derivatives are linear transformations, we can look at their determinants. Then we can ask whether the determinant has a positive or a negative sign. Geometrically, this sign will detect whether or not the linear transformation given by the derivative preserves or reverses the orientation of the respective tangent spaces. In order to make this idea work, we first have to define what an orientation of a vector space is, and then we need to transfer this to smooth manifolds. This is what we are going to do next.

14.2 Oriented Smooth Manifolds

We first define an orientation of a vector space and will then transfer the concept to manifolds using tangent spaces.

14.2.1 Orientation on Vector Spaces

An *orientation* for a finite-dimensional real vector space V is an equivalence class of ordered bases where the relation is defined as follows: the ordered basis $(v_1, \dots, v_n)$ has the same orientation as the basis $(v_1', \dots, v_n')$ if the matrix A with $v_i' = Av_i$ for all i has determinant $\det(A) > 0$. It has the opposite orientation if $\det(A) < 0$. That this is an equivalence relation follows from the fact that the determinant function is multiplicative. Thus each finite-dimensional vector space has precisely two orientations, corresponding to the two equivalence classes of ordered bases. We emphasise that the ordering of the basis elements is essential. Interchanging the positions of two basis vectors changes the sign of the orientation since the determinant of the corresponding permutation matrix is negative.

Hence an orientation of V is a choice of an equivalence class of ordered bases. To make it easier to talk about the choice of orientation, we assign a positive sigh to the chosen orientation and a negative sign to the other orientation. We say then that an ordered basis is *positively oriented* (respectively *negatively oriented*) if its equivalence class belongs to the orientation $+1$ (respectively -1). We often confuse orientations with their corresponding signs $+1$ or -1. In fact, in the case of the zero-dimensional vector space it is convenient to define an orientation as the symbol $+1$ or -1.

Example 14.1 (Standard orientation on $\mathbb{R}^n$) The vector space $\mathbb{R}^n$ has a *standard orientation* given by the ordered basis $(e_1, \dots, e_n)$. We always assign $+1$ to the standard orientation of $\mathbb{R}^n$.

Remark 14.2 Let $\varphi\colon V \to W$ be an isomorphism between two oriented real vector spaces V and W. Let β be a positively oriented ordered basis of V. Then $\varphi(\beta)$ is either a positively or negatively oriented basis of W. If $\varphi(\beta)$ is positively oriented, then φ preserves the orientations, i.e., it sends every positively oriented basis of V to a positively oriented basis of W. If $\varphi(\beta)$ is negatively oriented, then φ reverses the orientations, i.e., it sends every positively oriented basis of V to a negatively oriented basis of W. We can test whether φ preserves or reverses orientations by using the determinant. However, we have to be a bit careful. In order to compute $\det(\varphi)$, we have to choose a basis β_V of V and a basis β_W of W. Let $A_{\beta_V}^{\beta_W}(\varphi)$ denote the matrix which represents φ with respect to these bases. Then, by definition, we have $\det(\varphi) = \det(A_{\beta_V}^{\beta_W}(\varphi))$. Now, if we choose β_V and β_W both to be *positively* oriented bases of V and W, respectively, then φ preserves orientations if and only if $\det(\varphi)$ is positive.

14.2.2 Orientation on Manifolds

Now we translate orientations from vector spaces to orientations for manifolds. The idea is to orient each tangent space. However, this needs to be done in a compatible way.

Definition 14.3 (Orientation on manifolds) Let X be a smooth manifold. An *orientation* of X is a smooth choice of orientations for all the tangent spaces $T_x X$ at each point $x \in X$, i.e., around each point $x \in X$ there exists a local parametrisation $\phi\colon U \to X$ such that the isomorphism $d\phi_u\colon \mathbb{R}^k \to T_{\phi(u)}X$ preserves orientations at each point u of $U \subseteq \mathbb{H}^k$ where the orientation on $\mathbb{R}^k$ is always assumed to be the standard one. We use the *symbol* $\mathfrak{o}_X$ to denote the data of an orientation on X. We will sometimes write $(X, \mathfrak{o}_X)$ to express that X is oriented with orientation $\mathfrak{o}_X$.

We now introduce an important new terminology.

Definition 14.4 (Orientable and oriented manifolds) A smooth manifold X is called *orientable* if a smooth choice of orientations of its tangent spaces exists. A smooth manifold is called *oriented* if it is orientable and a choice of orientation has been made.

Hence an oriented manifold is a pair consisting of a manifold together with a chosen orientation. An important example is the n-dimensional sphere $\mathbb{S}^n$, which is orientable and has a standard orientation as we will explain in Example 14.19.

Example 14.5 For zero-dimensional manifolds, orientations are very simple. To each point $x \in X$, we assign an orientation number $+1$ or -1.

Definition 14.6 (Maps and orientations) A smooth map $f\colon X \to Y$ between oriented smooth manifolds is called *orientation preserving* if its derivative

$df_x\colon T_x(X) \to T_{f(x)}(Y)$ preserves orientations, i.e., sends positive bases to positive bases, at every point $x \in X$.

We show in Exercise 14.3 that if f is a diffeomorphism and if df_x preserves orientation at one point x, then f preserves orientation at every point. This implies the following result.

Lemma 14.7 (Diffeomorphisms and orientation) *Let X and Y be oriented smooth manifolds, and let $f\colon X \to Y$ be a diffeomorphism. Then f either preserves or reverses orientations.* □

Example 14.8 (Reflections and orientation) Let $r_i\colon \mathbb{R}^n \to \mathbb{R}^n$ be the reflection of the ith coordinate, i.e., $r_i(x_1,\ldots,x_n) = (x_1,\ldots,-x_i,\ldots,x_n)$. Then r_i is an isomorphism which reverses the orientation. The composition $r_i \circ r_j$ of two reflections is an isomorphism which preserves the orientation. This shows that $-\mathrm{Id}_{\mathbb{R}^n}$, the negative of the identity of $\mathbb{R}^n$, preserves the orientation if n is even, and $-\mathrm{Id}_{\mathbb{R}^n}$ reverses the orientation if n is odd. Now we can consider the reflection $r_i\colon \mathbb{S}^{n-1} \to \mathbb{S}^{n-1}$ on the n-sphere, i.e., the restriction of r_i to $\mathbb{S}^{n-1}$. Since the derivative of r_i is a reflection on a vector space, we see that r_i is a diffeomorphism which reverses the orientation.

Since the antipodal map on $\mathbb{S}^n$ is induced by a composition of $n+1$ reflections, the above observation yields the following result.

Proposition 14.9 (Orientation and the antipodal map) *The antipodal map $a\colon \mathbb{S}^n \to \mathbb{S}^n$, $a(x) = -x$, is a diffeomorphism which preserves the orientation if n is odd and reverses the orientation if n is even.*

Later we will use the following version of the Isotopy Theorem 11.3.

Lemma 14.10 (Linear Isotopy Lemma) *Let E be a linear isomorphism of $\mathbb{R}^k$. If E preserves orientation, then there exists a homotopy E_t consisting of linear isomorphisms such that $E_0 = E$ and E_1 is the identity. If E reverses orientation, then there exists a homotopy E_t consisting of linear isomorphisms such that $E_0 = E$ and E_1 is the reflection map*

$$E_1(x_1,x_2,\ldots,x_k) = (-x_1,x_2,\ldots,x_k).$$

Proof: See Exercise 14.5. □

Example 14.11 (Paths on manifolds) Let X be a smooth manifold without boundary. Let $\gamma\colon [0,1] \to X$ be a smooth path on X. The product $[0,1] \times X$ is an oriented manifold. Let $\Gamma \subset [0,1] \times X$ denote the graph of γ. We recall from Exercise 9.6 that Γ is a smooth one-dimensional manifold with boundary and that the map $g\colon [0,1] \to \Gamma$, defined by $g(t) = (t, \gamma(t))$, is a diffeomorphism. We can then use the diffeomorphism g to define an orientation on Γ such that g preserves orientations. More precisely, we may orient $T_{(t,\gamma(t))}\Gamma$ such that the basis $\{dg_t(e_1)\}$ is positive, where $e_1 \in \mathbb{R} = T_t([0,1])$ denotes the standard basis of $\mathbb{R}$.

We learned that a manifold may or may not be orientable. To assign $+1$ or -1 to the orientation of $T_x(X)$ for every point is a locally constant function. If X is orientable, this assignment is continuous. If X is in addition connected, then this assignment must be constant. Hence, on every connected component of an orientable manifold, the orientation is constant $+1$ or -1. Here is a rigorous proof of this fact.

Proposition 14.12 *A connected, orientable smooth manifold X with boundary admits exactly two orientations.*

Proof: Assume that we are given two orientations on X. In fact, we know that there are at least two, since given one, we can reverse signs everywhere and get another orientation. We need to show that there are not more than two. We do this by showing that the set of points at which two orientations agree and the set where they disagree are both open. Since X is connected, this will imply the assertion. Since X is orientable, we can choose local parametrisations $\phi\colon U \to X$ and $\psi\colon V \to X$ around a point $x \in X$ with $\phi(0) = x = \psi(0)$ such that $d\phi_u$ preserves the first orientation for all $u \in U$ and $d\psi_v$ preserves the second $v \in V$. After possibly shrinking, i.e., after possibly replacing U and V with $\phi^{-1}(\phi(U) \cap \psi(V))$ and $\psi^{-1}(\phi(U) \cap \psi(V))$, respectively, we can assume $\phi(U) = \psi(V)$. If the two orientations of $T_x(X)$ agree, then the map $d(\phi^{-1} \circ \psi)_0\colon \mathbb{R}^k \to \mathbb{R}^k$ is an orientation-preserving isomorphism. Thus the determinant of $d(\phi^{-1} \circ \psi)_0$ is positive. Hence the function $\varphi\colon V \to \mathbb{R}$, $v \mapsto \det(d(\phi^{-1} \circ \psi)_{u'})$, satisfies $\varphi(0) > 0$. Since the derivative depends continuously on v and the determinant function is continuous, φ is continuous. Hence, since $\varphi(0) > 0$, there is an open neighbourhood V' around 0 in V on which $\varphi_{|V'} > 0$. This implies that the orientations of $T_x(X)$ induced by ϕ and ψ, respectively, agree for all x in the open subset $\psi(V')$. Since every point on X has such an open neighbourhood, the set of points where the orientations agree is open. If the orientations on $T_x(X)$ induced by ϕ and ψ, respectively, do not agree, the same argument shows that the set of points where the orientations do not agree is open. □

If $(X, \mathfrak{o}_X)$ is an oriented manifold X, then we can talk about the manifold X with the *opposite orientation*. We write $-\mathfrak{o}_X$ for the opposite orientation. We will often just write $-X$ for the oriented manifold $(X, -\mathfrak{o}_X)$.

Remark 14.13 (Classification of manifolds) Not all manifolds are orientable. One of the most famous examples of a non-orientable manifold is the Möbius band. We will give a rigorous proof of the non-orientability in Theorem 14.16. Other examples of non-orientable manifolds include the Klein bottle and the real projective plane $\mathbb{R}P^2$. We will discuss the latter case in Theorem 14.15. As a consequence, we see that the question whether a smooth manifold is orientable or not helps us classify manifolds up to diffeomorphism. There is the class of orientable manifolds and the class of non-orientable manifolds.

Now we study how orientations behave with respect to smooth maps.

Theorem 14.14 (Pullback of an orientation) *Let X and Y be smooth manifolds with or without boundary. Assume that $Y = (Y, \mathfrak{o}_Y)$ is oriented and that there is a local diffeomorphism $f: X \to Y$. Then there is a unique orientation on X such that f preserves orientations. We call the induced orientation on X the* pullback orientation *induced by f and denote it by $f^*\mathfrak{o}_Y$. Moreover, if $g: Y \to Z$ is another smooth map and $(Z, \mathfrak{o}_Z)$ is an oriented manifold, then the pullback orientation on X induced by $g \circ f$ is the same as the orientation on X obtained by pulling back the orientation on Z via g to Y and then further to X, i.e., $(g \circ f)^*\mathfrak{o}_Z = f^*(g^*\mathfrak{o}_Z)$ as orientations on X. In other words, the pullback of orientations behaves functorially.*

Proof: Since f is a local diffeomorphism, at every point $x \in X$, there are open subsets $U \subset X$ and $V \subset Y$ with $x \in U$ and $f(x) \in V$ such that $f_{|U}: U \to V$ is a diffeomorphism. By choosing U and V small enough, we can assume that there are local parametrisations $\phi: W \to U \subset X$ and $\psi: W \to V \subset Y$ such that $f \circ \phi = \psi$ with $W \subset \mathbb{R}^n$ open. Since Y is oriented, we can choose ψ such that $d\psi_w$ determines the orientation on T_yY for all $y = \psi(w)$ in V. Since df_x is an isomorphism for all x by assumption, we can define an orientation on T_xX by requiring $\det(df_x) > 0$. This defines an orientation on T_xX for all $x \in U$. It remains to check that this orientation on the tangent spaces for points in $U \subset X$ does not depend on the choice of ϕ. So let $\phi': W' \to U' \subset X$ be another local parametrisation of X with $U \cap U' \neq \emptyset$. Then we can find a local parametrisation $\psi': W' \to V' \subset Y$ of Y such that $f_{|U'}: U' \to V'$ is a diffeomorphism. Since Y is oriented, the orientation of T_yY determined by $d\psi_w$ and by $d\psi_{w'}$, with $\psi(w) = y = \psi(w')$ are the same for all y in $V \cap V'$. In other words, we have $\det(d(\psi^{-1} \circ \psi')_{w'} > 0$ for all $w' \in W \cap W' \subset \mathbb{R}^n$. Since $\det(df_x) > 0$, this implies $\det(d(\phi^{-1} \circ \phi')_{w'} > 0$ for all $w' \in W \cap W' \subset \mathbb{R}^n$ as well. Hence the orientation of T_xX via $d\phi_w$ and $d\phi'_{w'}$ is the same for all $w, w' \in W \cap W$ with $\phi(w) = x = \phi'(w')$. This shows that we have defined an orientation on X. The facts that f preserves orientations, uniqueness, and functoriality follow from the construction of the induced orientation. □

Here is an interesting application of Theorem 14.14.

Theorem 14.15 (Orientability of real projective space) *Real projective space $\mathbb{RP}^n$ is orientable if and only if n is odd.*

Proof: Let $q\colon \mathbb{S}^n \to \mathbb{RP}^n$, $x \mapsto [x]$, be the canonical projection, and let $a\colon \mathbb{S}^n \to \mathbb{S}^n$ be the antipodal map. We have $q(x) = q(-x)$ for all $x \in \mathbb{S}^n$, i.e., we have $q = q \circ a$. Both q and a are local diffeomorphisms. Hence, if $\mathbb{RP}^n$ is oriented by an orientation $\mathfrak{o}_{\mathbb{RP}^n}$, then the orientation on $\mathbb{S}^n$ induced by q and by $q \circ a$ must be the same, i.e., we must have

$$q^*\mathfrak{o}_{\mathbb{RP}^n} = (q \circ a)^*\mathfrak{o}_{\mathbb{RP}^n} = a^*(q^*\mathfrak{o}_{\mathbb{RP}^n}),$$

where we used the functoriality of the pullback orientation. This shows that $\mathbb{RP}^n$ can be oriented only if a preserves orientation. By Proposition 14.9, this is the case if and only if n is odd. This proves that $\mathbb{RP}^n$ is not orientable if n is even.

We will now define an orientation of $\mathbb{RP}^n$ when n is odd. At every point $[x] \in \mathbb{RP}^n$, the fibre under q consists of two antipodal points $q^{-1}([x]) = \{x, -x\} \subset \mathbb{S}^n$. Given a point $x \in \mathbb{S}^n$, we can use stereographic projection from a point which is not $\pm x$ on $\mathbb{S}^n$ as one local parametrisation ϕ around both x and $-x$. Let β_x be a basis of $T_x\mathbb{S}^n$. We now define an orientation on $T_{[x]}\mathbb{RP}^n$ by requiring that the image of β_x in $T_{[x]}\mathbb{RP}^n$ under dq_x has the same sign as β_x in $T_x\mathbb{S}^n$, i.e., we set $\operatorname{sign}(dq_x(\beta_x)) := \operatorname{sign}(\beta_x)$. We have to check that this definition is independent of the choice of representing $[x] \in \mathbb{RP}^n$ by x or $-x$ in $\mathbb{S}^n$. By Proposition 14.9, the antipodal map $a\colon \mathbb{S}^n \to \mathbb{S}^n$ preserves orientations, since we assume that n is odd. Considering the commutative diagram (14.1), this implies that the orientations of $T_x\mathbb{S}^n$ and $T_{-x}\mathbb{S}^n$ are such that the respective images of the standard bases of $\mathbb{R}^n$ under $d\phi_{\phi^{-1}(x)}\colon \mathbb{R}^n \to T_x\mathbb{S}^n$ and $d\phi_{\phi^{-1}(-x)}\colon \mathbb{R}^n \to T_{-x}\mathbb{S}^n$ are either both positively or both negatively oriented.

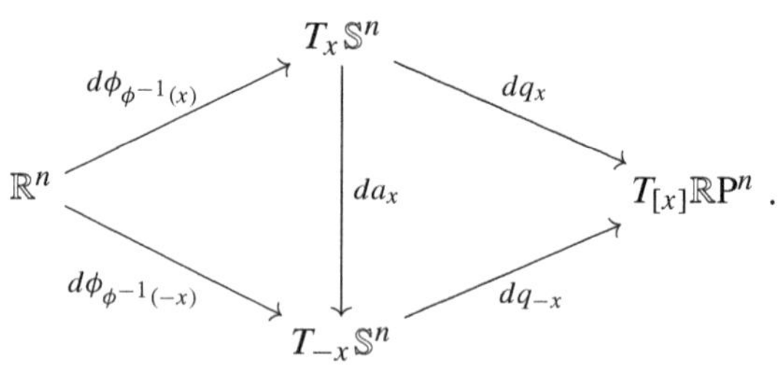

(14.1)

This shows $\operatorname{sign}(a_x(\beta_x)) = \operatorname{sign}(\beta_x)$ where $a_x(\beta_x) =: \beta_{-x}$ is a basis of $T_{-x}\mathbb{S}^n$. Thus, the orientations on $T_{[x]}\mathbb{RP}^n$ determined by $T_x\mathbb{S}^n$ and $T_{-x}\mathbb{S}^n$ via dq_x and dq_{-x}, respectively, are the same. Since $\mathbb{S}^n$ is oriented as explained in Example 14.19, this defines an orientation on $\mathbb{RP}^n$ for n odd. □

We now look at another important example of a non-orientable manifold. Recall the Möbius band which we met in Example 6.43, Exercise 9.4, Example 13.10, and Exercise 13.10.

Theorem 14.16 *The Möbius band is not orientable.*

Proof: Let X denote the Möbius band. Suppose that X is orientable and assume that we have chosen an orientation. We will now show that this leads to a contradiction to other results. Let $Z \subset X$ be the central circle. Then Z is orientable as a copy of $\mathbb{S}^1$, and it is a submanifold of codimension one. By Exercise 14.7, this implies that the normal bundle $N(Z, X)$ is trivial. By Corollary 12.22, this implies that there is a submersion $g\colon U \to \mathbb{R}$ defined on an open subset $U \subset X$ such that $Z = g^{-1}(0)$. By Exercise 13.10, such a g does not exist. This shows that X cannot be equipped with an orientation. □

Next we will study how orientations behave under the main constructions and concepts we have developed so far. The product of oriented manifolds inherits an orientation as follows.

Definition 14.17 (Product orientation) Assume that X and Y are oriented, and at least one of them is without boundary. Then $X \times Y$ is a manifold with boundary and inherits an orientation, which we call the *product orientation*, in the following way: at a point $(x, y) \in X \times Y$, let $\alpha = (v_1, \ldots, v_k)$ and $\beta = (w_1, \ldots, w_m)$ be ordered bases of $T_x(X)$ and $T_y(Y)$, respectively. We denote the ordered basis $((v_1, 0), \ldots, (v_k, 0), (0, w_1), \ldots, (0, w_m))$ of $T_x(X) \times T_y(Y) = T_{(x,y)}(X \times Y)$ by $(\alpha \times 0, 0 \times \beta)$. Now it comes in handy that we have related orientations of ordered bases to signs. For we can define the orientation of $T_x(X) \times T_y(Y)$ by determining a sign by setting $\operatorname{sign}(\alpha \times 0, 0 \times \beta) := \operatorname{sign}(\alpha) \cdot \operatorname{sign}(\beta)$.

14.3 Induced Orientation on the Boundary

Let X be an oriented smooth manifold with boundary. Then the boundary submanifold ∂X inherits an orientation as follows. At every point $x \in \partial X$, $T_x(\partial X)$ is a subspace of codimension one in $T_x(X)$. Its orthogonal complement in $T_x(X)$, is a line which contains exactly two unit vectors: one is pointing inwards into $T_x(X)$, the other one is pointing outwards away from $T_x(X)$. This can be made precise by choosing a local parametrisation $\phi\colon U \to X$ around x with $U \subset \mathbb{H}^k$ open and $\phi(0) = x$. The derivative $d\phi_0\colon \mathbb{R}^k \to T_x(X)$ is by definition of $T_x(X)$ an isomorphism. In $\mathbb{R}^k$, there are the two unit vectors

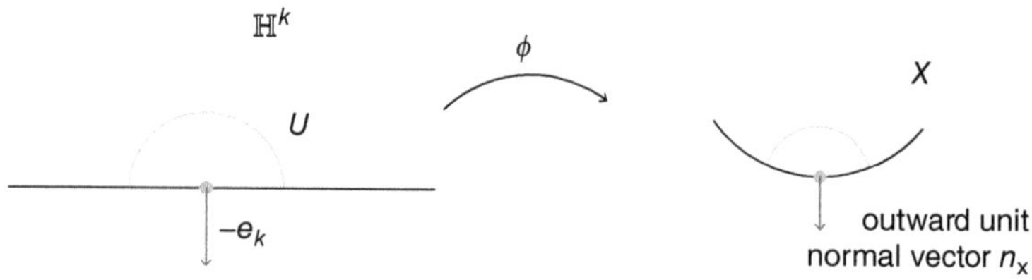

Figure 14.2 We define an outward normal vector n_x as the image $d\phi_0(-e_k)$ of the outward-pointing standard basis vector $-e_k$.

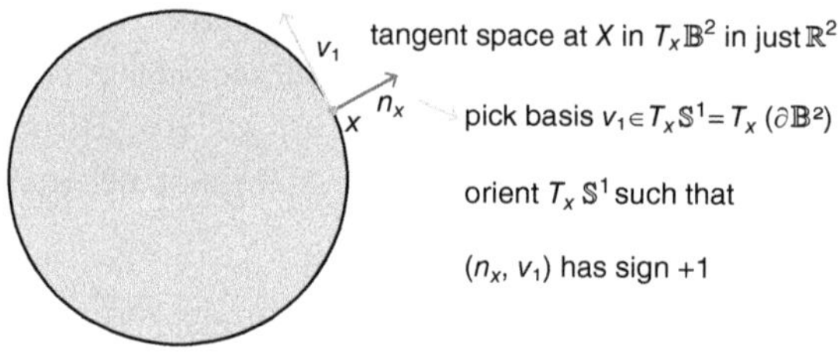

Figure 14.3 The outward-pointing unit vector n_x is the vector orthogonal to the tangent line at x to $\mathbb{S}^1$ and pointing away from the origin.

$e_k = (0, \ldots, 0, 1)$ pointing into $\mathbb{H}^k$, and $-e_k = (0, \ldots, 0, -1)$ pointing out of $\mathbb{H}^k$. See Figure 14.2. Using the Gram–Schmidt process we can orthonormalise the image of e_k under $d\phi_0$ with respect to $T_x(\partial X)$ and get the inward-pointing unit normal vector. The orthonormalisation with respect to $T_x(\partial X)$ of $d\phi_0(-e_k)$ is the *outward-pointing unit normal vector*. Here we recall that the inner product on $T_x(X)$ is induced by the standard inner product on $\mathbb{R}^N$, where $X \subset \mathbb{R}^N$ and hence $T_x(X) \subset \mathbb{R}^N$. We denote the outward-pointing unit normal vector by n_x. We checked in Exercise 9.5 that the construction of n_x does not depend on the choice of ϕ and that the assignment $x \mapsto n_x$ is a smooth map on ∂X. See Figure 14.3. Now we are ready to orient $T_x(\partial X)$.

Definition 14.18 (Boundary orientation) We define the sign of an ordered basis $(v_1, \ldots, v_{k-1})$ in $T_x(\partial X)$ to be the sign of the ordered basis $(n_x, v_1, \ldots, v_{k-1})$ in $T_x(X)$, i.e., we set

$$\text{sign}\,(v_1, \ldots, v_{k-1}) \text{ in } T_x(\partial X) := \text{sign}\,(n_x, v_1, \ldots, v_{k-1}) \text{ in } T_x(X).$$

Since both the assignment $x \mapsto n_x$ and the choice of sign for ordered bases on $T_x(X)$ vary smoothly, this defines an orientation on ∂X which is called the *boundary orientation*.

Example 14.19 (Orientation of the sphere) The n-dimensional sphere $\mathbb{S}^n$ is an orientable smooth manifold. The standard orientation of $\mathbb{S}^n$ is given as the boundary orientation induced from the orientation of the unit disk $\mathbb{D}^{n+1}$ defined by the standard orientation of $T_x(\mathbb{D}^{n+1}) = \mathbb{R}^{n+1}$. See Figure 14.3.

Remark 14.20 (Orientation of one-manifolds) Let us apply what we just learned to the case of a one-manifold with boundary. The boundary ∂X is zero dimensional. The orientation of the zero-dimensional vector space $T_x(\partial X)$ equals the sign of the basis of $T_x(X)$ consisting of the outward-pointing unit vector n_x.

Example 14.21 (Unit interval) Let us look at the compact interval $X = [0, 1]$ with its standard orientation inherited from being a subset of $\mathbb{R}$. Note that local parametrisations of $[0, 1]$ are given by $\phi\colon [0, 1) \to [0, 1]$, $x \mapsto x$, around $0 \in [0, 1]$ and $\psi\colon [0, 1) \to [0, 1]$, $x \mapsto 1 - x$, around $1 \in [0, 1]$. Hence, at $x = 1$, the outward-pointing normal vector is $1 \in \mathbb{R} = T_x(X)$. The basis consisting of this vector is positively oriented. At $x = 0$ the outward-pointing normal vector is the negatively oriented $-1 \in \mathbb{R} = T_0(X)$. Thus the orientation of $T_1(\partial X)$ is $+1$, and the orientation of $T_0(\partial X)$ is -1. Reversing the orientation on $[0, 1]$ simply reverses the orientations at each boundary point. Thus the sum of both orientation numbers at the boundary points of $[0, 1]$ is always zero.

Since, by Theorem 10.1, every compact one-manifold with boundary is diffeomorphic to the disjoint union of copies of $[0, 1]$, we conclude:

Proposition 14.22 (Boundary orientation of one-manifolds) *The sum of the orientation numbers at the boundary points of any compact oriented one-dimensional manifold with boundary is zero. In particular, the boundary points of a smooth path γ on an oriented manifold X, i.e., a smooth map $\gamma\colon [0, 1] \to X$, must have opposite orientation signs.* □

Remark 14.23 This is an important observation which makes it possible to define a homotopy invariant degree of a smooth map with values in $\mathbb{Z}$.

Example 14.24 (Orienting the domain of a homotopy) As an application of product and boundary orientations, we would like to orient the product $[0, 1] \times X$ for a boundaryless smooth oriented manifold X. This will be crucial for the homotopy invariance of the integer-valued Brouwer degree in Section 15. We learned that products and boundaries inherit orientations. For each $t \in [0, 1]$, $X_t := \{t\} \times X$ is diffeomorphic to X, and the orientation on X_t should be such that the diffeomorphism $X \to X_t$, $x \mapsto (t, x)$, preserves orientations.

For future applications, we are particularly interested in the orientation of the boundary $\partial([0,1] \times X) = \{0\} \times X \cup \{1\} \times X$. So let us try to understand the induced orientation on the boundary. We start with X_1. We see from the local parametrisation ψ above that along X_1 the outward-pointing normal vector is the vector $n_{(1,x)} = (1,0) = (1,0,\ldots,0)$ in $T_1([0,1]) \times T_x(X)$. If $\beta = (v_1,\ldots,v_k)$ is an ordered basis of $T_x(X)$, then $0 \times \beta = ((0,v_1),\ldots,(0,v_k))$ is an ordered basis of $T_x(X_1)$. By definition of the boundary orientation, $(n_{(1,0)}, (0 \times \beta))$ is positively oriented if and only if β is positively oriented. In terms of signs, this means $\operatorname{sign}(n_{(1,0)}, (0 \times \beta)) = \operatorname{sign}(\beta)$. If we calculate the orientation induced from the product structure, then we get

$$\operatorname{sign}((1,0),(0 \times \beta)) = (+1) \cdot \operatorname{sign}(\beta) = \operatorname{sign}(\beta).$$

We learn from these two equations that the boundary orientation of X_1 is just the orientation of X as a copy in the product $[0,1] \times X$. This may sound obvious, but we should be careful. Along X_0, the outward-pointing normal vector is the vector $n_{(0,x)} = (-1,0) = (-1,0,\ldots,0)$ in $T_0([0,1]) \times T_x(X)$. Hence the orientation on $T_0([0,1])$ is opposite to the standard orientation of $\mathbb{R}$. The formula for product orientations then yields

$$\operatorname{sign}((-1,0), 0 \times \beta)) = \operatorname{sign}(-1) \cdot \operatorname{sign}(\beta) = -\operatorname{sign}(\beta).$$

Thus the boundary orientation on X_0 is the reverse of its orientation as a copy of X in the product $[0,1] \times X$. Thus the orientation on the boundary is

$$\partial([0,1] \times X) = X_1 \cup (-X_0).$$

We will also express this fact by using the *notation* $\partial([0,1] \times X) = X_1 - X_0$. See Figure 14.4.

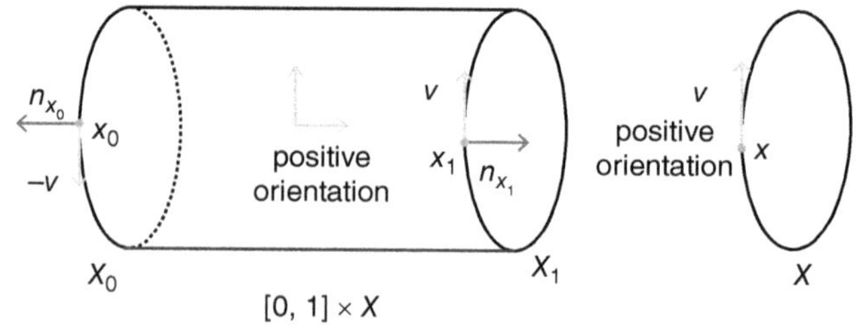

Figure 14.4 The vectors n_{x_0} and n_{x_1} point in opposite directions. This provides the two boundary copies of X with opposite orientations.

14.4 Orientation of Transverse Preimage

Our next goal is to orient preimages. In order to do so, we will have to look at direct sums of vector spaces.

Definition 14.25 (Orientation on a direct sum of vector spaces) Suppose that $V = V_1 \oplus V_2$ is a direct sum of vector spaces. Then the *direct sum orientation* on V is defined as follows. Choose ordered bases β_1 of V_1 and β_2 of V_2. Let $\beta = (\beta_1, \beta_2)$ be the combined ordered basis of V. We then set

$$\mathrm{sign}\,(\beta) := \mathrm{sign}\,(\beta_1) \cdot \mathrm{sign}\,(\beta_2).$$

Remark 14.26 It follows immediately from the way matrices on direct sums are put together that the above formula determines an orientation on the third space if two orientations are given. Note that this not only means that orientations on V_1 and V_2 determine an orientation on V, but also orientations on V and, say, V_2 determine an orientation on V_1. Note again that the order of the summands V_1 and V_2 is crucial.

14.4.1 Orientation of Transverse Preimages

Let $f: X \to Y$ be a smooth map and $Z \subset Y$ be a submanifold with $f \pitchfork Z$ and $\partial f \pitchfork Z$. We assume that X, Y, and Z are all oriented, and Y and Z are without boundary. We would like to define a preimage orientation on the manifold with boundary $S := f^{-1}(Z)$. If $f(x) = z \in Z$ and z is a regular value, then, by Proposition 5.5, we have $T_x(S) = (df_x)^{-1}(T_z(Z)) \subset T_x(X)$. Let $N_x(S; X)$ be the orthogonal complement to $T_x(S)$ in $T_x(X)$. By definition, we then have a direct sum decomposition

$$N_x(S; X) \oplus T_x(S) = T_x(X).$$

Hence, by Remark 14.26, we only need to choose an orientation on $N_x(S; X)$ to obtain an orientation on $T_x(S)$. Since $f \pitchfork Z$, we have

$$\begin{aligned} T_z(Y) &= df_x(T_x(X)) + T_z(Z) = df_x(N_x(S; X) \oplus T_x(S)) + T_z(Z) \\ &= df_x(N_x(S; X)) \oplus T_z(Z), \end{aligned}$$

where we use $df_x(T_x(S)) = T_z(Z)$ for the last step. Thus the orientations on $T_z(Y)$ and $T_z(Z)$ induce a direct sum orientation on $df_x(N_x(S; X))$. It remains to show that this also induces an orientation on $N_x(S; X)$. Since $\mathrm{Ker}\,(df_x)$ is contained in $(df_x)^{-1}(T_z(Z)) = T_x(S)$, the restriction of df_x to $N_x(S; X)$ is in fact an isomorphism onto its image. Therefore, the induced orientation on $df_x(N_x(S; X))$ defines an orientation on $N_x(S; X)$ via the isomorphism

$(df_x)_{|N_x(S;X)}$. Since the orientations on X, Y, and Z vary smoothly and df_x also depends smoothly on x, the induced orientation on $T_x(S)$ varies smoothly with x. We put this into a formula in the following definition.

Definition 14.27 (Preimage orientation) Let $f\colon X \to Y$ be a smooth map and $Z \subset Y$ be a submanifold with $f \pitchfork Z$ and $\partial f \pitchfork Z$. Let X, Y, and Z be oriented and Y and Z be without boundary. We choose an ordered basis β_V for the vector space V which runs through the spaces $T_x f^{-1}(Z)$, $T_x X$, $T_{f(x)}Z$, $T_{f(x)}Y$, and $N_x(f^{-1}(Z); X)$. We define the sign of $\beta_{N_x(f^{-1}(Z);X)}$, and thereby the orientation of $N_x(f^{-1}(Z); X)$, such that

$$\operatorname{sign}\left(df_x\left(\beta_{N_x(f^{-1}(Z);X)}\right)\right)\cdot \operatorname{sign}\left(\beta_{T_{f(x)}Z}\right) = \operatorname{sign}\left(\beta_{T_{f(x)}Y}\right).$$

Then we define the sign of $\beta_{T_x f^{-1}(Z)}$ and hence the orientation of $f^{-1}(Z)$, called the *preimage orientation*, such that

$$\operatorname{sign}\left(\beta_{N_x(f^{-1}(Z);X)}\right)\cdot \operatorname{sign}\left(\beta_{T_x f^{-1}(Z)}\right) = \operatorname{sign}\left(\beta_{T_x X}\right).$$

Remark 14.28 (Orthogonality is not important) Note that we did not really use that $N_x(S; X)$ is orthogonal to $T_x(S)$. All we needed was a direct sum decomposition $H \oplus T_x(S) = T_x(X)$ with a space H with an orientation induced by the orientation of X. We will exploit this fact below.

14.4.2 Orientation on Boundary of Preimage

Let $f\colon X \to Y$ be a smooth map with $f \pitchfork Z$ and $\partial f \pitchfork Z$, where X, Y, and Z are all oriented, Y and Z are boundaryless, and X has a boundary. Then the manifold $\partial f^{-1}(Z)$ acquires two orientations:

- an orientation as the boundary of the manifold $f^{-1}(Z)$, and
- an orientation as the preimage of Z under the map $\partial f\colon \partial X \to Y$.

It turns out that these two orientations may not agree. However, there is a formula that relates them.

Theorem 14.29 (Orientation on boundary of preimage) *Let $f\colon X \to Y$ be a smooth map with $f \pitchfork Z$ and $\partial f \pitchfork Z$, where X, Y, and Z are all oriented, Y and Z are without boundary, while X has a boundary. Then the orientation of the boundary of $f^{-1}(Z)$ satisfies the formula:*

$$\partial(f^{-1}(Z)) = (-1)^{\operatorname{codim} Z}(\partial f)^{-1}(Z).$$

This means the orientations of $\partial f^{-1}(Z)$, induced by being a boundary or by being a preimage, are the same if codim$Z$ *is even, and opposite if* codim$Z$ *is odd.*

Proof: Denote $f^{-1}(Z)$ again by S. Let H be a subspace of $T_x(\partial X)$ complementary to $T_x(\partial S)$, i.e., $H \oplus T_x(\partial S) = T_x(\partial X)$. Note that H is also complementary to $T_x(S)$ in $T_x(X)$, i.e., $H \oplus T_x(S) = T_x(X)$. For we have

$$H \cap T_x(S) = \{0\} \text{ and } T_x(S) \cap T_x(\partial X) = T_x(\partial S),$$

and $\dim H = \dim T_x(\partial X) - \dim T_x(\partial S) = \dim T_x(X) - \dim T_x(S)$. Hence we may use H to define the direct sum orientation of both S and ∂S at x. Since $H \subset T_x(\partial X) \subset T_x(X)$, the maps df_x and $d(\partial f)_x$ agree on H, i.e., $df_x(H) = d(\partial f)_x(H)$. As in the case of $N_x(S; X)$, since

$$\{0\} \subset T_z(Z) \Rightarrow \text{Ker}\,(df_x) \subset f^{-1}(T_z(Z)) = T_x(S),$$

the intersection $\text{Ker}\,(df_x) \cap H$ is $\{0\}$. Hence the restrictions of df_x and $d(\partial f)_x$ to H are isomorphisms onto their common image. Thus, $f \pitchfork Z$ and $\partial f \pitchfork Z$ imply that we have two direct sum decompositions $df_x(H) \oplus T_z(Z) = T_z(Y) = d(\partial f)_x(H) \oplus T_z(Z)$, and the two orients of H via these direct sums agree.

To conclude, we obtained that H has a well-defined orientation. Hence we can use this unique orientation on H to orient

$$S \text{ via } H \oplus T_x(S) = T_x(X) \text{ and } \partial S \text{ via } H \oplus T_x(\partial S) = T_x(\partial X).$$

It remains to check how this orientation of $T_x(\partial S)$ relates to the orientation of the boundary induced from the orientation of $T_x(S)$. Let n_x be the outward unit vector to ∂S in $T_x(S)$, and let $\mathbb{R} \cdot n_x$ represent the one-dimensional subspace spanned by n_x. We orient this space by assigning the sign $+1$ to the basis (n_x). Even though n_x need not be perpendicular to all of $T_x(\partial X)$, it suffices to know that n_x lies in the half-space pointing away from $T_x(X)$ to know that the orientations of $\mathbb{R} \cdot n_x$, $T_x(\partial X)$, and $T_x(X)$ are related by the direct sum

$$T_x(X) = \mathbb{R} \cdot n_x \oplus T_x(\partial X).$$

Now we use that H is complementary to both $T_x(S)$ in $T_x(X)$ and $T_x(\partial S)$ in $T_x(\partial X)$ and plug this into the above direct sum to get

$$H \oplus T_x(S) = \mathbb{R} \cdot n_x \oplus H \oplus T_x(\partial S).$$

This equation is already almost what we need, since we would like to compare the orientations $T_x(S)$ and $\mathbb{R} \cdot n_x \oplus T_x(\partial S)$. For doing so, we need to move $\mathbb{R} \cdot n_x$ past H. If $\dim H = m$, H has m basis vectors $(w_1, \dots, w_m)$. Remembering the rule for orienting direct sums, this means we have to apply exactly m transpositions to the ordered set $(n_x, w_1, \dots, w_m)$ to get to $(w_1, \dots, w_m, n_x)$. This results in m shifts of signs. Hence we get

$$H \oplus T_x(S) = (-1)^{\text{codim} Z} H \oplus \mathbb{R} \cdot n_x \oplus T_x(\partial S).$$

Since H appears on both sides as the first summand, we can disregard it for the computation and get that if ∂S is oriented as a preimage under ∂f, then its orientation relates to that of $T_x(S)$ by $T_x(S) = (-1)^{\operatorname{codim} Z}\mathbb{R} \cdot n_x \oplus T_x(\partial S)$. Now, if ∂S is oriented as a boundary, then $T_x(S) = \mathbb{R} \cdot n_x \oplus T_x(\partial S)$. Thus, sign (∂S) as a boundary $= (-1)^{\operatorname{codim} Z} \cdot$ sign (∂S) as a preimage. □

14.5 Fibres of the Hopf Map

Recall the *Hopf map* $\hbar$ from Example 2.10 and Exercise 2.10. We consider $\mathbb{S}^3$ as a subset of $\mathbb{C}^2$, i.e., $\mathbb{S}^3 = \{(z_0, z_1) \in \mathbb{C}^2 : |z_0|^2 + |z_1|^2 = 1\}$, and $\mathbb{S}^2$ as a subset of $\mathbb{C} \times \mathbb{R}$, i.e., $\mathbb{S}^2 = \{(z, x) \in \mathbb{C} \times \mathbb{R} : |z|^2 + x^2 = 1\}$. Then $\hbar$ is the map $\mathbb{S}^3 \to \mathbb{S}^2$ given by $\hbar(z_0, z_1) = (2z_0\bar{z}_1, |z_0|^2 - |z_1|^2)$. Consider the point $a = (0, 0, 1)$ on $\mathbb{S}^2 \subset \mathbb{R}^3 \cong \mathbb{C} \times \mathbb{R}$. In Exercise 4.12, we determined the fibre $\hbar^{-1}(a)$. Now we would like to compute its orientation as a preimage under $\hbar$. Recall that the fibre over a is $\hbar^{-1}(a) = \{(z_0, 0) \in \mathbb{S}^3 \subset \mathbb{C}^2 : |z_0|^2 = 1\}$. Let $q = (x_0, y_0, 0, 0) \in \hbar^{-1}(a)$ be a point in the fibre over a. The tangent space $T_q\mathbb{S}^3$ is the vector space

$$T_q\mathbb{S}^3 = \{\mathbf{u} \in \mathbb{R}^4 : \mathbf{u} \perp q\} = \operatorname{Span}\left\{ q^\perp = \begin{pmatrix} -y_0 \\ x_0 \\ 0 \\ 0 \end{pmatrix}, \mathbf{e}_3^4 = \begin{pmatrix} 0 \\ 0 \\ 1 \\ 0 \end{pmatrix}, \mathbf{e}_4^4 = \begin{pmatrix} 0 \\ 0 \\ 0 \\ 1 \end{pmatrix} \right\}.$$

The orientation of $T_q\mathbb{S}^3$ as a boundary of the unit ball is such that the outward-pointing vector q together with the basis vectors of $T_q\mathbb{S}^3$ form a positively oriented basis of $\mathbb{R}^4$. The determinant of the matrix

$$\begin{pmatrix} x_0 & -y_0 & 0 & 0 \\ y_0 & x_0 & 0 & 0 \\ 0 & 0 & 1 & 0 \\ 0 & 0 & 0 & 1 \end{pmatrix},$$

which expresses the basis $(q, q^\perp, \mathbf{e}_3^4, \mathbf{e}_4^4)$ in the standard basis of $\mathbb{R}^4$ equals $x_0^2 + y_0^2 = 1 > 0$. In particular, it is positive, and the basis $(q^\perp, \mathbf{e}_3^4, \mathbf{e}_4^4)$ is a positively oriented basis of $T_q\mathbb{S}^3$. The tangent space $T_q\hbar^{-1}(a)$ equals the kernel of $d\tilde{\hbar}_q$, where we consider the map $\tilde{\hbar}\colon \mathbb{R}^4 \cong \mathbb{C}^2 \to \mathbb{C} \times \mathbb{R} \cong \mathbb{R}^3$ using $\hbar = \tilde{\hbar}_{|\mathbb{S}^3}$. In Exercise 4.12, we computed this map as represented by the matrix

$$d\tilde{\hbar}_q = 2 \cdot \begin{pmatrix} 0 & 0 & x_0 & y_0 \\ 0 & 0 & y_0 & -x_0 \\ x_0 & y_0 & 0 & 0 \end{pmatrix}.$$

The kernel of this map is the span of the vector $q^\perp$ that we have just seen above. The normal space $N_q(\hbar^{-1}(a);\mathbb{S}^3) \subset T_q\mathbb{S}^3$ of vectors which are orthogonal to $T_q\hbar^{-1}(a)$ is the span of $(\mathbf{e}_3^4, \mathbf{e}_4^4)$. The map $d\tilde{\hbar}_q$ sends $\mathbf{e}_3$ and $\mathbf{e}_4$ to $d\tilde{\hbar}_q(\mathbf{e}_3^4) = 2(x_0, y_0, 0, 0)$ and $d\tilde{\hbar}_q(\mathbf{e}_4^4) = 2(y_0, -x_0, 0, 0)$. These two vectors form a basis $(d\tilde{\hbar}_q(\mathbf{e}_3^3), d\tilde{\hbar}_q(\mathbf{e}_4^4))$ of $T_a\mathbb{S}^2$. We need to check the orientation of this basis. The tangent space $T_a\mathbb{S}^2$ has a basis $(\mathbf{e}_1^3, \mathbf{e}_2^3)$ as a subspace in $\mathbb{R}^3$. This basis is positively oriented since, together with the outward-pointing vector $a = \mathbf{e}_3^3$, the basis $(\mathbf{e}_3^3, \mathbf{e}_1^3, \mathbf{e}_2^3)$ is a positively oriented basis of $\mathbb{R}^3$ since we make two permutations which lead to multiplying with $(-1)^2 = +1$. The matrix A which expresses $(d\tilde{\hbar}_q(\mathbf{e}_3^3), d\tilde{\hbar}_q(\mathbf{e}_4^4))$ in terms of the basis $(\mathbf{e}_1^3, \mathbf{e}_2^3)$ is given by

$$A = 2 \cdot \begin{pmatrix} x_0 & y_0 \\ y_0 & -x_0 \end{pmatrix}.$$

We see that $\det A = 4(-x_0^2 - y_0^2) = -4 < 0$ is negative. Hence the basis $(d\tilde{\hbar}_q(\mathbf{e}_3^3), d\tilde{\hbar}_q(\mathbf{e}_4^4))$ is a negatively oriented basis of $T_a\mathbb{S}^2$. This defines an orientation on the normal space $N_q(\hbar^{-1}(a);\mathbb{S}^3)$ by declaring the orientation of the basis $(\mathbf{e}_3^4, \mathbf{e}_4^4)$ to be negative. Finally, the orientation of $T_q\hbar^{-1}(a)$ is such that the direct sum $N_q(\hbar^{-1}(a);\mathbb{S}^3) \oplus T_q\hbar^{-1}(a) = T_q\mathbb{S}^3$ induces the given orientation on $T_q\mathbb{S}^3$. We make this explicit by looking at the basis $(\mathbf{e}_3^4, \mathbf{e}_4^4, q^\perp)$ of $N_q(\hbar^{-1}(a);\mathbb{S}^3) \oplus T_q\hbar^{-1}(a)$. As a basis of $T_q\mathbb{S}^3$, $(\mathbf{e}_3^4, \mathbf{e}_4^4, q^\perp)$ is positively oriented since it arises by two permutations from the positively oriented basis $(q^\perp, \mathbf{e}_3^4, \mathbf{e}_4^4)$. Since the sign of $(\mathbf{e}_3^4, \mathbf{e}_4^4)$ is negative as a basis of $N_q(\hbar^{-1}(a);\mathbb{S}^3)$, we need that $q^\perp$ also has negative sign. Hence the vector $q^\perp$ provides a negatively oriented basis of $T_q\hbar^{-1}(a)$. Comparing this orientation with the standard orientation of $\mathbb{S}^1$, we see that $\hbar^{-1}(a)$ has the opposite orientation. We will compute the orientation of the fibre of $\hbar$ at another point in Exercise 14.9.

14.6 Simply Connected Manifolds Are Orientable

The following theorem shows that a large class of manifolds is orientable. Recall that a manifold X is called simply connected if it is connected and every continuous map $\mathbb{S}^1 \to X$ is homotopic to a constant map.

Theorem 14.30 *Every simply connected manifold is orientable.*

The idea of the proof is to choose an orientation at one point and then to transport this orientation to all other points along smooth paths. Since the manifold is connected, we can connect any two points via a path. This gives us an orientation at every point. However, the orientations may depend on the choices of paths. Then we use that the manifold is simply connected, i.e., any two paths are homotopic. This will allow us to show that everything only depends on the choice at the initial point. We will now explain the details of the proof. It is a bit lengthy, but we think it is worth the effort as the proof sheds light on many of the technicalities related to orientations.

Proof of Theorem 14.30: Let $k = \dim X$. We start by picking a point $x \in X$ and choosing an orientation of the tangent space $T_x(X)$. We will refer to it as the positive orientation class. Now let $y \in X$ be any other point in X. Since X is connected and a smooth manifold, X is in fact path-connected by Exercise 8.5. Hence there is a smooth map $\gamma\colon [0,1] \to X$ with $\gamma(0) = x$ and $\gamma(1) = y$. Now we choose an open covering of X by local parametrisations $\phi_\alpha\colon V_\alpha \to U_\alpha$. After shrinking V_α if necessary, we can assume that each V_α is an open ball in $\mathbb{R}^k$. If there is a single U_α such that $\gamma([0,1] \subset U_\alpha$, then we can define an orientation on $T_y(X)$ as follows. Let $\mathfrak{b}^k$ denote the standard basis of $\mathbb{R}^k$. Then $d(\phi_\alpha)_{\phi_\alpha^{-1}(x)}(\mathfrak{b}^k)$ is a basis of $T_x(X)$. If this basis is not in the chosen orientation class for $T_x(X)$, we first replace ϕ_α with $\phi_\alpha \circ r_1\colon V_\alpha \to U_\alpha$, where $r_1\colon \mathbb{R}^k \to \mathbb{R}^k$ is the reflection in the first coordinate. This new map is also a local parametrisation with domain V_α and image U_α since V_α is an open ball in $\mathbb{R}^k$ and hence $v \in V_\alpha$ if and only if $r_1(v) \in V_\alpha$. By Example 14.8, r_1 reverses the orientation. Hence, after replacing ϕ_α with $\phi_\alpha \circ r_1$ if necessary, we can assume that $d(\phi_\alpha)_{\phi_\alpha^{-1}(x)}(\mathfrak{b}^k)$ is a basis in the chosen orientation class for $T_x(X)$. Now we orient $T_y(X)$ such that the basis $(d(\phi_\alpha^{-1})_y \circ d(\phi_\alpha)_{\phi_\alpha^{-1}(x)})(\mathfrak{b}^k)$ is positively oriented.

If $\gamma([0,1])$ is not contained in a single U_α, then we proceed as follows. Since $[0,1]$ is compact and γ is continuous, the image $\gamma([0,1])$ is compact by Lemma 1.13 which was proven in Exercise 1.1. Hence finitely many of the U_α suffice to cover $\gamma([0,1])$. We label these open sets $U_1, \ldots, U_n$, and we order them such that $x \in U_1$ and $y \in U_n$. The collection of subsets $\gamma^{-1}(\gamma([0,1]) \cap U_i) \subset [0,1]$ is an open cover of $[0,1]$. By the Lebesgue number lemma proven in Exercise 1.4, we can find a partition of $[0,1]$ into n subintervals $[t_{i-1}, t_i]$ with $t_0 = 0$ and $t_n = 1$ such that, for each $i = 1, \ldots, n$, and after possibly relabelling the U_i, we have $\gamma([t_{i-1}, t_i])$ is contained in U_i. In particular, $\gamma([0,t_1])$ is contained in U_1, and we use the above procedure to orient $T_{\gamma(t)}(X)$ for all $t \in [0,t_1]$. Next, we apply the procedure to $\gamma([t_1,t_2]) \subset U_2$. That is, we first check that $d(\phi_2)_{\phi_1^{-1}(\gamma(t_1))}(\mathfrak{b}^k)$ has the same orientation as $d(\phi_1)_{\phi_1^{-1}(\gamma(t_1))}(\mathfrak{b}^k)$, and, if not, we first compose ϕ_2 with a reflection. Then we orient $T_{\gamma(t)}(X)$ for all $t \in [t_1,t_2]$ such that $d(\phi_2)_{\phi_1^{-1}(\gamma(t))}(\mathfrak{b}^k)$ is in the positive orientation class. After n steps, we obtain an orientation of $T_{\gamma(t)}(X)$ for all $t \in [0,1]$, in particular, for $T_{\gamma(t_n)}(X) = T_y(X)$. We write $\mathfrak{o}_t$ for the orientation of $T_{\gamma(t)}(X)$.

Remark 14.31 We note that, at each $t \in [t_{i-1}, t_i]$, we chose the orientation for $T_{\gamma(t)}(X)$ such that the image $d(\phi_i)_{\phi_i^{-1}(\gamma(t))}(\mathfrak{b}^k)$ of the standard basis $\mathfrak{b}^k$ of $\mathbb{R}^k$ is positively oriented. And we can do so for all points z in U_i. The path γ and the partition of $[0,1]$ give us reference points which ensure that, at least along γ, our choices are compatible. The latter would not be guaranteed if we only had the local parametrisations.

Now we show that the orientation of $T_y(X)$ does not depend on the choice of local parametrisations. We choose another open covering by local parametrisations $\phi'_\beta\colon V'_\beta \to U'_\beta$. By repeating the above process, with a potentially different partition of $[0,1]$ into m subintervals $[t'_{j-1}, t'_j]$ such that $\gamma([t'_{j-1}, t'_j]) \subset U'_j$, we obtain an orientation $\mathfrak{o}'_t$ of $T_{\gamma(t)}(X)$ for all $t \in [0,1]$.

Lemma 14.32 *We have $\mathfrak{o}_t = \mathfrak{o}'_t$ for all $t \in [0,1]$.*

Proof: For every $t \in [0,1]$, there are indices i, j such that $t \in [t_{i-1}, t_i] \cap [t'_{j-1}, t'_j]$, and we consider the parametrisations ϕ_i and ϕ_j. To simplify the notation we write $z = \gamma(t)$. The local parametrisations ϕ_i and ϕ'_j then define a matrix $A_{i,j}(t)$ via the composition

$$\begin{array}{ccc} & T_z(X) & \\ {\scriptstyle d(\phi_i)_{\phi_i^{-1}(z)}} \nearrow & & \searrow {\scriptstyle d((\phi'_j)^{-1})_z} \\ \mathbb{R}^k & \xrightarrow[A_{i,j}(t)]{\quad\cdots\cdots\quad} & \mathbb{R}^k \end{array} \quad . \tag{14.2}$$

That is, $A_{i,j}(t)$ is the transition matrix between the bases $d(\phi_i)_{\phi_i^{-1}(\gamma(t))}(\mathfrak{b}^k)$ and $d(\phi'_j)_{\phi_j^{-1}(\gamma(t))}(\mathfrak{b}^k)$ of $T_{\gamma(t)}(X)$. Since ϕ_i and ϕ'_j are local parametrisations, we have $\det A_{i,j}(t) \neq 0$ for all t. By construction of the orientations, we have $\mathfrak{o}_t = \mathfrak{o}'_t$ if and only if $\det A_{i,j}(t) > 0$. We note that by construction of the orientations, the sign of $\det A_{i,j}(t)$, i.e., the number $\det A_{i,j}(t)/|\det A_{i,j}(t)|$, is independent of the choice of i or $i+1$ if t equals one of the t_i for $i = 1, \ldots, n-1$ and hence $t \in [t_{i-1}, t_i]$ and $t \in [t_i, t_{i+1}]$, and similarly for t equal t'_j for $j = 1, \ldots, m-1$. This implies that the number $\det A_{i,j}(t)/|\det A_{i,j}(t)|$ only depends on t, and we obtain a well-defined map $\sigma\colon [0,1] \to \{\pm 1\}$ defined by $\sigma(t) := \det A_{i,j}(t)/|\det A_{i,j}(t)|$. Since $d(\phi_i)_{\phi_i^{-1}(z)}$ and $d((\phi'_j)^{-1})_z$ vary smoothly with z, the map σ is continuous where $\{\pm 1\}$ has the discrete topology. This implies that σ is locally constant. Since $[0,1]$ is connected, σ is constant by Lemma 1.30. As $\sigma(0) = 1$, we must have $\sigma(t) = 1$ and hence $\mathfrak{o}_t = \mathfrak{o}'_t$ for all $t \in [0,1]$. This proves the lemma. □

This shows that the orientation of the $T_{\gamma(t)}(X)$ for all $t \in [0,1]$ does not depend on the choice of local parametrisations. Actually, the argument of the proof of Lemma 14.32 also shows that the orientation of $T_{\gamma(t)}(X)$ for all $[0,1]$ does not depend on the choice of partition of $[0,1]$. It remains to use the assumption that X is simply connected to prove the following lemma.

Lemma 14.33 *The orientation of $T_y(X)$ does not depend on the choice of the path γ.*

Proof: Let γ' be another path from x to y. Since X is simply connected, there is a homotopy $\Gamma\colon [0,1] \times [0,1] \to X$ between γ and γ'. To simplify the notation, we will also write $\gamma_s(t) = \Gamma(s,t)$. Let $\phi_\alpha\colon V_\alpha \to U_\alpha$ be a collection of local parametrisations such that the U_α form an open cover of X and the V_α are open balls in $\mathbb{R}^k$. Since Γ is continuous and $[0,1] \times [0,1]$ is compact, the image $\Gamma([0,1] \times [0,1])$ is a compact subset of X by Lemma 1.13. Hence finitely many of the U_α suffice to cover $\Gamma([0,1] \times [0,1])$. By the Lebesgue Number Lemma proven in Exercise 1.5, we can find $s_i, t_j \in [0,1]$ for $i = 1, \ldots, n$ and $j = 1, \ldots, m$ with $s_0 = t_0 = 0$ and $s_n = t_m = 1$ such that each $\Gamma([s_{i-1}, s_i] \times [t_{j-1}, t_j])$ is contained in some $U_{\alpha(i,j)}$. We write $U_{i,j}$ for $U_{\alpha(i,j)}$. The collection of the $U_{i,j}$ then forms an open cover of $\Gamma([0,1] \times [0,1])$. We write $\phi_{i,j}\colon V_{i,j} \to U_{i,j}$ for the corresponding local parametrisation. Starting with the chosen orientation of $T_x(X)$ we can now orient each $T_{\gamma_s(t)}(X)$ inductively as above. That is, assuming $T_{\gamma_{s_i}(t_j)}(X)$ is already oriented, we first check that $d(\phi_{i+1,j+1})_{\phi_{i+1,j+1}^{-1}(\gamma_{s_i}(t_j))}(\mathfrak{b}^k)$ has the same orientation as $d(\phi_{i,j})_{\phi_{i,j}^{-1}(\gamma_{s_i}(t_j))}(\mathfrak{b}^k)$, and, if not, we first compose $\phi_{i+1,j+1}$ with a reflection. Then we orient $T_{\gamma_s(t)}(X)$ for all $(s,t) \in [s_i, s_{i+1}] \times [t_j, t_{j+1}]$ such that $d(\phi_{i+1,j+1})_{\phi_{i+1,j+1}^{-1}(\gamma_s(t))}(\mathfrak{b}^k)$ is in the positive orientation class.

By construction, at each $(s,t) \in [s_{i-1}, s_i] \times [t_{j-1}, t_j]$, the transition matrix $A_{i,j}(s,t)$ for the composition

$$\begin{array}{ccc} & T_{\gamma_s(t)}(X) & \\ {\scriptstyle d(\phi_i)_{\phi_i^{-1}(\gamma_s(t))}} \nearrow & & \searrow {\scriptstyle d((\phi'_j)^{-1})_{\gamma_s(t)}} \\ \mathbb{R}^k & \xrightarrow[A_{i,j}(s,t)]{\qquad\qquad} & \mathbb{R}^k \end{array} \qquad (14.3)$$

has strictly positive determinant. Moreover, if $(s,t) = (s_i, t_j)$ for some i and j, the sign of $\det A_{i,j}(s,t)/|\det A_{i,j}(s,t)|$ is independent of the choice of local parametrisation. Hence the function $[0,1] \times [0,1] \to \{\pm 1\}$, which sends (s,t) to $\det A_{i,j}(s,t)/|\det A_{i,j}(s,t)|$ is constant. This shows that the orientation of $T_y(X)$ is the same for every path from x to y. □

Thus, we have shown that starting with a choice of a point x and an orientation of $T_x(X)$ we can construct a smoothly varying orientation of $T_y(X)$

for every $y \in X$ which is compatible with the orientation on $T_x(X)$. In other words, we have proven that X is orientable and thereby proven Theorem 14.30. □

14.7 Exercises

14.1 Let $\beta = (v_1, \ldots, v_k)$ be an ordered basis of a vector space V.

(a) Show that replacing one v_i by a multiple cv_i yields an equivalently oriented ordered basis if $c > 0$, and an oppositely oriented one if $c < 0$.

(b) Show that transposing two elements, i.e., interchanging the places of v_i and v_j for $i \neq j$, yields an oppositely oriented ordered basis.

(c) Show that subtracting from one v_i a linear combination of the others yields an equivalently oriented ordered basis.

(d) Suppose that V is the direct sum of V_1 and V_2. Show that the direct sum orientation of V from $V_1 \oplus V_2$ equals $(-1)^{(\dim V_1)(\dim V_2)}$ times the orientation from $V_2 \oplus V_1$.

14.2 The upper half space $\mathbb{H}^k$ is oriented by the standard orientation of $\mathbb{R}^k$. Thus $\partial\mathbb{H}^k$ acquires a boundary orientation. But $\partial\mathbb{H}^k$ may also be identified with $\mathbb{R}^{k-1}$. Show that the boundary orientation agrees with the standard orientation of $\mathbb{R}^{k-1}$ if and only if k is even.

14.3 Suppose that $f\colon X \to Y$ is a diffeomorphism of connected oriented manifolds with boundary. Show that if $df_x\colon T_x(X) \to T_{f(x)}(Y)$ preserves orientation at one point x, then f preserves orientation at every point.

14.4 In this exercise, we study the orientation of spheres:

(a) Write down the orientation of $\mathbb{S}^2$ as the boundary of the closed unit ball $\mathbb{B}^3$ in $\mathbb{R}^3$, by specifying a positively oriented ordered basis for the tangent space at each $(a, b, c) \in \mathbb{S}^2$.

(b) Show that the boundary orientation of $\mathbb{S}^k$ equals the orientation of $\mathbb{S}^k = g^{-1}(1)$ as the preimage of the map $g\colon \mathbb{R}^{k+1} \to \mathbb{R}$, $g(x) = |x|^2$.

14.5 In this exercise, we will prove the Linear Isotopy Lemma 14.10:

(a) Show that it suffices to treat the case that E preserves orientations. So let E be a linear isomorphism of $\mathbb{R}^k$ that preserves orientations. The proof now proceeds by induction on the dimension k. We need to check two initial cases:

(b) Prove the assertion for $k = 1$.

(c) Prove the assertion for $k = 2$ and assume that E has only complex eigenvalues.

(d) Now show the induction step.

14.6 Let X and Z be transverse submanifolds in Y and assume X, Z, and Y are oriented. Let $i\colon X \hookrightarrow Y$ be the inclusion of X into Y, $j\colon Z \hookrightarrow Y$ be the inclusion of Z into Y. We orient the intersection $X \cap Z$ as the preimage $i^{-1}(Z)$, and the intersection $Z \cap X$ as the preimage $j^{-1}(X)$. Show that the orientations of $X \cap Z$ and $Z \cap X$ are related by

$$X \cap Z = (-1)^{(\mathrm{codim} X)(\mathrm{codim} Z)} Z \cap X.$$

Hint: Show that the orientation of $S = X \cap Z$ at any y is induced by the direct sum $(N_y(S, X) \oplus N_y(S, Z)) \oplus T_y(S) = T_y(Y)$. What happens when you consider $Z \cap X$ instead?

14.7 Let X be an oriented manifold and let $Z \subset X$ be a submanifold of codimension one. Show that Z is orientable if and only if the relative normal bundle $N(Z, X)$ is trivial.

14.8 (a) Let V be a vector space. Show that both orientations on V define the same product orientation on $V \times V$.

(b) Let X be an orientable manifold. Show that the product orientation on $X \times X$ is the same for all choices of orientation on X.

(c) Suppose that X is not orientable. Show that $X \times Y$ is never orientable, no matter what manifold Y may be. In particular, $X \times X$ is not orientable.

Hint: First show that $X \times \mathbb{R}^m$ is not orientable, and then use that every Y has an open subset diffeomorphic to $\mathbb{R}^m$.

(d) Prove that there exists a natural orientation on some neighbourhood of the diagonal Δ in $X \times X$, whether or not X can be oriented. But note that Δ itself is orientable if and only if $X \times X$ is orientable. Why?

Hint: Cover a neighbourhood of Δ by local parametrisations $\phi \times \phi\colon U \times U \to X \times X$, where $\phi\colon U \to X$ is a local parametrisation of X, then apply the previous observations.

14.9 Let $\hbar\colon \mathbb{S}^3 \to \mathbb{S}^2$ be the Hopf map. Consider the point $b = (0, 1, 0)$ on $\mathbb{S}^2 \subset \mathbb{R}^3 \cong \mathbb{C} \times \mathbb{R}$. Recall that we computed the fibre $\hbar^{-1}(b)$ in Exercise 4.12. Determine the orientation of the fibre $\hbar^{-1}(b)$ as a preimage under $\hbar$.

15

The Integer-Valued Brouwer Degree

We have seen in Chapter 11 that the mod 2 degree is a powerful invariant of smooth maps between manifolds. Now we remedy the defect that it was merely an element in $\mathbb{Z}/2$ and define an integer-valued degree. The price we pay for the extended set of values of the degree is that the latter will only be defined for smooth maps between oriented manifolds. However, we will see from the applications of the degree that the new construction is very useful.

15.1 A Well-Defined Invariant

Let X and Y be *oriented* n-dimensional smooth manifolds without boundary. We assume that X is *compact* and that Y is *connected.* Let $f: X \to Y$ be a smooth map. Then the degree of f is defined as follows. Let $x \in X$ be a regular point of f. With our assumptions, this means that $df_x: T_xX \to T_{f(x)}Y$ is a linear isomorphism between oriented vector spaces. The sign of df_x is defined to be $+1$ if df_x preserves orientations, and it is defined to be -1 if df_x reverses orientations. See Figure 15.1. Now, for a regular value $y \in Y$ of f, we define the integer

$$\deg(f; y) := \sum_{x \in f^{-1}(y)} \operatorname{sign}(df_x).$$

Lemma 15.1 *The function $Y \to \mathbb{Z}$, $y \mapsto \deg(f; y)$ is locally constant on the set of regular values of f.*

Proof: Let y be a regular value of f. First assume $f^{-1}(y) = \emptyset$. Since X is compact, $f(X)$ is compact and hence closed in Y. Thus $Y \setminus f(X)$ is open in Y. Since $y \in Y \setminus f(X)$, the subset $U := Y \setminus f(X)$ is an open neighbourhood around y such that $\#f^{-1}(y) = \#f^{-1}(y')$ for all $y' \in U$. Now we assume that

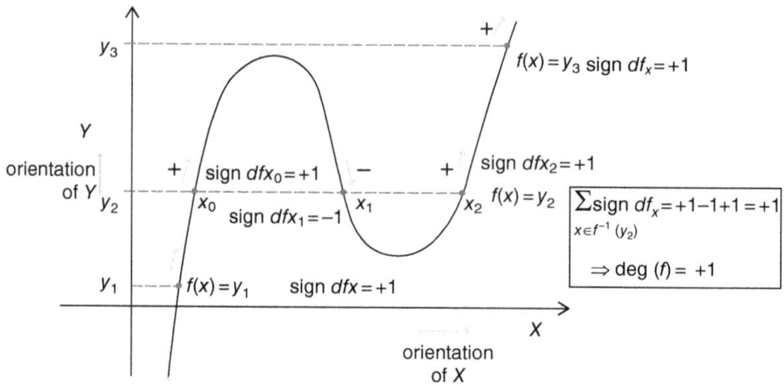

Figure 15.1 We calculate the degree by counting the number of points in the fibre of a regular value. The number is well defined if we take signs determined by orientation into account. In dimension one, this boils down to looking at the sign of the derivative or by checking whether the graph hits the horizontal line corresponding to a value from below or above.

$f^{-1}(y) \neq \emptyset$. By the Stack of Records Theorem 4.28 and Proposition 4.9, we can find a neighbourhood U of y which only consists of regular values such that the preimage $f^{-1}(U)$ is a disjoint union $V_1 \cup \cdots \cup V_n$, where each V_i is an open set in X mapped by f diffeomorphically onto U. Hence, for all points $z \in U$, we have $\#f^{-1}(\{z\}) = n$. It remains to take orientations into account. Since $f_{|V_i}\colon V_i \to U$ is a diffeomorphism, we know that $df_{x_i}\colon T_{x_i}(X) \to T_y(Y)$ is an isomorphism. Now both $T_{x_i}(X)$ and $T_y(Y)$ are oriented, and hence df_{x_i} is either orientation preserving or reversing. By our definition of orientations on manifolds, we have either $\det(df_{x_i}) > 0$ and hence for all $z \in U$ we have $\det(df_{w_i}) > 0$, or $\det(df_{x_i}) < 0$ and hence for all $z \in U$ we have $\det(df_{w_i}) < 0$, where w_i is the unique point in V_i with $f(w_i) = z$. Thus, the orientation number is the same for all points in V_i. Hence the sum of orientation numbers of the points in $f^{-1}(z)$ is the same for all points $z \in U$. □

Now we show an analogue of Theorem 11.14.

Theorem 15.2 (Boundary Theorem for deg) *Let Y be an oriented smooth manifold. Assume that $X = \partial W$ is the boundary of a compact oriented manifold W and that X is oriented as the boundary of W. If $f\colon X \to Y$ can be extended to a smooth map $F\colon W \to Y$, then $\deg(f; y) = 0$ for every regular value y of f.*

Proof: First we suppose that y is a regular value for both F and f. By Theorem 9.18, $F^{-1}(y)$ is a compact submanifold of W of dimension $\dim W - \dim Y = 1$

with boundary $\partial F^{-1}(y) = \partial W \cap F^{-1}(y) = X \cap (F_{|X})^{-1}(y) = f^{-1}(y)$. We also see that only the boundary points of $F^{-1}(y)$ lie on X. By Theorem 10.1, the one-dimensional manifold $F^{-1}(y)$ is the disjoint union of finitely many connected components which are diffeomorphic to either $\mathbb{S}^1$ or $[0,1]$. Since $\mathbb{S}^1$ does not have boundary points, only the boundary points of the components diffeomorphic to $[0,1]$ lie on $X = \partial W$. Let A be one such component diffeomorphic to $[0,1]$, and let a and b denote the boundary points of A, i.e., $\partial A = \{a\} \cup \{b\} \subset X$. We will show that

$$\operatorname{sign} df_a + \operatorname{sign} df_b = 0. \tag{15.1}$$

Since $f^{-1}(y) = (F_{|\partial W})^{-1}(y)$ consists of finitely many boundary points, this will imply that the sum of the signs of all df_x for $x \in f^{-1}(y)$ is zero as claimed.

To show (15.1), we look at how the tangent spaces at the boundary points of A are oriented. So let $p \in A$ and let $(v_1, v_2, \ldots, v_n, v)$ be a positively oriented basis for T_pW such that v is tangent to the one-dimensional submanifold A, i.e., v forms a basis of the one-dimensional vector subspace $T_pA \subset T_pW$, while $(v_1, \ldots, v_n)$ is an ordered basis of the normal space $N_p(A; W) \subset T_pW$. Since we have a direct sum decomposition $N_p(A; W) \oplus T_pA = T_pW$, we get the equation $\operatorname{sign}(v_1, \ldots, v_n) \cdot \operatorname{sign}(v) = \operatorname{sign}(v_1, v_2, \ldots, v_n, v) = +1$. Thus we have

$$\operatorname{sign}(v) = +1 \text{ in } T_pA \iff \operatorname{sign}(v_1, \ldots, v_n) = +1 \text{ in } N_p(A; W).$$

Since A is a component of $F^{-1}(y)$, it inherits an orientation as a preimage as explained in Section 14.4. This means that $\operatorname{sign}(v_1, \ldots, v_n)$ in $N_p(A; W)$ is determined by the effect of dF_p on $(v_1, \ldots, v_n)$, i.e., as in Definition 14.27 we have $\operatorname{sign}(v_1, \ldots, v_n) = \operatorname{sign}(dF_p(v_1, \ldots, v_n))$ where the right-hand side is determined by the given orientation of T_yY. Thus, we conclude that v is a positively oriented basis of T_pA if and only if dF_p sends the basis $(v_1, \ldots, v_n)$ to a positively oriented basis in T_yY.

The vector $v(p)$ depends smoothly on p. Now let $\varphi\colon [0,1] \to A$ be a diffeomorphism with $\varphi(0) = a$ and $\varphi(1) = b$. Since 0 and 1 have opposite orientations in $[0,1]$, the points a and b have opposite orientations, i.e., we have $\operatorname{sign} v(a) + \operatorname{sign} v(b) = 0$. Now, at the boundary points $p \in X$, dF_x restricts to df_x. We just learned that the sign of $v(x)$ is positive if and only if $\operatorname{sign} df_x = +1$, i.e., $\operatorname{sign} v(x) = \operatorname{sign} df_x$ at boundary points. Hence we conclude $\operatorname{sign} df_a + \operatorname{sign} df_b = 0$. By taking the sum over all arcs in $F^{-1}(y)$, we have proved the assertion when y is a regular value for F.

Now suppose that y_0 is a regular value for f, but not for F. The function $y \mapsto \deg(f; y)$ is constant within some open neighbourhood U of y_0 by

Lemma 15.1. By Sard's Theorem 9.21, the open subset U contains a regular value for F. Hence we can choose a regular value y for F within U and get

$$\deg(f; y_0) = \deg(f; y).$$

The previous case then implies the assertion of the theorem. □

Lemma 15.3 (Homotopy Lemma for deg) *Let $f, g\colon X \to Y$ be two smooth maps. Assume f and g are smoothly homotopic. If $y \in Y$ is a regular value for both f and g, then* $\deg(f; y) = \deg(g; y)$.

Proof: Let $F\colon [0,1] \times X \to Y$ be a smooth homotopy between $F_0 = f$ and $F_1 = g$. Recall that the boundary of $[0,1] \times X$ is given by $\{0\} \times X \cup \{1\} \times X$, where the orientation of $\{1\} \times X$ is that of X and the orientation of $\{0\} \times X$ is the opposite one. Since the degree is additive on connected components, we deduce that the degree at y of the restriction $\partial F := F_{|\partial([0,1]\times X}\colon \{0\} \times X \cup \{1\} \times X \to Y$ of F to the boundary of $[0,1] \times X$ equals $\deg(g; y) - \deg(f; y)$. Since ∂F can be extended to the compact oriented manifold $[0,1] \times X$, this difference must be zero by Theorem 15.2. □

Now we can prove the following key result.

Theorem 15.4 (The Brouwer Degree is well defined) *Let X and Y be oriented n-dimensional smooth manifolds without boundary, with X compact and Y connected. Let $f\colon X \to Y$ be a smooth map. The number* $\deg(f; y)$ *does not depend on the choice of the regular value y, and we denote it by* $\deg f$. *Moreover,* $\deg f$ *only depends on the homotopy class of f, i.e., if f_0 and f_1 are smoothly homotopic then* $\deg f_0 = \deg f_1$.

Proof: Let y and z be two regular values of f. By the Isotopy Theorem 11.3, we can choose a diffeomorphism $h\colon X \to Y$ which is isotopic to the identity and with $h(y) = z$. Every diffeomorphism either preserves or reverses the orientation at every point. Since h is isotopic to the identity, we must have $\operatorname{sign} dh_x = +1$ for all x. Thus h preserves orientation, and we get $\deg(f; y) = \deg(h \circ f; h(y))$. Since h is homotopic to the identity, f is homotopic to $h \circ f$. Hence $\deg(h \circ f; z) = \deg(f; z)$ by Lemma 15.3. Thus we get $\deg(f; y) = \deg(f; z)$. Now let $f_1\colon X \to Y$ be a smooth map which is smoothly homotopic to $f_0 = f$. By Sard's Theorem 9.21, we can choose $y \in Y$ which is a regular value for both f_0 and f_1. By the first assertion, we can use y to calculate $\deg(f_0) = \deg(f_0; y)$ and $\deg(f_1) = \deg(f_1; y)$. Then Lemma 15.3 implies that $\deg(f_0) = \deg(f_1)$. □

Remark 15.5 The degree $\deg(f)$ is well defined since we proved that the sum of preimage points counted with orientation signs is constant. The number $\#f^{-1}(y)$ itself, however, given by counting the finite number of points in the fibre without signs may vary.

In the exercises, we are going to show that the degree is multiplicative in the following sense.

Lemma 15.6 (The degree is multiplicative) *Let $f\colon X \to Y$ and $g\colon Y \to Z$ be smooth maps between manifolds with X and Y compact, Y and Z connected. Assume that all three manifolds are oriented, without boundary, and that $\dim X = \dim Y = \dim Z$. Then we have $\deg(g \circ f) = \deg(g) \cdot \deg(f)$.*

Proof: This is Exercise 15.6. □

In the following examples, we keep the assumption that X and Y are smooth, oriented manifolds without boundary, X is compact, Y is connected, and $\dim X = \dim Y$.

Example 15.7 (Constant maps have degree zero) Let $f\colon X \to Y$ be a constant map with value y_0. Then $\deg(f) = 0$ since any y with $y \neq y_0$ is a regular value and $f^{-1}(y) = \emptyset$.

Example 15.8 (Degree of a diffeomorphism) Let $f\colon X \to Y$ be a diffeomorphism. Then we have $\#f^{-1}(y) = 1$ for every $y \in Y$, and $\deg(f) = +1$ if f preserves the orientation and $\deg(f) = -1$ if f reverses the orientation.

As a direct consequence of the previous example, we get the following result.

Proposition 15.9 *An orientation-reversing diffeomorphism of a compact boundaryless manifold is not smoothly homotopic to the identity.* □

Example 15.10 (Reflection on $\mathbb{S}^n$) An example of an orientation-reversing diffeomorphism is provided by the ith reflection $r_i\colon \mathbb{S}^n \to \mathbb{S}^n$ defined by $r_i(x_1, \ldots, x_{n+1}) = (x_1, \ldots, -x_i, \ldots, x_{n+1})$. By Examples 14.8 and 15.8, we have $\deg(r_i) = -1$.

Example 15.11 (Degree of self-maps of $\mathbb{S}^1$) Recall that the restriction of complex multiplication $z \to z^m$ defines a smooth map $f_m\colon \mathbb{S}^1 \to \mathbb{S}^1$ for every $m \in \mathbb{Z}$. For $m \neq 0$, let us calculate the derivative $d(f_m)_z\colon T_z(\mathbb{S}^1) \to T_{f_m(z)}(\mathbb{S}^1)$. We use the parametrisation $\phi\colon t \mapsto (\cos t, \sin t)$. We have the commutative diagram

$$\begin{array}{ccc} \mathbb{S}^1 & \xrightarrow{\quad f_m \quad} & \mathbb{S}^1 \\ {\scriptstyle \phi}\uparrow & & \uparrow{\scriptstyle \phi} \\ \mathbb{R} & \xrightarrow[t \mapsto mt]{} & \mathbb{R} \end{array} . \tag{15.2}$$

Taking derivatives yields, where we note that $t \mapsto mt$ is a linear map and therefore equal to its derivative:

$$\begin{array}{ccc} T_z(\mathbb{S}^1) & \xrightarrow{\quad d(f_m)_z \quad} & T_{z^m}(\mathbb{S}^1) \\ {\scriptstyle d\phi_t}\uparrow & & \uparrow{\scriptstyle d\phi_{mt}} \\ \mathbb{R} & \xrightarrow[t \mapsto mt]{} & \mathbb{R} \end{array} . \tag{15.3}$$

In order to determine $d(f_m)_z$ we recall that $d\phi_t$ has the form

$$d\phi_t : \mathbb{R} \to \mathbb{R}^2,\ s \mapsto (-\sin t,\, \cos t) \cdot s,$$

and hence at $z = \phi(t)$ we have $T_z(\mathbb{S}^1) = (-\sin t,\, \cos t) \cdot \mathbb{R}$. Thus, we obtain

$$\begin{aligned} d(f_m)_z \colon T_z(\mathbb{S}^1) &\to T_{z^m}(\mathbb{S}^1), \\ (-\sin t,\, \cos t) \cdot s &\mapsto m \cdot (-\sin(mt),\, \cos(mt)) \cdot s, \end{aligned}$$

which means that $d(f_m)_z$ is the linear map given by multiplication by m. Hence, when $m > 0$, f_m wraps the circle uniformly around itself m times preserving orientation. The map is everywhere regular and orientation preserving, so its degree is the number of preimages of any point. And this number is m. Similarly, when $m < 0$, the map is everywhere regular but orientation reversing. As each point has $|m|$ preimages, the degree is $-|m| = m$. Finally, when $m = 0$ the map is constant, so its degree is zero.

Remark 15.12 (One homotopy class $\mathbb{S}^1 \to \mathbb{S}^1$ for each integer) One immediate consequence of the calculation of Example 15.11 is the interesting fact that the circle admits an infinite number of homotopically distinct maps since $\deg(z^m) = m$ implies that f_n and f_m are not homotopic if $n \neq m$. We provide a more complete picture of the self-maps of $\mathbb{S}^1$ in Hopf's Theorem 15.19.

Example 15.13 (Self-maps of $\mathbb{S}^2$ and complex polynomials) Let $p(z) = z^m + a_1 z^{m-1} + \cdots + a_m$ be a monic polynomial with complex coefficients $a_1, \ldots, a_m$. As in (4.6), we may consider p as a smooth map $\mathbb{C} \to \mathbb{C}$ which in addition sends ∞ to ∞ to get an induced smooth map $f \colon \mathbb{S}^2 \to \mathbb{S}^2$ which sends the north pole to the north pole. We claim that the map f has degree m. To show our claim, we first observe that the homotopy

$$H(z,t) = tz^m + (1-t)(a_1 z^{m-1} + \cdots + a_m)$$

induces a smooth homotopy $F: \mathbb{S}^2 \times [0,1] \to \mathbb{S}^2$ between the maps induced by p and the monomial z^m. The invariance of the degree under homotopy then shows that we can assume that $p(z) = z^m$ is a monomial. Then the construction of f and the computation of Example 15.11 yield the claim. We recommend to look at Exercises 15.3–15.5 which study this situation further. Hopf's Degree Theorem 16.26 yields a more general statement.

Theorem 15.14 (Self-maps of $\mathbb{S}^n$) *Let $k \geq 1$ and $m \in \mathbb{Z}$ be an integer. Then there is a smooth map $f_m: \mathbb{S}^n \to \mathbb{S}^n$ with $\deg(f_m) = m$.*

Proof: For $m = 0$, it suffices to take any constant map. For $n = 1$ and $m \in \mathbb{Z}$, we proved the assertion in Example 15.11. We can generalise this case as follows. First, we assume that m is positive, i.e., we assume $m \geq 1$. We can describe the coordinates of any point in $\mathbb{S}^n$ as the tuple $\left(x, \sqrt{1-|x|^2}\cos t, \sqrt{1-|x|^2}\sin t\right)$ where $x = (x_1, \ldots, x_{n-1})$ denotes the first $n-1$ coordinates and $|x|^2$ denotes the sum $x_1^2 + \cdots + x_{n-1}^2$. We define the smooth map $f_m: \mathbb{S}^n \to \mathbb{S}^n$ by

$$\left(x, \sqrt{1-|x|^2}\cos t, \sqrt{1-|x|^2}\sin t\right) \mapsto \left(x, \sqrt{1-|x|^2}\cos(mt), \sqrt{1-|x|^2}\sin(mt)\right).$$

The map f_m is surjective, and by Sard's Theorem 5.17 we can find a regular value $p \in \mathbb{S}^n$. The preimage of p consists of exactly m points. It remains to check the sign of the orientation at each preimage point $q \in f^{-1}(p)$. With respect to the bases of $T_q\mathbb{S}^n$ and $T_{f_m(q)}\mathbb{S}^n$ that are induced by the standard basis in $\mathbb{R}^n$, the derivative $d(f_m)_q$ at q is given by the $(n \times n)$-matrix which consists of the $(n-1 \times n-1)$-identity matrix and the integer m in the bottom right-hand corner. This matrix has determinant $m > 0$. Thus, the orientation is preserved at each point in $q \in f^{-1}(p)$, and we get $\deg(f_m) = m$.

Second, we assume that $m < 0$ is negative. Then we may consider the composition $f_{|m|} \circ r_1$, where r_1 denotes the reflection in the first coordinate. Since the degree is multiplicative by Lemma 15.6, this map has degree m by the first case and Example 15.10. □

Remark 15.15 (Homotopy groups of spheres – surjectivity) Let $[\mathbb{S}^n, \mathbb{S}^n]$ denote the set of smooth maps $\mathbb{S}^n \to \mathbb{S}^n$ modulo homotopy. By Theorem 15.4, we can think of the degree as a map $\deg: [\mathbb{S}^n, \mathbb{S}^n] \longrightarrow \mathbb{Z}$. By Theorem 15.14, this map is surjective. We will later prove Hopf's Theorem 16.26, which tells us that this map is injective as well.

In the exercises, we are going to prove the following two applications of the degree, which may be considered as partial converses of the Borsuk–Ulam Theorem 11.20.

Theorem 15.16 (Fixed points of self-maps of the sphere) *Let* $f: \mathbb{S}^n \to \mathbb{S}^n$ *be a smooth map. Assume that* $\deg(f) \neq (-1)^{n+1}$. *Then* f *must have a fixed point.*

Proof: We prove the assertion in Exercise 15.7. □

Theorem 15.17 (Self-map of the sphere of odd degree) *Let* $f: \mathbb{S}^n \to \mathbb{S}^n$ *be a smooth map. Assume* $\deg(f)$ *is odd. Then* f *must send some pair of antipodal points to antipodal points, i.e., there is at least one pair of antipodal points* x_0 *and* $-x_0$ *such that* $f(-x_0) = -f(x_0)$.

Proof: We prove the assertion in Exercise 15.9. □

Example 15.18 (Self-maps of the torus) Let $\mathbb{T}^2 = \mathbb{S}^1 \times \mathbb{S}^1$ be the two-dimensional torus. Another presentation of $\mathbb{T}^2$ is as the quotient $\mathbb{R}^2/\mathbb{Z}^2$ where two points $p, q \in \mathbb{R}^2$ are identified if $p - q \in \mathbb{Z}^2$. We equip this space with the quotient topology induced by the quotient map $\mathbb{R}^2 \to \mathbb{R}^2/\mathbb{Z}^2$. Now let n be a fixed positive integer. We consider the two linear maps $\mathbb{R}^2 \to \mathbb{R}^2$ represented by the matrices $A = \begin{pmatrix} n & 0 \\ 0 & 1 \end{pmatrix}$ and $B = \begin{pmatrix} 1 & 0 \\ 0 & n \end{pmatrix}$. Since A and B send $\mathbb{Z}^2$ to $\mathbb{Z}^2$, they induce well-defined maps $\mathbb{T}^2 = \mathbb{R}^2/\mathbb{Z}^2 \longrightarrow \mathbb{R}^2/\mathbb{Z}^2 = \mathbb{T}^2$. We denote the induced maps on $\mathbb{T}^2$ by the same symbols A and B, respectively. The effect of A on $\mathbb{T}^2$ is to revolve n times around the circle $\mathbb{S}^1 \times \{y\}$ in $\mathbb{T}^2$, for some $y \in \mathbb{S}^1$. while the effect of B on $\mathbb{T}^2$ is to revolve n times around the circle $\{x\} \times \mathbb{S}^1$ in $\mathbb{T}^2$, for some $x \in \mathbb{S}^1$. Hence, for any point $q \in \mathbb{T}^2$, the preimages $A^{-1}(q)$ and $B^{-1}(q)$ consist of exactly n points. The derivative dA_p of A at any point $p \in \mathbb{T}^2$ equals the linear map $\mathbb{R}^2 \to \mathbb{R}^2$ represented by A, i.e., $dA_p = A$, and similarly $dB_p = B$. In particular, we get $\det(dA_p) = \det(A) = n = \det(B) = \det(dB_p)$. Since $n > 0$, the determinant of dA_p and dB_p is positive at every point p, which is sent to q. Thus we can conclude $\deg(A) = n = \deg(B)$.

15.2 Hopf Degree Theorem in Dimension One

We return to self-maps of $\mathbb{S}^1$. We learned in Example 15.11 that there is a homotopy class of maps $\mathbb{S}^1 \to \mathbb{S}^1$ for every integer m. This shows that the map $\deg: [\mathbb{S}^1, \mathbb{S}^1] \to \mathbb{Z}$ given by the degree $f \mapsto \deg(f)$ is surjective. degree Hopf showed that this map is also injective.

Theorem 15.19 (Hopf Degree Theorem in dimension one) *Two smooth maps $f_0, f_1 : \mathbb{S}^1 \to \mathbb{S}^1$ are homotopic if and only if they have the same degree.*

Remark 15.20 We generalise the assertion to all n in Theorem 16.26. Note that the statement also holds for continuous maps as we now explain. Since we know that every continuous map between smooth manifolds is homotopic to a smooth map by Whitney's Approximation Theorem 12.26. Hence all we need to extend the assertion is a definition of the degree for merely continuous maps. We can do this using Whitney's Approximation Theorem 12.26, or using the techniques of algebraic topology, e.g., singular homology. See also Section 17.2.

Proof of Theorem 15.19: We already know that if f_0 and f_1 are homotopic, then we have $\deg(f_0) = \deg(f_1)$. So assume that $\deg(f_0) = \deg(f_1)$, and we need to show $f_0 \sim f_1$. By Lemma 8.23, we know that there exists a smooth map $g : \mathbb{R} \to \mathbb{R}$ such that $f \circ p = p \circ g$, and there is a unique $q \in \mathbb{Z}$ such that $g(t+1) = g(t) + q$ for all $t \in \mathbb{R}$.

Lemma 15.21 $q = \deg(f)$.

Proof of Lemma 15.21: If f is not surjective, then we can pick a point $y \notin f(\mathbb{S}^1)$. Thus y is automatically a regular value. Since $\#f^{-1}(y) = 0$, we must have $\deg(f) = 0$. In this case, we need to have $q = 0$, i.e., $g(t+1) = g(t)$. For otherwise, $p \circ g$ was surjective and hence f would be surjective. Note that, since the stereographic projection map $\mathbb{S}^1 \setminus \{y\} \to \mathbb{R}$ is a diffeomorphism and $\mathbb{R}$ is contractible, this shows that $\mathbb{S}^1 \setminus \{y\}$ is contractible. Hence f is a map to a contractible space and therefore homotopic to a constant map and has degree zero.

Now we assume that f is surjective. Let $y \in \mathbb{S}^1$ be a regular value of f, and let $z \in f^{-1}(y)$. Since p is surjective, there is a $t \in \mathbb{R}$ with $p(t) = z$. Since y is a regular value, f is a local diffeomorphism around z. Its derivative is related to that of g by the chain rule $df_z \circ dp_t = dp_{g(t)} \circ dg_t$. The derivative of $p : \mathbb{R} \to \mathbb{S}^1$ at any t is $dp_t : \mathbb{R} \to T_{p(t)}(\mathbb{S}^1)$, $w \mapsto 2\pi \cdot (-\sin(2\pi t), \cos(2\pi t)) \cdot w$. Hence the determinant of dp_t is 2π, and hence it is positive at every t. Thus the sign of the determinant of df_z equals the sign of dg_t, i.e.,

$$\operatorname{sign} df_z = \operatorname{sign} dg_t.$$

As above, let $y \in \mathbb{S}^1$ be a regular value of f and $z \in f^{-1}(y)$. We fix a $t_0 \in \mathbb{R}$ with $p(t_0) = z$. When we walk from t_0 to $t_0 + 1$, we need to count how many preimages of y we collect along the way, *with their orientation*.

We start with the case $q = 0$, i.e., $g(t+1) = g(t)$. It will actually teach us the key ideas we need to remember from this proof. We need to count how

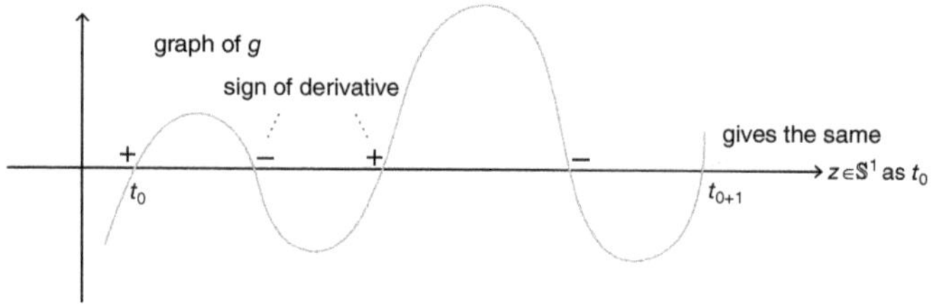

Figure 15.2 We count the intersection points together with the sign of the derivative. If $q = 0$, there are as many crossing points from below with positive derivative as there are crossing points from above with negative derivative.

often $g(s) = g(t_0)$ with $dg_s = g'(s) > 0$ and how often $g(s) = g(t_0)$ with $dg_s = g'(s) < 0$. Note that since y is regular, dg_s is always $\neq 0$ for such an s. Since g is a smooth function $\mathbb{R} \to \mathbb{R}$, this is now just an exercise in calculus. Using the periodicity of g, i.e., that $g'(t_0)$ must have the same sign as $g'(t_0+1)$, we see that there are exactly as many points s with $g(s) = g(t_0)$ and $dg_s = g'(s) > 0$ as there are points with $g(s) = g(t_0)$ and $dg_s = g'(s) < 0$. Thus $\deg(f) = 0$. See Figure 15.2.

Now assume $q > 0$, and $g(t+1) = g(t) + q$. Again, we walk from t_0 to $t_0 + 1$ and sum up the orientation numbers of all the preimages of y that we collect along the way. This corresponds to counting how often we have $g(s) = g(t_0) + i$ for some $i = 0, 1, \dots, q-1$ and $s \in [t_0, t_0+1]$. Let us look at one interval $[g(t_0)+i, g(t_0)+i+1]$ at a time. We would like to know how many $s \in [t_0, t_0+1]$ are sent to either $g(t_0)+i$ or $g(t_0)+i+1$ together with the sign of the derivative. Therefore we look at the preimage $g^{-1}([g(t_0)+i, g(t_0)+i+1])$. This set is a disjoint union of closed intervals. For each of these intervals, the start and endpoints are sent to either $g(t_0) + i$ or $g(t_0) + i + 1$. Let us think of the graph of g passing $g(t_0) + i$ with a positive sign of the derivative as going in with $+1$ and passing $g(t_0) + i + 1$ with a positive sign of the derivative as going out with $+1$, and the other two alternatives as the ones with -1. Then we see that the graph has to go in with $+1$ for the first time and has to go out with $+1$ for the last time, since the graph starts at $g(t_0) \leq g(t_0) + i$ and ends at $g(t_0) + q \geq g(t_0) + i + 1$. In between these two points, the graph goes out with -1 as often as it goes in with $+1$, and it goes in with -1 as often as it goes out with $+1$. See Figure 15.3.

Thus in total the orientation numbers for $g^{-1}([g(t_0) + i, g(t_0) + i + 1])$ add up to $+2$. Repeating this for all $i = 0, 1, \dots, q-1$ gives a sum of orientation numbers equal to q, since we have to account for that we counted the inner points twice. Since the sum of orientation numbers of f equals that of g, this

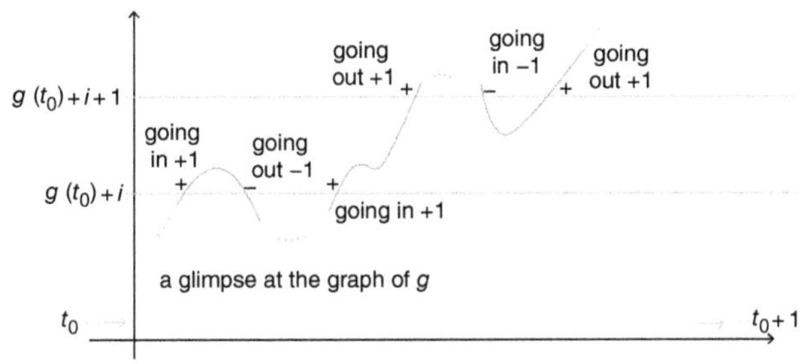

Figure 15.3 We count the intersection points with the sign of the derivative. If $q > 0$, there are exactly q more crossing points from below with positive derivative than there are crossing points from above with negative derivative.

shows $\deg(f) = q$. Finally, if $q < 0$, the same argument works with signs and directions reversed. This finishes the proof of the lemma. □

Now assume we have two maps f_0 and f_1 with $\deg(f_0) = q = \deg(f_1)$. Let g_0 and g_1 be smooth maps $\mathbb{R} \to \mathbb{R}$ which satisfy $g_0(t+1) = g_0(t)+q$ and $g_1(t+1) = g_1(t)+q$ such that $f_0(p(t)) = p(g_0(t))$, $f_1(p(t)) = p(g_1(t))$. Then the smooth map $g_s(t) := sg_1 + (1-s)g_0$ also satisfies $g_s(t+1) = g_s(t)+q$. Note $g_s(t)$ defines a homotopy G from g_0 to g_1 by $G(t,s) = g_s(t)$. Then G defines a homotopy $G\colon \mathbb{R}\times[0,1] \to \mathbb{R}$ with $G(t+1,s) = G(t,s)+q$ for all t, s which induces a well-defined homotopy

$$F\colon \mathbb{S}^1 \times [0,1] \to \mathbb{S}^1, (z,s) \mapsto p(G(t,s)) \text{ for any } t \in p^{-1}(z).$$

Hence F is a homotopy from $f_0 = p\circ g_0$ to $p \circ g_1 = f_1$. □

15.3 Linking Number and the Hopf Invariant

We now study some basic properties of the *Hopf invariant*. Recall that we have already defined the linking number and the Hopf invariant in a mod 2 version in Section 11.5. Now we improve the constructions to get $\mathbb{Z}$-valued invariants. In particular, we show that the Hopf invariant of the Hopf map $\hbar$ equals one. This is a much stronger result than the one we proved previously in Theorem 11.34. This section is rather brief, and most proofs are exercises for which you find solutions at the end of the book. As in Definition 11.25, we define the linking number of submanifolds in Euclidean space as follows.

Definition 15.22 (Linking number) For $k \geq 1$, let $X, Y \subset \mathbb{R}^{k+1}$ be two disjoint smooth manifolds. We assume that X and Y are compact and without

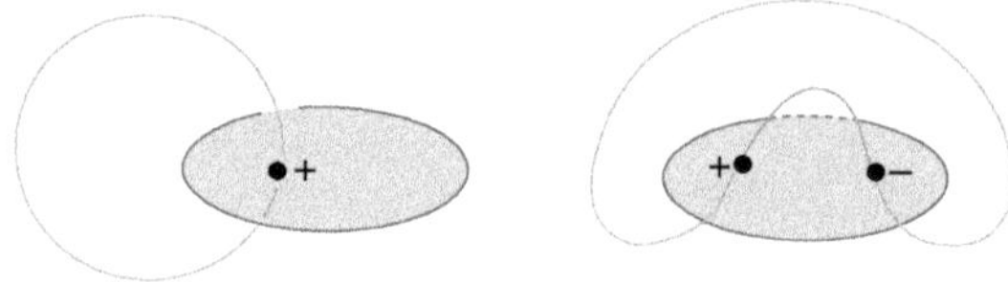

Figure 15.4 The shaded disk is a compact manifold D with boundary. In both cases, the curve intersects the manifold D. The linking number detects the intersection. The circles on the left-hand side are linked and cannot be moved apart. However, multiple intersections may occur. The curves on the right-hand side are not linked and can be moved. The linking number detects this by counting the linking points with signs. See also Example 11.26.

boundary, of dimensions $\dim X = m$ and $\dim Y = n$ such that $m + n = k$. We assume that X and Y are oriented. We define the map

$$\lambda\colon X \times Y \to \mathbb{S}^k,\ (x, y) \mapsto \frac{x - y}{|x - y|}.$$

The *linking number* $L(X,Y)$ of X and Y is defined to be the degree of λ, i.e., $L(X,Y) := \deg(\lambda)$.

Lemma 15.23 *For X and Y as above, $L(Y,X) = (-1)^{(m+1)(n+1)}L(X,Y)$.*

Proof: This is Exercise 15.10. □

Since it is defined via the degree, the linking number satisfies a version of the boundary theorem, which justifies its name.

Proposition 15.24 *Assume in addition that X is the boundary of an oriented manifold W which is disjoint from Y. Then $L(X,Y) = 0$.*

Proof: This is Exercise 15.11. □

We extend the definition of the linking number to submanifolds X, Y in $\mathbb{S}^{k+1}$.

Definition 15.25 Assume that X and Y are compact, oriented, and boundaryless, and $\dim X + \dim Y = k$. Since the sphere is connected and X and Y are closed subsets, there must be a point p which is not contained in either X or Y. We identity $\mathbb{S}^{k+1} \setminus \{p\}$ with $\mathbb{R}^{k+1}$ via the diffeomorphism defined by stereographic projection from p. Then we consider X and Y as submanifolds of $\mathbb{R}^{k+1}$ and define the *linking number* $L(X,Y)$ as in Definition 15.22. The linking number of empty submanifolds of $\mathbb{S}^{k+1}$ is defined to be zero.

We can now define the Hopf invariant for maps between spheres of certain dimensions.

Definition 15.26 For $n \geq 1$, consider a smooth map $f: \mathbb{S}^{2n-1} \to \mathbb{S}^n$. Let $w \neq z \in \mathbb{S}^n$ be two distinct regular values for f. Then $f^{-1}(w)$ and $f^{-1}(z)$ are compact, oriented, and boundaryless submanifolds of $\mathbb{S}^n$, and their linking number is defined. The number $H(f) := L(f^{-1}(w), f^{-1}(z))$ is called the *Hopf invariant* of f.

Remark 15.27 The number $H(f)$ is a famous invariant that played a crucial role in the development of mathematics. As in Section 11.5.2, one can show that $H(f)$ does not depend on the choice of w and z and only depends on the homotopy class of f. We skip this verification here.

In the exercises, we verify the following properties of the Hopf invariant.

Lemma 15.28 *Let f be as above. Assume that n is odd. Then $H(f) = 0$.*

Proof: This is Exercise 15.12. □

Lemma 15.29 *Let f be as above and $g: \mathbb{S}^n \to \mathbb{S}^n$ be a smooth map. Consider the composition*

$$\mathbb{S}^{2n-1} \xrightarrow{f} \mathbb{S}^n \xrightarrow{g} \mathbb{S}^n.$$

Then we have $H(g \circ f) = H(f) \cdot \deg(g)^2$.

Proof: This is Exercise 15.13 □

Now we compute the Hopf invariant for the *Hopf map* $\hbar: \mathbb{S}^3 \to \mathbb{S}^2$. We have met $\hbar$ several times before and will first recall some of its properties which we have mostly proven in exercises in previous chapters. We recommend doing these exercises again as preparation. The calculation of $H(\hbar) = 1$ in Theorem 15.30 below is a key new step which goes beyond the computation modulo 2. Recall that we may consider $\mathbb{S}^3$ as a subset of $\mathbb{C}^2$, i.e., $\mathbb{S}^3 = \{(z_0, z_1) \in \mathbb{C}^2 : |z_0|^2 + |z_1|^2 = 1\}$, and $\mathbb{S}^2$ as a subset of $\mathbb{C} \times \mathbb{R}$, i.e., $\mathbb{S}^2 = \{(z, x) \in \mathbb{C} \times \mathbb{R} : |z|^2 + x^2 = 1\}$. Then $\hbar$ is the map $\mathbb{S}^3 \to \mathbb{S}^2$ given by

$$\hbar(z_0, z_1) = \left(2z_0\bar{z}_1, |z_0|^2 - |z_1|^2\right).$$

Theorem 15.30 *The Hopf invariant of the Hopf map $\hbar: \mathbb{S}^3 \to \mathbb{S}^2$ is one, i.e., $H(\hbar) = 1$.*

Proof: By definition of $H(\hbar)$, we need to choose two distinct regular values a and b of $\hbar$ and calculate the linking number of $\hbar^{-1}(a)$ and $\hbar^{-1}(b)$. We choose the points $a = (0,0,1)$ and $b = (0,1,0)$ on $\mathbb{S}^2 \subset \mathbb{R}^3 \cong \mathbb{C} \times \mathbb{R}$. In Section 14.5 and Exercise 14.9, we determined the fibres $\hbar^{-1}(a)$ and $\hbar^{-1}(b)$ together with their orientations as preimages under $\hbar$. In the proof of Theorem 11.34, we computed the mod 2 linking number of $\hbar^{-1}(a)$ and $\hbar^{-1}(b)$. We will use the same notation as in the proof of Theorem 11.34 and will recall the key steps of the computation. For the proof of Theorem 15.30, it remains to understand the effect of the linking map λ on the orientations to get the orientation number at the preimage points.

To calculate the linking number of $\hbar^{-1}(a)$ and $\hbar^{-1}(b)$, we need to choose a point on $\mathbb{S}^3$ disjoint from these two subsets and stereographically project $\mathbb{S}^3$ from this point onto $\mathbb{R}^3$. By our choice of a and b, we get that the north pole $N = (0,0,0,1)$ is neither on $\hbar^{-1}(a)$ nor on $\hbar^{-1}(b)$. We then consider the spaces $S_a := \phi_N^{-1}(\hbar^{-1}(a)) = \{\mathbf{v} = (v_0, v_1, v_2) \in \mathbb{R}^3 : v_0^2 + v_1^2 = 1 \text{ and } v_2 = 0\}$ and $S_b := \phi_N^{-1}(\hbar^{-1}(b))$, i.e.,

$$S_b = \{\mathbf{w} = (w_0, w_1, w_2) \in \mathbb{R}^3 : w_1 = w_2 \text{ and } w_0^2 + w_1^2 = (1 - w_0)^2/2\}.$$

In the proof of Theorem 11.34, we showed that $p = (1,0,0)$ is a regular value for $\lambda\colon \mathbb{S}_a \times \mathbb{S}_b \to \mathbb{S}^2$ defined by $\lambda(\mathbf{x},\mathbf{y}) = (\mathbf{x} - \mathbf{y})/|\mathbf{x} - \mathbf{y}|$. We showed that there are three points $(\mathbf{v},\mathbf{w})$ in $S_a \times S_b$ with $\lambda(\mathbf{v},\mathbf{w}) = p$: one with $v_0 = 1$, $w_0 = \sqrt{2} - 1$, one with $v_0 = 1$, $w_0 = -\sqrt{2} - 1$, and one with $v_0 = -1$, $w_0 = -\sqrt{2} - 1$.

We have to show that the sum of the signs of the determinants of the derivatives of λ at these points is $+1$. For then we get $\deg(\lambda) = +1$ as claimed. Since S_a is the unit circle in the xy-plane, the tangent space of S_a at a point $\mathbf{v}$ is given by $T_{\mathbf{v}} S_a = \{\mathbf{u} = (u_0, u_1, u_2) \in \mathbb{R}^3 : u_2 = 0 \text{ and } u_0 v_0 + u_1 v_1 = 0\}$. In the proof of Theorem 11.34, we showed that

$$T_{\mathbf{w}} S_b = \{\mathbf{u} = (u_0, u_1, u_2) \in \mathbb{R}^3 : u_2 = u_1 \text{and } u_0(w_0 + 1) + u_1 w_1 = 0\}.$$

Now we study the effect of the linear map $d\lambda_{(\mathbf{v},\mathbf{w})}\colon T_{\mathbf{v}} S_a \times T_{\mathbf{w}} S_b \to T_p \mathbb{S}^2$ on oriented bases. A basis of $T_p \mathbb{S}^2$ is given by the vectors $\mathbf{e}_2 = (0,1,0)$ and $\mathbf{e}_3 = (0,0,1)$. At each of the points $(\mathbf{x},\mathbf{y})$ with $\lambda(\mathbf{x},\mathbf{y}) = \mathbf{p}$, the vector $\mathbf{a} = (0,1,0)$ is a basis of $T_{\mathbf{x}} \mathbb{S}_a$, and the vector $\mathbf{b} = (0,1,1)$ provides a basis of $T_{\mathbf{y}} \mathbb{S}_b$. The map $d\lambda_{(\mathbf{x},\mathbf{y})}$ sends $\mathbf{a}$ to $\frac{1}{|x_0 - y_0|} \cdot (0,1,0)$ and $\mathbf{b}$ to $\frac{1}{|x_0 - y_0|} \cdot (0,-1,-1)$. Hence we have $d\lambda_{(\mathbf{x},\mathbf{y})}(\mathbf{a}) = \frac{1}{|x_0 - y_0|} \cdot \mathbf{e}_2$ and $d\lambda_{(\mathbf{x},\mathbf{y})}(\mathbf{b}) = -\frac{1}{|x_0 - y_0|} \cdot \mathbf{e}_2 - \frac{1}{|x_0 - y_0|} \cdot \mathbf{e}_3$. These two vectors form a basis of $T_p \mathbb{S}^2$. To check the orientation of the basis, we calculate the determinant of $d\lambda_{(\mathbf{v},\mathbf{w})} = \frac{1}{|v_0 - w_0|} \begin{pmatrix} 1 & -1 \\ 0 & -1 \end{pmatrix}$, which is also the transition matrix from $(\mathbf{e}_2, \mathbf{e}_3)$ to the new basis. The determinant of this matrix is negative. Hence the basis $(d\lambda_{(\mathbf{v},\mathbf{w})}(\mathbf{a}), d\lambda_{(\mathbf{v},\mathbf{w})}(\mathbf{b}))$ of $T_p \mathbb{S}^2$ is negatively oriented. In order to understand the effect of $d\lambda_{(\mathbf{v},\mathbf{w})}$ on the orientations, we must determine whether $(\mathbf{a}, \mathbf{b})$ is a positively or negatively oriented basis of $T_{\mathbf{v}}S_a \times T_{\mathbf{w}}S_b$:

- $(\mathbf{v}, \mathbf{w})$ with $v_0 = 1$, $w_0 = \sqrt{2} - 1$ and $v_1 = v_2 = w_1 = w_2 = 0$: as we checked in a previous point, the orientation of S_a is opposite to the standard orientation of the circle. Hence the basis $\mathbf{a}$ is a negatively oriented basis of $T_{\mathbf{v}}S_a$. The basis $\mathbf{b}$, however, is a positively oriented basis of $T_{\mathbf{w}}S_b$.[1] For, at this $\mathbf{w}$, the corresponding vectors are $q = (1/\sqrt{2}, 0, 0, -1/\sqrt{2})$ and $q_0^{\perp} = (0, 1/\sqrt{2}, 1/\sqrt{2}, 0)$, and $\mathbf{b}$ is a positive multiple of $d(\phi_N^{-1})_q(q_0^{\perp}) = (0, 1/\sqrt{2}, 1/\sqrt{2})$. Since $q_0^{\perp}$ has sign $+1$ and ϕ_N preserves orientations, $\mathbf{b}$ also has sign $+1$. Hence $(\mathbf{a} \times 0, 0 \times \mathbf{b})$ is a negatively oriented basis of $T_{\mathbf{v}}S_a \times T_{\mathbf{w}}S_b$. Thus $d\lambda_{(\mathbf{v},\mathbf{w})}$ sends a negatively oriented basis to a negatively oriented basis and therefore preserves orientations. Hence the orientation number at $(\mathbf{v}, \mathbf{w})$ is $+1$.
- $(\mathbf{v}, \mathbf{w})$ with $v_0 = 1$, $w_0 = -\sqrt{2} - 1$, and $v_1 = v_2 = w_1 = w_2 = 0$: the vector $\mathbf{a}$ is a negatively oriented basis of $T_{\mathbf{v}}S_a$. The basis $\mathbf{b}$, however, is now also a negatively oriented basis of $T_{\mathbf{w}}S_b$. For, at this $\mathbf{w}$, the corresponding vector are $q = (-1/\sqrt{2}, 0, 0, 1/\sqrt{2})$ and $q_0^{\perp} = (0, -1/\sqrt{2}, -1/\sqrt{2}, 0)$, and $\mathbf{b}$ is a negative multiple of $d(\phi_N^{-1})_q(q_0^{\perp}) = (0, -1/\sqrt{2}, -1/\sqrt{2})$. Since $q_0^{\perp}$ has sign $+1$ and ϕ_N preserves orientations, $\mathbf{b}$ has sign -1. Hence $(\mathbf{a} \times 0, 0 \times \mathbf{b})$ is a positively oriented basis of $T_{\mathbf{v}}S_a \times T_{\mathbf{w}}S_b$. Thus $d\lambda_{(\mathbf{v},\mathbf{w})}$ sends a positively oriented basis to a negatively oriented basis and therefore reverses orientations. Hence the orientation number at $(\mathbf{v}, \mathbf{w})$ is -1.
- $(\mathbf{v}, \mathbf{w})$ with $v_0 = -1$, $w_0 = -\sqrt{2} - 1$ and $v_1 = v_2 = w_1 = w_2 = 0$: now the vector $\mathbf{a}$ is a positively oriented basis of $T_{\mathbf{v}}S_a$ and $\mathbf{b}$ is a negatively oriented basis of $T_{\mathbf{w}}S_b$. Hence $(\mathbf{a} \times 0, 0 \times \mathbf{b})$ is a negatively oriented basis of $T_{\mathbf{v}}S_a \times T_{\mathbf{w}}S_b$. Thus $d\lambda_{(\mathbf{v},\mathbf{w})}$ sends a negatively oriented basis to a negatively oriented basis and therefore preserves orientations. Hence the orientation number at $(\mathbf{v}, \mathbf{w})$ is $+1$.

Thus, in total, we get that the sum of the orientation numbers is $+1 - 1 + 1 = +1$. Hence we have shown $H(\hbar) = \deg(\lambda) = 1$. □

[1] Imagine S_b as a deformation of the unit circle lying in the plane $y = z$ in $\mathbb{R}^3$ and the vector $\mathbf{b}$ points upwards at a point with positive x-coordinate. We checked that the orientation of S_b is opposite to the standard orientation of the circle.

15.4 Exercises

15.1 Let $a\colon \mathbb{S}^k \to \mathbb{S}^k$, $a(x) = -x$, be the antipodal map on $\mathbb{S}^k$.

(a) Show that a has degree $(-1)^{k+1}$. Hint: Consider the reflection maps $r_i(x_1, \dots, x_{k+1}) = (x_1, \dots, -x_i, \dots, x_{k+1})$.
Recall that reflections are diffeomorphisms which reverse orientations.

(b) Show that a is homotopic to the identity if and only if k is odd. Hint: Use that we showed in Exercise 8.4 that a is homotopic to the identity if k is odd.

15.2 Prove the following theorem: the n-dimensional sphere $\mathbb{S}^n$ admits a vector field $\mathbf{v}$ *without zeros* if and only if n is *odd*. Hint: Combine Exercises 8.10 and 15.1.

15.3 Use the degree to prove the Fundamental Theorem of Algebra: let $p(z) = z^m + a_1 z^{m-1} + \cdots + a_m$ be a monic polynomial with complex coefficients. Show that there exists a $w \in \mathbb{C}$ such that $p(w) = 0$.

15.4 Explain why the degree cannot be used to prove that there is a zero in $\mathbb{R}$ for every monic real polynomial.

15.5 Let $p(z) = z^m + a_1 z^{m-1} + \cdots + a_m$ be a monic complex polynomial. At every point $z_0 \in \mathbb{C}$, there is an integer $m \geq 0$ such that we can factor $p(z) = (z - z_0)^m q(z)$ with $q(z_0) \neq 0$. Note that z_0 is a zero of p if and only if $m > 0$. In this case, we call m the *multiplicity* of the zero z_0. In this exercise, we relate multiplicities and degrees as follows. Let $\mathbb{D}_0 \subset \mathbb{C}$ be a disk with centre z_0 and radius r small enough such that there is no other zero of p in $\mathbb{D}_0$ except z_0. Recall that we showed that this is possible when we discussed Milnor's proof of the FTA in Section 4.5.

(a) Check that the formula $g(z) = z_0 + rz$ defines an orientation-preserving diffeomorphism $\mathbb{S}^1 \to \partial\mathbb{D}_0$ where $\mathbb{D}_0$ inherits an orientation from $\mathbb{C} = \mathbb{R}^2$.

(b) Explain why $\frac{p}{|p|}\colon \partial\mathbb{D}_0 \to \mathbb{S}^1$ and $\frac{p\circ g}{|p\circ g|}\colon \mathbb{S}^1 \to \mathbb{S}^1$ have the same degree.

(c) Find a smooth homotopy $h_t\colon \mathbb{S}^1 \to \mathbb{S}^1$ between $h_1 = \frac{p\circ g}{|p\circ g|}$ and $h_0(z) = c \cdot z^m$ with c being the constant $c = \frac{q(z_i)}{|q(z_i)|}$.

(d) Conclude that $\deg\left(\frac{p}{|p|}\right) = m$.

15.6 Let $X \xrightarrow{f} Y \xrightarrow{g} Z$ be a sequence of smooth maps between manifolds with X and Y compact and Y and Z connected. Assume that all three manifolds are oriented and boundaryless and $\dim X = \dim Y = \dim Z$. Show that

$$\deg(g \circ f) = \deg(g) \cdot \deg(f).$$

15.7 Let $\mathbb{S}^k \xrightarrow{f} \mathbb{S}^k$ be a smooth map. Show that if $\deg(f) \neq (-1)^{k+1}$, then f must have a fixed point. Hint: Assume that f had no fixed point. Show that then the composition of f with the antipodal map would be homotopic to the identity map.

15.8 Let $k \geq 1$ be an *odd* integer. Let $q\colon \mathbb{S}^k \to \mathbb{RP}^k$, $x \mapsto [x]$, be the canonical projection. Recall from Theorem 14.15 that $\mathbb{RP}^k$ is compact and orientable, so that $\deg(q)$ is defined. Show that the degree of q is ± 2, and that it is $+2$ if the orientation on $\mathbb{RP}^k$ is chosen such that q preserves orientation. Hint: Use that q does not change after composition with the antipodal map.

15.9 In this exercise, we prove Theorem 15.17: let $\mathbb{S}^k \xrightarrow{f} \mathbb{S}^k$ be a smooth map. If $\deg(f)$ is *odd*, then f must send *some* pair of antipodal points to antipodal points. To prove the claim, we assume that $f\colon \mathbb{S}^k \to \mathbb{S}^k$ sends no pair of antipodal points to antipodal points. We will show that this implies that $\deg(f)$ must be even.

(a) Assume that $f\colon \mathbb{S}^k \to \mathbb{S}^k$ sends no pair of antipodal points to antipodal points. Show that then f and $f \circ a$ are both smoothly homotopic to the map $g\colon \mathbb{S}^k \to \mathbb{S}^k$ defined by $g(x) := \frac{f(x)+f(-x)}{|f(x)+f(-x)|}$, and deduce that $\deg(f) = \deg(g) = \deg(f \circ a)$.

(b) Now further assume that k is *even*. Show that this implies $\deg(g) = 0$.

(c) Now assume that k is *odd*. Use Exercises 15.6 and 15.8 to deduce that this implies $\deg(g)$ is *even*.

(d) Deduce the claim and thereby prove Theorem 15.17.

15.10 Assume that X and Y are such that their linking number is defined Show that $L(Y, X) = (-1)^{(m+1)(n+1)} L(X, Y)$ where m and n denote the dimensions of X and Y, respectively. Hint: Check what happens with the orientation numbers at points when we switch the order of X and Y, and think of our computation of the degree of the antipodal map on $\mathbb{S}^k$.

15.11 Assume that X is the boundary of an oriented manifold W which is disjoint from Y. Show that this implies $L(X, Y) = 0$. Hint: Use the Boundary Theorem 15.2 for degrees.

15.12 Let $f\colon \mathbb{S}^{2n-1} \to \mathbb{S}^n$ be a smooth map. Show that if n is odd, then $H(f) = 0$.

15.13 Let $f\colon \mathbb{S}^{2n-1} \to \mathbb{S}^n$ and $g\colon \mathbb{S}^n \to \mathbb{S}^n$ be smooth maps. Consider the composition $\mathbb{S}^{2n-1} \xrightarrow{f} \mathbb{S}^n \xrightarrow{g} \mathbb{S}^n$. Show that $H(g \circ f) = H(f) \cdot \deg(g)^2$.

16

Pontryagin Construction and Hopf's Degree Theorem

We have now seen in many examples that the degree is a powerful invariant which can be used in a variety of settings. However, it does have a drawback. The degree is only defined under quite strict assumptions, most notably the manifolds must have the same dimension, and, for the integral-valued version, both have to be oriented. In this chapter, we will study an invariant which generalises the degree and reduces the required assumptions. This is due to the work of Pontryagin [41]. See also [35, chapter 7], and for example [18], [28], and [49].

16.1 Framed Manifolds and Cobordisms

We recall the following important concept, which we met in slightly different form in Exercise 13.8.

Definition 16.1 (Cobordism) A *cobordism* between two k-dimensional submanifolds Z and Z' of $X \subset \mathbb{R}^N$ is a compact smooth $(k+1)$-dimensional submanifold $W \subset X \times [0,1]$ such that

(a) $(Z \times [0, \varepsilon)) \cup (Z' \times (1-\varepsilon, 1]) \subset W$ for an $\varepsilon > 0$, and
(b) $\partial W = W \cap (X \times \{0,1\}) = (Z \times \{0\}) \cup (Z' \times \{1\})$.

If such a cobordism exists, we say that Z and Z' are *cobordant*. We refer to the subset $(Z \times [0, \varepsilon)) \cup (Z' \times (1-\varepsilon, 1]) \subset W$ as a *collar of* W.

Remark 16.2 (Cobordism to the empty manifold) We allow that one of the manifolds is empty, say $Z' = \emptyset$. Then a cobordism between a k-dimensional submanifold Z and the empty set is a compact smooth $(k+1)$-dimensional submanifold $W \subset X \times [0,1]$ such that $Z \times [0, \varepsilon) \subset W$ for an $\varepsilon > 0$ and $\partial W = W \cap (X \times \{0,1\}) = Z \times \{0\}$.

We will be interested in cobordisms of submanifolds with the following additional structure.

Definition 16.3 (Framed submanifolds) Let X be a smooth n-dimensional manifold without boundary. A *framing* of a k-dimensional submanifold $Z \subset X$ is a smooth assignment $\mathfrak{n}\colon x \mapsto \mathfrak{n}(x) = (\mathfrak{n}_1(x), \ldots, \mathfrak{n}_{n-k}(x))$ which assigns to each $x \in Z$ a basis for the space $N_x(Z,X) = (T_xZ)^\perp \subset T_xX$ of normal vectors to Z in X at x. A pair $(Z,\mathfrak{n})$ consisting of a submanifold Z and a framing $\mathfrak{n}$ is called a *framed submanifold* of X.

Example 16.4 (Framing on zero-dimensional submanifolds) Let X be a smooth n-dimensional manifold. We can consider a point x in X as a submanifold of codimension n. The normal space to the submanifold $Z = \{x\}$ is the whole tangent space T_xX. More generally, let $Z = \{x_1, \ldots, x_m\}$ be a collection of points. The normal space $N_{x_i}(Z,X)$ is given by $N_{x_i}(Z,X) = N_{x_i}(\{x_i\},X) = T_{x_i}X$. A framing of Z is then just a choice of a basis for each $T_{x_i}X$.

Recall from Definition 12.15 that the collection of spaces $N_x(Z,X)$ of normal vectors to Z together with the canonical map $N(Z,X) \to Z$ forms a vector bundle over Z, which is called the *normal bundle to* Z.

Lemma 16.5 *The normal bundle of a framed submanifold is trivial.*

Proof: The framing $\mathfrak{n}$ on Z allows us to define a diffeomorphism

$$Z \times \mathbb{R}^p \to N(Z,X),\ (x,\lambda_1,\ldots,\lambda_p) \mapsto (x,\lambda_1\mathfrak{n}_1(x),\ldots,\lambda_p\mathfrak{n}_p(x)),$$

where $p = n - k$ denotes the codimension of Z in X. □

As the following example demonstrates, not all submanifolds can be framed.

Example 16.6 (An unframable submanifold) Let X denote the Möbius band, and let $Z \subset X$ denote the central circle. By Lemma 16.5, the existence of a framing of Z in X would imply that the normal bundle $N(Z,X)$ was trivial. However, by Corollary 12.22, this implies that there is a submersion $g\colon U \to \mathbb{R}$ defined on an open subset $U \subset X$ such that $Z = g^{-1}(0)$. By Exercise 13.10, such a g does not exist. This shows that Z cannot be equipped with a framing.

We now add framings to cobordisms.

Definition 16.7 (Framed cobordism) A *framed cobordism* between two k-dimensional framed submanifolds $(Z,\mathfrak{n})$ and $(Z',\mathfrak{n}')$ of $X \subset \mathbb{R}^N$ is a cobordism W between Z and Z' together with a framing $\mathfrak{w}$ of W which restricts to the given framings of $\mathfrak{n}$ and $\mathfrak{n}'$, i.e., for each $i = 1, \ldots, n-k$,

$$\mathfrak{w}_i(x,t) = \begin{cases} (\mathfrak{n}_i(x), 0) & \text{for } (x,t) \in Z \times [0,\varepsilon) \\ (\mathfrak{n}_i'(x), 0) & \text{for } (x,t) \in Z' \times (1-\varepsilon, 1] \end{cases}$$

as vectors in the normal spaces

$$N_{(x,t)}(W, X \times [0,1]) = (T_{(x,t)}W)^{\perp} \subset T_{(x,t)}(X \times [0,1]) \subset \mathbb{R}^N \times \mathbb{R}.$$

If such a framed cobordism exists, we say that Z and Z' are *framed cobordant*.

Lemma 16.8 *Being framed cobordant is an equivalence relation.*

Proof: This follows from gluing together framed cobordisms, which agree on the boundary. □

Remark 16.9 We may think of a cobordism as a generalisation of the notion of a homotopy. We will explore this perspective in the following special cases which will play an important role in Section 16.3. In Section 16.2, we will see other types of examples of framed manifolds and framed cobordisms.

Let X be an oriented smooth n-dimensional manifold. Recall from Example 16.4 that we can consider a point $x \in X$ with a basis $\mathfrak{b}_x$ on the tangent space $T_x(X)$ as a framed submanifold $(x, \mathfrak{b}_x)$ of codimension n. The given orientation $\mathfrak{o}$ on X then provides an *orientation number for* $\mathfrak{b}_x$ as follows. We set $\operatorname{sign}(x) = +1$ if $\mathfrak{b}_x$ is positively oriented with respect to $\mathfrak{o}$, and we set $\operatorname{sign}(x) = -1$ if $\mathfrak{b}_x$ is negatively oriented.

Proposition 16.10 (Framed cobordisms between points with equal orientations) *Let $X \subset \mathbb{R}^N$ be a connected, oriented, smooth n-dimensional manifold. Let y and z be two points in X, and let $\mathfrak{b}_y$ and $\mathfrak{b}_z$ be bases of the tangent space T_yX and T_zX, respectively. We assume that* $\operatorname{sign}(y) = \operatorname{sign}(z)$. *Then there is a framed cobordism between $(y, \mathfrak{b}_y)$ and $(z, \mathfrak{b}_z)$.*

Proof: Let us assume $\operatorname{sign}(y) = +1 = \operatorname{sign}(z)$. The case -1 is similar. Since X is a connected smooth manifold, X is path-connected by Exercise 8.5. Thus, there is a smooth path $\gamma\colon [0,1] \to X$ with $\gamma(0) = y$ and $\gamma(1) = z$. After possibly composing with a suitable bump function, we can assume that there is an $\varepsilon > 0$ such that γ is constant equal y on $[0,\varepsilon)$ and constant equal z on $(1-\varepsilon, 1]$. The graph Γ of such a path is a one-dimensional submanifold in $X \times [0,1]$ by Exercise 9.6, and Γ is a cobordism between y and z.

It remains to construct a framing on Γ. For each $(x,t) \in X \times [0,1]$, the tangent space $T_{(x,t)}(X \times [0,1]) = T_x(X) \times T_t([0,1])$ is a vector subspace of $\mathbb{R}^N \times \mathbb{R}$. The tangent space $T_{(x,t)}(\Gamma)$ of Γ at (x,t) is a one-dimensional subspace of the $(n+1)$-dimensional vector space $T_{(x,t)}(X \times [0,1])$. The normal space $N_{(x,t)}(\Gamma, X \times [0,1])$ is the n-dimensional orthogonal complement, with respect to the standard inner product in $\mathbb{R}^{N+1}$, of $T_{(x,t)}(\Gamma)$ in $T_{(x,t)}(X \times [0,1])$. At $(y,0)$, we have $N_{(y,0)}(\Gamma, X \times [0,1]) = T_y(X)$ with basis $\mathfrak{b}_y$. Moving the basis of $T_{(x,t)}(X \times [0,1])$ along Γ produces a smooth rotation of the basis such that the last vector is tangent to Γ and the first n vectors form a basis of $N_{(x,t)}(\Gamma, X \times [0,1])$. More precisely, we proceed as follows. Since Γ is compact as the continuous image of $[0,1]$, we can cover Γ by finitely many local parametrisations $\phi_i \colon U_i \to V_i \subset X \times [0,1]$ with $U_i \subset \mathbb{H}^{n+1}$ and $V_i \subset X \times [0,1]$ open for $i = 0, \ldots, k$ such that $(y,0) \in V_0$ and $(z,1) \in V_k$. Let $e_1, \ldots, e_n, e_{n+1}$ denote the standard basis of $\mathbb{R}^{n+1}$. For $(\gamma(t),t) \in \Gamma \cap V_i$, we can assume that $d\phi_i(e_{n+1})$ is tangent to Γ, and the vectors $d\phi_i(e_1), \ldots, d\phi_i(e_n)$ form a basis of a vector subspace $W_{(\gamma(t),t)}$ of $T_{(\gamma(t),t)}(X \times [0,1])$ such that $W_{(\gamma(t),t)} \oplus T_{(\gamma(t),t)}(\Gamma) = T_{(\gamma(t),t)}(X \times [0,1])$. The vector space $W_{(\gamma(t),t)}$ is isomorphic to $N_{(\gamma(t),t)}(\Gamma, X \times [0,1])$ via a Gram–Schmidt process, and framing $W_{(\gamma(t),t)}$ is equivalent to framing $N_{(\gamma(t),t)}(\Gamma, X \times [0,1])$. We will therefore, by slight abuse of notation, identify $W_{(\gamma(t),t)}$ and $N_{(\gamma(t),t)}(\Gamma, X \times [0,1])$. Then the basis $d\phi_i(e_1), \ldots, d\phi_i(e_n)$ defines a framing $\mathfrak{w}_{(\gamma(t),t)}$ of Γ at all points $(\gamma(t),t) \in \Gamma \cap V_i$. Since the $\Gamma \cap V_i$s cover Γ, this process provides a framing $\mathfrak{w}$ of Γ.

It remains to check the orientation and to adjust the framing of the collar of Γ. We equip $X \times [0,1]$ with the product orientation of Definition 14.17. Then, for $(x,t) \in V_i$, $T_{(x,t)}(X \times [0,1])$ is oriented such that the basis $d\phi_i(e_1), \ldots, d\phi_i(e_n), d\phi_i(e_{n+1})$ is positive. At each point $(\gamma(t),t) \in \Gamma \cap V_i$, the basis $d\phi_i(e_1), \ldots, d\phi_i(e_n), d\phi_i(e_{n+1})$ is positively oriented in $T_{(\gamma(t),t)}(X \times [0,1])$, and hence $d\phi_i(e_1), \ldots, d\phi_i(e_n)$ is positively oriented in $T_x X$ for all $(x,t) \in \Gamma$. Here we assume that all local parametrisations are chosen such that they induce the chosen orientation of X. This implies that the function $\operatorname{sign} \colon \Gamma \to \{\pm 1\}$, which sends a point $(\gamma(t),t) \in \Gamma \cap V_i$ to the sign of the basis $d\phi_i(e_1), \ldots, d\phi_i(e_n)$, is constant equal $+1$. In order to define a framing of the whole manifold Γ, we need to check the behaviour at the endpoints. Since we chose γ constant on $[0,\varepsilon)$ and $(1-\varepsilon, 1]$, the normal spaces $N_{(y,0)}(\Gamma, X \times [0,1])$ and $N_{(z,1)}(\Gamma, X \times [0,1])$ are actually equal $T_y X \times \{0\}$ and $T_z X \times \{0\}$, respectively. Hence, at $(y,0)$, the framing $\mathfrak{w}_{(y,t)}$ equals $(\mathfrak{b}_y, 0)$ on $\{y\} \times [0,\varepsilon)$. At $(z,1)$, $\mathfrak{w}_{(z,t)}$ is constant equal to a basis $(\mathfrak{b}'_z, 0)$ on $\{z\} \times (1-\varepsilon, 1]$. See Figure 16.1. Since $\mathfrak{w}_{(z,t)}$ is positively oriented, there is a matrix $A \in \mathrm{GL}(n,\mathbb{R})^+$ such that $\mathfrak{b}_z = A \cdot \mathfrak{w}(z,1)$. Now, by Proposition 1.41, we can choose a smooth path $A(t)$ in $\mathrm{GL}(n,\mathbb{R})^+$ such that $A(0) = \mathrm{Id}$ and $A(1) = A$. In fact, after reparametrisation, we can choose the smooth path

$A(t)$ to be constant equal to the identity matrix on $[0, \varepsilon)$ and to be constant equal to the matrix A on $(1-\varepsilon, 1]$. Then we replace the framing $\mathfrak{w}_{(\gamma(t),t)}$ with $A(t) \cdot \mathfrak{w}_{(\gamma(t),t)}$. This defines the desired framed cobordism between $(y, \mathfrak{b}_y)$ and $(z, \mathfrak{b}_z)$. □

We will learn in Section 16.3 that, in an oriented manifold, it is not possible to find a framed cobordism between $(y, \mathfrak{b}_y)$ and $(z, \mathfrak{b}_z)$ when $\operatorname{sign}(y) \neq \operatorname{sign}(z)$. However, the pair of two points with opposite signs is framed cobordant to the empty manifold as we will now show.

Proposition 16.11 (Union of two points) *Let X be a connected, oriented smooth manifold. Let y and z be two points in X, and let $\mathfrak{b}_y$ and $\mathfrak{b}_z$ be bases of the tangent space T_yX and T_zX, respectively. We assume that* $\operatorname{sign}(y) \neq \operatorname{sign}(z)$. *Then there is a framed cobordism between the framed submanifold of codimension n given by the union $(Z, \mathfrak{b}) = \{(y, \mathfrak{b}_y), (z, \mathfrak{b}_z)\}$ and the empty manifold.*

Proof: Since X is a connected smooth manifold, X is path-connected by Exercise 8.5. We choose a smooth path $\gamma\colon [0,1] \to X$ from y to z in X such that $\gamma(t) = y$ for $t \in [0, \varepsilon)$ and $\gamma(t) = z$ for $t \in (1-\varepsilon, 1]$. We then define the subspace

$$W := \{(\gamma(t), t(1-t)) : t \in [0,1]\} \subset X \times [0,1].$$

Since t varies smoothly in both components, this is a one-dimensional compact smooth manifold with boundary $\partial W = \{y, z\} \times \{0\}$. In particular, W is a cobordism between $\{y, z\}$ and the empty manifold. Now we can determine a framing on W similar to the proof of Proposition 16.10. Since we walk along W from $(y, 0)$ towards $(z, 0)$ pointing into $X \times [0,1]$ at $(y, 0)$ and pointing out of $X \times [0,1]$ at $(z, 0)$, the orientations of $N_{(y,0)}(W, X \times [0,1]) \cong T_yX$ and $N_{(z,0)}(W, X \times [0,1]) \cong T_zX$ have opposite signs. Thus, W is the desired framed cobordism between $(Z, \mathfrak{b}) = (y, \mathfrak{b}_y) \cup (z, \mathfrak{b}_z)$ and the empty manifold. See also Figure 16.1. □

Remark 16.12 We note that, in the proof of Proposition 16.11, the function $t(1-t)$ is strictly increasing for $0 < t < 1/2$ and strictly decreasing for $1/2 < t < 1$. At $t = 1/2$, the vector tangent to the path is parallel to X, and it is at this inflection point that the orientation number of the framing changes.

Proposition 16.13 (Framed cobordisms in non-orientable manifolds) *Let X be a connected smooth manifold. We assume that X is not orientable. Let y and z be two points in X, and let $\mathfrak{b}_y$ and $\mathfrak{b}_z$ be bases of the tangent space*

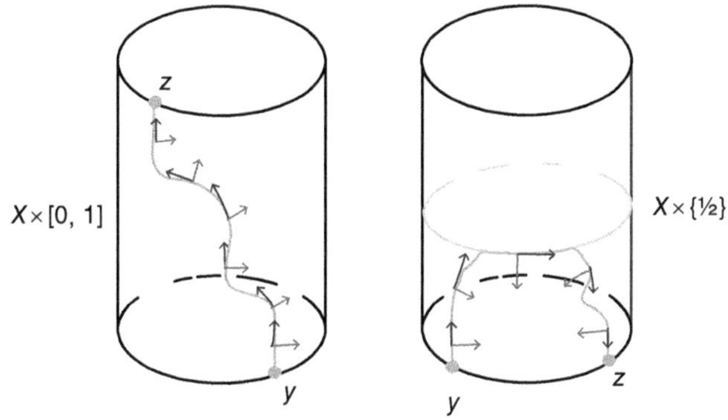

Figure 16.1 Two framed cobordisms: on the left-hand side, a framed path from one point to another. One vector points in the direction of the path, and the orthogonal complement is the normal space. The sign of the orientation is preserved along the path. On the right-hand side, a framed cobordism between the union of two points. The cobordism does not intersect $X \times \{1\}$. The switch in the sign of the orientation of the normal space happens inside $X \times \{1/2\}$, where the path is parallel to X.

T_yX and T_zX, respectively. Then there is a framed cobordism between the framed submanifolds $(y, \mathfrak{b}_y)$ and $(z, \mathfrak{b}_z)$. Moreover, there is a framed cobordism between the framed submanifold $(Z, \mathfrak{b}) = \{(y, \mathfrak{b}_y), (z, \mathfrak{b}_z)\}$ and the empty manifold.

Proof: Since X is a connected smooth manifold, X is path-connected by Exercise 8.5. We choose a smooth path $\gamma\colon [0, 1] \to X$ with $\gamma(0) = y$ and $\gamma(1) = z$ and such that γ is constant equal y on $[0, \varepsilon)$ and constant equal z on $(1 - \varepsilon, 1]$ for some $0 < \varepsilon < 1$. The graph Γ of γ is a cobordism between y and z. We can then construct a framing on Γ as in the proof of Proposition 16.10. It remains to show that we can turn Γ into a framed cobordism between $(y, \mathfrak{b}_y)$ and $(z, \mathfrak{b}_z)$, i.e., that we can choose the framing on Γ equal to the given framings $\mathfrak{b}_y$ and $\mathfrak{b}_z$ on the collar.

The proof of Proposition 16.10 shows that the framing on Γ comes equipped with a smooth assignment of an orientation of each normal space $N_{(x,t)}(\Gamma, X \times [0, 1])$. If we say that $\mathfrak{b}_y$ is a positively oriented basis of $N_{(y,0)}(\Gamma, X \times [0, 1]) = T_yX$, then $\mathfrak{b}_z$ is either a positively oriented basis of $N_{(z,0)}(\Gamma, X \times [0, 1]) = T_zX$ or not. If it is positive, we can adjust the framing on the collar $z \times (1 - \varepsilon, 1]$ as in the proof of Proposition 16.10. If $\mathfrak{b}_z$ is negatively oriented though, then we cannot perform the same smooth operation. Now we use that X is not orientable. This implies that there is at least one point $x_0 \in X$ such that we can find a loop γ_0 based at x_0 such that the above process equips $T_{\gamma(0)}X = T_{x_0}X$

and $T_{\gamma(1)}X = T_{x_0}X$ with opposite orientations. If $\mathfrak{b}_z$ is negatively oriented, then we find a smooth path γ_1 from y to x_0 and a smooth path γ_2 from x_0 to z. Then, the above procedure produces framed submanifolds Γ_0, Γ_1, and Γ_2. If the framing equips T_yX and $T_{x_0}X$ with opposite orientations and $T_{x_0}X$ and T_zX with opposite orientations, then we define a new path γ as the concatenation of γ_1 and γ_2. If only one of γ_1 and γ_2 switches the orientation at the endpoints, then we define a new path γ as the concatenation of γ_1 followed by γ_0 and then followed by γ_2. In all cases, we obtain a smooth path γ from y to z such that both $\mathfrak{b}_y$ and $\mathfrak{b}_z$ are positively oriented under the induced framing on Γ. Thus, we have shown that we always find a path such that we can adjust the framing on the collar $z \times (1-\varepsilon, 1]$ to turn Γ into a framed cobordism between $(y, \mathfrak{b}_y)$ and $(z, \mathfrak{b}_z)$. This proves the first assertion. For the second assertion, we construct a cobordism W as in the proof of Proposition 16.11. To ensure that W is compatible with the framing $\mathfrak{b}_y$ and $\mathfrak{b}_z$, we use the same trick as above. □

16.2 The Pontryagin Construction

From now on, we are going to assume that X is compact. We consider a smooth map $f : X \to \mathbb{S}^p$, and let $y \in \mathbb{S}^p$ be a regular value for f. Then $f^{-1}(y)$ is a submanifold by the Preimage Theorem. When we consider $\mathbb{S}^p$ with its usual orientation, then f induces a framing on the manifold $f^{-1}(y)$ as follows. We have the positively oriented basis $\mathfrak{o} = (\mathfrak{o}_1, \ldots, \mathfrak{o}_p)$ for the tangent space $T_y\mathbb{S}^p$. For each $x \in f^{-1}(y)$, by Proposition 4.13, the subspace $T_x f^{-1}(y) \subset T_xX$ is the kernel of the derivative $df_x : T_xX \to T_y\mathbb{S}^p$. Moreover, df_x maps the orthogonal complement $(T_x f^{-1}(y))^{\perp}$ isomorphically onto $T_y\mathbb{S}^p$. Hence, there is a unique basis $\mathfrak{n}(x)$ of $(T_x f^{-1}(y))^{\perp} \subset T_xX$ which maps to $\mathfrak{o}$ under df_x. This defines an orientation $\mathfrak{n}$ on the normal bundle $N(f^{-1}(y), X)$ which we also denote by $\mathfrak{n} = f^*\mathfrak{o}$.

Definition 16.14 (Pontryagin manifold) The framed manifold $(f^{-1}(y), f^*\mathfrak{o})$ is called the *Pontryagin manifold* associated with f.

Our next goal is to show that the framed cobordism class of the Pontryagin manifold associated with f is independent of the choices we have made. We will achieve this goal by showing independence step by step in the following sequence of lemmas.

Lemma 16.15 *Let $\mathfrak{o}$ and $\mathfrak{o}'$ be two different positively oriented bases of $T_y\mathbb{S}^p$. Then the Pontryagin manifold $(f^{-1}(y), f^*\mathfrak{o})$ is framed cobordant to $(f^{-1}(y), f^*\mathfrak{o}')$.*

Proof: Since $\mathfrak{o}$ and $\mathfrak{o}'$ are both positively oriented, there is a matrix A with $\det(A) > 0$ such that $A \cdot \mathfrak{o} = \mathfrak{o}'$. By Proposition 1.41, we can choose a smooth path $A(t)\colon [0,1] \to \mathrm{GL}(n,\mathbb{R})^+$ with $A(0) = I$ and $A(1) = A$. In fact, for a given $\varepsilon > 0$, we can choose the smooth path $A(t)$ to be constantly equal to the identity matrix on $[0,\varepsilon)$ and to be constantly equal to the matrix A on $(1-\varepsilon, 1]$. For each $t \in [0,1]$, we then get an orientation $A(t) \cdot \mathfrak{o}$ at the point $(y,t) \in \{y\} \times [0,1]$. This defines a framing on $\{y\} \times [0,1]$ within $\mathbb{S}^p \times [0,1]$. The submanifold $f^{-1}(y) \times [0,1] \subset X \times [0,1]$ is a cobordism between $f^{-1}(y) \times \{0\}$ and $f^{-1}(y) \times \{1\}$, and, by Theorem 14.14, the pullback of the framing on $\{y\} \times [0,1]$ along $f \times \mathrm{id}$ defines a framed cobordism between $(f^{-1}(y), f^*\mathfrak{o})$ and $(f^{-1}(y), f^*\mathfrak{o}')$. □

Lemma 16.15 shows that, up to framed cobordism, we can drop the orientation $\mathfrak{o}$ from the notation of a Pontryagin manifold. Next, we are going to show that the framed cobordism class of $f^{-1}(y)$ is independent of the choice of regular value y, at least in a small neighbourhood.

Lemma 16.16 *Let y be a regular value for the smooth map $f\colon X \to \mathbb{S}^p$. If $z \in \mathbb{S}^p$ is sufficiently close to y, then $f^{-1}(z)$ is framed cobordant to $f^{-1}(y)$.*

Proof: Assume that df_{x_0} is surjective at a point x_0. It follows from Theorem 4.7 that there is an open neighbourhood U around x_0 such that df_x is surjective for all $x \in U$. Hence the set of regular points for f is open in X, and hence the set C of critical points is closed. Since X is compact, this implies that C is compact by Lemma 1.14. Since f is continuous, this implies that the set $f(C) \subset \mathbb{S}^p$ of critical values is compact by Lemma 1.13. Since $\mathbb{S}^p$ is Hausdorff, this implies that $f(C)$ is closed in $\mathbb{S}^p$ by Proposition 1.21. This shows that we can choose an $\varepsilon > 0$ such that $\mathbb{B}_\varepsilon(y) \cap \mathbb{S}^p$ contains only regular values. Now let $z \in \mathbb{S}^p$ be such that $|z - y| < \varepsilon$. We can choose a family of smooth rotations $r_t\colon \mathbb{S}^n \to \mathbb{S}^n$ such that $r_1(y) = z$, and such that, for some $0 < \delta < 1$,

(a) r_t is the identity for $0 \le t < \delta$, r_t equals r_1 for $1 - \delta < t \le 1$, and
(b) each $r_t^{-1}(z)$ lies on the great circle from y to z, which implies that $r_t^{-1}(z)$ is a regular value for f.

Such a family r_t is then an isotopy as defined in Definition 11.2. Now we can define a homotopy $F\colon X \times [0,1] \to \mathbb{S}^n$ by $F(x,t) := r_t(f(x))$. By our choice of r_t and z, note that, for each fixed t, z is a regular value for the composition $r_t \circ f\colon X \to \mathbb{S}^n$. This implies that z also is a regular value for the map F. Hence the preimage $F^{-1}(z) \subset X \times [0,1]$ is a framed manifold which provides a framed cobordism between the framed submanifolds $f^{-1}(z)$ and $(r_1 \circ f)^{-1}(z) = f^{-1}(r_1^{-1}(z)) = f^{-1}(y)$. □

Now we show that, up to framed cobordism, a Pontryagin manifold only depends on the homotopy class of f.

Lemma 16.17 *Let f and g be smooth maps $X \to \mathbb{S}^p$. Let y be a regular value for both f and g. If f and g are smoothly homotopic, then $f^{-1}(y)$ is framed cobordant to $g^{-1}(y)$.*

Proof: Using a suitable bump function, we can choose a smooth homotopy $F\colon X \times [0,1] \to \mathbb{S}^p$ with

$$\begin{cases} F(x,t) = f(x) & \text{for } 0 \leq t < \varepsilon, \\ F(x,t) = g(x) & \text{for } 1 - \varepsilon < t \leq 1. \end{cases}$$

By Sard's Theorem 9.21, we can choose a regular value z for F which is close enough to y such that, by Lemma 16.16, $f^{-1}(z)$ is framed cobordant to $f^{-1}(y)$ and such that $g^{-1}(z)$ is framed cobordant to $g^{-1}(y)$. Then $F^{-1}(z)$ is a framed cobordism between $f^{-1}(z)$ and $g^{-1}(z)$. Since framed cobordism is an equivalence relation, we get that $f^{-1}(y)$ and $g^{-1}(y)$ are framed cobordant as well. □

We can prove our first main result of this section.

Theorem 16.18 (A well-defined cobordism class) *Let $X \subset \mathbb{R}^N$ be a compact smooth manifold, and let $f\colon X \to \mathbb{S}^p$ be a smooth map. The framed cobordism class of a Pontryagin manifold associated with f only depends on the homotopy class of f. We write N_f for the framed cobordism class of f.*

Proof: By the previous lemmas, the only remaining point is to show the independence of the choice of regular value y. So let y, y' be two arbitrary regular values for f. Then we can choose a family of rotations $r_t\colon \mathbb{S}^p \to \mathbb{S}^p$ such that r_0 is the identity and $r_1(y) = y'$, and f is smoothly homotopic to $r_1 \circ f$. By Lemma 16.17, the Pontryagin manifold $f^{-1}(y')$ is then framed cobordant to the Pontryagin manifold $(r_1 \circ f)^{-1}(y') = f^{-1}(r_1^{-1}(y')) = f^{-1}(y)$. □

Next, we show that the Pontryagin manifold construction yields a member in every framed cobordism class.

Theorem 16.19 (Surjectivity) *Let $X \subset \mathbb{R}^N$ be a compact, smooth manifold. Every compact, boundaryless, framed submanifold $(Z, \mathfrak{n})$ of codimension p in X occurs as the Pontryagin manifold N_f for some smooth map $f\colon X \to \mathbb{S}^p$.*

Proof: Let $(Z, \mathfrak{n})$ be a compact framed submanifold of codimension p in X. By the Tubular Neighbourhood Theorem 12.21, there is a diffeomorphism $N^{\varepsilon}(Z, X) \to U \subset X$ to an open neighbourhood U of Z in X. Since the normal bundle to Z is trivial by Lemma 16.5, there is a diffeomorphism $Z \times \mathbb{R}^p \xrightarrow{\cong} Z \times \mathbb{B}_{\varepsilon'}(0) \to N^{\varepsilon}(Z, X)$ for a sufficiently small $\varepsilon' > 0$. Composing these maps gives us a diffeomorphism $g\colon Z \times \mathbb{R}^p \to U$ which restricts to the identity on $Z \times \{0\}$, i.e., $g((x,0)) = x$ for all $x \in Z$. We can then define a smooth map

$$\pi\colon U \to \mathbb{R}^p,\ \pi((g(x,\mathbf{v})) = \mathbf{v}.$$

By the construction of π and the Tubular Neighbourhood Theorem 12.21, 0 is a regular value for π and $\pi^{-1}(0)$ equals Z. Moreover, by the construction of the diffeomorphism $Z \times \mathbb{R}^p \to N(Z, X)$, the framing on $\pi^{-1}(Z) = Z$ is the initial framing $\mathfrak{n}$ on Z.

We can now extend π to a smooth map on X as follows. Now let $y_0 \in \mathbb{S}^p$ be a point, and let $\phi_{y_0}^{-1}\colon \mathbb{S}^p \setminus \{y_0\} \to \mathbb{R}^p$ denote the stereographic projection from y_0. Using the arguments in the proof of Lemma 7.5, we can find a smooth monotone decreasing function $\lambda\colon \mathbb{R} \to \mathbb{R}$ with

$$\begin{cases} \lambda(t) > 0 & \text{for } t < 1 \\ \lambda(t) = 0 & \text{for } t \geq 1. \end{cases}$$

We can then define a smooth map $\varphi\colon \mathbb{R}^p \to \mathbb{S}^p$ by $\varphi(x) = \phi_{y_0}(x/\lambda(|x|^2))$. The map φ maps x with $|x| \geq 1$ to $y_0 \in \mathbb{S}^p$ and maps the open unit ball in $\mathbb{R}^p$ diffeomorphically onto $\mathbb{S}^p \setminus \{y_0\}$. We then define $f\colon X \to \mathbb{S}^p$ by

$$f(x) = \begin{cases} \varphi(\pi(x)) & \text{for } x \in U \\ y_0 & \text{for } x \notin U. \end{cases}$$

The map f is smooth as a composition of smooth maps. In fact, we may consider f as an extension of π to a smooth map $X \to \mathbb{S}^p = \mathbb{R}^p \cup \{y_0\}$. Moreover, $\varphi(0) = -y_0 \in \mathbb{S}^p$ is a regular value for f since φ is a local diffeomorphism around 0, and 0 is a regular value for π. Since the Pontryagin manifold $\pi^{-1}(0)$ is $(Z, \mathfrak{n})$ and φ is a local diffeomorphism around 0, the Pontryagin manifold

$$f^{-1}(\varphi(0)) = \pi^{-1}(0)$$

also equals $(Z, \mathfrak{n})$. Thus, we have found a smooth map $f\colon X \to \mathbb{S}^p$ with $N_f = (Z, \mathfrak{n})$, which proves the theorem. □

Theorem 16.20 (Injectivity) *Let $X \subset \mathbb{R}^N$ be a compact smooth manifold. Two smooth maps from X to $\mathbb{S}^p$ are smoothly homotopic if and only if the associated Pontryagin manifolds are framed cobordant.*

For the proof of Theorem 16.20, we need some preparation.

Lemma 16.21 *Let f and g be smooth maps $X \to \mathbb{S}^p$. Let y be a regular value for both f and g. If the framed manifolds $(f^{-1}(y), f^*\mathfrak{o})$ and $(g^{-1}(y), g^*\mathfrak{o})$ are equal, then f and g are smoothly homotopic.*

Proof: We write $Z = f^{-1}(y) = g^{-1}(y)$. Since $T_x Z$ is the kernel of both df_x and dg_x for an x with $f(x) = y = g(x)$, the assumption that $f^*\mathfrak{o} = g^*\mathfrak{o}$ implies that $df_x = dg_x$ for all $x \in Z$.

- Claim: It suffices to show that f and g coincide on an open neighbourhood V of Z.

Proof of the claim: For the given point $y \in \mathbb{S}^p$, let $\phi_y^{-1} \colon \mathbb{S}^p \setminus \{y\} \to \mathbb{R}^p$ denote the stereographic projection from y. Then, after removing the open neighbourhood V, we can use a linear homotopy in $\mathbb{R}^p$, i.e., we can define a smooth map $F \colon X \times [0,1] \to \mathbb{S}^p$ by

$$F(x,t) := \begin{cases} f(x) & \text{for } x \in V \text{ and all } t \\ \phi_y\left(t \cdot \phi_y^{-1}(f(x)) + (1-t) \cdot \phi_y^{-1}(g(x))\right) & \text{for } x \in X \setminus Z. \end{cases}$$

Since we assume that $f_{|V} = g_{|V}$, F defines a smooth homotopy between f and g. This proves the claim. □

Now we want to smoothly deform f so that it coincides with g in some open neighbourhood of Z. But we have to make sure that we do not map new points to y, since this would alter the fibre $f^{-1}(y)$. As in the proof of Theorem 16.19, we can find an open neighbourhood V of Z in X together with a diffeomorphism $Z \times \mathbb{R}^p \to V$. Since both f and g map Z to y, we can choose V small enough such that neither $f(V)$ nor $g(V)$ contains the antipodal point $-y$ of y. We then get two maps $F, G \colon Z \times \mathbb{R}^p \to \mathbb{R}^p$ as the compositions

$$F, G \colon Z \times \mathbb{R}^p \xrightarrow{\cong} V \xrightarrow[g_{|V}]{f_{|V}} \mathbb{S}^p \setminus \{-y\} \xrightarrow{\cong} \mathbb{R}^p.$$

By construction, we have $F^{-1}(0) = G^{-1}(0) = Z \times \{0\}$ and both $dF_{(x,0)}$ and $dG_{(x,0)}$ equal the linear projection map $T_x(Z \times \mathbb{R}^p) = T_x Z \times \mathbb{R}^p \to \mathbb{R}^p$ for all $x \in Z$.

- Claim: We can find a small constant $c > 0$ such that $F(x,u) \cdot u > 0$ and $G(x,u) \cdot u > 0$ for all $x \in Z$ and $u \in \mathbb{B}_c^p(0)$.

Proof of the claim: First we find such a suitable constant for F. For a fixed $x \in Z$, we consider the linear map $F(x,-) \colon \mathbb{R}^p \to \mathbb{R}^p$ defined by $u \mapsto F(x,u)$. Since $T_x(Z)$ is the kernel of df_x, the derivative of $dF(x,-)_0$ equals $dF_{(x,0)}$

which we recall is the linear projection $T_x Z \times \mathbb{R}^p \to \mathbb{R}^p$. Hence we may consider as a linear map $\mathbb{R}^p \to \mathbb{R}^p$. Since $F(x,0) = 0$, we may then apply Taylor's Theorem from calculus to $F(x,-)$ to get a constant $c_x > 0$ such that $|F(x,u) - u| \leq c_x \cdot |u|^2$ for $|u| < 1$. This implies

$$\begin{aligned} 2F(x,u) \cdot u &= |F(x,u)|^2 + |u|^2 - |F(x,u) - u|^2 \\ &> |F(x,u)|^2 + |u|^2 - c_x^2 |u|^4 \\ &> |u|^2 - c_x^2 |u|^4 > 0 \end{aligned}$$

for $0 < |u| < \min(c_x^{-1}, 1)$. By the Cauchy–Schwartz inequality, the dot product is continuous, which implies that there is an open neighbourhood $U_x \subset Z$ of x such that $F(x,u) \cdot u > 0$ for $0 < |u| < \min(c_x^{-1}, 1)$. Then the U_x, for every $x \in Z$, define an open cover of Z. Since Z is compact, finitely many of the U_x suffice to cover Z. Labelling them $U_{x_1}, \ldots, U_{x_n}$, we can then define $c_F := \min_i(c_{x_i})$ as the minimum of the c_{x_i} to get $F(x,u) \cdot u > 0$ for $0 < |u| < c_F$. By repeating the argument for G, we get another constant c_G such that $G(x,u) \cdot u > 0$ for $0 < |u| < c_G$. Now we set $c := \min(c_F, c_G)$. This proves the claim. □

After fixing a constant $c > 0$ as in the claim, we now know that the points $F(x,u)$ and $G(x,u)$ belong to the same open half-space of $\mathbb{R}^p$ for all $x \in Z$ and $u \in \mathbb{B}_c^p(0)$. In particular, the line segment between $F(x,u)$ and $G(x,u)$ does not pass through the origin. Thus, the linear homotopy $t \mapsto (1-t) \cdot F(x,u) + t \cdot G(x,u)$ between F and G does not map any new points to 0 as long as $u \in \mathbb{B}_c^p(0)$. To avoid moving points outside of $\mathbb{B}_c^p(0)$ we choose, as in Lemma 7.5, a smooth function $\varphi\colon \mathbb{R}^p \to [0,1]$ with

$$\varphi(u) = \begin{cases} 1 & \text{for } |u| \leq c/2 \\ 0 & \text{for } |u| \geq c. \end{cases}$$

Then we define the smooth map

$$F_t(x,u) = (1 - \varphi(u)t) \cdot F(x,u) + \varphi(u)t \cdot G(x,u).$$

Then F_t is a smooth deformation from $F_0 = F$ to the smooth map F_1, which has the following properties:

$$F_1(x,u) = \begin{cases} G(x,u) & \text{for } |u| \leq c/2 \\ F(x,u) & \text{for } |u| \geq c, \end{cases}$$

and F_1 has the same zeros as F. Now we define the new map $f_1\colon V \to \mathbb{S}^p$ such that the diagram

$$\begin{array}{ccc} V & \overset{f_1}{\dashrightarrow} & \mathbb{S}^p \setminus \{y\} \\ \downarrow & & \uparrow \phi_y^{-1} \\ Z \times \mathbb{R}^p & \underset{F_1}{\longrightarrow} & \mathbb{R}^p \end{array} \tag{16.1}$$

commutes. The map f_1 is a smooth deformation of f since F_1 is a smooth deformation of $F = F_0$. Moreover, since F_1 coincides with G on $Z \times \mathbb{B}^p_{c/2}(0)$, f_1 coincides with g on the open neighbourhood $V_{c/2}$ of Z, where $V_{c/2}$ is the image of $Z \times \mathbb{B}^p_{c/2}(0)$ under the diffeomorphism $Z \times \mathbb{R}^p \xrightarrow{\cong} V$. By the first claim, this shows $f_1 \simeq g$, and since $f_1 \simeq f$, we get $f \simeq g$ since homotopy is a transitive relation. This proves the lemma. □

Proposition 16.22 (Tubular neighbourhood for framed cobordisms) *Let $X \subset \mathbb{R}^N$ be a compact smooth n-dimensional manifold without boundary, and let $(W, \mathfrak{w})$ be a framed cobordism between two framed submanifolds of X of codimension p. Then there is a tubular neighbourhood W^ε of W in $X \times [0,1]$ which restricts to a tubular neighbourhood of ∂W in $\partial(X \times [0,1]) = X \times \{0\} \cup X \times \{1\}$.*

Proof: Since W is a cobordism, the normal spaces near the boundary ∂W are parallel to the normal spaces of ∂W in $X \times \{0\} \cup X \times \{1\}$. In particular, for normal vectors v to ∂W in $X \times \{0\} \cup X \times \{1\}$, $h(w,v) = w + v$ remains in the boundary $X \times \{0\} \cup X \times \{1\}$. We can therefore argue as in the proof of Theorem 12.21. Let $\widetilde{X} \subset \mathbb{R}^{N+1}$ be a levelwise defined tubular neighbourhood of $X \times [0,1]$. Then we get a submersion $\tilde{\pi}\colon \widetilde{X} \to X \times [0,1]$. We define the smooth map $h\colon N(W, X \times [0,1]) \to \mathbb{R}^{N+1}$ by $(w,v) \mapsto w + v$. The inverse image $\widetilde{U} := h^{-1}(\widetilde{X}) \subset N(W, X \times [0,1])$ is an open neighbourhood of the zero section W_0 in $N(W, X \times [0,1])$. Since $h(w,0) = w$ for all $w \in W$, the composition $\varphi\colon \widetilde{U} \xrightarrow{h} \widetilde{X} \xrightarrow{\tilde{\pi}} X \times [0,1]$ is the identity when we restrict it to W_0. By the same argument as in the proof of Theorem 12.11, we can show that $dh_{(w,0)}$ is an isomorphism at every point of W_0 in $N(W, X \times [0,1])$. Since $d\tilde{\pi}_w$ is the identity for all $w \in W \subset \widetilde{X}$, $d\varphi_{(w,0)}$ is an isomorphism for every $(w,0) \in W_0 \subset N(W, X \times [0,1])$. Hence the assumptions of the generalised Inverse Function Theorem 12.1 are satisfied, and we can conclude as before that there is an open neighbourhood $N^\varepsilon(W, X \times [0,1])$ of W_0 in $N(W, X \times [0,1])$ which is mapped diffeomorphically onto an open neighbourhood W^ε of W in $X \times [0,1]$ by $\varphi = \tilde{\pi} \circ h$. The composition $W^\varepsilon \xrightarrow{\cong} N^\varepsilon(W, X \times [0,1]) \to W$

is a submersion. Since W is a cobordism, W^ε restricts to a neighbourhood $(\partial W)^\varepsilon$ of ∂W in $X \times \{0\} \cup X \times \{1\}$. □

Using Proposition 16.22, we can now proceed exactly as in the proof of Theorem 16.19 to show the following result.

Theorem 16.23 (Framed cobordisms are Pontryagin manifolds) *Let $X \subset \mathbb{R}^N$ be a compact smooth n-dimensional manifold without boundary, and let $(W, \mathfrak{w})$ be a framed cobordism between two framed submanifolds of X of codimension p. Then there is a smooth map $F\colon X \times [0,1] \to \mathbb{S}^p$ such that $(W, \mathfrak{w})$ is the Pontryagin manifold associated to F.* □

Now we can finish the proof of Theorem 16.20.

Proof of Theorem 16.20: Let $f, g\colon X \to \mathbb{S}^p$ be smooth maps, and let $y \in \mathbb{S}^p$ be a regular value for both f and g. If f and g are smoothly homotopic, then Lemma 16.17 shows that the Pontryagin manifolds $f^{-1}(y)$ and $g^{-1}(y)$ are framed cobordant. Now assume that there exists a framed cobordism $(W, \mathfrak{w})$ between $f^{-1}(y)$ and $g^{-1}(y)$. By Theorem 16.23, there is a smooth homotopy $F\colon X \times [0,1] \to \mathbb{S}^p$ between f and g with regular value y such that $(F^{-1}(y), F^*\mathfrak{v})$ is equal to $(W, \mathfrak{w})$. Setting $F_t(x) = F(x,t)$, the Pontryagin manifolds associated to f and F_0, respectively, are equal by construction of F. Then Lemma 16.21 implies that f and F_0 are smoothly homotopic. Similarly, we get that F_1 and g are smoothly homotopic. Thus, f and g are smoothly homotopic since smooth homotopy is an equivalence relation. This proves the theorem. □

16.3 Hopf Degree Theorem

As an application, we will deduce Hopf's Degree Theorem generalising Theorem 15.19 to higher dimensions. We then consider the case of non-orientable manifolds. Before we can prove the main results of this section, we need some more preparation.

Lemma 16.24 (Disentangling cobordisms) *Let X be a smooth manifold of dimension $n \geq 2$. Let W_1 and W_2 be two one-dimensional framed cobordisms in $X \times [0,1]$. Then there is an open subset $U \subset X \times [0,1]$ which contains $W_1 \cap W_2$, and there is a one-dimensional framed cobordism $\widetilde{W}_1$ which coincides with W_1 outside U and such that $\widetilde{W}_1 \cap W_2 = \emptyset$.*

Proof: The assertion follows from Proposition 12.51 after some adjustments. The latter result was proven for a manifold without boundary, whereas $X \times$

$[0,1]$ has a boundary. However, we can remedy this as follows. Let $\varepsilon > 0$ be such that both W_1 and W_2 are constant on their respective collars of height ε. Then we set $Y := X \times (\varepsilon/4, 1-\varepsilon/4)$. This is a smooth manifold without boundary of dimension $n+1 \geq 3$. Now we also have to modify W_1 and W_2 since their restrictions to Y are not compact anymore. To do so, we can shorten their respective collars a bit and set $W_1' = W_1 \cap X \times [\varepsilon/2, 1-\varepsilon/2]$ and $W_2' = W_2 \cap X \times [\varepsilon/2, 1-\varepsilon/2]$. Then $W_1' \cap W_2' = W_1 \cap W_2$ and this intersection happens away from the boundaries of W_1' and W_2'. Now we can apply Proposition 12.51 to get $\widetilde{W}_1'$. We note that $\widetilde{W}_1'$ is still a compact one-dimensional manifold since it is the image of a compact one-dimensional manifold under an embedding. The collar of $\widetilde{W}_1'$ coincides with the collar of W_1'. Now it remains to stretch the collar of $\widetilde{W}_1'$ along the collar of W_1 to obtain the desired one-dimensional cobordism $\widetilde{W}_1$, which does not intersect W_2. □

Proposition 16.25 *Let X be a compact, connected, oriented smooth manifold without boundary. We assume that X has dimension $n \geq 2$. Let $(Z, \mathfrak{n})$ and $(Z', \mathfrak{n}')$ be two framed submanifolds of codimension n in X. If $\sum_{x \in Z} \operatorname{sign}(x) = \sum_{x' \in Z'} \operatorname{sign}(x')$, then $(Z, \mathfrak{n})$ and $(Z', \mathfrak{n}')$ are framed cobordant.*

Proof: Both Z and Z' are finite collections of points. First we pair points of opposite signs in Z and construct a framed cobordism using Proposition 16.11 between a pair $\{x_i, x_j\}$ of points in Z with $\operatorname{sign}(x_i) \neq \operatorname{sign}(x_j)$ with the empty manifold until we end with points of the same sign. Then we perform the same process for Z'. Assume now that $\sum_{x \in Z} \operatorname{sign}(x) = \sum_{x' \in Z'} \operatorname{sign}(x')$, i.e., we have $\sum_{x \in Z} \operatorname{sign}(x) = \sum_{x' \in Z'} \operatorname{sign}(x')$ many points in both Z and Z' with the same sign. We can then construct a cobordism between pairs of points $x \in Z$ and $x' \in Z'$ as in the proof of Proposition 16.10. However, before we equip them with a framing, we first use that $\dim X \geq 2$ and Lemma 16.24 to disentangle these cobordisms. Then we equip the resulting disjoint cobordisms with a framing as in the proof of Proposition 16.10. The union of these framed cobordisms is then a framed cobordism between $(Z, \mathfrak{n})$ and $(Z', \mathfrak{n}')$. □

We can now prove the following famous theorem of Hopf [20].

Theorem 16.26 (Hopf Degree Theorem) *Let X be a compact, connected, oriented smooth n-dimensional manifold without boundary. Then two smooth maps $X \to \mathbb{S}^n$ are smoothly homotopic if and only if they have the same degree. In other words, the degree defines an injective map $[X, \mathbb{S}^n] \to \mathbb{Z}$.*

Proof: For $n = 1$, the classification of one-manifolds in Theorem 10.1 implies that X is diffeomorphic to $\mathbb{S}^1$. The case $n = 1$ is therefore already proven in Theorem 15.19. Hence we now assume $n \geq 2$. We already know that two smooth maps f and g from X to $\mathbb{S}^n$ have the same degree if they are homotopic. Now we assume that $\deg(f) = \deg(g)$. By Theorem 16.20, it then suffices to show that the associated Pontryagin manifolds are framed cobordant. For $f: X \to \mathbb{S}^p$, let $(f^{-1}(y), \mathfrak{n})$ be the associated Pontryagin manifold. The degree of f then equals $\sum_{x \in Z} \operatorname{sign}(x)$ since $\mathfrak{n} = f^*\mathfrak{o}$ where $\mathfrak{o}$ is the canonical orientation of $\mathbb{S}^p$. Hence, the assumption implies

$$\sum_{x \in f^{-1}(y)} \operatorname{sign}(x) = \deg(f) = \deg(g) = \sum_{x \in g^{-1}(y)} \operatorname{sign}(x).$$

The assertion then follows from Theorem 16.20. □

Theorem 16.20 shows that the degree defines a map from $[X, \mathbb{S}^n]$ to $\mathbb{Z}$ which is injective. We will now explain that it is also surjective. For any non-zero integer k, we can choose a framed manifold $(Z, \mathfrak{n})$ consisting of $|k|$ many points where each point is framed either by a positively oriented basis if k is positive, or by a negatively oriented basis if k is negative. By Theorem 16.19, there is then a smooth map $f: X \to \mathbb{S}^n$ with $(Z, \mathfrak{n})$ as its associated Pontryagin manifold. Since every continuous map $X \to \mathbb{S}^n$ is homotopic to a smooth map, we get the following result.

Corollary 16.27 (Homotopy groups of spheres) *Let X be a compact, connected, oriented smooth n-dimensional manifold without boundary. The degree defines a bijection $[X, \mathbb{S}^n] \xrightarrow{\cong} \mathbb{Z}$. In particular, if $\pi_n(\mathbb{S}^n)$ denotes the nth homotopy group of $\mathbb{S}^n$, the degree defines a bijection $\pi_n(\mathbb{S}^n) \xrightarrow{\cong} \mathbb{Z}$.*

Remark 16.28 We emphasise that Hopf's theorem requires that we consider maps into the sphere, and not other smooth manifolds. For instance, in Example 15.18 we have seen two smooth maps A and B from the torus $\mathbb{T}^2$ to itself such that $\deg(A) = \deg(B)$. However, the maps A and B are not homotopic. One can show this for example by studying the effect of these maps on the singular homology of $\mathbb{T}^2$. See also Example 17.14. In particular, this example shows that the homotopy class of a smooth map from a compact, connected, oriented smooth two-dimensional manifold without boundary into $\mathbb{T}^2$ is not determined by the degree.

Now we consider the case that X is not orientable. In this case, there are much fewer homotopy classes. First, we prove the following lemma.

Proposition 16.29 *Let X be a compact, connected smooth manifold without boundary. We assume that X is not orientable and that X has dimension $n \geq 2$.*

Let $(Z, \mathfrak{n})$ *and* $(Z', \mathfrak{n}')$ *be two framed zero-dimensional submanifolds of* X*. Then* $(Z, \mathfrak{n})$ *and* $(Z', \mathfrak{n}')$ *are framed cobordant if* $\#Z = \#Z'$ *modulo* 2*.*

Proof: Both Z and Z' are finite collections of points. By the second assertion of Proposition 16.13, we can construct cobordisms between pairs of points in Z and the empty manifold, and cobordisms between pairs of points in Z' and the empty manifold. We then use that $\dim X \geq 2$ and Lemma 16.24 to disentangle these cobordisms. Then we equip the resulting disjoint cobordisms with a framing and take their union. We can then assume that both $(Z, \mathfrak{n}) = (\{x\}, \mathfrak{n}_x)$ and $(Z', \mathfrak{n}') = (\{x'\}, \mathfrak{n}'_{x'})$ consist of a single point. By the first assertion of Proposition 16.13, we can then construct a framed cobordism between $(\{x\}, \mathfrak{n}_x)$ and $(\{x'\}, \mathfrak{n}'_{x'})$. This proves the claim. □

Now we can prove Hopf's Theorem for non-orientable manifolds.

Theorem 16.30 (Hopf Degree Theorem for non-orientable manifolds) *Let* X *be a compact, connected, smooth* n*-dimensional manifold without boundary. Assume that* X *is not orientable. Then two smooth maps* $X \to \mathbb{S}^n$ *are homotopic if and only if they have the same mod* 2 *degree.*

Proof: It follows from the classification of all connected smooth one-manifolds that the latter are orientable. Thus, we may assume $n \geq 2$. We already know that two smooth maps f and g from X to $\mathbb{S}^n$ have the same degree if they are homotopic. So we assume that $\deg(f) = \deg(g)$ modulo 2. By Whitney's Approximation Theorem 12.26, we can assume that f and g are smooth maps. By Theorem 16.20, it then suffices to show that the associated Pontryagin manifolds are framed cobordant. Let y be a regular value for both f and g. Since the associated Pontryagin manifolds $f^{-1}(y)$ and $g^{-1}(y)$ are finite collections of points and $\#f^{-1}(y) = \deg(f) = \deg(g) = \#g^{-1}(y)$ modulo 2, the assertion follows from Proposition 16.29. □

17
Vector Fields and the Poincaré–Hopf Index Theorem

Let $X \subset \mathbb{R}^N$ be a smooth manifold. A *vector field* on X is a smooth assignment of a vector tangent to X at each point x, i.e., a smooth map $\mathbf{v}: X \to \mathbb{R}^N$ such that $\mathbf{v}(x) \in T_x(X)$ for every x. We have studied properties of vector fields in the exercises before. Vector fields play a crucial role in many applications, for example, in mathematical physics. Providing efficient tools to understand such fields is an important motivation for differential topology. A priori, one might expect that vector fields are free to move in every way they want. It turns out, however, that the topology of X provides quite strong restrictions for what is allowed and what is not. This is the content of the Poincaré–Hopf Index Theorem 17.19 that we are going to prove in this chapter.

17.1 Index of a Vector Field at a Zero

We have already seen some examples of vector fields in the exercises. See Exercises 8.10 and 15.2. We first look at some more examples.

17.1.1 Examples of Vector Fields

Example 17.1 (Vector field on an open subset) Let $U \subset \mathbb{R}^k$ be an open subset. Recall that the tangent space $T_u U$ is just $\mathbb{R}^k$ for every $u \in U$. Hence a vector field $\mathbf{v}$ on U is just a smooth map $U \to \mathbb{R}^k$. Let $e_1, \dots, e_k$ be the standard basis of $\mathbb{R}^k$. Then $\mathbf{v}$ is of the form $\mathbf{v} = \sum_{i=1}^k v_i e_i$ for smooth functions $v_i : U \to \mathbb{R}$.

Example 17.2 (A vector field on the sphere) On the n-dimensional sphere $\mathbb{S}^n$, let $p_N = (0, \dots, 0, 1) \in \mathbb{S}^n$ denote the north pole. We can construct a vector field $\mathbf{v}$ on $\mathbb{S}^n$ by setting $\mathbf{v}(x) := p_N - (p_N \cdot x)x$, where $\cdot$ denotes the inner product defined by considering p_N and x as vectors in $\mathbb{R}^{n+1}$. To check that this actually defines a vector field, it suffices to observe that $\mathbf{v}(x)$ is perpendicular to every x since $\mathbf{v}(x) \cdot x = p_N \cdot x - (p_N \cdot x)(x \cdot x) = 0$, where we use $x \cdot x = 1$ for every $x \in \mathbb{S}^n$.

Example 17.3 (Gradient vector field of a function) Let $X \subset \mathbb{R}^N$ be a smooth k-dimensional manifold, and let $f : X \to \mathbb{R}$ be a smooth real-valued function on X. For each $x \in X$, the derivative $df_x : T_x X \to \mathbb{R}$ defines a linear functional on $T_x X$. Recall from linear algebra that for any such functional there is a vector $\mathbf{v}(x) \in T_x X$ such that

$$df_x(w) = \mathbf{v}(x)^\top \cdot w,$$

where $\mathbf{v}(x)^\top$ denotes the transpose of $\mathbf{v}(x)$ and $\cdot$ is matrix multiplication. We may think of this assignment $x \mapsto \mathbf{v}(x) \in T_x X$ as a vector field. This is called the *gradient field of* f, and we denote it by $\mathbf{grad}(f)$. We note that if $z \in X$ is a zero of $\mathbf{grad}(f)$, i.e., $\mathbf{grad}(f)(z) = 0$, then $df_z(w) = (\mathbf{grad}(f)(z))^\top \cdot w = 0$ for all w. Hence $df_z = 0$ and z is a critical point of f. Conversely, if z is a critical point of f, then $df_z = 0$ and $\mathbf{grad}(f)(z)$ must be zero. Thus, z is a zero the gradient field of f if and only if z is a critical point.

In the special case $X = \mathbb{R}^k$, the gradient field of f can be represented by the Jacobian $(1 \times k)$-matrix of f at x, i.e.,

$$\mathbf{grad}(f)(x) = \left(\frac{\partial f}{\partial x_1}(x), \ldots, \frac{\partial f}{\partial x_k}(x) \right).$$

That is, for the standard basis vectors $e_1, \ldots, e_k$ of $\mathbb{R}^k$, we have

$$\mathbf{grad}(f)(x) = \sum_{i=1}^{k} \frac{\partial f}{\partial x_i}(x) e_i.$$

17.1.2 Pullback of Vector Fields

Next we introduce a useful construction which allows us to transport vector fields along smooth maps.

Definition 17.4 (Pullback of vector fields) Let X and Y be smooth manifolds and let $\mathbf{v}$ be a vector field on Y. Assume that there is a diffeomorphism $f : X \to Y$. Then we define the *pullback vector field*, denoted $f^*\mathbf{v}$, by assigning to x the unique vector which corresponds to the value of $\mathbf{v}$ at $f(x)$ under the derivative of f, i.e.,

$$f^*\mathbf{v}(x) := df_x^{-1}(\mathbf{v}(f(x))).$$

Example 17.5 (Gradient vector field - pullback) Let $X \subset \mathbb{R}^N$ be a smooth k-dimensional manifold, and let $f : X \to \mathbb{R}$ be a real-valued function on X.

Recall from Example 17.3 the gradient field of f. Let $\phi\colon U \to X$ be a local parametrisation of X. We will now have a closer look at the pullback vector field $\phi^*\mathbf{grad}(f)\colon U \to \mathbb{R}^k$ on U. By definition, we have

$$\phi^*\mathbf{grad}(f)(u) = d\phi_u^{-1}(\mathbf{grad}(f)(\phi(u)).$$

To compute the right-hand side, let $(e_1, \ldots, e_k)$ denote the standard basis of $\mathbb{R}^k$. For $u \in U$, the vectors $d\phi_u(e_1), \ldots, d\phi_u(e_k)$ form a basis of $T_{\phi(u)}X$. We need to determine how $f_{\phi(u)}$ acts on each $d\phi_u(e_i)$. By the definition of the induced derivative, we have

$$df_{\phi(u)}(d\phi_u(e_i)) = d(f \circ \phi)_u(d\phi_u^{-1} \circ d\phi_u(e_i)) = d(f \circ \phi)_u(e_i),$$

where $d(f \circ \phi)_u$ is the derivative of the smooth map $U \xrightarrow{\phi} X \xrightarrow{f} \mathbb{R}$. Hence, with respect to the basis $(d\phi_u(e_1), \ldots, d\phi_u(e_k))$, we can represent $df_{\phi(u)}$ by the Jacobian $(1 \times k)$-matrix

$$df_{\phi(u)} = \left(\frac{\partial(f \circ \phi)}{\partial u_1}(u), \ldots, \frac{\partial(f \circ \phi)}{\partial u_k}(u)\right).$$

The vector $\mathbf{grad}(f)(\phi(u)) \in T_{\phi(u)}X$ such that

$$df_{\phi(u)}(w) = (\mathbf{grad}(f)(\phi(u)))^\top \cdot w,$$

for all $w \in T_{\phi(u)}X$, is then given by

$$\mathbf{grad}(f)(\phi(u)) = \sum_{i=1}^{k} \frac{\partial(f \circ \phi)}{\partial u_i}(u) d\phi_u(e_i) \in T_{\phi(u)}X.$$

For a basis vector $d\phi_u(e_j) \in T_{\phi(u)}X$, we then get

$$\begin{aligned} df_{\phi(u)}(d\phi_u(e_j)) &= (\mathbf{grad}(f)(\phi(u)))^\top \cdot d\phi_u(e_j) \\ &= \left(\sum_{i=1}^{k} \frac{\partial(f \circ \phi)}{\partial u_i}(u) d\phi_u(e_i)\right)^\top \cdot d\phi_u(e_j) \in \mathbb{R}. \qquad (17.1) \end{aligned}$$

In order to obtain $\phi^*\mathbf{grad}(f)$ it remains to apply $d\phi_u^{-1}$ and to express the effect of $\phi^*\mathbf{grad}(f)(u)$ in terms of the standard basis $e_1, \ldots, e_k$ of $\mathbb{R}^k$. For the latter, we observe that we have

$$\left(\phi^*\mathbf{grad}(f)(u)\right)^\top \cdot e_j = (\mathbf{grad}(f)(\phi(u)))^\top \cdot d\phi_u(e_j) \in \mathbb{R}.$$

Moreover, (17.1) tells us which real number that is. Thus, in terms of the standard basis of $\mathbb{R}^k$, we can express $\phi^*\mathbf{grad}(f)(u)$ as follows. We define smooth functions $g_{ij}\colon U \to \mathbb{R}$ by the formula

$$g_{ij}(u) := (d\phi_u(e_i))^\top \cdot d\phi_u(e_j) \in \mathbb{R}.$$

Then we get, for all $u \in U$,

$$(\phi^*\mathbf{grad}(f))(u) = \sum_{j=1}^{k} \left(\sum_{i=1}^{k} \frac{\partial (f \circ \phi)}{\partial u_i}(u) g_{ij}(u) \right) \cdot e_j \in \mathbb{R}^k. \tag{17.2}$$

Finally, recall from Example 17.1 that every vector field $\mathbf{v}$ on $U \subset \mathbb{R}^k$ is of the form $\mathbf{v} = \sum_j v_j e_j$ for smooth functions $v_j \colon U \to \mathbb{R}$. Then we can write $\phi^*\mathbf{grad}(f)$ as $\phi^*\mathbf{grad}(f) = \sum_j v_j e_j$ for the smooth functions $v_j = \sum_{i=1}^{k} \frac{\partial (f \circ \phi)}{\partial u_i} g_{ij}$.

17.1.3 Index at a Zero

Let $X \subset \mathbb{R}^N$ be a smooth manifold, and let $\mathbf{v} \colon X \to \mathbb{R}^N$ be a vector field. Since $\mathbf{v}(x)$ varies smoothly with x, $\mathbf{v}$ behaves mostly quite regularly, except at the points where $\mathbf{v}(x) = 0$. For, in a small neighbourhood of such a point, $\mathbf{v}$ can change directions radically. Hence the main points we need to investigate are the zeros of the vector field. We will now use the degree to characterise the behaviour of vector fields at their zeros. We begin with Euclidean space, i.e., we assume first $X = \mathbb{R}^n$.

Definition 17.6 (Index at the origin) Let $\mathbf{v}$ be a vector field on $\mathbb{R}^n$. Assume that $\mathbf{v}$ has an *isolated zero* at the origin, i.e., there is a small radius $\varepsilon > 0$ such that $\mathbf{v}$ has no other zeros in the closed ball $\overline{\mathbb{B}^n_\varepsilon}$ around the origin with radius ε. Then we can define the map

$$\bar{\mathbf{v}} \colon \mathbb{S}^{n-1}_\varepsilon \to \mathbb{S}^{n-1},\ y \mapsto \frac{\mathbf{v}(y)}{|\mathbf{v}(y)|},$$

where $\mathbb{S}^{n-1}_\varepsilon$ denotes the sphere of radius ε which bounds $\overline{\mathbb{B}^n_\varepsilon}$. We equip both spheres with the standard orientation, i.e., they are oriented as the boundary of $\overline{\mathbb{B}^n_\varepsilon}$ and $\overline{\mathbb{B}^n_1}$, respectively. We define the *index* of $\mathbf{v}$ at 0, denoted $\mathrm{ind}_0(\mathbf{v})$, to be the degree of $\bar{\mathbf{v}}$, i.e.,

$$\mathrm{ind}_0(\mathbf{v}) := \deg(\bar{\mathbf{v}}).$$

We can think of the map $y \mapsto \mathbf{v}(y)/|\mathbf{v}(y)|$ as a measure of the variation of the direction of $\mathbf{v}$ around the origin. The choice of the radius does not matter for the definition of ind_0, since if we choose another radius, say $\varepsilon' < \varepsilon$, then $\bar{\mathbf{v}}$ can be extended to a compact manifold $W = \overline{\mathbb{B}^n_\varepsilon} \setminus \mathbb{B}^n_{\varepsilon'}$ where $\overline{\mathbb{B}}$ denotes the closed ball and $\mathbb{B}$ the open ball. The boundary of W is ∂W is the union of $\mathbb{S}^{n-1}_\varepsilon$ and $\mathbb{S}^{n-1}_{\varepsilon'}$. However, the two spheres inherit opposite orientations as boundaries. Hence the Boundary Theorem 15.2 implies that the two degrees that arise from using $\mathbb{S}^{n-1}_\varepsilon$ and $\mathbb{S}^{n-1}_{\varepsilon'}$, respectively, agree.

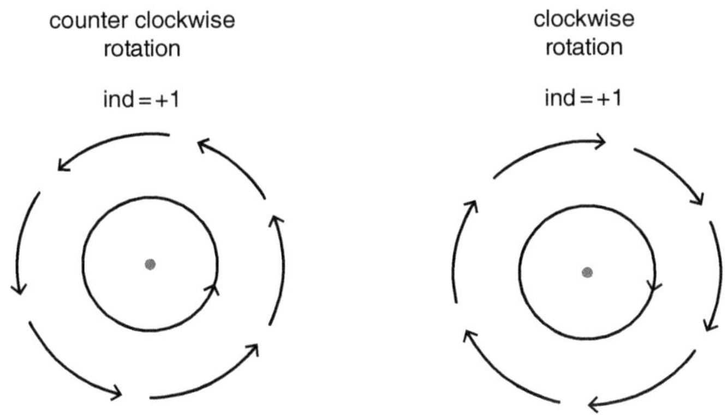

Figure 17.1 Vector fields arising from the maps $z \mapsto z^1$ and $z \mapsto \bar{z}$ with indices 1 and -1, respectively.

Example 17.7 On $\mathbb{R}^2$, we can easily construct vector fields with a zero of arbitrary index as follows. In the plane of complex numbers, the polynomial $z \mapsto z^k$ defines a smooth vector field with a zero of index k at the origin, and the function $z \mapsto \bar{z}^k$ defines a vector field with a zero of index $-k$. See Figure 17.1. For more examples of vector fields and for some curves to which the fields are tangent, see Figures 17.2 and 17.3.

Now let X be a smooth manifold and let $\mathbf{v}$ be a vector field on X. We would like to transfer the definition of the index of $\mathbf{v}$ at a point $x \in X$ from Euclidean space to X. The idea of how to do this is familiar: we use a local parametrisation to transfer information between X and Euclidean space.

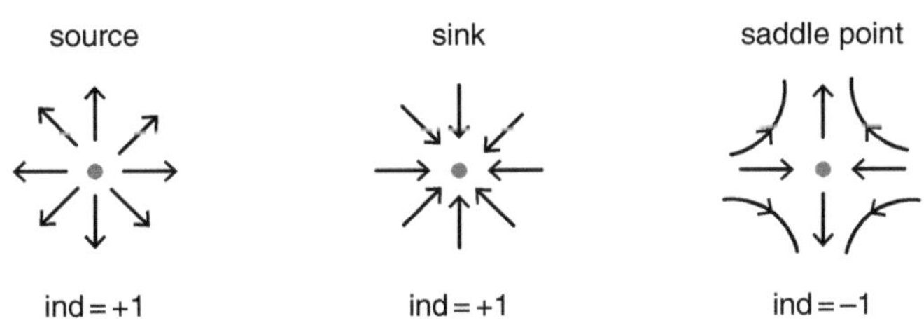

Figure 17.2 Source is the vector field on $\mathbb{R}^2$ given by $(x, y) \mapsto (x, y)$. The induced map on $\mathbb{S}^1$ is the identity. Hence the index is $+1$. The sink is the vector field on $\mathbb{R}^2$ given by $(x, y) \mapsto (-x, -y)$. The induced map on $\mathbb{S}^1$ is the antipodal map, which is homotopic to the identity in this case. Hence the index of the sink is also $+1$. The saddle point field is the vector field on $\mathbb{R}^2$ given by $(x, y) \mapsto (y, x)$. The induced map on $\mathbb{S}^1$ reverses the orientation. Hence the index of the saddle point at the origin is -1.

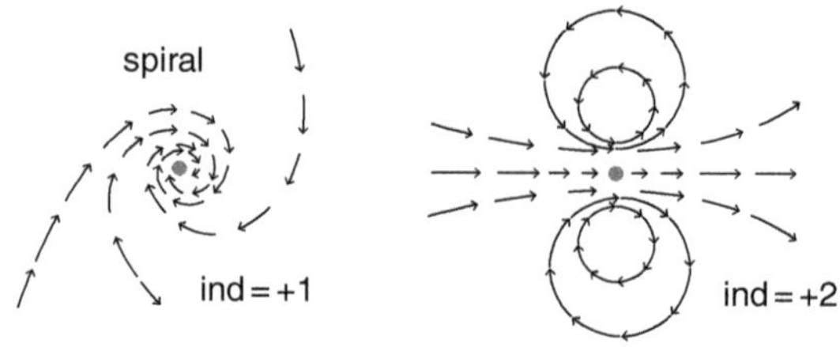

Figure 17.3 Two types of spiral movements. The left-hand spiral is just a scaled circular movement. Since the length of the vectors does not matter, the index is $+1$. The right-hand vector field on $\mathbb{R}^2$ is given by $(x,y) \mapsto (x^2 - y^2, 2xy)$. Considering the first coordinate as the real and the second as the imaginary part of points $z = (x,y)$ in the complex plane, this map is just $z \mapsto z^2$. Hence the index is $+2$.

Definition 17.8 (Index of a zero) Let X be a smooth manifold, and let $\mathbf{v}$ be a vector field on X with an isolated zero at $z \in X$. Let $\phi \colon U \to X$ be a local parametrisation with $\phi(0) = z$. We define the index of $\mathbf{v}$ at z to be

$$\mathrm{ind}_z(\mathbf{v}) := \mathrm{ind}_0(\phi^*\mathbf{v}).$$

We need to show that this definition does not depend on the choice of the local parametrisation ϕ. We do this in two steps via the following two lemmas.

Lemma 17.9 *Every orientation-preserving diffeomorphism $f \colon \mathbb{R}^n \to \mathbb{R}^n$ is smoothly isotopic to the identity.*

Proof: We may assume that $f(0) = 0$. Since the evaluation at a point x in $T_x\mathbb{R}^n = \mathbb{R}^n$ of the derivative of f at 0 can be defined by $df_0(x) = \lim_{t\to 0} f(tx)/t$, we define an isotopy $F \colon \mathbb{R}^n \times [0,1] \to \mathbb{R}^n$ by the formula

$$\begin{cases} F(x,t) := f(tx)/t & \text{for } 0 < t \leq 1 \\ F(x,0) := df_0 & \text{for } t = 0. \end{cases}$$

We need to check that F is smooth even when $t \to 0$. To do so we write $f(x) = x_1 g_1(x) + \cdots + x_n g_n(x)$ for smooth functions $g_i \colon \mathbb{R}^n \to \mathbb{R}$ defined by

$$g_i(x_1, \ldots, x_n) := \int_0^1 \frac{\partial f}{\partial x_i}(tx_1, \ldots, tx_n)dt.$$

We then have $F(x,t) = x_1 g_1(tx) + \cdots + x_n g_n(tx)$ for all $t \in [0,1]$. For all t, this map is a diffeomorphism and smooth when $t \to 0$. Thus, f is smoothly isotopic to the linear map df_0. Since f is an orientation-preserving diffeomorphism, we know $\det df_0 > 0$, i.e., $df_0 \in \mathrm{GL}_n(\mathbb{R})^+$. Since $\mathrm{GL}_n(\mathbb{R})^+$ is path-connected by Proposition 1.41, this implies that df_0 is smoothly isotopic to the identity. Since the relation of being smoothly isotopic is transitive, this proves the assertion. □

Note that the assertion of Lemma 17.9 is quite different from the situation for maps $\mathbb{S}^n \to \mathbb{S}^n$ on the sphere $\mathbb{S}^n$, where there may exist an orientation-preserving diffeomorphism which is not smoothly homotopic to the identity. That the index of a zero of a vector field on a smooth manifold is well defined will now follow from the next lemma.

Lemma 17.10 *Let U and U' be open neighbourhoods of the origin in $\mathbb{R}^n$. Let $\mathbf{v}$ be a vector field on U with an isolated zero at the origin. Assume there is a diffeomorphism $f: U' \to U$ with $f(0) = 0$. Then the index of $\mathbf{v}$ at 0 is equal to the index of $f^*\mathbf{v}$ at 0, i.e., $\operatorname{ind}_0(\mathbf{v}) = \operatorname{ind}_0(f^*\mathbf{v})$.*

Proof: Since we only need to study the zeros in a small open neighbourhood of the origin, we may assume that U and U' are open balls around the origin with sufficiently small radii ε and ε', respectively. First we assume that f preserves orientation. Then, by Lemma 17.9, we can find a smooth isotopy $f_t: U' \times [0,1] \to \mathbb{R}^n$ with $f_0 = \mathrm{id}$, $f_1 = f$, and $f_t(0) = 0$ for all t. For each t, we set $\mathbf{v}_t := f_t^*\mathbf{v}$ on $f_t^{-1}(U) \subset \mathbb{R}^n$. Since f_t is a smooth isotopy, each f_t is a diffeomorphism and the vector field $\mathbf{v}_t$ is well defined with an isolated zero at the origin for every $t \in [0,1]$. For each $t \in [0,1]$, let $\varepsilon_t > 0$ be the maximal radius such that $\mathbf{v}_t$ has no zeros other than the origin in $\mathbb{B}^n_{\varepsilon_t}$. The function $[0,1] \to \mathbb{R}^+$, $t \mapsto \varepsilon_t$ is continuous and hence takes a minimum when t varies in $[0,1]$. Let $\varepsilon > 0$ be that minimum. Then we have a well-defined smooth map $\bar{\mathbf{v}}_t: \mathbb{S}^{n-1}_\varepsilon \to \mathbb{S}^{n-1}$ defined by $y \mapsto \mathbf{v}_t(y)/|\mathbf{v}_t(y)|$. As $\mathbf{v}_0 = \mathbf{v}$ and $\mathbf{v}_1 = f^*\mathbf{v}$, we deduce $\operatorname{ind}_0(\mathbf{v}) = \deg(\bar{\mathbf{v}}_0) = \deg(\bar{\mathbf{v}}_1) = \operatorname{ind}_0(f^*\mathbf{v})$ from the invariance of the degree under homotopy. This proves the assertion for an orientation-preserving diffeomorphism.

Now we assume that f reverses orientations. Then we can write f as the composition of an orientation-preserving diffeomorphism φ and a reflection $r: \mathbb{R}^n \to \mathbb{R}^n$ which switches the sign of exactly one coordinate, i.e., r sends $(x_1, \ldots, x_i, \ldots, x_n)$ to $(x_1, \ldots, -x_i, \ldots, x_n)$. By the first case and since $(\varphi \circ r)^*\mathbf{v} = r^*(\varphi^*\mathbf{v})$, we may therefore assume that $f = r$. We write $\mathbf{v}' = f^*\mathbf{v}$. Then the associated map $\bar{\mathbf{v}}'$ on the ε-sphere $\mathbb{S}^{n-1}_\varepsilon \to \mathbb{S}^{n-1}$ with $\bar{\mathbf{v}}'(y) = \mathbf{v}'(y)/|\mathbf{v}'(y)|$ satisfies $\bar{\mathbf{v}}' = r^{-1} \circ \bar{\mathbf{v}} \circ r$ since r is a linear isomorphism and therefore $dr_0 = r$. Using the definition of the degree, this shows $\deg(\bar{\mathbf{v}}) = \deg(\bar{\mathbf{v}}')$ and completes the proof of the lemma. □

Proposition 17.11 (Index of a zero is well defined) *Let X be a smooth manifold, and let $\mathbf{v}$ be a vector field on X with an isolated zero at $z \in X$. Then the index of $\mathbf{v}$ at z does not depend on the choice of a local parametrisation at z.*

Proof: Let $\phi\colon U \to X$ and $\psi\colon U' \to X$ be local parametrisations with $\phi(0) = z = \psi(0)$. By choosing U and U' small enough, the composition $f := \psi^{-1} \circ \phi$ defines a diffeomorphism $f\colon U' \to U$. We check using the definition of the pullback that we have $\phi^*\mathbf{v} = (\psi^{-1} \circ \phi)^*(\psi^*\mathbf{v})$. By Lemma 17.10, we then get

$$\mathrm{ind}_0(\phi^*\mathbf{v}) = \mathrm{ind}_0\left((\psi^{-1} \circ \phi)^*(\psi^*\mathbf{v})\right) = \mathrm{ind}_0(\psi^*\mathbf{v}). \qquad \square$$

Example 17.12 (Index sum on spheres) Recall the vector field $\mathbf{v}$ on the n-dimensional sphere $\mathbb{S}^n$ from Example 17.2 given by $\mathbf{v}(x) = p_N - (p_N \cdot x)x$ where $p_N \in \mathbb{S}^n$ denotes the north pole and $\cdot$ denotes the inner product defined by considering p_N and x as vectors in $\mathbb{R}^{n+1}$. We compute the effect of $\mathbf{v}$ on $x = (x_1, \ldots, x_{n+1}) \in \mathbb{S}^n$ as

$$\mathbf{v}(x) = (-x_1x_{n+1}, -x_2x_{n+1}, \ldots, -x_nx_{n+1}, 1 - x_{n+1}^2). \tag{17.3}$$

This shows that $\mathbf{v}$ has exactly two zeros: the north pole $p_N = (0, \ldots, 0, 1)$ and the south pole $p_S = (0, \ldots, 0, -1)$. Now we compute the index of $\mathbf{v}$ at both zeros. By definition, we obtain the index at a zero by choosing a local parametrisation and pulling back the vector field to $\mathbb{R}^n$. For this example, it is convenient to use the stereographic projections from the poles of $\mathbb{S}^n$. First, we look at the zero at p_S. Let $\mathbb{R}^n \to \mathbb{S}^n \setminus p_N$ be the local parametrisation which is the inverse of the stereographic projection from the north pole. We then let ϕ_N be the restriction to the open ball $U_S := \mathbb{B}_1^n \subset \mathbb{R}^n$ of radius one around the origin. Then ϕ_N is a diffeomorphism onto the lower hemisphere V_S on $\mathbb{S}^n$, which is an open neighbourhood of p_S which does not contain any other zeros of $\mathbf{v}$. Thus the boundary of the closure $\overline{U}_S$ of U_S is just the unit sphere $\mathbb{S}^{n-1} \in \mathbb{R}^n$. Hence ϕ_N sends a point $u \in \partial\overline{U}_S$ to $\phi_N(u) = (u_1, \ldots, u_n, 0) \in \mathbb{S}^n \subset \mathbb{R}^{n+1}$. Thus, by (17.3), we get

$$\mathbf{v}(\phi_N(u)) = (0, \ldots, 0, 1) \text{ for all } u \in \partial\overline{U}_S.$$

This does not mean that $\phi_N^*\mathbf{v}$ is constant, since we still have to apply the derivative $(d(\phi_N)_u)^{-1}$. We computed this derivative in the suggested solution for Exercise 2.9. The formula there yields

$$\phi_N^*\mathbf{v}(u) = \begin{pmatrix} 1 & 0 & \ldots & 0 & u_1 \\ 0 & 1 & \ldots & 0 & u_2 \\ \vdots & \vdots & \ldots & \vdots & \vdots \\ 0 & \ldots & 0 & 1 & u_n \end{pmatrix} \cdot \begin{pmatrix} 0 \\ \vdots \\ 0 \\ 1 \end{pmatrix} = \begin{pmatrix} u_1 \\ u_2 \\ \vdots \\ u_n \end{pmatrix} \quad \text{for all } u \in \partial\overline{U}_S.$$

Thus, on the boundary of the closure of the neighbourhood around the zero p_S, $\phi_N^*\mathbf{v}$ acts as the identity. This shows that the index of $\mathbf{v}$ at p_S is $+1$:

$$\mathrm{ind}_S(\mathbf{v}) = \mathrm{ind}_0(\phi_N^*\mathbf{v}) = \deg(\mathrm{id}_{\mathbb{S}^{n-1}}) = +1.$$

To compute the index at the north pole p_N, we use the stereographic projection from the south pole and the upper hemisphere V_N as an open neighbourhood of p_N. We then get again $\mathbf{v}(\phi_S(u)) = (0, \ldots, 0, 1)$ for all $u \in \partial\overline{U}_N$. However, the vector field $\phi_S^*\mathbf{v}$ now acts as

$$\phi_S^*\mathbf{v}(u) = \begin{pmatrix} 1 & 0 & \ldots & 0 & -u_1 \\ 0 & 1 & \ldots & 0 & -u_2 \\ \vdots & \vdots & \ldots & \vdots & \vdots \\ 0 & \ldots & 0 & 1 & -u_n \end{pmatrix} \cdot \begin{pmatrix} 0 \\ \vdots \\ 0 \\ 1 \end{pmatrix} = \begin{pmatrix} -u_1 \\ \vdots \\ -u_n \end{pmatrix} \quad \text{for all } u \in \partial\overline{U}_N.$$

This shows that $\overline{\phi_S^*\mathbf{v}}$ acts as the antipodal map $a_{\mathbb{S}^{n-1}}$ on $\mathbb{S}^{n-1}$. Hence the index of $\mathbf{v}$ at N equals the degree of the antipodal map on $\mathbb{S}^{n-1}$, which is $(-1)^n$ by Exercise 15.1:

$$\mathrm{ind}_N(\mathbf{v}) = \mathrm{ind}_0(\phi_S^*\mathbf{v}) = \deg(a_{\mathbb{S}^{n-1}}) = (-1)^n.$$

We observe that, in this example, the sum of the indices of the zeros of the vector field $\mathbf{v}$ is 0 if n is odd and is 2 if n is even. We will see in the next section that this sum is actually independent of the choice of vector field on $\mathbb{S}^n$.

17.2 The Euler Characteristic: Algebraic Topology in a Nutshell

In order to state the main result of this chapter, we take a brief detour to algebraic topology and introduce the Euler characteristic. There are several ways to define the Euler characteristic of a smooth manifold X. However, the definitions that are accessible with the techniques developed in this book, for example, as the self-intersection number of the diagonal in $X \times X$, do not convey the property that we are interested in here. Namely, the Euler characteristic is a purely topological invariant, i.e., it only depends on the topology of X and not the structure as a smooth manifold. Other definitions, for example via triangulation, are more intuitive and may be more elementary. To make them precise and independent of choices, however, requires a lot of work. Here we will define the Euler characteristic via singular homology, a key tool in algebraic topology. We will briefly introduce the main idea and the construction of singular homology groups. We refer the reader to any general introduction to algebraic topology for further details, explanation, and motivation. See for example [15].

The underlying idea may roughly be summarised as follows. In order to understand a complicated space X, we study all the ways we can map certain model spaces with precisely understood properties into X. There are different choices for these model spaces, for example, one can use spheres and define homotopy groups. For singular homology, one uses the following spaces, which have a very nice combinatorial behaviour. For $n \geq 0$, the *standard n-simplex* Δ^n is the set $\Delta^n \subset \mathbb{R}^{n+1}$ defined by

$$\Delta^n = \left\{ (t_0, \ldots, t_n) \in \mathbb{R}^{n+1} : \sum_{i=0}^{n} t_i = 1, t_i \geq 0 \text{ for all } i \right\}.$$

The standard simplices for varying n are related by face maps for $0 \leq i \leq n$ which can be described as $\phi_i^n(t_0, \ldots, t_{n-1}) = (t_0, \ldots, t_{i-1}, 0, t_i, \ldots, t_{n-1})$ with the 0 inserted at the ith coordinate, where t_0 is the 0th coordinate. Let X be a topological space. A *singular n-simplex* in X is a continuous map $\sigma : \Delta^n \to X$. We denote by $\mathrm{Sing}_n(X)$ the set of all singular n-simplices in X. For example, $\mathrm{Sing}_0(X)$ is just the set of points of X. For $0 \leq i \leq n$, we can use the face maps ϕ_i^n to define maps

$$d_i^n : \mathrm{Sing}_n(X) \to \mathrm{Sing}_{n-1}(X),\ \sigma \mapsto \sigma \circ \phi_i^n,$$

by sending a singular n-simplex σ to the $n-1$-simplex defined by precomposition with the ith face inclusion. The image $d_i^n(\sigma) = \sigma \circ \phi_i^n$ is called the *ith face of* σ. See Figure 17.4.

To make this construction accessible to algebraic tools, we define the group $S_n(X)$ of *singular n-chains* in X as the free abelian group generated by singular n-simplices, i.e., $S_n(X) := \mathbb{Z}\mathrm{Sing}_n(X)$. Thus a singular n-chain is a finite $\mathbb{Z}$-linear combination of singular n-simplices. We define the *boundary operator* as the homomorphism of abelian groups given by

$$\partial_n : S_n(X) \to S_{n-1}(X),\ \partial_n(\sigma) := \sum_{i=0}^{n} (-1)^i d_i^n \sigma.$$

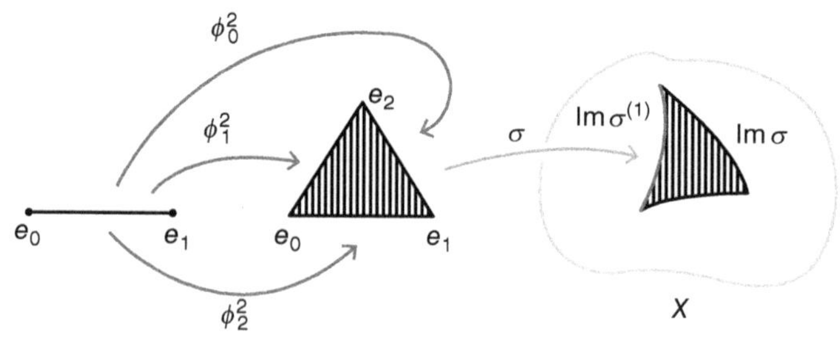

Figure 17.4 The face maps from the one-dimensional simplex into the two-dimensional simplex. There are three maps corresponding to the three edges of Δ^2. Then the singular two-simplex σ maps Δ^2 to X.

A key property of the operators ∂_n is that their composition vanishes, i.e., $\partial_{n+1} \circ \partial_n = 0$, which we can check via a simple computation. This implies that the sequence of pairs $\{S_n(X), \partial_n\}$ forms a *chain complex*:

$$\cdots \xrightarrow{\partial_{n+1}} S_n(X) \xrightarrow{\partial_n} S_{n-1}(X) \xrightarrow{\partial_{n-1}} \cdots \xrightarrow{\partial_2} S_1(X) \xrightarrow{\partial_1} S_0(X) \xrightarrow{\partial_0} 0.$$

This allows us to make the following definition.

Definition 17.13 (Singular homology) The nth *singular homology group* of X with coefficients in $\mathbb{Z}$ is defined to be the quotient group

$$H_n(X;\mathbb{Z}) = \frac{\operatorname{Ker}(\partial_n \colon S_n(X) \to S_{n-1}(X))}{\operatorname{Im}(\partial_{n+1} \colon S_{n+1}(X) \to S_n(X))}.$$

The construction of singular homology is functorial, i.e., a continuous map $f \colon Y \to Y$ induces a homomorphism of abelian groups $f_* \colon H_n(X;\mathbb{Z}) \to H_n(Y;\mathbb{Z})$. Moreover, the assignment $f \mapsto f_*$ is invariant under homotopy, i.e., $f \simeq g$ implies $f_* = g_*$. For each n, the group $H_n(X;\mathbb{Z})$ therefore only depends on the homotopy type of X, i.e., $f \colon Y \to Y$ is a homotopy equivalence, then $f_* \colon H_n(X;\mathbb{Z}) \to H_n(Y;\mathbb{Z})$ is an isomorphism of groups. Intuitively, singular homology measures the following property of X. A one-cycle γ, i.e., an element in the kernel of ∂_1, is a closed loop on X. The cycle γ vanishes in homology if it is a boundary, i.e., if there is an element β in $S_2(X)$ such that $\partial_2(\beta) = \gamma$. In other words, γ vanishes if we can continuously shrink γ to a point inside X. More generally, an n-cycle α, i.e., an element in the kernel of ∂_n, vanishes in homology if it is a boundary, i.e., if there is an element β in $S_{n+1}(X)$ such that $\partial_{n+1}(\beta) = \alpha$. That is, α vanishes if it can be filled by a chain of one dimension higher. Hence we may therefore think of a non-zero element in $H_n(X;\mathbb{Z})$ as a tool to detect an n-dimensional hole in X. In particular, the group $H_0(X;\mathbb{Z})$ is determined by the number m of connected components of X, i.e., $H_0(X;\mathbb{Z})$ is a sum of m copies of $\mathbb{Z}$. The $H_1(X;\mathbb{Z})$ is the abelianisation of the fundamental group of X.

Examples 17.14 The homology groups of the n-dimensional sphere are given by $H_0(\mathbb{S}^n;\mathbb{Z}) = \mathbb{Z}$, $H_n(\mathbb{S}^n;\mathbb{Z}) = \mathbb{Z}$, and $H_i(\mathbb{S}^n;\mathbb{Z}) = 0$ for all $i \neq 0, n$. In particular, for $n = 1$, we have $\pi_1(\mathbb{S}^1) \cong H_1(\mathbb{S}^1;\mathbb{Z})$. For the two-dimensional torus $\mathbb{T}^2$, we have $H_0(\mathbb{T}^2;\mathbb{Z}) = \mathbb{Z}$, $H_2(\mathbb{T}^2;\mathbb{Z}) = \mathbb{Z}$, $H_1(\mathbb{T}^2;\mathbb{Z}) = \mathbb{Z} \times \mathbb{Z}$, and $H_i(\mathbb{T}^2;\mathbb{Z}) = 0$ for all $i \neq 0, 1, 2$. Moreover, the maps A and B from $\mathbb{T}^2$ to itself in Example 15.18 have the effect on $H_1(\mathbb{T}^2;\mathbb{Z}) = \mathbb{Z} \times \mathbb{Z}$ that A_* sends (x, y) to (nx, y) and B_* sends (x, y) to (x, ny). Thus, for $n \neq 1$, A_* and B_* are different maps, and hence A and B cannot be homotopic. See also Remark 16.28.

The Euler characteristic may now be viewed as a very rough summary of the information contained in the homology groups. First we reduce the information of homology groups to their rank, i.e., we let b_i denote the rank of $H_i(X;\mathbb{Z})$ as an abelian group. The number b_i is called the ith *Betti number of* X. If X is a k-dimensional manifold, then we have $b_i = 0$ for all $i \geq k+1$. Hence we may take the alternating sum of the Betti numbers.

Definition 17.15 (Euler characteristic) Let X be a smooth k-dimensional manifold. The *Euler characteristic* of X, denoted by $\chi(X)$, is defined as the alternating sum of the Betti numbers of X, i.e.,

$$\chi(X) := \sum_{i=1}^{k}(-1)^i b_i = \sum_{i=1}^{k}(-1)^i \cdot (\text{rank of } H_i(X;\mathbb{Z})).$$

For all i, the groups $H_i(X;\mathbb{Z})$ only depend on the topology of X. Hence the Euler characteristic of X only depends on the topology of X. In fact, $\chi(X)$ only depends on the homotopy type of X. Before we compute some examples of the Euler characteristic, we first consider a useful variation of homology groups.

Remark 17.16 (Relative homology and attaching cells) Given a subspace $Y \subset X$, there are *relative homology* groups denoted $H_n(X,Y;\mathbb{Z})$ which, roughly speaking, measure only the features of the complement of Y in X. They are defined as the homology of the induced chain complex $\{S_n(X)/S_n(Y), \bar{\partial}_n\}_n$, where we use that $S_n(Y)$ is a subgroup of $S_n(X)$ and $\bar{\partial}_n$ is the homomorphism induced by ∂_n on the quotient.

The following situation is of particular interest in Morse theory, which will play a role again later in this chapter.

Example 17.17 (Cell complexes) A certain kind of relative homology is situation is of particular interest for us. Let Y be a topological space, and suppose we are given a map $f\colon \mathbb{S}^{n-1} \to Y$. Then we can attach $\overline{\mathbb{B}^n}$ to Y via f by forming the pushout diagram

$$\begin{array}{ccc} \mathbb{S}^{n-1} & \xrightarrow{\;f\;} & Y \\ \big\downarrow & & \big\downarrow \\ \overline{\mathbb{B}^n} & \longrightarrow & Y \cup_f \overline{\mathbb{B}^n} \end{array} \quad . \tag{17.4}$$

We think of this process as obtaining X by gluing an n-dimensional cell to Y. In this case, the relative homology groups are given by

$$H_j(X,Y;\mathbb{Z}) = H_j(Y \cup_f \overline{\mathbb{B}^n}, Y;\mathbb{Z}) = \begin{cases} \mathbb{Z} & \text{if } j = n \\ 0 & \text{otherwise.} \end{cases}$$

A space X, which is built by inductively attaching cells in different dimensions, is called a *cell complex*. For example, $\mathbb{S}^n$ is a cell complex with one cell in dimension zero and one cell in dimension n. The torus $\mathbb{T}^2$ is a cell complex with one cell in dimension zero, two cells in dimension one, and one cell in dimension two.

Example 17.18 (Euler characteristic of a cell complex) Assume that the manifold X is homotopy equivalent to a cell complex Y, i.e., a space which is obtained by successively gluing i-dimensional cells together for $i = 1, \ldots, k$. Let c_i denote the number of i-dimensional cells we attach. In this case, the Euler characteristic of X can be computed as

$$\chi(X) = \sum_{i=1}^{k} (-1)^i c_i. \tag{17.5}$$

Equation (17.5) comes quite close to the familiar formula of Euler which says that the alternating sum $v - e + f$ is an invariant of a surface S, where v, e, and f denote the number of vertices, edges, and faces, respectively, of a polygon which covers S. For example, the Euler characteristic of the n-sphere is two if n is even, and it is zero if n is odd. The Euler characteristic of the torus is zero. See also Examples 17.22 and 17.26.

17.3 The Poincaré–Hopf Index Theorem

We are now ready to state a famous theorem due to Poincaré and Hopf, which we will prove in the following sections.

Theorem 17.19 (Poincaré–Hopf Index Theorem) *Let X be a compact smooth manifold and let $\mathbf{v}$ be a vector field on X with only finitely many isolated zeros. Then the sum of the indices of $\mathbf{v}$ at its zeros is equal to the Euler characteristic of X.*

Theorem 17.19 consists of two remarkable statements. First, the sum of the indices is independent of the vector field, i.e., the sum is entirely determined by the manifold X. Second, it is just the topology of X that matters since the latter determines the Euler characteristic of X. The Euler characteristic is a topological invariant while the nature of the index is analytic. Results of this nature provide a bridge between different types of phenomena and are very influential in mathematics. The theorem can be extended to manifolds with boundary by requiring that the vector field has to point outwards at every boundary point. We omit the proof of this extended result here.

Remark 17.20 We note that the Eisenbud–Levine–Khimshiashvili signature formula provides an algebraic method to compute the index at isolated zeros by assigning a quadratic form to the germ of a smooth $f: \mathbb{R}^n \to \mathbb{R}^n$ with $f(0) = 0$ and an isolated zero at the origin. See [12] and [13].

We will provide the proof of Theorem 17.19 in Sections 17.3.1 and 17.3.2. But first we look at some immediate consequences of Theorem 17.19.

Corollary 17.21 *Let X be a compact smooth manifold. If X admits a vector field without zeros, then the Euler characteristic of X is zero.*

Example 17.22 We know from Exercise 8.10 that odd-dimensional spheres admit non-vanishing vector fields. In particular, this shows $\chi(\mathbb{S}^{2k+1}) = 0$.

Corollary 17.23 *Let X be a compact smooth manifold. If $\chi(X) \neq 0$, then every vector field on X must have at least one zero.*

Example 17.24 (Vector fields on even spheres) We have seen in Example 17.12 that the index sum for some vector field on $\mathbb{S}^{2n}$ is 2. Hence the index sum for *every* vector field on $\mathbb{S}^{2n}$ must be 2. In particular, every vector field on $\mathbb{S}^{2n}$ must have at least one zero. For vector fields on $\mathbb{S}^2$, this is often referred to as the hairy ball theorem, as we may think of a tangent vector in $T_x\mathbb{S}^2$ as an arrow pointing from x.

In fact, the dimension of the manifold has a strong influence on the Euler characteristic.

Corollary 17.25 (Odd-dimensional manifolds have vanishing Euler characteristic) *Let X be a compact smooth manifold. Assume that the dimension of X is odd. Then the index sum of any vector field on X is zero, and hence the Euler characteristic of X is zero.*

Proof: Let $\mathbf{v}$ be a vector field on X, and let n denote the dimension of X. Then we can multiply each $\mathbf{v}(x)$ by -1 to get a new field which we denote by $-\mathbf{v}$. If z is a zero of $\mathbf{v}$, z is a zero of $-\mathbf{v}$ as well. The induced maps $\mathbb{S}^{n-1}_{\varepsilon} \to \mathbb{S}^{n-1}$ we use to compute the indices at z of $\mathbf{v}$ and $-\mathbf{v}$, respectively, differ by a composition with the antipodal map $a_{\mathbb{S}^{n-1}}$ on $\mathbb{S}^{n-1}$. This changes the degree by multiplication by $(-1)^n$ which is -1, since n is odd. Thus, by Theorem 17.19, the index sum of $\mathbf{v}$ and $-\mathbf{v}$ satisfy

$$\sum_{z\in\mathbf{v}^{-1}(0)} \mathrm{ind}_z(\mathbf{v}) = \sum_{z\in(-\mathbf{v})^{-1}(0)} \mathrm{ind}_z(-\mathbf{v}) = -\sum_{z\in\mathbf{v}^{-1}(0)} \mathrm{ind}_z(\mathbf{v}).$$

Hence the index sum must be zero. By Theorem 17.19, this implies that the Euler characteristic is zero as well. □

Example 17.26 (Nowhere-vanishing vector field on the torus) Let $\mathbb{T} = \mathbb{S}^1 \times \mathbb{S}^1 \subset \mathbb{R}^4$ be the two-dimensional torus. We define a vector field $\mathbf{v}$ on $\mathbb{T}$ by

$$\mathbf{v}\colon \mathbb{T} \to \mathbb{R}^4,\ x = (x_1, y_1, x_2, y_2) \mapsto (-y_1, x_1, -y_2, x_2).$$

We can verify that $\mathbf{v}(x) \in T_x\mathbb{T} = (T_{(x_1,y_1)}\mathbb{S}^1) \times (T_{(x_2,y_2)}\mathbb{S}^1) \subset \mathbb{R}^4$. Moreover, we see that $\mathbf{v}(x)$ is never zero. Hence there are nowhere-vanishing vector fields on the torus. We deduce from this observation and Theorem 17.19 that the index sum of every vector field on the torus and the Euler characteristic of the torus are both zero.

Remark 17.27 (Hopf's Theorem on non-vanishing vector fields) In fact, it is a theorem of Hopf [21] that the vanishing of the index sum is not just a necessary condition for the existence of a non-vanishing vector field but also sufficient. That is, if the index sum of some vector field on a connected compact oriented smooth manifold X is zero, then there exists a vector field on X with no zeros at all. We will discuss a proof of this result in Section 17.4.

17.3.1 Independence

We split the proof of Theorem 17.19 into two steps: the assertion on the independence of the index sum of the vector field and the identification with the Euler characteristic. We begin with independence and will prove the following theorem.

Theorem 17.28 (Poincaré–Hopf Index Theorem – Independence) *Let X be a compact smooth manifold and let $\mathbf{v}$ be a vector field on X with only finitely many isolated zeros. If X has a boundary, we require that $\mathbf{v}$ points outwards at all boundary points of X. Then the index sum does not depend on the choice of the vector field.*

We prove Theorem 17.28 in several steps. The key idea is to show that the index sum equals the degree of a particular map that only depends on X.

Definition 17.29 (Gauss map) Let $X \subset \mathbb{R}^n$ be a compact smooth n-dimensional manifold with boundary. The *Gauss map*

$$g\colon \partial X \to \mathbb{S}^{n-1}$$

is defined by sending $x \in \partial X$ to the outward-pointing unit normal vector at x.

Remark 17.30 In the following, we are going to use the following observation without mentioning it. The assumption that $X \subset \mathbb{R}^n$ is an n-dimensional manifold implies that its tangent space, which is an n-dimensional subspace of $\mathbb{R}^n$, is equal to $\mathbb{R}^n$ at every point. Hence a vector field on X is a smooth map $\mathbf{v}\colon X \to \mathbb{R}^n$.

Now we prove the first case of Theorem 17.28.

Proposition 17.31 *Let $X \subset \mathbb{R}^n$ be a compact smooth n-dimensional manifold with boundary. Let $\mathbf{v}\colon X \to \mathbb{R}^n$ be a vector field on X with finitely many isolated zeros and such that $\mathbf{v}$ points outwards at all boundary points of X. Then the index sum is equal to the degree of the Gauss map. In particular, the index sum does not depend on the choice of $\mathbf{v}$.*

Proof: Let $z_1, \ldots, z_k$ be the zeros of $\mathbf{v}$ which are not on the boundary of X. After removing an ε-ball around each zero for a sufficiently small ε, we obtain again a smooth manifold with boundary which we denote by Y. Since there are no zeros of $\mathbf{v}$ on the boundary of X by assumption, the assignment $y \mapsto \bar{\mathbf{v}}(y) = \mathbf{v}(y)/|\mathbf{v}(y)|$ defines a smooth map $Y \to \mathbb{S}^{n-1}$. Since $\bar{\mathbf{v}}$ is defined on all of Y, the Boundary Theorem 15.2 for the degree implies that the degree of the restriction of $\bar{\mathbf{v}}$ to ∂Y is zero. Since the degree is additive with respect to connected components, the degree of the restriction of $\bar{\mathbf{v}}$ to the boundary ∂Y equals the sum of the degrees of the restrictions of $\bar{\mathbf{v}}$ to the components of ∂Y. One of the components of ∂Y is ∂X. By assumption, $\mathbf{v}(x)$ points outwards at all points $x \in \partial X$ on the boundary. By smoothly normalising the outward-pointing vector $\mathbf{v}(x)$, this implies $\bar{\mathbf{v}}_{|\partial X}$ is homotopic to the Gauss map g, which sends $x \in \partial X$ to the outward unit normal vector at x. Since the degree is invariant under homotopy, this implies $\deg(\bar{\mathbf{v}}_{|\partial X}) = \deg(g)$.

The other boundary components of ∂Y are $n-1$-dimensional ε-spheres with the boundary orientation induced by the orientation of X. By the construction of Y, the boundary orientation is such that the normal vector pointing into the ε-sphere completes a positively oriented basis. That is, the orientation of each of the ε-spheres, which are the components of ∂Y, is the opposite of the standard orientation. Thus, by definition of the index of a zero of $\mathbf{v}$, the degree of the restriction of $\bar{\mathbf{v}}$ to the other boundary components of ∂Y equals $-\sum_{i=0}^{k} \mathrm{ind}_{z_i}(\mathbf{v})$. Hence, in total, we have shown

$$\deg(g) - \sum_{i=0}^{k} \mathrm{ind}_{z_i}(\mathbf{v}) = \deg(\bar{\mathbf{v}}_{|\partial Y}) = 0. \qquad \square$$

In order to extend Proposition 17.31 to manifolds $X \subset \mathbb{R}^n$ of dimension less than n, we will now study the derivative of the vector field.

Definition 17.32 (Non-degenerate zeros - Euclidean case) Let $U \subset \mathbb{R}^n$ be an open subset. Let $\mathbf{v}\colon U \to \mathbb{R}^n$ be a vector field on U, and let $z \in U$ be a zero of $\mathbf{v}$. Then the vector field $\mathbf{v}$ is said to be *non-degenerate at z* if the linear transformation $d\mathbf{v}_z\colon \mathbb{R}^n \to \mathbb{R}^n$ is an isomorphism.

Note that it follows from the Inverse Function Theorem 12.1 that, if $\mathbf{v}$ is non-degenerate at z, then z is an isolated zero.

Example 17.33 (Gradient vector field - non-degenerate zeros - Euclidean case) Let $U \subset \mathbb{R}^n$ be an open subset. Let $f\colon U \to \mathbb{R}$ be a smooth function and let $\mathbf{v} := \mathbf{grad}(f)\colon U \to \mathbb{R}^n$ denote its gradient vector field. Recall from Example 17.3 that $\mathbf{v}(u)$ is represented, with respect to the standard basis of $\mathbb{R}^n$, by the $(1 \times k)$-Jacobian matrix of f. Hence the derivative $d\mathbf{v}_u$ is represented by the Hessian matrix of f. This shows that a zero of $\mathbf{grad}(f)$ is non-degenerate if and only if z is a non-degenerate critical point of f, i.e., the Hessian matrix of f at z is invertible.

Lemma 17.34 *Let $U \subset \mathbb{R}^n$ be an open subset. Let $\mathbf{v}\colon U \to \mathbb{R}^n$ be a vector field on X and let $u \in U$ be a zero of $\mathbf{v}$. Assume that $\mathbf{v}$ is non-degenerate at u. Then we have* $\operatorname{ind}_z(\mathbf{v}) = \det d\mathbf{v}_z/|\det d\mathbf{v}_z|$. *That is, the index of $\mathbf{v}$ at u is either $+1$ or -1 according to whether the determinant of $d\mathbf{v}_z$ is positive or negative.*

Proof: We may assume $z = 0$. We can find a sufficiently small open ball $\mathbb{B}$ around the origin inside U. We may then consider $\mathbf{v}_{|\mathbb{B}}$ as a diffeomorphism into $\mathbb{R}^n$. Now we can apply Lemma 17.9 and the method used in the proof of Lemma 17.10. If $\mathbf{v}$ preserves orientation, then we can smoothly deform it into the identity without introducing any new zeros. In this case the index equals $+1$. If $\mathbf{v}$ reverses orientation, then we can smoothly deform it into a reflection without introducing any new zeros. In this case the index equals -1. □

Lemma 17.35 (Derivative of a vector field) *Let $X \subset \mathbb{R}^N$ be a smooth n-dimensional manifold, $\mathbf{w}\colon X \to \mathbb{R}^N$ a vector field on X, and let z be a zero of $\mathbf{w}$. Then the derivative $d\mathbf{w}_z\colon T_zX \to \mathbb{R}^N$ has image $T_zX \subset \mathbb{R}^N$, and we may consider $d\mathbf{w}_z$ as a linear transformation from T_zX into itself. If this linear transformation has determinant $\neq 0$, then z is an isolated zero of $\mathbf{w}$ with index equal* $\det d\mathbf{w}_z/|\det d\mathbf{w}_z|$, *i.e., with index $+1$ or -1 according to whether the determinant of $d\mathbf{w}_z$ is positive or negative.*

Proof: We choose a local parametrisation $\phi\colon U \to X$ around z. Let e^i denote the ith basis vector of $\mathbb{R}^n$. For a point $u = (u_1, \ldots, u_n) \in U$ and writing $\phi = (\phi^1, \ldots, \phi^N)$, we set

$$t^i(u) := d\phi_u(e^i) = \begin{pmatrix} \frac{\partial \phi^1}{\partial u_i}(u) \\ \vdots \\ \frac{\partial \phi^N}{\partial u_i}(u) \end{pmatrix} =: \frac{\partial \phi}{\partial u_i}(u). \tag{17.6}$$

By definition of the tangent space and the derivative, the $t^1(u), \dots, t^n(u)$ then form a basis of the tangent space $T_{\phi(u)}X$. We need to determine the image of the $t^i(u)$ under the linear transformation $d\mathbf{w}_{\phi(u)}$. We fix an i and then have

$$d\mathbf{w}_{\phi(u)}(t^i(u)) = d(\mathbf{w} \circ \phi)_u(e^i) = \frac{\partial(\mathbf{w} \circ \phi)}{\partial u_i}(u), \tag{17.7}$$

using the shortened notation of (17.6) on the right-hand side. Now we let $\mathbf{v} = \phi^*\mathbf{w}$ denote the pullback vector field on U along ϕ. Since U is an open subset of $\mathbb{R}^n$, the tangent space T_uU at any $u \in U$ is just $\mathbb{R}^n$. We may then write $\mathbf{v}$ as $\mathbf{v} = \sum_{i=1}^n v_i e^i$ with suitable smooth functions $v_1, \dots, v_n : U \to \mathbb{R}$. By definition, we have $\mathbf{v}(u) = d\phi_u^{-1} \circ \mathbf{w} \circ \phi(u)$ and $t^i(u) = d\phi_u(e^i)$ so that

$$\mathbf{w}(\phi(u)) = d\phi_u(\mathbf{v}(u)) = \sum_i v_i(u)t^i(u).$$

Taking partial derivatives with respect to u_i and using the product rule then yields

$$\frac{\partial(\mathbf{w} \circ \phi)}{\partial u_i}(u) = \sum_j \frac{\partial v_j}{\partial u_i}(u)t^j(u) + \sum_j v_j(u)\frac{\partial t^j}{\partial u_i}(u). \tag{17.8}$$

Now we apply (17.8) at the zero $\bar{u} = \phi^{-1}(z)$ of $\mathbf{v}$, i.e., $\mathbf{v}(\bar{u}) = \sum_i v_i(\bar{u})e^i = 0$. Since the $e^1, \dots, e^n$ form a basis of $\mathbb{R}^n$, we must have $v_1(\bar{u}) = \dots = v_n(\bar{u}) = 0$. Thus, at the zero $\bar{u}$, we get

$$d\mathbf{w}_z(t^i(\bar{u})) = \frac{\partial(\mathbf{w} \circ \phi)}{\partial u_i}(\bar{u}) = \sum_j \frac{\partial v_j}{\partial u_i}(\bar{u})t^j(\bar{u}). \tag{17.9}$$

This shows that, at the zero z, the image of the basis vector t^i under $d\mathbf{w}_z$ is a linear combination of the basis vectors of T_zX. Thus, the image of $d\mathbf{w}_z$ in $\mathbb{R}^N$ is contained in the subspace T_zX. This proves the first assertion of the lemma.

Equation (17.9) also shows that the determinant of the linear transformation

$$d\mathbf{w}_z : T_zX \to T_zX$$

equals the determinant of the matrix with (i, j)th entry $\frac{\partial v_j}{\partial u_i}(\bar{u})$. Thus we get $\det d\mathbf{w}_z = \det d\mathbf{v}_{\bar{u}}$. Since ϕ is a diffeomorphism between open neighbourhoods of the zero $\bar{u}$ of $\mathbf{v}$ and z of $\mathbf{w}$, the indices $\operatorname{ind}_z(\mathbf{w})$ and $\operatorname{ind}_{\bar{u}}(\mathbf{v})$ are equal. Moreover, by the Inverse Function Theorem 12.1, the zero z is isolated if $\det d\mathbf{w}_z \neq 0$, since then there is a small open neighbourhood of z such that z is the only point where $\mathbf{w}(x)$ vanishes. By Lemma 17.34, we know $\mathrm{ind}_{\bar{u}}(\mathbf{v})$ is $+1$ or -1 according to whether $\det d\mathbf{v}_{\bar{u}}$ is positive or negative. Thus, $\mathrm{ind}_z(\mathbf{w})$ is $+1$ or -1 according to whether $\det d\mathbf{w}_z$ is positive or negative. This proves the second assertion of the lemma. □

Lemma 17.35 allows us, in particular, to extend the definition of non-degenerate zeros to arbitrary vector fields.

Definition 17.36 (Non-degenerate zeros – general case) Let $X \subset \mathbb{R}^n$ be a smooth manifold and $\mathbf{v}$ be a vector field on X. Let $z \in X$ be a zero of $\mathbf{v}$. Then the zero z is said to be *non-degenerate* if the linear transformation $d\mathbf{v}_z : T_zX \to T_zX$ is an isomorphism.

Non-degenerate zeros of the gradient field of a function can be detected by studying the Hessian matrix.

Lemma 17.37 (Gradient vector field – non-degenerate zeros) *Let $X \subset \mathbb{R}^n$ be a smooth k-dimensional manifold. Let $f: X \to \mathbb{R}$ be a smooth real-valued function on X and $\mathbf{v} := \mathbf{grad}(f)$ be its gradient field defined in Example 17.3. A zero z of $\mathbf{grad}(f)$ is non-degenerate if and only if z is a non-degenerate critical point of f, i.e., the Hessian matrix of f at z, computed in any local coordinate system, is invertible.*

Proof: We already know that $z \in X$ is a zero of $\mathbf{grad}(f)$ if and only if z is a critical point of f. Let $z \in X$ be a zero of $\mathbf{grad}(f)$. Let $\phi: U \to X$ be a local parametrisation of X with $\phi(u) = z$. Then it follows from (17.2) that the derivative $d(\phi^*\mathbf{grad}(f))_u$ is represented, with respect to the standard basis of $\mathbb{R}^k$, by a matrix which is the product of the Hessian matrix of $f \circ \phi$ at u and the real $(k \times k)$-matrix $G(u)$ with (i, j)-entry $g_{ij}(u)$. By definition, z is a non-degenerate critical point of f if and only if the Hessian matrix of $f \circ \phi$ at u is invertible. Thus, to prove the assertion, it suffices to show $\det G(u) \neq 0$. Let A_u denote the matrix which represents $d\phi_u$ with respect to the bases $e_1, \dots, e_k$ of $\mathbb{R}^k$ and $d\phi_u(e_1), \dots, d\phi_u(e_k)$ of $T_{\phi(u)}X$. Then $d\phi_u(e_i)^\top$ is the ith row and $d\phi_u(e_j)$ is the jth column of A_u. Thus, by definition of $G(u)$ and the $g_{ij}(u)$, we have $G(u) = A_u^\top \cdot A_u$. Since A_u is invertible, this shows that $G(u)$ is an invertible real symmetric matrix. Hence it is positive definite and all its eigenvalues are strictly positive. In particular, we see that $\det G(u) > 0$. □

Remark 17.38 (Existence of vector fields with finitely many isolated zeros on compact manifolds) Since smooth functions with only non-degenerate critical points are generic by Theorem 5.35, it follows from Lemma 17.37 that there are vector fields on every smooth manifold with only non-degenerate zeros. In particular, every smooth manifold admits a vector field with only isolated zeros. Moreover, if X is compact, then the set of isolated zeros is closed and hence finite. Thus, on every compact manifold, there exists a vector field with only finitely many zeros, which are all isolated.

Now we let $X \subset \mathbb{R}^n$ be a compact k-dimensional smooth manifold without boundary. We recall from the ε-Neighbourhood Lemma 12.9 that we can find an $\varepsilon > 0$ small enough such that

$$X^{\varepsilon} = \left\{y \in \mathbb{R}^n : |y - x| < \varepsilon \text{ for some } x \in X\right\}$$

is an open subspace X^{ε} in $\mathbb{R}^n$. We then write

$$\overline{X^{\varepsilon}} = \left\{y \in \mathbb{R}^n : |y - x| \leq \varepsilon \text{ for some } x \in X\right\}$$

for the closure of X^{ε}. Note that if we choose $\varepsilon > 0$ small enough, then $\overline{X^{\varepsilon}}$ is a smooth n-dimensional manifold with boundary.

Theorem 17.39 (Index sum and Gauss map on ε-neighbourhood) *Let X and $\overline{X^{\varepsilon}}$ be as above. Let $\mathbf{w}$ be a vector field on X with only finitely many zeros, all of which are non-degenerate. Then the index sum of $\mathbf{w}$ is equal to the degree of the Gauss map $g\colon \partial\overline{X^{\varepsilon}} \to \mathbb{S}^{k-1}$. In particular, the index sum does not depend on the choice of vector field.*

Proof of Theorem 17.39: Given a point $y \in \overline{X^{\varepsilon}}$. We know from Exercise 12.2 that, if we choose ε small enough, we can find the unique point $\pi(y) \in X$ which is closest to y and such that the vector $y - \pi(y)$ is orthogonal to the tangent space of X at $\pi(y)$. Moreover, we can consider the assignment $y \mapsto \pi(y)$ as a smooth map $\pi : \overline{X^{\varepsilon}} \to X$. For a point y on the boundary $\partial\overline{X^{\varepsilon}}$, we have $|y - \pi(y)| = \varepsilon$ and the vector $y - \pi(y)$ points outwards from $\pi(y)$ to y. Thus, the outward unit normal vector at $y \in \partial\overline{X^{\varepsilon}}$ is $g(y) = (y - \pi(y))/\varepsilon$. This shows that the Gauss map on $\partial\overline{X^{\varepsilon}}$ is given by

$$g\colon \partial\overline{X^{\varepsilon}} \to \mathbb{S}^{n-1},\ y \mapsto (y - \pi(y))/\varepsilon.$$

Now we extend the vector field $\mathbf{v}$ to a vector field $\mathbf{w}$ on $\overline{X^{\varepsilon}}$ by setting

$$\mathbf{w}(y) := (y - \pi(y)) + \mathbf{v}(\pi(y)).$$

We will now show that $\overline{X^{\varepsilon}}$ and $\mathbf{w}$ satisfy the hypotheses of Proposition 17.31. First, since X is compact, $\overline{X^{\varepsilon}}$ is a bounded and closed subset of $\mathbb{R}^n$ and therefore a compact subspace in $\mathbb{R}^n$ by Theorem 1.16. Hence $\overline{X^{\varepsilon}}$ is a compact n-dimensional submanifold of $\mathbb{R}^n$ with boundary.

- Claim: $\mathbf{w}$ points outwards along the boundary.

Proof of the claim: Let $y \in \partial\overline{X^\varepsilon}$ be a boundary point. Since $g(y)$ points outwards and the inner product of $\mathbf{w}(y)$ and $g(y)$, considered as vectors in $\mathbb{R}^n$, is given by

$$\begin{aligned}\mathbf{w}(y) \cdot g(y) &= [(y - \pi(y)) + \mathbf{v}(\pi(y))] \cdot [(y - \pi(y))/\varepsilon] \\ &= [(y - \pi(y)) \cdot (y - \pi(y))]/\varepsilon + [\mathbf{v}(\pi(y) \cdot (y - \pi(y))]/\varepsilon \\ &= \varepsilon^2/\varepsilon + 0 = \varepsilon > 0,\end{aligned}$$

where we use $|y - \pi(y)| = \varepsilon$ and that $y - \pi(y)$ is orthogonal to the tangent vector $\mathbf{v}(\pi(y) \in T_{\pi(y)}X$. □

- Claim: $\mathbf{v}$ and $\mathbf{w}$ have the same zeros.

Proof of the claim: Since the vectors $y - \pi(y)$ and $\mathbf{v}(\pi(y))$ are orthogonal, they cannot cancel each other whenever they are non-zero. Thus, $\mathbf{w}(y)$ can only be zero if both $y - \pi(y)$ and $\mathbf{v}(\pi(y))$ are the zero vector. Since $y - \pi(y) = 0$ only if $y \in X$, we see that $\mathbf{w}$ vanishes exactly when $\mathbf{v}$ does. □

- Claim: $\mathrm{ind}_z(\mathbf{w}) = \mathrm{ind}_z(\mathbf{v})$ at every zero $z \in X$.

Proof of the claim: Let $z \in X$ be a zero of $\mathbf{w}$ and therefore of $\mathbf{v}$ by the previous claim. By Lemma 17.35, we can compute both $\mathrm{ind}_z(\mathbf{w})$ and $\mathrm{ind}_z(\mathbf{v})$ using the derivatives $d\mathbf{w}_z$ and $d\mathbf{v}_z$, respectively. Moreover, we know that $d\mathbf{v}_z$ is a linear transformation of T_zX into itself, and $d\mathbf{w}_z$ is a linear transformation of $T_z\overline{X^\varepsilon}$ into itself. Since $T_zX \subset T_z\overline{X^\varepsilon}$ and hence $T_z\overline{X^\varepsilon} = T_zX \oplus T_zX^\perp$, we can determine the effect of $d\mathbf{w}_z$ by computing its effect on vectors in T_zX and the orthogonal complement $T_zX^\perp$ separately. Since the restriction of π to X is the identity, the derivative of $y \mapsto y - \pi(y)$ in the direction of T_zX acts trivially. Thus we get $d\mathbf{w}_z(h) = d\mathbf{v}_z(h)$ for all $h \in T_zX$. In the direction orthogonal to T_zX, however, $d\mathbf{v}_z$ acts trivially, and π is constant along a fixed line orthogonal to X. Thus we get $d\mathbf{w}_z(h) = h$ for all $h \in T_zX^\perp$. We can then compute the determinant of $d\mathbf{w}_z$ as the product

$$\det d\mathbf{w}_z = \det(d\mathbf{w}_z)_{|T_zX} \cdot \det(d\mathbf{w}_z)_{|T_zX^\perp} = \det d\mathbf{v}_z \cdot 1 = \det d\mathbf{v}_z.$$

This proves the claim. □

Hence, in particular, the index sums of $\mathbf{w}$ and of $\mathbf{v}$ are equal. Moreover, since $\det d\mathbf{w}_z = \det d\mathbf{v}_z \neq 0$, all zeros of $\mathbf{w}$ are isolated by Lemma 17.35. Thus we can apply Proposition 17.31, which shows that the index sum of $\mathbf{w}$ and hence the index sum of $\mathbf{v}$ is equal to the degree of the Gauss map. □

It remains to remove the assumption that the zeros of the vector field have to be non-degenerate.

Theorem 17.40 (Index sums of fields with degenerate zeros) *Let $X \subset \mathbb{R}^n$ be a compact k-dimensional smooth manifold without boundary. Let $\mathbf{v}$ be a vector field on X with finitely many isolated zeros. Then there is a vector field $\mathbf{w}$ on X with finitely many zeros, all of which are non-degenerate, such that the index sums of $\mathbf{v}$ and $\mathbf{w}$ are equal.*

Before we prove the theorem, we note that Theorems 17.39 and 17.40 together imply Theorem 17.28.

Proof of Theorem 17.40: First we assume that $\mathbf{v}$ is a vector field defined on an open subset $U \subset \mathbb{R}^k$, and assume that z is the only zero of $\mathbf{v}$. We choose a smooth bump function $\varphi\colon U \to [0,1]$ which has value 1 on a small ball $\mathbb{B}_{\varepsilon'}$ around z and has value 0 outside a ball $\mathbb{B}_\varepsilon$ around z with slightly larger radius ε. Since $\mathbf{v}$ is a smooth map and has no other zeros than z, we can choose a regular value y of $\mathbf{v}$. We then define a new vector field $\mathbf{w}$ on U by

$$\mathbf{w}(u) := \mathbf{v}(u) - \varphi(u)y.$$

Then $\mathbf{w}(u)$ can only be zero if $\mathbf{v}(u) = \varphi(u)y$. Since z is the only zero of $\mathbf{v}$, our choice of φ implies that $\mathbf{w}$ does not have any zeros outside of $\mathbb{B}_\varepsilon$. After possibly rescaling by multiplying with a constant, we can choose y small enough such that $|\mathbf{v}(u) - \varphi(u)y| > 0$ on $\mathbb{B}_\varepsilon \setminus \mathbb{B}_{\varepsilon'}$. Moreover, within $\mathbb{B}_{\varepsilon'}$, we have $\varphi(u) = 1$, i.e., $\mathbf{w}(u) = \mathbf{v}(u) - y$ for all $u \in \mathbb{B}_{\varepsilon'}$. Thus, inside $\mathbb{B}_{\varepsilon'}$ we can only have $\mathbf{w}(u) = 0$ if $\mathbf{v}(u) = y$. Since y is a regular value of $\mathbf{v}$, this implies that $d\mathbf{w}_u = d\mathbf{v}_u\colon T_uU = \mathbb{R}^k \to \mathbb{R}^k$ is an isomorphism at every $u \in \mathbf{v}^{-1}(y)$. This proves that $\mathbf{w}$ has only non-degenerate zeros. Since the closed ball $\overline{\mathbb{B}}_{\varepsilon'}$ is compact, $\mathbf{w}$ can only have finitely many zeros on U. Since $\overline{\mathbb{B}}_{\varepsilon'}$ is compact, we can apply Proposition 17.31 to see that the index sum of $\mathbf{v}$ and the index sum of $\mathbf{w}$ on $\overline{\mathbb{B}}_{\varepsilon'}$ equal the degree of the Gauss map $g\colon \partial\overline{\mathbb{B}}_{\varepsilon'} = \mathbb{S}^{k-1}_{\varepsilon'} \to \mathbb{S}^{k-1}$ Thus, the index sums of $\mathbf{v}$ and $\mathbf{w}$ are equal.

Now we let $X \subset \mathbb{R}^n$ and $\mathbf{v}$ be as assumed in the theorem. Let z_i be one of the finitely many isolated zeros $z_1, \dots, z_m$ of $\mathbf{v}$. Then we can choose a local parametrisation $\phi_i : U_i \to X$ around z. Let $\varphi_i : U_i \to [0,1]$ be a smooth bump function and $y_i \in \mathbb{R}^k$ be a regular value of $\phi_i^*\mathbf{v}$ as in the first case above. Then we can form the new vector field $\mathbf{w}_i$ on X by setting

$$\mathbf{w}_i(x) := \mathbf{v}(x) - \varphi_i(\phi_i^{-1}(x))d\phi_{\phi_i^{-1}(x)}(y_i).$$

We can check as above that $\mathbf{w}_i$ has finitely many zeros and the zeros of $\mathbf{w}_i$ which lie in the image of ϕ_i are all non-degenerate. Moreover, the index sums of $\mathbf{v}$ and $\mathbf{w}_i$ are equal. Choosing the local parametrisations around all the z_i small enough so that they do not overlap, we can perform this process for each zero and define a vector field $\mathbf{w}$ on X by

$$\mathbf{w}(x) := \mathbf{v}(x) - \sum_{i=1}^{m} \varphi_i(\phi_i^{-1}(x)) d\phi_{\phi_i^{-1}(x)}(y_i).$$

Then the vector field $\mathbf{w}$ satisfies the properties claimed in the theorem. □

17.3.2 Index Sum of a Gradient Field and Morse Theory

Theorem 17.28 reduces the proof of Theorem 17.19 to finding at least one vector field on a given manifold X with index sum equal to $\chi(X)$. We will now explain why such a vector field exists. The idea is to find a suitable Morse function $f: X \to \mathbb{R}$ such that the index sum of its gradient field equals the Euler characteristic of X. Given a smooth real-valued function $f: X \to \mathbb{R}$. By Lemma 17.37 we know that the non-degenerate zeros of $\mathbf{grad}(f)$ are exactly the non-degenerate critical points of f. We now need to express the index of a zero of $\mathbf{grad}(f)$ in terms of the behaviour of the critical points of f.

Definition 17.41 (Functions – index of non-degenerate critical points) Let $X \subset \mathbb{R}^n$ be a smooth manifold. Let $f: X \to \mathbb{R}$ be a smooth real-valued function on X with a non-degenerate critical point $z \in X$. The *index* of f at the critical point z is defined as the number of negative eigenvalues of the Hessian matrix of f at z computed in a local coordinate system of X around z. We denote the index by $\mathrm{ind}_z(f)$.

One can show that the index is well defined, i.e., the number of negative eigenvalues of the Hessian of f at z is the same for every local coordinate system.

Remark 17.42 (Index of a critical point and bilinear forms) The index of a bilinear form B on a vector space V is defined to be the maximal dimension of a vector subspace of V on which B is negative definite. The Hessian matrix $H_z(f)$ of a smooth function $f: X \to \mathbb{R}$ at z, computed in a local coordinate system around z and with respect to a suitable basis, is a symmetric matrix and hence defines a symmetric bilinear form on T_zX by setting $w \mapsto w^\top \cdot H_z(f) \cdot w$. It then follows that the index of f at a critical point z equals the index of the symmetric bilinear form of the Hessian of f at z.

Now we can relate the index of a critical point to the index of a zero.

Proposition 17.43 (Gradient field – index of a non-degenerate zero) *Let $X \subset \mathbb{R}^n$ be a smooth k-dimensional manifold. Let $f\colon X \to \mathbb{R}$ be a smooth real-valued function on X with a non-degenerate critical point $z \in X$. Let $s = \mathrm{ind}_z(f)$ be the index of f at z. Then the index of* $\mathbf{grad}(f)$ *at the zero z is $(-1)^s$, i.e.,* $\mathrm{ind}_z(\mathbf{grad}(f)) = (-1)^s$.

Proof: Let $\phi\colon U \to X$ be a local parametrisation of X around z with $\phi(0) = z$. We may choose U small enough such that U does not contain any other critical points of f. By definition, the index of $\mathbf{grad}(f)$ at z is the index of $\phi^*\mathbf{grad}(f)$ at 0. Thus, we may assume that $X = U$ is an open subset of $\mathbb{R}^k$ and 0 is the only critical point of $f\colon U \to \mathbb{R}$. Moreover, by Morse's Lemma 5.33, we can choose ϕ such that we may assume that f is of the form

$$f(x) = f(0) - x_1^2 - \cdots - x_s^2 + x_{s+1}^2 + \cdots + x_k^2 \tag{17.10}$$

for all $x \in U$, where s is the number of negative eigenvalues of the Hessian $H(f)_0$ of f at 0. From (17.10) we deduce that

$$\mathbf{grad}(f)(x) = (-2x_1, \ldots, -2x_s, 2x_{s+1}, \ldots, 2x_k) \in \mathbb{R}^k.$$

Writing $\mathbf{v} = \mathbf{grad}(f)$, this shows that the induced map $\bar{\mathbf{v}}\colon \mathbb{S}^{k-1} \to \mathbb{S}^{k-1}$ is the composition of the reflection in each of the first s coordinates. Since the reflection of an individual coordinate has degree -1, this implies that $\bar{\mathbf{v}}$ has degree $(-1)^s$. □

Proposition 17.44 (Gradient vector field – index sum) *Let $X \subset \mathbb{R}^n$ be a compact smooth k-dimensional manifold. Let $f\colon X \to \mathbb{R}$ be a Morse function on X, i.e., a smooth real-valued function with only non-degenerate critical points. For $\lambda \in \mathbb{Z}$, let c_λ denote the number of critical points with index λ. Then the index sum of* $\mathbf{grad}(f)$ *is equal to $\sum_\lambda (-1)^\lambda c_\lambda$.*

Proof: Let z be a zero of $\mathbf{grad}(f)$ and let λ be its index as a critical point. By Proposition 17.43, z contributes the summand $(-1)^\lambda$ to the index sum of $\mathbf{grad}(f)$. Hence, if there are c_λ many such zeros, they contribute the summand $(-1)^\lambda c_\lambda$ to the index sum. Since the zeros of $\mathbf{grad}(f)$ are the critical points of f, this proves the claim. □

By Morse's Theorem 5.35, every smooth manifold admits a Morse function. Hence we can always find a Morse function, and, by Theorem 17.28 and Proposition 17.44, we know that the corresponding sum $\sum_\lambda (-1)^\lambda c_\lambda$ is independent of our choice of function. It remains to relate this sum to the Euler characteristic, i.e., it remains to show $\sum_\lambda (-1)^\lambda c_\lambda = \chi(X)$. The details of this identification require some more theory on cell complexes and singular homology than we have discussed in this book. We may, however, give a sketch of the argument and refer to [34, part I] for the details. For a smooth function $f\colon X \to \mathbb{R}$ and $a \in \mathbb{R}$, we set $X^a = f^{-1}((-\infty, a]) = \{x \in X : f(x) \leq a\}$. Recall that we refer to the equivalence class of a space under the relation of homotopy equivalence as the *homotopy type* of the space. For a proof of the following theorem, we refer to [34, theorem 3.2].

Theorem 17.45 *Let X be a compact smooth manifold. Let $f\colon X \to \mathbb{R}$ be a smooth function, and let p be a non-degenerate critical point with index λ. Setting $f(p) = c$, suppose that $f^{-1}([c-\varepsilon, c+\varepsilon])$ is compact, and contains no critical point of f other than p, for some $\varepsilon > 0$. Then, for all sufficiently small $\varepsilon > 0$, the set $X^{c+\varepsilon}$ has the homotopy type of $X^{c-\varepsilon}$ with a λ-dimensional cell attached.*

The idea of the proof is based on Morse's Lemma 5.33, which says that we can choose local coordinates on X around p such that f is locally of the form

$$f(x) = c - x_1^2 - \cdots - x_\lambda^2 + x_{\lambda+1}^2 + \cdots + x_k^2.$$

Then the preimage under the sum $-x_1^2 - \cdots - x_\lambda^2$ provides a λ-dimensional disk which is glued in between $X^{c-\varepsilon}$ and $X^{c+\varepsilon}$. We illustrate this process by the following standard example.

Example 17.46 (Height function on the torus) We consider the torus as a subspace in $\mathbb{R}^3$ and look at the preimages of intervals $(-\infty, a]$ under a height function h, i.e., the projection onto one of the three coordinates. In a rather informal discussion, we study what happens at the critical points of h. See Figure 17.5.

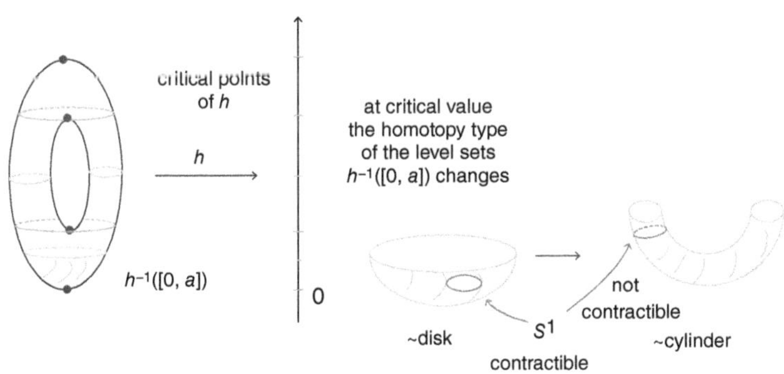

Figure 17.5 The critical points of the height function yield a decomposition of the torus.

For $a < 0$, $h^{-1}((-\infty, a])$ is empty. The first critical value is $a = 0$. For sufficiently small $a > 0$, the preimage $h^{-1}((-\infty, a])$ has changed and looks like a disk, just a bit punched in at the centre. In particular, the preimage is contractible in this range. Thus, the homotopy type near the first critical point is that of a zero-dimensional cell. When we pass the next critical value, another significant change happens as $h^{-1}((-\infty, a])$ looks like a bent cylinder. On this cylinder, there are loops which are not homotopic to a constant map. Thus, above the critical value, $h^{-1}((-\infty, a])$ is not contractible anymore, and the homotopy type of $h^{-1}((-\infty, a])$ has changed at a critical value. Another change of the homotopy type happens when we pass the next critical value. The preimage $h^{-1}((-\infty, a])$ becomes homeomorphic to a compact surface with one hole and a circle as boundary. Finally, after passing the last critical value, the preimage is the whole torus, which is a compact surface with one hole and without boundary.

In general, we can use Theorem 17.45 to find a cell complex which is homotopy equivalent to X. Let X be a compact smooth k-dimensional manifold and let $f\colon X \to \mathbb{R}$ be a Morse function. Let $a_1 < \cdots < a_n$ be real numbers such that X^{a_i} contains exactly i critical points, and $X^{a_n} = X$. We then have $\emptyset \subset X^{a_1} \subset \ldots \subset X^{a_n} = X$. Let $\bar{\mathbb{B}}^\lambda$ denote the closed unit ball in $\mathbb{R}^\lambda$. We think of $\bar{\mathbb{B}}^\lambda$ as a λ-dimensional cell. In X^{a_i} there is exactly one critical point which is not contained in $X^{a_{i-1}}$. Let λ_i denote the index of f at this critical point. We then get

$$
\begin{aligned}
H_j(X^{a_i}, X^{a_{i-1}}; \mathbb{Z}) &\cong H_j(X^{a_{i-1}} \cup \bar{\mathbb{B}}^{\lambda_i}, X^{a_{i-1}}; \mathbb{Z}) \\
&\cong H_j(\bar{\mathbb{B}}^{\lambda_i}, \mathbb{B}^{\lambda_i}; \mathbb{Z}) \cong \begin{cases} \mathbb{Z} & \text{if } j = \lambda_i \\ 0 & \text{otherwise,} \end{cases}
\end{aligned}
$$

where the second isomorphism arises from a technique in singular homology, which is called *excision*. See for example [15, chapter 2.1]. For a given λ, let c_λ denote the number of critical points of f of index λ, and let $b_\lambda(X^{a_i}, X^{a_{i-1}})$ denote the *relative Betti number*, i.e., the rank of the abelian group $H_\lambda(X^{a_i}, X^{a_{i-1}}; \mathbb{Z})$. We then get

$$
\chi(X^{a_i}, X^{a_{i-1}}) = \sum_{i=1}^{k} b_\lambda(X^{a_i}, X^{a_{i-1}}) = c_\lambda.
$$

Since the Euler characteristic is additive (see [34, part I, §5]), this shows

$$
\chi(X) = \sum_{i=1}^{k} \chi(X^{a_i}, X^{a_{i-1}}) = c_0 - c_1 + c_2 - + \cdots \pm c_k = \sum_{i=1}^{k} (-1)^\lambda c_\lambda.
$$

Together with Proposition 17.44, this shows that the index sum of the gradient field equals the Euler characteristic and finishes the proof of the Poincaré–Hopf Index Theorem 17.19. □

17.4 Existence of Vector Fields with No Zeros

Given a compact oriented smooth manifold X, it is a natural question whether we can find a nowhere-vanishing vector field $\mathbf{v}$ on X. Theorem 17.19 provides a necessary condition: if $\mathbf{v}$ has no zeros, then the index sum is zero and hence the Euler characteristic of X must be zero, too. Hopf showed in [21] that this is also a *sufficient* condition.

Theorem 17.47 (Existence of nowhere-vanishing vector fields) *Let X be a compact, connected smooth manifold. Then X possesses a nowhere-vanishing vector field if and only if the Euler characteristic of X is zero.*

Before we prove this theorem, we point out the following consequence, which follows from the fact that smooth manifolds of odd dimension have Euler characteristic zero by Corollary 17.25.

Corollary 17.48 *Let X be a compact smooth manifold without boundary. If the dimension of X is odd, then X possesses a nowhere-vanishing vector field.*

To prove Theorem 17.47, we will use the following lemma.

Lemma 17.49 *Let Y be a smooth manifold. Let $\mathbb{B}$ be an open ball in $\mathbb{R}^k$, and let $\overline{\mathbb{B}}$ be its closure. Let $f\colon \mathbb{R}^k \setminus \mathbb{B} \to Y$ be a smooth map defined outside $\mathbb{B}$, and let $\partial f\colon \partial\mathbb{B} \to Y$ be the restriction to the boundary. Assume that ∂f is homotopic to a constant map. Then f extends to a smooth map defined on all of $\mathbb{R}^k$ into Y.*

Proof: We assume that $\mathbb{B}$ is centred at 0. Every non-zero point $x \in \overline{\mathbb{B}}$ can be written uniquely as $x = ty$ for some $y \in \partial\overline{\mathbb{B}}$ and some $t \in [0, 1]$. By assumption, there is a homotopy $g_t\colon \partial\mathbb{B} \to Y$ with $g_1 = \partial f$ and g_0 being a constant map. Now we define the map $F\colon \mathbb{R}^k \to Y$ by setting

$$F(x) := \begin{cases} f(x) & \text{if } x \in \mathbb{R}^k \setminus \mathbb{B} \\ g_t(x) & \text{if } x \in \overline{\mathbb{B}} \text{ and } x = ty \text{ for some } y \in \partial\overline{\mathbb{B}} \text{ and } t \in [0, 1]. \end{cases}$$

Note that F is well defined on $\mathbb{R}^k \setminus \mathbb{B}$, since f and g_t agree on $\partial\mathbb{B} = \overline{\mathbb{B}} \cap (\mathbb{R}^k \setminus \mathbb{B})$ where we have $f = \partial f = g_1$. Note also that $F(0)$ is well defined as the constant value of g_0. Now, F is smooth except, possibly, on $\partial\overline{\mathbb{B}}$. We therefore use, if necessary, a smooth bump function to turn F into a smooth homotopy. □

Proof of Theorem 17.47: First we assume that $X = \mathbb{R}^k$ and that we have a vector field $\mathbf{v}$ on X with only finitely many zeros. Then the vector field is just a smooth map $\mathbf{v}\colon \mathbb{R}^k \to \mathbb{R}^k$. Since $\mathbf{v}$ has only finitely many zeros, there is a closed ball $\overline{\mathbb{B}} = \bar{\mathbb{B}}^k_r$ containing all the zeros of $\mathbf{v}$ in its interior. Using the technique of the proof of Theorem 17.40, we can assume that the zeros of $\mathbf{v}$ are non-degenerate. Moreover, we can assume that the origin in $\mathbb{R}^k$ is not a zero of $\mathbf{v}$. Since there are no zeros of $\mathbf{v}$ on the boundary of $\overline{\mathbb{B}}$, $\mathbf{v}$ points either inwards or outwards at all boundary points of $\overline{\mathbb{B}}$. Since the index sum of $\mathbf{v}$ is zero, we can replace $\mathbf{v}$ with $-\mathbf{v}$ if $\mathbf{v}$ points inwards. Hence we can assume that $\mathbf{v}$ points outwards at all boundary point of $\overline{\mathbb{B}}$. Since $\overline{\mathbb{B}}$ is compact, we can apply Proposition 17.31 and obtain that the index sum of $\mathbf{v}$ equals the degree of the Gauss map $g\colon \partial\overline{\mathbb{B}} = \mathbb{S}^{k-1}_r \to \mathbb{S}^{k-1}$. By assumption, the index sum of $\mathbf{v}$ is zero, and hence the degree of g is zero. Since g and $\bar{\mathbf{v}} = \mathbf{v}/|\mathbf{v}|$ are homotopic, this show that $\bar{\mathbf{v}}$ has degree zero. By Theorem 16.26, this implies that $\partial\mathbf{v}_{|\overline{\mathbb{B}}} = \mathbf{v}_{|\partial\overline{\mathbb{B}}}\colon \partial\overline{\mathbb{B}} \to \mathbb{R}^k \setminus \{0\}$ is homotopic to a constant map. Hence $\mathbf{v}_{|\mathbb{R}^k\setminus\mathbb{B}}\colon \mathbb{R}^k \setminus \mathbb{B} \to \mathbb{R}^k \setminus \{0\}$ is a map to which we can apply Lemma 17.49. This shows that $\mathbf{v}_{|\mathbb{R}^k\setminus\mathbb{B}}$ extends to a smooth map

$$\mathbf{w}\colon \mathbb{R}^k \to \mathbb{R}^k \setminus \{0\}$$

with $\mathbf{w} = \mathbf{v}$ outside the compact space $\overline{\mathbb{B}}$. Hence we have constructed a vector field $\mathbf{w}$ on $\mathbb{R}^k$ without zeros which equals $\mathbf{v}$ outside a compact subset. This proves the theorem for the case $X = \mathbb{R}^k$.

Now we let $X \subset \mathbb{R}^N$ be a compact, oriented, connected k-dimensional smooth manifold with $\chi(X) = 0$. For $k = 1$, the assertion follows from the classification of compact one-manifolds in Theorem 10.1 and from the fact that $\mathbb{S}^1$ admits a nowhere-zero vector field by Exercise 8.10. Hence we may assume that $k \geq 2$. As pointed out in Remark 17.38, there exists a vector field $\mathbf{v}$ on X with only finitely many non-degenerate zeros. Since $\chi(X) = 0$, Theorem 17.19 implies the index sum of $\mathbf{v}$ is zero. Let $\phi\colon U \to V \subset X$ be a local parametrisation of X. By the Isotopy Theorem 11.18, we can find a diffeomorphism of X to itself such that $f^*\mathbf{v}$ has finitely many non-degenerate zeros, all of which lie in V. Hence we may assume that the zeros of $\mathbf{v}$ all lie in an open subset which is diffeomorphic to $\mathbb{R}^k$. Thus, we can apply the case $X = \mathbb{R}^k$ we considered above to replace $\mathbf{v}$ with a vector field that has no zeros. This proves the theorem. □

Appendix

Solutions to Selected Exercises

A.1 Recollection on Topological Spaces

Solution (to Exercise 1.1) By definition of the subspace topology, an open cover of $f(A)$ consists of a collection $\{V_\alpha\}_\alpha$ of open sets $V_\alpha \subset Y$ such that $f(A) = \cup_\alpha (f(A) \cap V_\alpha)$. Since f is continuous, the subsets $f^{-1}(V_\alpha)$ are open in X. Hence each $A \cap f^{-1}(V_\alpha)$ is open in A. Since $f(A) = \cup_\alpha (f(A) \cap V_\alpha)$, we have $A = f^{-1}(\cup_\alpha (f(A) \cap V_\alpha)) = \cup_\alpha (A \cap f^{-1}(V_\alpha))$. Since A is compact, we can choose finitely many $\alpha_1, \ldots, \alpha_n$ such that $A = (A \cap f^{-1}(V_{\alpha_1})) \cup \cdots \cup (A \cap f^{-1}(V_{\alpha_n}))$. This implies $f(A) = (f(A) \cap V_{\alpha_1}) \cup \cdots \cup (f(A) \cap V_{\alpha_n})$. This shows that $f(A)$ is compact.

Solution (to Exercise 1.2) Assume $f \colon X \to Y$ is closed. Let $W \subset X$ be an open subset with $f^{-1}(y) \subset W$. Then the complement $X \setminus W$ is closed in X. Since f is closed, the image $f(X \setminus W)$ is closed in Y. Hence the complement $U := Y \setminus f(X \setminus W)$ is open in Y. As $f^{-1}(y) \subset W$, we know $y \notin f(X \setminus W)$ and hence $y \in U$ and $f^{-1}(y) \subseteq f^{-1}(U) \subseteq W$.

Solution (to Exercise 1.3) Assume S is not finite. By Theorem 1.16, compact subsets of $\mathbb{R}^n$ are bounded. Hence there is an $r > 0$ such that S is contained in the n-dimensional box with edges of length r centred at the origin. Divide this box into 2^n many n-dimensional boxes of equal size. The length of their edges is $r/2$. If S is infinite, there must be at least one small box which still contains infinitely many points of S. We take this box and divide it again into 2^n many n-dimensional boxes of equal size. The length of their edges is now $r/4$. Again, if S is infinite there must be at least one of the smaller boxes which still contains infinitely many points of S. After k steps of this process, we get an n-dimensional box with edges of length $r/2^k$ which contains infinitely many points of S. By repeating the process, we see that we can find an infinite sequence of points in S which forms a Cauchy sequence and therefore converges. We denote the limit by s. But then it is not possible to find a small open ball $\mathbb{B}^n_\varepsilon(s)$ around s such that $\mathbb{B}^n_\varepsilon(s)$ and $S \setminus \{s\}$ are disjoint. Hence S would not be discrete.

Solution (to Exercise 1.4) Assume there is no such λ. Then, for every n, we can choose an interval $J_n \subset [0, 1]$ of length $1/n$ such that J_n is not contained in any of the intervals I_α. This defines an infinite sequence $(J_n)_n$ of intervals in $[0, 1]$. The sequence of midpoints of the intervals J_n forms an infinite sequence in $[0, 1]$. Then there is a subsequence of this sequence which forms a Cauchy sequence and converges to a point x in $[0, 1]$. Since the I_α cover X, there is an α_0 such that $x \in I_{\alpha_0}$. Since I_{α_0} is open, we can find an N large enough such that $J_n \subset I_{\alpha_0}$ for all $n \geq N$. This contradicts our choice of the sequence $(J_n)_n$. Thus, the Lebesgue number must exist. We then deduce that $[0, 1]$ is compact, since we can then cover $[0, 1]$ by finitely many intervals of length λ and therefore by finitely many of the U_α.

Solution (to Exercise 1.5) For every $x \in X$, there is an $\alpha(x)$ such that $x \in U_{\alpha(x)}$. Since $U_{\alpha(x)}$ is open, there is a small $\delta(x) > 0$ such that the open ball $\mathbb{B}^n_{\delta(x)}(x)$ is contained in $U_{\alpha(x)}$. The collection of open balls $\mathbb{B}^n_{\delta(x)}(x)$ for all $x \in X$ defines a new open cover of X. Since X is compact, there are finitely many points $x_1, \ldots, x_n$ such that $\{\mathbb{B}^n_{\delta(x_i)}(x_i) : i = 1, \ldots, n\}$ is a cover of X as well. Now we set $\lambda := \min_{1 \leq i \leq n} \delta(x_i)/2$. We will now show that λ is a Lebesgue number. So, let C be a subset of X with diameter less than λ. Pick a point $y \in C$. Then $y \in \mathbb{B}^n_{\delta(x_i)/2}(x_i)$ for some i. Then, for every $x \in C$, the triangle inequality implies $|x - x_i| \leq |x - y| + |y - x_i| < \lambda + \delta(x_i)/2$. By definition of λ, we have $\lambda \leq \delta(x_i)/2$, and hence $|x - x_i| < \lambda + \delta(x_i)/2 \leq \delta(x_i)$. This shows $C \subset \mathbb{B}^n_{\delta(x_i)}(x_i) \subset U_{\alpha(x_i)}$ as required.

Solution (to Exercise 1.6) First, assume that Δ is closed. For $x \neq y$ in X, we have $(x, y) \notin \Delta$, and hence $(x, y) \in (X \times X) \setminus \Delta$. Since Δ is closed, $(X \times X) \setminus \Delta$ is open. By definition of the product topology, there are open subsets U and V in X such that $(x, y) \in U \times V \subset (X \times X) \setminus \Delta$. Thus, we have $x \in U$ and $y \in V$ with $U \cap V = \emptyset$. This shows that X is Hausdorff. Now assume that X is Hausdorff. Let $(x, y) \in (X \times X)$ with $x \neq y$. Since X is Hausdorff, we can find disjoint open subsets U and V in X such that $x \in U$ and $y \in V$. Since $U \cap V = \emptyset$, we have $(x, y) \in U \times V \subset (X \times X) \setminus \Delta$, and $(X \times X) \setminus \Delta$ is open. This implies that Δ is closed.

Solution (to Exercise 1.7) We write $\Gamma := \Gamma(R)$. We define the map $\widetilde{q} := q \times q \colon X \times X \to (X/R) \times (X/R)$. Since $\widetilde{q}(\Gamma)$ is the diagonal of $(X/R) \times (X/R)$, it suffices to show that $\widetilde{q}(\Gamma)$ is closed by Exercise 1.6. We have $\widetilde{q}((X \times X) \setminus \Gamma) = ((X/R) \times (X/R)) \setminus \widetilde{q}(\Gamma)$. Since q is an open map, the map $\widetilde{q}$ is open as well. Since Γ is closed by assumption, we see that $((X/R) \times (X/R)) \setminus \widetilde{q}(\Gamma)$ is open. Thus, $\widetilde{q}(\Gamma)$ is closed.

Solution (to Exercise 1.8) The determinant map $\det \mathrm{M}(n) \to \mathbb{R}$ is continuous. Hence the preimage $\mathrm{SL}(n) = \det^{-1}(\{1\})$ of the closed subset $\{1\} \subset \mathbb{R}$ is closed both in $\mathrm{M}(n)$ and in $\mathrm{GL}(n, \mathbb{R})$.

Solution (to Exercise 1.9) Given $A = (a_{ij}) \in \mathrm{O}(n)$. By definition of matrix-multiplication, $\sum_j a_{ij}^2$ is the ith diagonal entry in $AA^\top$. Moreover, $\sum_j a_{ij}^2$ is also the square of the norm of the ith row vector of A. Since $A \in \mathrm{O}(n)$, we have $AA^\top = I$ and the ith diagonal entry in $AA^\top$ is equal 1. This shows that $\mathrm{O}(n)$ is contained in the product of n spheres $\prod \mathbb{S}^{n-1}$ in $\prod \mathbb{R}^n = \mathbb{R}^{n^2} = M(n)$. Hence $\mathrm{O}(n)$ is bounded. Moreover, $\mathrm{O}(n)$ is closed in $\mathbb{R}^{n^2}$ since we can define it as the inverse image of the closed subset$\{I\} \subset \mathrm{M}(n)$ under the map $\mathrm{M}(n) \to \mathrm{M}(n)$ sending A to $AA^\top$. We note that this map is continuous, since multiplication of matrix entries and reshuffling of the entries in a matrix is continuous. Thus $\mathrm{O}(n)$ is closed and bounded in $\mathbb{R}^{n^2}$ and therefore compact by Theorem 1.16. Finally, $\mathrm{SO}(n)$ is a closed subspace of $\mathrm{O}(n)$. By Lemma 1.14, $\mathrm{SO}(n)$ is therefore compact since $\mathrm{O}(n)$ is compact.

Solution (to Exercise 1.10) Since $\mathrm{GL}(n, \mathbb{R})$ is an open subspace of $\mathrm{M}(n) \cong \mathbb{R}^{n^2}$, it suffices by Proposition 1.36 to show that $\mathrm{GL}(n, \mathbb{R})$ is not path-connected. Let γ be a path in $\mathrm{GL}(n, \mathbb{R})$, i.e., a continuous map $\gamma\colon [0,1] \to \mathrm{GL}(n, \mathbb{R})$. Since γ and det are continuous, so is their composite

$$\det \circ \gamma\colon [0,1] \xrightarrow{\gamma} \mathrm{GL}(n, \mathbb{R}) \xrightarrow{\det} \mathbb{R}.$$

Now assume there is a path γ with $\det(\gamma(0)) > 0$ and $\det(\gamma(1)) < 0$. Then the Intermediate Value Theorem from calculus implies that there is a real number $t_0 \in (0,1)$ such that $\det(\gamma(t_0)) = 0$. However, this means $\gamma(t_0) \notin \mathrm{GL}(n, \mathbb{R})$. Hence γ would have to leave $\mathrm{GL}(n, \mathbb{R})$, and $\mathrm{GL}(n, \mathbb{R})$ is not path-connected.

A.2 Smooth Manifolds

Solution (to Exercise 2.2) (a) The assertion is true if X, Y, and Z are open subsets of $\mathbb{R}^N$, $\mathbb{R}^M$, and $\mathbb{R}^L$, respectively. For in this case, this is just the chain rule from calculus. Now let $x \in X$ and $y = f(x)$. Since g is smooth at y, there is an open subset $V \subset \mathbb{R}^M$ with $y \in V$ and a smooth map

$G: V \to \mathbb{R}^L$ such that $G_{|V\cap Y} = g_{|V\cap Y}$. Since f is smooth at x, there is an open subset $U \subset \mathbb{R}^N$ with $x \in U$ and a smooth map $F: U \to \mathbb{R}^M$ such that $F_{|U\cap X} = f_{|U\cap X}$. After replacing U with $U \cap f^{-1}(V)$ if necessary (to make sure that the image of F lies in V), we have $G \circ F = g \circ f$ on $X \cap U$, since $f(X \cap U) \subset Y \cap V$. By the chain rule, we know that $G \circ F: U \to \mathbb{R}^L$ is smooth, since U is open in $\mathbb{R}^N$. This shows that $g \circ f$ is smooth.

(b) By assumption f and g are smooth and have smooth inverses f^{-1} and g^{-1} respectively. Since f and g are bijective, so is $g \circ f$. By the previous point, both $g \circ f$ and $(g \circ f)^{-1} = f^{-1} \circ g^{-1}$ are smooth.

Solution (to Exercise 2.3) (a) The inverse is $g: \mathbb{R}^k \to \mathbb{B}_r$, $y \mapsto \frac{ry}{\sqrt{r^2+|y|^2}}$. To show this, we compute both compositions and show that we get the respective identities. Both f and g are smooth, since they are the composites of smooth maps. Hence Exercise 2.2 shows that they are both smooth.

(b) Let $x \in X$. By definition of a smooth manifold, there exist an open subset $V \subset X$ with $x \in V$, an open subset $U \subset \mathbb{R}^k$, and a diffeomorphism $\phi: U \to V$. Let $u \in U$ be the inverse $\phi^{-1}(x)$. If u is not the origin in $\mathbb{R}^k$, we compose ϕ with the translation $T_u: \mathbb{R}^k \to \mathbb{R}^k$ defined by $T_u(y) = y + u$, which satisfies $T_u(0) = u$. Note that T_u is a diffeomorphism, since it is invertible with inverse T_{-u} and both T_u and T_{-u} have the identity matrix as their Jacobian matrix at any point. In particular, all higher partial derivatives exist. Thus, after composing ϕ with T_u, we can assume $\phi(0) = x$. Now it suffices to choose a small enough radius r such that $\phi(\mathbb{B}_r^k(0)) \subset V$. Then $\phi_{|\mathbb{B}_r^k(0)}: \mathbb{B}_r^k(0) \to \phi(\mathbb{B}_r^k(0)) \subset X$ is the desired local parametrisation.

(c) By the previous point, for every $x \in X$, there is a diffeomorphism $\phi: \mathbb{B}_r^k(0) \to V$ for some open subset $V \subset X$ with $x \in V$. Now it suffices to precompose ϕ with the diffeomorphism $g: \mathbb{R}^k \to \mathbb{B}_r^k(0)$ of point (a) in this exercise.

Solution (to Exercise 2.5) Every linear map $\mathbb{R}^n \to \mathbb{R}^m$ is smooth, and the derivative is equal to the map itself. Hence, given a k-dimensional vector subspace V of $\mathbb{R}^N$, it suffices to choose a basis in V to get a linear isomorphism $\phi: V \to \mathbb{R}^k$. This map serves as a parametrisation, since it is a diffeomorphism. Now, given a linear map $f: V \to \mathbb{R}^m$, the composite $\phi \circ f: \mathbb{R}^k \to \mathbb{R}^m$ is linear and therefore smooth. Since ϕ is a diffeomorphism, this implies that f must be smooth, too.

Solution (to Exercise 2.6) Let $E_0 \in M(p,n;n)$. After performing row operations if necessary, we may assume that E_0 is of the form $E_0 = \begin{pmatrix} A_0 \\ B_0 \end{pmatrix}$, where A_0 is an invertible $n \times n$ matrix and B is a $(p-n) \times n$ matrix. Since $\det A_0 \neq 0$ and $\det\colon M(n,n) \to \mathbb{R}$ is a function which depends continuously on the entries in a matrix, there is an $\varepsilon > 0$ such that, if the absolute values of all the entries of the matrix $A - A_0$ are less than ε, then $\det A \neq 0$ as well. Hence the subset $U \subset M(p,n)$ consisting of all matrices of the form $E = \begin{pmatrix} A \\ B \end{pmatrix}$ for which the absolute values of all the entries of $A - A_0$ are less than ε is an open subset of $M(p,n)$. This shows that every point $E_0 \in M(p,n;n)$ has an open neighbourhood $U \cap M(p,n;n)$. Hence $M(p,n;n)$ is an open subset of $M(p,n)$. Since $M(p,n) = \mathbb{R}^{pn}$ is a smooth manifold, $M(p,n;n)$ is a smooth manifold of dimension $p \cdot n$.

Solution (to Exercise 2.9) (a) Let $N = (0,\ldots,0,1) \in \mathbb{S}^k$ be the north pole on the k-dimensional sphere. The stereographic projection ϕ_N^{-1} from $\mathbb{S}^k \setminus \{N\}$ onto $\mathbb{R}^k$ is the map which sends a point p to the point at which the line through N and p intersects the subspace in $\mathbb{R}^{k+1}$ defined by $x_{k+1} = 0$, and we identify $\mathbb{R}^k$ with $(\mathbb{R}^k \times \{0\}) \subset \mathbb{R}^{k+1}$. In order to get from N to p, we walk in the direction of the vector $v := p - N = (x_1,\ldots,x_{k+1} - 1)$. To find $\phi_N^{-1}(p)$ we need to find the real number λ such that the $(k+1)$-st coordinate of $N + \lambda \cdot v$ is 0. Hence we need to solve $1 + \lambda(x_{k+1} - 1) = 0$ for λ and get $\lambda = \frac{1}{1-x_{k+1}}$. Thus

$$\phi_N^{-1}(x_1,\ldots,x_{k+1}) = \frac{1}{1-x_{k+1}}(x_1,\ldots,x_k).$$

(b) We compute the inverse ϕ_N. Given the point $x = (x_1,\ldots,x_k) \in \mathbb{R}^k$. To find $\phi_N(x)$, we walk from N in the direction of the vector $w = x - N$ until we reach the sphere, i.e., we need to find $\lambda \in \mathbb{R}$ such that $N + \lambda \cdot w = (\lambda x_1,\ldots,\lambda x_k, 1-\lambda)$ lies on $\mathbb{S}^k$. Hence we need to solve $\lambda^2(x_1^2 + \cdots + x_k^2) + (1-\lambda)^2 = 1$, i.e., $\lambda^2(1+|x|^2) - 2\lambda = 0$ for λ. Since 0 is not a valid solution, we get $\lambda = \frac{2}{1+|x|^2}$. Thus,

$$\phi_N(x) = \frac{1}{1+|x|^2}\left(2x_1,\ldots,2x_k,|x|^2 - 1\right).$$

We need to check that both ϕ_N and ϕ_N^{-1} are smooth. All entries of ϕ_N are smooth functions, since they are polynomials on the domain and hence infinitely differentiable functions in each variable. This implies that ϕ_N is smooth. But let us calculate the derivative anyway, since we are going to use it later. We get that $d(\phi_N)_x$ is given by

$$\frac{2}{(1+|x|^2)^2}\begin{pmatrix} 1+|x|^2-x_1^2 & -2x_1x_2 & -2x_1x_3 & \dots & -2x_1x_k \\ -2x_2x_1 & 1+|x|^2-x_2^2 & -2x_2x_3 & \dots & -2x_2x_k \\ \vdots & \vdots & \vdots & \vdots & \vdots \\ -2x_kx_1 & \dots & \dots & \dots & 1+|x|^2-x_k^2 \\ 2x_1 & 2x_2 & \dots & \dots & 2x_k \end{pmatrix}.$$

Similarly, the derivative of ϕ_N^{-1} at x is given by

$$d(\phi_N^{-1})_x = \begin{pmatrix} \frac{1}{1-x_{k+1}} & 0 & \dots & 0 & \frac{x_1}{(1-x_{k+1})^2} \\ 0 & \frac{1}{1-x_{k+1}} & \dots & 0 & \frac{x_2}{(1-x_{k+1})^2} \\ \vdots & \vdots & \dots & \vdots & \vdots \\ 0 & 0 & \dots & \frac{1}{1-x_{k+1}} & \frac{x_k}{(1-x_{k+1})^2} \end{pmatrix}.$$

(c) The formulae for the projection from the south pole $S = (0, \dots, 0, -1) \in \mathbb{S}^k$ are similar. We get $\phi_S^{-1}(x_1, \dots, x_{k+1}) = \frac{1}{1+x_{k+1}}(x_1, \dots, x_k)$ and $\phi_S(x) = \frac{1}{1+|x|^2}\left(2x_1, \dots, 2x_k, 1-|x|^2\right)$. To check that both ϕ_S and ϕ_S^{-1} are both smooth is completely analogous. Since all points are covered by these two parametrisations, $\mathbb{S}^k$ is a smooth manifold. Finally, we note that the formula for the derivative of ϕ_S^{-1} is given by

$$d(\phi_S^{-1})_x = \begin{pmatrix} \frac{1}{1+x_{k+1}} & 0 & \dots & 0 & \frac{-x_1}{(1+x_{k+1})^2} \\ 0 & \frac{1}{1+x_{k+1}} & \dots & 0 & \frac{-x_2}{(1+x_{k+1})^2} \\ \vdots & \vdots & \dots & \vdots & \vdots \\ 0 & 0 & \dots & \frac{1}{1+x_{k+1}} & \frac{-x_k}{(1+x_{k+1})^2} \end{pmatrix}.$$

Solution (to Exercise 2.10) (a) For $(z_0, z_1) \in \mathbb{S}^3$, we need to check that $\hbar(z_0, z_1)$ is actually a point in $\mathbb{S}^2$. To do this, we recall that $|z| = z\bar{z}$ and $\bar{\bar{z}} = z$ for any complex number z. Now we compute, using that $(z_0, z_1) \in \mathbb{S}^3$:

$$\begin{aligned} |2z_0\bar{z}_1|^2 + \left(|z_0|^2 - |z_1|^2\right)^2 &= 4|z_0|^2|z_1|^2 + |z_0|^4 - 2|z_0|^2|z_1|^2 + |z_1|^4 \\ &= \left(|z_0|^2 + |z_1|^2\right)^2 = 1. \end{aligned}$$

(b) First we assume $\hbar(z_0, z_1) = \hbar(w_0, w_1)$. Then we get

$$\begin{aligned} &\left(2w_0\bar{w}_1, |w_0|^2 - |w_1|^2\right) = \left(2z_0\bar{z}_1, |z_0|^2 - |z_1|^2\right) \\ \iff &w_0\bar{w}_1 = z_0\bar{z}_1 \text{ and } |w_0|^2 - |w_1|^2 = |z_0|^2 - |z_1|^2. \end{aligned}$$

The equations $|w_0|^2 - |w_1|^2 = |z_0|^2 - |z_1|^2$ and $|w_0|^2 + |w_1|^2 = 1 = |z_0|^2 + |z_1|^2$ together imply $z_0\bar{z}_0 = |z_0|^2 = |w_0|^2 = w_0\bar{w}_0$ and $z_1\bar{z}_1 =$

$z_1^2 = w_1^2 = w_1\bar{w}_1$. Remembering that neither of the numbers can be zero, and by rewriting these equations, we get $w_0/z_0 = \overline{z}_0/\overline{w}_0$ and $w_1/z_1 = \overline{z}_1/\overline{w}_1$. Hence, looking at the left-hand side, there is a complex number $\alpha \neq 0$ such that $\alpha = w_0/z_0 = \bar{z}_0/\bar{w}_0 = 1/\bar{\alpha}$, and thus $w_0 = \alpha z_0$ with $\alpha\bar{\alpha} = 1$. On the other hand, we also know $w_0/z_0 = \bar{z}_1/\bar{w}_1$. Combining these equations yields $\alpha = w_0/z_0 = \bar{z}_1/\bar{w}_1 = w_1/z_1$, and thus $w_1 = \alpha z_1$. This shows that the desired α with $\alpha\bar{\alpha} = 1$ exists. Now we assume that $(w_0, w_1) = (\alpha z_0, \alpha z_1)$ with $|\alpha|^2 = \alpha\bar{\alpha} = 1$. Then we compute

$$\hbar(w_0, w_1) = \left(2w_0\bar{w}_1, |w_0|^2 - |w_1|^2\right) = \left(2\alpha z_0\bar{\alpha}\bar{z}_1, |\alpha|^2|z_0|^2 - |\alpha|^2|z_1|^2\right)$$
$$= \left(2z_0\bar{z}_1, |z_0|^2 - |z_1|^2\right) = \hbar(z_0, z_1).$$

(c) Let $p \in \mathbb{S}^2$ be a fixed point and fix a point (z_0, z_1) in $\mathbb{S}^3$ with $\hbar(z_0, z_1) = p$. By the previous point, we know that the points in $\hbar^{-1}(p)$ are parametrised by the complex numbers α with $|\alpha|^2 = 1$. The latter condition means $\alpha \in \mathbb{S}^1 \subset \mathbb{C}$. Hence we get a bijective map $f\colon \mathbb{S}^1 \to \hbar^{-1}(p)$ defined by $\alpha \mapsto (\alpha z_0, \alpha z_1)$. Since this map just consists of multiplication with complex numbers, we can conclude that it is a smooth map where we consider $\mathbb{S}^1 \subset \mathbb{C} = \mathbb{R}^2$ and $\hbar^{-1}(p) \subset \mathbb{R}^4$ as subsets in real Euclidean space. The previous point implies that f is onto. Since not both z_0 and z_1 can be zero, we see that f is one-to-one. For, say $z_0 \neq 0$, then $\alpha z_0 = \beta z_0$ implies $\alpha = \beta$. The inverse is given by sending $(w_0, w_1) \in \hbar^{-1}(p)$ to $\alpha \in \mathbb{S}^1$ such that $w_0 = \alpha z_0$ and $w_1 = \alpha z_1$. In every open subset of $\hbar^{-1}(p) \subset \mathbb{S}^3$ where $z_0 \neq 0$, the map $w_0 \mapsto w_0 z_0 = \alpha$ is smooth. Similarly, in every open subset of $\hbar^{-1}(p) \subset \mathbb{S}^3$ where $z_1 \neq 0$, the map $w_1 \mapsto w_1 z_1 = \alpha$ is smooth. Hence f^{-1} is a smooth map as well, and f is a diffeomorphism.

Solution (to Exercise 2.14) (a) The inverse of F is the projection π onto the first factor since we have $\pi \circ F = \mathrm{Id}_X$ and $F \circ \pi = \mathrm{Id}_{\Gamma(f)}$. Let $X \subset \mathbb{R}^N$, $Y \subset \mathbb{R}^M$, and let f be smooth. Then, for any point $x \in X$, there is an open subset $U \subset \mathbb{R}^N$ and a smooth map $\tilde{f}\colon U \to \mathbb{R}^M$ with $\tilde{f}_{X\cap U} = f_{X\cap U}$. Then $(\tilde{f} \times \mathrm{Id})\colon U \to \mathbb{R}^N \times \mathbb{R}^M$ is a smooth extension of $F_{X\cap U}$. Hence F is smooth. The inverse π is a smooth map since it extends to the smooth projection on all of $\mathbb{R}^N \times \mathbb{R}^M$. Hence F is a diffeomorphism. Hence every local parametrisation $\phi\colon V \to X$ can be extended to a local parametrisation $F \circ \phi\colon V \to \Gamma(f)$. Thus the graph $\Gamma(f)$ is a manifold.

(b) For $\in X$, let $\phi\colon U \to X$ be a local parametrisation around x with $\phi(0) = x$ and open $U \subseteq \mathbb{R}^k$. Let $\psi\colon W \to Y$ be a local parametrisation around $f(x)$ with $\psi(0) = f(x)$ and open $W \subseteq \mathbb{R}^l$. Then $\phi \times \psi\colon U \times W \to X \times Y$ is a local parametrisation around $(x, f(x))$ with $U \times W \subset \mathbb{R}^{k+l}$ open. Then we have the commutative diagram

$$\begin{array}{ccc} X & \xrightarrow{\quad F \quad} & X \times Y \\ {\scriptstyle \phi}\uparrow & & \uparrow{\scriptstyle \phi\times\psi} \\ U & \xrightarrow[G=\mathrm{Id}_U\times(\psi^{-1}\circ f\circ\phi)]{} & U \times W \end{array}, \tag{A.1}$$

where G is the map defined by $u \mapsto (u, \psi^{-1}(f(\phi(u))))$. Thus G is the map $\mathrm{Id}_U \times (\psi^{-1} \circ f \circ \phi)$, and $dG_0\colon \mathbb{R}^k \to \mathbb{R}^k \times \mathbb{R}^l$ is given by the linear map $dG_0 = \mathrm{Id}_{\mathbb{R}^k} \times (d\psi^{-1}_{f(x)} \circ df_x \circ d\phi_0)$. Thus dF_x has to be defined as $\mathrm{Id}_{T_x(X)} \times df_x$:

$$\begin{array}{ccc} T_x(X) & \xrightarrow{dF_x=\mathrm{Id}_{T_x(X)}\times df_x} & T_x(X) \times T_y(Y) \\ {\scriptstyle d\phi_0}\uparrow & & \uparrow{\scriptstyle d\phi_0\times d\psi_0} \\ \mathbb{R}^k & \xrightarrow[dG_0=\mathrm{Id}_{\mathbb{R}^k}\times(d\psi^{-1}_{f(x)}\circ df_x\circ d\phi_0)]{} & \mathbb{R}^k \times \mathbb{R}^l \end{array}. \tag{A.2}$$

Hence $dF_x(v) = (v, df_x(v))$.

(c) For $\in X$, let $\phi\colon U \to X$ be a local parametrisation around x with $\phi(0) = x$ and open $U \subseteq \mathbb{R}^k$. Since F is a diffeomorphism, $\psi = F \circ \phi\colon U \to \Gamma(f)$ is then a local parametrisation around $(x, f(x))$ with $\psi(0) = F(\phi(0)) = (x, f(x))$. The tangent space of $\Gamma(f)$ at $(x, f(x))$ is by definition the image of $d\psi_0\colon \mathbb{R}^k \to \mathbb{R}^{N+M}$. By the chain rule, we have $d\psi_0 = dF_x \circ d\phi_0$ as linear maps $\mathbb{R}^k \to \mathbb{R}^{N+M}$. Hence, by our definition of tangent spaces, we get

$$T_{(x,f(x))}(\Gamma(f)) = d\psi_0(\mathbb{R}^k) = dF_x(d\phi_0(\mathbb{R}^k)) = dF_x(T_x(X)).$$

Finally, by the previous point, we know $dF_x = \mathrm{Id}_{T_x(X)} \times df_x$ and get

$$T_{(x,f(x))}(\Gamma(f)) = (\mathrm{Id}_{T_x(X)} \times df_x)(T_x(X)) = \Gamma(df_x) \subset T_x(X) \times T_{f(x)}(Y),$$

which is the graph of df_x in $T_x(X) \times T_{f(x)}(Y)$.

Solution (to Exercise 2.15) For matrices $A \in \mathrm{GL}(n)$ and $B \in \mathrm{M}(n)$, we have $\det(B) = (\det A) \cdot \det(A^{-1}B) = (\det A) \cdot \det(L_{A^{-1}}B)$, where we denote by

$L_{A^{-1}}\colon \mathrm{GL}(n) \to \mathrm{GL}(n)$ the left multiplication map defined by $B \mapsto A^{-1} \cdot B$. Taking the derivative at A and remembering the chain rule yields

$$\begin{aligned} d(\det)_A(B) &= d((\det A) \cdot \det \circ L_{A^{-1}})(B) = (\det A) \cdot d(\det \circ L_{A^{-1}})B \\ &= (\det A) \cdot d(\det)_{(L_{A^{-1}}A)} \circ d(L_{A^{-1}})_A(B) \\ &= (\det A) \cdot d(\det)_I(A^{-1}B) = (\det A) \cdot \operatorname{tr}(A^{-1}B), \end{aligned}$$

where we have used $d(L_{A^{-1}})_A(B) = A^{-1}B$ which can be easily checked, since matrix multiplication is linear. Hence, after replacing B with AB, we get $d(\det)_A(AB) = (\det A) \cdot (\operatorname{tr} B)$ for all $B \in \mathrm{M}(n)$.

Solution (to Exercise 2.16) (a) Given a smooth map $c\colon I \to \mathbb{R}^k$ with $c = (c_1, \dots, c_k)$ and $c_i : I \to \mathbb{R}$ all smooth. The derivative of c at $t_0 \in I$ is a linear map $T_{t_0}I = \mathbb{R} \to \mathbb{R}^k = T_{x_0}(\mathbb{R}^k)$ with $dc_{t_0}(v) = (c_1'(t_0), \dots, c_k'(t_0)) \cdot v$. Since $v \in \mathbb{R}$ is just a real number, we get $dc_{t_0}(1) = (c_1'(t_0), \dots, c_k'(t_0)) \in \mathbb{R}^k = T_{x_0}(\mathbb{R}^k)$.

(b) First, assume $X = \mathbb{R}^k$ and let $w = (w_1, \dots, w_k)$ be a vector in $T_xX = \mathbb{R}^k$. We define the curve $c_w : \mathbb{R} \to \mathbb{R}^k$ by $t \mapsto t \cdot w$. The derivative of c_w at any t_0 is $d(c_w)_{t_0} : \mathbb{R} \to \mathbb{R}^k, t \mapsto (w_1, \dots, w_k) \cdot t$. Thus we have $d(c_w)_{t_0}(1) = w$. Now let X be an arbitrary k-dimensional smooth manifold, $x \in X$, and let v be a vector in $T_x(X)$. Let $\phi\colon U \to X$ be a local parametrisation around x with $\phi(0) = x$. By definition, $T_x(X) = d\phi_0(\mathbb{R}^k)$ and there is a unique vector $w \in \mathbb{R}^k$ with $d\phi_0(w) = v$. Since any open ball around the origin in $\mathbb{R}^k$ is diffeomorphic to $\mathbb{R}^k$ by Exercise 2.3, we can assume $U = \mathbb{R}^k$ and have $w \in U$. Let $c_w : \mathbb{R} \to \mathbb{R}^k$ be the linear curve in $\mathbb{R}^k$ defined in the previous point. Then we define $c\colon \mathbb{R} \to X$ by $c = \phi \circ c_w$, i.e., $c(t) = \phi(t \cdot w)$. The derivative of c at $t_0 = 0 \in \mathbb{R}$ is $d(c)_{t_0} = d\phi_0 \circ d(c_w)_{t_0}$. Thus, $d(c)_{t_0}(1) = d\phi_0(d(c_w)_{t_0}(1)) = d\phi_0(w) = v$.

A.3 The IFT, Immersions, and Embeddings

Solution (to Exercise 3.2) Since f is bijective, there is an inverse map $f^{-1}\colon Y \to X$. Since f is a local diffeomorphism, we can find, for every $x \in X$, an open subset $U \subset X$ such that $f_{|U} : U \to f(U)$ is a diffeomorphism and $f(U)$ is open in Y. Hence there is a smooth inverse $(f_{|U})^{-1}$ of $f_{|U}$. Since the inverse of a map is unique, we must have $(f_{|U})^{-1} = (f^{-1})_{|f(U)}$. This shows that f^{-1} is smooth on the open subset $f(U) \subset Y$. Since f is bijective, the open subsets $f(U)$, for all $x \in X$, cover Y. Hence, at every point $y \in Y$, we can find an open subset on which the restriction of f^{-1} is smooth. This shows that f^{-1} is a smooth map. Thus, f is a diffeomorphism.

Solution (to Exercise 3.3) Since f is one-to-one, it is a bijection from X onto its image $\operatorname{Im}(f) \subseteq Y$. Since it is a local diffeomorphism, f is an open map. Hence, $\operatorname{Im}(f)$ is open in Y, and a subset $U \subset \operatorname{Im}(f)$ is open in $\operatorname{Im}(f)$ if and only if it is open in Y. Hence $f\colon X \to \operatorname{Im}(f)$ is a bijective local diffeomorphism. By the previous point, it is a diffeomorphism.

Solution (to Exercise 3.6) The map f is not an embedding, since it is not injective. But we can check it is an immersion by showing that the derivative is injective everywhere. The derivative of f at (s,t) is

$$df_{(s,t)}\colon \mathbb{R}^2 \to \mathbb{R}^3,\ df_{(s,t)} = \begin{pmatrix} -\sin s \cos t & -(2+\cos s)\sin t \\ -\sin s \sin t & (2+\cos s)\cos t \\ \cos s & 0 \end{pmatrix}.$$

In order to show that $df_{(s,t)}$ is injective, we need to check that it has full or maximal rank, i.e., rank 2. Hence we need to check that the two column vectors are always linearly independent. To simplify notation, we set $x = \sin s$, $y = \cos s$, $u = \sin t$, and $v = \cos t$. Now assume there are two real numbers λ and μ such that $\lambda(-xv) + \mu(-(2+y)u) = 0$, $\lambda(-xu) + \mu((2+y)v) = 0$, and $\lambda y = 0$. We distinguish two cases: $y = \cos s = 0$ and $y = \cos s \neq 0$. If $y \neq 0$, then we must have $\lambda = 0$, $\mu(2+y)u = 0$, and $\mu(2+y)v = 0$. Since $|y| \leq 1$, we know $2 + y \neq 0$. Hence we can divide by $2 + y$. Moreover, we know that not both u and v can be 0. This implies $\mu = 0$. If $y = 0$, then $x = \sin s = \pm 1$. Hence we get the system $\pm\lambda v - 2\mu u = 0$ and $\pm\lambda u + 2\mu v = 0$. But since u and v are never both 0, we know that the vectors (v,u) and (u,v) are linearly independent, unless $u = v$. If $u \neq v$, we get $\lambda = 0$ and $\mu = 0$. If $u = v \neq 0$, we get, for $a = \pm 1$, a matrix of the form au

$$\begin{pmatrix} 1 & -2 \\ 1 & 2 \\ 0 & 0 \end{pmatrix},$$

which has rank two as well.

Solution (to Exercise 3.7) Let a and b be two relatively prime integers with $a \neq 0$. Recall that, for $t_1, t_2 \in \mathbb{R}$, we have $e^{2\pi i t_1} = e^{2\pi i t_2}$ if and only if $t_1 - t_2 \in \mathbb{Z}$. Since a and b are integers, $t_1 - t_2 \in \mathbb{Z}$ implies $at_1 - at_2 \in \mathbb{Z}$ and $bt_1 - bt_2 \in \mathbb{Z}$. Hence $e^{2\pi i t_1} = e^{2\pi i t_2}$ implies $e^{2\pi i a t_1} = e^{2\pi i a t_2}$ and $e^{2\pi i b t_1} = e^{2\pi i b t_2}$. This shows that we have a well-defined map $g_{a,b}\colon \mathbb{S}^1 \to \mathbb{S}^1 \times \mathbb{S}^1$

given by $e^{2\pi it} \mapsto (e^{2\pi iat}, e^{2\pi ibt})$. Hence, for $f: \mathbb{R} \to \mathbb{S}^1$, $t \mapsto e^{2\pi it}$, we get $\gamma_{a,b} = g_{a,b} \circ f$. Since $\mathbb{S}^1$ is compact and $g_{a,b}$ is a one-to-one immersion, $g_{a,b}$ is an embedding.

Solution (to Exercise 3.9) Assume $X \subseteq \mathbb{R}^M$ and $Y \subseteq \mathbb{R}^N$, $\dim X = m$, $\dim Y = n$, and set $r := \operatorname{rank}(f)$. By definition of the rank, we have $m \geq r$ and $n \geq r$. We need to show $r = m$. By assumption, for any point $x \in X$, the linear map df_x has rank r. As in the proof of the Local Immersion Theorem 3.11, we can choose local parametrisations $\phi: U \to X$ around x and $\psi: V \to Y$ around y such that the map $\theta: U \to V$ in the commutative diagram

$$\begin{array}{ccc} X & \xrightarrow{\quad f \quad} & Y \\ {\scriptstyle \phi}\uparrow & & \uparrow{\scriptstyle \psi} \\ U & \xrightarrow[\theta = \psi^{-1} \circ f \circ \phi]{} & V \end{array} \tag{A.3}$$

has the form $\theta(x_1, \ldots, x_m) = (x_1, \ldots, x_r, 0) \in \mathbb{R}^n$. (Note that the 0s at the end of $\theta(x)$ only occur if $r < n$.) If $m > r$, then for a sufficiently small $\varepsilon > 0$, $\theta(x_1, \ldots, x_r, \varepsilon, 0) = (x_1, \ldots, x_r, 0)$ and θ is not injective. Since ϕ and ψ are diffeomorphisms, this would imply that f is not injective, which contradicts that f is bijective. Hence we get $m = r$ and f is an immersion.

Solution (to Exercise 3.10) By the Local Immersion Theorem 3.11, we can choose local parametrisations $\phi: V \to Z$ and $\psi: W \to X$ around z with $V \subset \mathbb{R}^k$ and $W = V \oplus V' \subset \mathbb{R}^n$ such that

$$\begin{array}{ccc} Z & \xrightarrow{\quad \text{inclusion} \quad} & X \\ {\scriptstyle \phi}\uparrow & & \uparrow{\scriptstyle \psi} \\ \mathbb{R}^k \supseteq V & \xrightarrow[\text{immersion}]{\text{canonical}} & W \subseteq \mathbb{R}^n \end{array} \tag{A.4}$$

commutes. The map ψ is a diffeomorphism onto its image $\psi(W) \subset X$. The inverse map $\psi^{-1}: \psi(W) \to W$ is a local coordinate system on the open neighbourhood $\psi(W)$ around $z \in X$. We write $x_i: \psi(W) \to \mathbb{R}$ for the ith component of ψ^{-1}, i.e., a point $p \in \psi(W)$ has the local coordinates $(x_1(p), \ldots, x_n(p)) = (\psi_1^{-1}(p), \ldots, \psi_n^{-1}(p))$. Since the diagram (A.4) commutes and ϕ is a diffeomorphism onto its image, we have $\phi(V) = Z \cap \psi(W)$. Hence, since the lower horizontal map is the canonical immersion, the points in $Z \cap \psi(W)$ are exactly those on which the coordinate functions $x_{k+1}, \ldots, x_n$ vanish. We then set $U := \psi(W)$ and get $Z \cap U = \{p \in U : x_{k+1}(p) = \cdots = x_n(p) = 0\}$.

A.4 Submersions and Regular Values

Solution (to Exercise 4.1) It suffices to show that $f(X) \subset Y$ is open in Y since for an arbitrary open subset $U \subset X$ we may consider the map $U \subseteq X \xrightarrow{f} Y$. Let y be any point in $f(X)$, and let $x \in X$ be a point in X with $f(x) = y$. By the Local Submersion Theorem 4.7, we can choose local parametrisations $\phi\colon V \to X$ around x with $V \subset \mathbb{R}^n$ open and $\psi\colon V' \to Y$ around y with $V' \subset \mathbb{R}^m$ open such that the induced map $V \to V'$ is the restriction of the canonical submersion

$$\begin{array}{ccc} X & \xrightarrow{\quad f \quad} & Y \\ {\scriptstyle \phi}\uparrow & & \uparrow{\scriptstyle \psi} \\ V & \xrightarrow[\text{submersion}]{\text{canonical}} & V' \end{array} . \tag{A.5}$$

By possibly shrinking V and V', we can assume that $V = \mathbb{B}_\varepsilon(0) \subset \mathbb{R}^n$ and $V' = \mathbb{B}_{\varepsilon'}(0) \subset \mathbb{R}^m$. Then the canonical submersion maps V surjectively onto V'. Since the diagram (A.5) commutes, we see that $\psi(V')$ is contained in $f(X)$. Since ψ is a local parametrisation, $W := \psi(V')$ is open in Y. Thus, every point in $f(X)$ has an open neighbourhood W in Y with $W \subset f(X)$. This shows that $f(X)$ is open in Y.

Solution (to Exercise 4.3) Let $f\colon X \to Y$ be a submersion. We have $Y = f(X) \cup (Y \setminus f(X))$. Since X is compact and f is continuous, we know $f(X)$ is compact and therefore closed in Y. Since X is open in X and f is a submersion, Exercise 4.1 implies that $f(X)$ is open in Y. Hence $f(X)$ is both open and closed in Y. Hence $f(X)$ must be either Y or $\emptyset$. Assuming f is non-trivial, $f(X)$ must be all of Y.

Solution (to Exercise 4.4) Given a compact smooth manifold X. Assume we had a submersion $f\colon X \to \mathbb{R}^n$ with for some n. By Exercise 4.3, we would have $f(X) = \mathbb{R}^n$. However, since X is compact, $f(X)$ is compact by Lemma 1.13. But $\mathbb{R}^n$ is not compact. Hence such a submersion cannot exist.

Solution (to Exercise 4.6) Assume $X \subseteq \mathbb{R}^M$ and $Y \subseteq \mathbb{R}^N$, $\dim X = m$, $\dim Y = n$, and set $r := \operatorname{rank}(f)$. By definition of the rank, we have $m \geq r$ and $n \geq r$.

(a) As in the proof of the Local Submersion Theorem 4.7, we can choose local parametrisations $\phi\colon U \to X$ around x and $\psi\colon V \to Y$ around y such that the map $\theta\colon U \to V$ in the commutative diagram

$$\begin{array}{ccc} X & \xrightarrow{\quad f \quad} & Y \\ \phi \uparrow & & \uparrow \psi \\ U & \xrightarrow[\theta=\psi^{-1}\circ f\circ\phi]{} & V \end{array}$$

has the form $\theta(x_1, \ldots, x_m) = (x_1, \ldots, x_r, 0) \in \mathbb{R}^n$. Note that the 0s at the end of $\theta(x)$ only occur if $r < n$. We need to show $r = n$. Assume we had $r < n$. Then, after possibly shrinking U, we can assume that U is a small open $\mathbb{B}_\varepsilon(0)$ around 0 in $\mathbb{R}^m$ and that $\theta(\overline{\mathbb{B}}_\varepsilon(0)) \subseteq V$, where $\overline{\mathbb{B}}_\varepsilon(0) = \{x \in \mathbb{R}^m : |x| \leq \varepsilon\}$ denotes the closed ball of radius ε. Since $\overline{\mathbb{B}}_\varepsilon(0)$ is compact, so is $\theta(\overline{\mathbb{B}}_\varepsilon(0))$ by Lemma 1.13. Since V is Hausdorff, $\theta(\overline{\mathbb{B}}_\varepsilon(0))$ is closed in V by Proposition 1.21. Moreover, $\theta(\overline{\mathbb{B}}_\varepsilon(0))$ is contained in $V \cap (\mathbb{R}^r \times \{0\})$. This shows that $\theta(\overline{\mathbb{B}}_\varepsilon(0))$ does not contain any open subsets of V. Since ϕ and ψ are diffeomorphisms, this implies that $f(\phi(\overline{\mathbb{B}}_\varepsilon(0)))$ is closed and does not contain any non-empty open subset of Y. Since we can cover X by such local parametrisations, we see that $f(X)$ is the union of subsets which do not contain any non-empty open subset of Y. Now if we assume that X is compact, then $f(X)$ is compact, and $f(X)$ can be covered by finitely many closed subsets which do not contain any non-empty open subset of Y. That implies that $f(X)$ is itself a closed subset which does not contain any non-empty open subset of Y. Hence $f(X)$ cannot be all of Y, and f is not surjective, which contradicts the assumption.

(c) For every open cover of a subspace in $\mathbb{R}^M$, we can always choose a countable subcover. This implies that $f(X)$ is the countable union of subsets which do not contain any non-empty open subset of Y. By Baire's Category Theorem, this implies that $f(X)$ does not contain any non-empty open subset of Y. Hence $f(X)$ cannot be equal Y and f would not be surjective.

(d) By Exercise 3.9 and point (b), we know $m = r = n$. Then the Inverse Function Theorem 3.4 implies that f is a local diffeomorphism. Since f is bijective, f is a diffeomorphism by Lemma 3.7 and Exercise 3.2.

Solution (to Exercise 4.7) We consider $\mathrm{O}(n)$ as a subspace in $\mathrm{M}(n) = \mathbb{R}^{n^2}$. Let $f\colon \mathrm{M}(n) \to S(n)$ be the smooth map given by $f(A) = AA^\top$. We know $\mathrm{O}(n) = f^{-1}(I)$. We have checked in the proof of Theorem 4.20 that I is a regular value of f. By Proposition 4.13, we know that $T_I(\mathrm{O}(n))$ equals the kernel of $df_I : \mathrm{M}(n) = T_I(\mathrm{M}(n)) \to T_I(S(n)) = S(n)$. We calculated the derivative df_A for any $A \in \mathrm{O}(n)$ in the proof of Theorem 4.20 as

$df_A(B) = BA^\top + AB^\top$. For $A = I$, this gives $df_I(B) = B + B^\top$. Hence the kernel of df_I is the space of matrices satisfying $B + B^\top = 0$, i.e., $B^\top = -B$.

Solution (to Exercise 4.8) (a) Since $d(\det)_A$ is a linear map $\mathbb{R}^4 \to \mathbb{R}$, it suffices to show $d(\det)_A$ is non-zero. Therefore, it suffices to show that $d(\det)_A(B) \neq 0$ for some matrix B. Since $A \neq 0$, there is at least one entry in A which is non-zero. Assume that $a_{11} \neq 0$, and for the other cases the argument is similar. Then we take the matrix $E_{22} = \begin{pmatrix} 0 & 0 \\ 0 & 1 \end{pmatrix}$:

$$\begin{aligned} d(\det)_A(E_{22}) &= \lim_{s\to 0} \frac{\det(A + sE_{22}) - \det A}{s} = \lim_{s\to 0} \frac{\det(A + sE_{22}) - \det A}{s} \\ &= \lim_{s\to 0} \frac{(a_{11})(a_{22} + s) - a_{12}a_{21} \ - \det A}{s} \\ &= \lim_{s\to 0} \frac{a_{11}s + (a_{11}a_{22} - a_{12}a_{21}) - \det A}{s} \\ &= \lim_{s\to 0} \frac{a_{11}s + \det A - \det A}{s} = \lim_{s\to 0} \frac{a_{11}s}{s} = a_{11} \neq 0. \end{aligned}$$

(b) A 2×2-matrix A has rank 0 if and only if it is the zero matrix. Thus $A \in \mathrm{M}(2) \setminus \{0\}$ has rank 1 if and only if it does not have rank 2, i.e., if and only if it is not invertible. Hence $A \in \mathrm{M}(2) \setminus \{0\}$ has rank 1 if and only if $\det A = 0$. By the previous point, the determinant function is a submersion $\mathrm{M}(2) \setminus \{0\} \to \mathbb{R}$. Hence $R_1 = \det^{-1}(0)$ is a submanifold of dimension $4 - 1 = 3$ by the Preimage Theorem 4.12.

Solution (to Exercise 4.9) We consider the function Q defined in the hint. Since P is homogeneous, we know Q is always 0. Hence its derivative with respect to t is zero as well. Hence we get

$$0 = \partial Q/\partial t = \sum_i x_i \partial P/\partial x_i(tx_1, \ldots, tx_k) - mt^{m-1} P(tx_1, \ldots, tx_k), \quad \text{(A.6)}$$

where we apply the chain rule to the first summand of Q, which is the composite $t \mapsto tx \mapsto P(tx)$. Setting $t = 1$ in (A.6) yields Euler's identity (4.7).

Solution (to Exercise 4.10) (a) The derivative of P at a point $(x_1, \ldots, x_k)$ is the linear map $dP_x \colon \mathbb{R}^k \to \mathbb{R}$ given by

$$(z_1, \ldots, z_k) \mapsto (\partial P/\partial x_1(x) \ldots \partial P/\partial x_k(x)) \cdot \begin{pmatrix} z_1 \\ \vdots \\ z_k \end{pmatrix} = \sum_i z_i \partial P/\partial x_i(x).$$

To show that dP_x is surjective, it suffices to show that dP_x is non-trivial. Applying dP_x to x and using Euler's identity yields

$$dP_x(x) = \sum_i x_i \partial P/\partial x_i(x_1, \ldots, x_k) = mP(x_1, \ldots, x_k).$$

Hence if $x = (x_1, \ldots, x_k)$ is not a zero of P, then $dP_x(x)$ is non-zero. Hence all non-zero real numbers are regular values of P. The Preimage Theorem now implies that $P^{-1}(a)$ is a $k-1$-dimensional submanifold of $\mathbb{R}^k$ for all $a \neq 0$. In fact, we should add that, if $m = 1$ and $k = 1$, then P is the identity map on $\mathbb{R}$ and all values $a \in \mathbb{R}$ are regular.

(b) Given two real numbers a, b with $ab > 0$. Then $(b/a)^{1/m}$ exists, and, if $P(x) = a$, then $P((b/a)^{1/m}x_1, \ldots, (b/a)^{1/m}x_k) = b/aP(x_1, \ldots, x_k) = b$. Multiplying each coordinate with $(b/a)^{1/m}$ corresponds to multiplication with the diagonal matrix with $(b/a)^{1/m}$ on the diagonal. This map is a linear isomorphism of $\mathbb{R}^k$ to itself. Hence we have a diffeomorphism $P^{-1}(a) \to P^{-1}(b)$ defined by $(x_1, \ldots, x_k) \mapsto ((b/a)^{1/m}x_1, \ldots, (b/a)^{1/m}x_k)$.

Solution (to Exercise 4.11) (a) If we think of the entries in an $n \times n$-matrix A as variables, then $\det A$ is a homogeneous polynomial of degree n given by Leibniz' formula. Hence we can apply Exercise 4.10 to the polynomial $P = \det\colon \mathrm{M}(n) = \mathbb{R}^{n^2} \to \mathbb{R}$ and conclude that 0 is the only critical value of det. Alternatively: using the formula $\det A = \sum_{i=1}^n (-1)^{i+j} a_{ij} \det A_{ij}$, we can compute the partial derivatives of det with respect to each variable x_{ij}. Since we remove the jth column from A for each of the A_{ij}, the entry a_{ij} occurs in this formula exactly once. Hence the partial derivative with respect to x_{ij} is just the factor of a_{ij}, i.e., $\frac{\partial \det}{\partial x_{ij}}(A) = (-1)^{i+j} \det A_{ij}$. The total derivative of det as a function of its entries can be represented by the $(n \times 1)$-matrix with these partial derivatives as entries. This linear map fails to be surjective if and only if all entries are zero, i.e., if and only if $\det A_{ij} = 0$ for all i, j. The latter happens if and only if the rank of A is $< n$, i.e., if and only if $\det A = 0$. Hence 0 is the only critical value.

(b) By the previous point, 1 is a regular value for det. Hence, by the Preimage Theorem 4.12, $\mathrm{SL}(n) = \det^{-1}(1)$ is a smooth manifold of dimension $\dim \mathrm{M}(n) - \dim \mathbb{R} = n^2 - 1$.

(c) By the previous point and Proposition 4.13, we can determine the tangent space as the kernel of the derivative of det at the identity matrix. Hence we need to calculate the derivative of det at the identity. To do this, we are going to use Leibniz' formula of equation (4.8). Given a matrix A, in

Leibniz' formula for the determinant of $B := I + sA$, every summand contains at least a factor s^2 unless it is the product of at least $n-1$ diagonal entries $b_{ii} = 1 + sa_{ii}$. For we need $n-1$ factors not containing s which is only possible when we multiply $n-1$ times 1. Moreover, if a permutation $\{1, \ldots, n\}$ leaves $n-1$ numbers fixed, it also has to leave the remaining one fixed. Hence the only summand in (4.8) which does not contain a factor s^2 is the summand

$$\prod_{i=1}^{n}(1 + sa_{ii}) = (1 + sa_{11}) \cdots (1 + sa_{nn}) = 1 + s \cdot \operatorname{tr}(A) + O(s^2).$$

The derivative $d(\det)_I : T_I(\mathrm{M}(n)) = \mathrm{M}(n) \to T_1(\mathbb{R}) = \mathbb{R}$ of the determinant at the identity is then given by

$$\begin{aligned} d(\det)_I(A) &= \lim_{s\to 0} \frac{\det(I + sA) - \det I}{s} = \lim_{s\to 0} \frac{1 + s \cdot \operatorname{tr}(A) + O(s^2) - 1}{s} \\ &= \lim_{s\to 0} \frac{s \cdot \operatorname{tr}(A) + O(s^2)}{s} = \lim_{s\to 0} \operatorname{tr}(A) + O(s) = \operatorname{tr}(A). \end{aligned}$$

Thus, $T_I(\mathrm{SL}(n)) = \operatorname{Ker}(d(\det)_I) = \{A \in \mathrm{M}(n) : \operatorname{tr}(A) = 0\}$. In other words, the tangent space to $\mathrm{SL}(n)$ at the identity is the space of matrices whose trace vanishes.

Solution (to Exercise 4.12) (a) We write $(z_0, z_1) = (x_0 + iy_0, x_1 = iy_1)$ for real coordinates x_0, y_0, x_1, y_1. First we get

$$\begin{aligned} \tilde{\hbar}(z_0, z_1) &= \left(2z_0\bar{z}_1, |z_0|^2 - |z_1|^2\right) \\ &= \left(2(x_0x_1 + y_0y_1) + i2(-x_0y_1 + y_0x_1), (x_0^2 + y_0^2) - (x_1^2 + y_1^2)\right). \end{aligned}$$

Then we can compute $d\tilde{\hbar}_q = 2 \cdot \begin{pmatrix} x_1 & y_1 & x_0 & y_0 \\ -y_1 & x_1 & y_0 & -x_0 \\ x_0 & y_0 & -x_1 & -y_1 \end{pmatrix}$.

(b) Let $g_4 : \mathbb{R}^4 \to \mathbb{R}$ and $g_3 : \mathbb{R}^3 \to \mathbb{R}$ be the smooth maps given by $x \mapsto |x|^2$. Then $\mathbb{S}^3 = g_4^{-1}(1)$ and $\mathbb{S}^2 = g_3^{-1}(1)$. Proposition 4.13, we have $T_q\mathbb{S}^3 = \operatorname{Ker}(d(g_4)_q) \subset T_q\mathbb{R}^4 = \mathbb{R}^4$ and $T_p\mathbb{S}^2 = \operatorname{Ker}(d(g_3)_p) \subset T_p\mathbb{R}^3 = \mathbb{R}^3$. We know that $\tilde{\hbar}(q) \in \mathbb{S}^2$ if $q \in \mathbb{S}^3$. Actually, our previous calculation shows that $\tilde{\hbar}(q) \in \mathbb{S}^2$ if and only if $q \in \mathbb{S}^3$. This implies $\mathbb{S}^3 = \tilde{\hbar}^{-1}(g_3^{-1}(1)) = (g_3 \circ \tilde{\hbar})^{-1}(1)$. In particular, $g_3(\tilde{\hbar}(q)) = 1$ is constant on $\mathbb{S}^3$. Hence, for every $q \in \mathbb{S}^3$, the image of the restriction $(d\tilde{\hbar}_q)_{|T_q\mathbb{S}^3}$ is contained in the kernel of $d(g_3)_{\tilde{\hbar}(q)}$ which is $T_{\tilde{\hbar}(q)}\mathbb{S}^2$.

(c) The fibre over a is $\hbar^{-1}(a) = \{(z_0, 0) \in \mathbb{S}^3 \subset \mathbb{C}^2 : |z_0|^2 = 1\}$. Let $q = (x_0, y_0, 0, 0) \in \hbar^{-1}(a)$ be a point in the fibre over a. The tangent space $T_q\mathbb{S}^3$ is the vector space

$$T_q\mathbb{S}^3 = \{\mathbf{u} \in \mathbb{R}^4 : \mathbf{u} \perp q\} = \text{Span}\left\{ q^\perp := \begin{pmatrix} -y_0 \\ x_0 \\ 0 \\ 0 \end{pmatrix}, \mathbf{e}_3^4 = \begin{pmatrix} 0 \\ 0 \\ 1 \\ 0 \end{pmatrix}, \mathbf{e}_4^4 = \begin{pmatrix} 0 \\ 0 \\ 0 \\ 1 \end{pmatrix} \right\}.$$

The vectors $q^\perp$, $\mathbf{e}_3^4$, and $\mathbf{e}_4^4$, are linearly independent and hence form a basis of $T_q\mathbb{S}^3$. Now we consider the map $d\tilde{\hbar}_q$. We computed this map as represented by the matrix

$$d\tilde{\hbar}_q = 2 \cdot \begin{pmatrix} 0 & 0 & x_0 & y_0 \\ 0 & 0 & y_0 & -x_0 \\ x_0 & y_0 & 0 & 0 \end{pmatrix}.$$

The kernel of this map is the span of the vector $q^\perp$. This implies that $d\hbar_q$ is surjective onto $T_a\mathbb{S}^2$. More concretely, the tangent space $T_a\mathbb{S}^2$ consists of the vectors which are orthogonal to a in $\mathbb{R}^3$. Hence it has a basis $(\mathbf{e}_1^3, \mathbf{e}_2^3)$ as a subspace in $\mathbb{R}^3$. The map $d\tilde{\hbar}_q : T_q\mathbb{R}^4 \to T_a\mathbb{R}^3$ sends

$$q^\perp \mapsto \mathbf{0}, \mathbf{e}_3^4 \mapsto \begin{pmatrix} 2x_0 \\ 2y_0 \\ 0 \end{pmatrix}, \mathbf{e}_4^4 \mapsto \begin{pmatrix} 2y_0 \\ -2x_0 \\ 0 \end{pmatrix}.$$

Hence the map $d\hbar_q : T_q\mathbb{S}^3 \to T_a\mathbb{S}^2$ can be represented in the chosen bases by the matrix $d\hbar_q = 2 \cdot \begin{pmatrix} 0 & x_0 & y_0 \\ 0 & y_0 & -x_0 \end{pmatrix}$. This map is surjective, since $-x_0^2 - y_0^2 = -1 \neq 0$. Hence q is a regular point for $\hbar$. Since q was any point in the fibre over a, we have shown that a is a regular value for $\hbar$.

To determine the fibre over b, we write $z_0 = x_0 + iy_0$ and $z_1 = x_1 + iy_1$. Then we get $\hbar(z_0, z_1) = (0, 1, 0)$ implies $2z_0\bar{z}_1 = i$ and $|z_0|^2 = |z_1|^2 = 1/2$, i.e., $y_0 = x_1$, $y_1 = -x_0$, and $x_0^2 + x_1^2 = 1/2$. Thus the fibre over b has the form

$$\begin{aligned} \hbar^{-1}(b) &= \{(z_0, z_1) \in \mathbb{S}^3 : \bar{z}_1 = \frac{i}{2z_0}\} \\ &= \{(x_0, y_0, x_1, y_1) \in \mathbb{S}^3 : y_0 = x_1, y_1 = -x_0\}. \end{aligned}$$

Let $q = (x_0, x_1, x_1, -x_0) \in \hbar^{-1}(b)$ be a point in the fibre over b. Since not both x_0 and x_1 can be zero, we assume that $x_0 \neq 0$. The tangent space $T_q\mathbb{S}^3$ is the vector space

$$T_q\mathbb{S}^3 = \{\mathbf{u} \in \mathbb{R}^4 : \mathbf{u} \perp q\} = \text{Span}\left\{q_1^\perp = \begin{pmatrix} -x_1 \\ x_0 \\ 0 \\ 0 \end{pmatrix}, q_2^\perp = \begin{pmatrix} -x_1 \\ 0 \\ x_0 \\ 0 \end{pmatrix}, q_3^\perp = \begin{pmatrix} x_0 \\ 0 \\ 0 \\ x_0 \end{pmatrix}\right\}.$$

Now we consider the map $d\tilde{\hbar}_q$. We computed this map as represented by the matrix $d\tilde{\hbar}_q = 2 \cdot \begin{pmatrix} x_1 & -x_0 & x_0 & x_1 \\ x_0 & x_1 & x_1 & -x_0 \\ x_0 & x_1 & -x_1 & x_0 \end{pmatrix}$. The kernel of this map is the span of the vector $q_0^\perp = \begin{pmatrix} -x_1 \\ x_0 \\ x_0 \\ x_1 \end{pmatrix}$. This implies that $d\hbar_q$ is surjective onto $T_a\mathbb{S}^2$. More concretely, the tangent space $T_b\mathbb{S}^2$ consists of the vectors which are orthogonal to b in $\mathbb{R}^3$. Hence it has a basis $(\mathbf{e}_1^3, \mathbf{e}_3^3)$ as a subspace in $\mathbb{R}^3$. The map $d\tilde{\hbar}_q : T_q\mathbb{R}^4 \to T_b\mathbb{R}^3$ sends

$$q_1^\perp \mapsto \begin{pmatrix} -2(x_0^2 + x_1^2) \\ 0 \\ 0 \end{pmatrix}, q_2^\perp \mapsto \begin{pmatrix} 2(x_0^2 - x_1^2) \\ 0 \\ -2x_0x_1 \end{pmatrix}, q_3^\perp \mapsto \begin{pmatrix} 2x_0x_1 \\ 0 \\ 2x_0^2 \end{pmatrix}.$$

Hence the map $d\hbar_q : T_q\mathbb{S}^3 \to T_b\mathbb{S}^2$ can be represented in the chosen bases by the matrix

$$d\hbar_q = 2 \cdot \begin{pmatrix} -x_0^2 - x_1^2 & x_0^2 - x_1^2 & x_0x_1 \\ 0 & -x_0x_1 & x_0^2 \end{pmatrix}.$$

This map is surjective, as one can check by using the conditions we have on x_0 and x_1. Hence q is a regular point for $\hbar$. Since q was any point in the fibre over b, we have shown that b is a regular value for $\hbar$.

(d) By our computation in point (a), we need to show that $d\tilde{\hbar}_q$ restricted to $T_q\mathbb{S}^3$ is surjective onto $T_{\hbar(q)}\mathbb{S}^2$ at every $q \in \mathbb{S}^3$. Since the tangent space $T_{\hbar(q)}\mathbb{S}^2$ of $\mathbb{S}^2$ is two-dimensional, we need to check that the image of $d\tilde{\hbar}_q$ restricted to Ker $(d(g_4)_q)$ spans a two-dimensional subspace. Since Ker $(d(g_4)_q)$ is of dimension 3, it suffices to show that $d\tilde{\hbar}_q$ has rank 3, which implies that the kernel of $d\tilde{\hbar}_q$ has dimension 1. Hence we need to show that $d\tilde{\hbar}_q$ always has three linearly independent columns. We can show this for example by calculating the determinants of appropriate 3×3-minors. Ignoring the factor 2 in our formula for $d\tilde{\hbar}_q$, we look at the minors A_j of the remaining matrix where we omit the jth column. We have $\det(A_4) = -x_1(x_0^2 + y_0^2 + x_1^2 + y_1^2) = -x_1$, $\det(A_3) = -y_0(x_0^2 + y_0^2 + x_1^2 + y_1^2) = -y_0$, $\det(A_2 = -x_0(x_0^2 + y_0^2 + x_1^2 + y_1^2) = -x_0$, and $\det(A_1) = -y_1(x_0^2 + y_0^2 + x_1^2 + y_1^2) = -y_1$. For every point

$q \in \mathbb{S}^3$, at least one of the coordinates x_0, y_0, x_1, y_1 is non-zero. Hence the matrix always has three linearly independent columns and $d\tilde{\hbar}_q$ has rank 3. This shows that each point in $\mathbb{S}^3$ is a regular point for $\hbar$, and hence every point in $\mathbb{S}^2$ is a regular value.

A.5 Transversality

Solution (to Exercise 5.5) By assumption, g is transverse to W, i.e.,

$$\operatorname{Im}(dg_y) + T_{g(y)}(W) = T_{g(y)}(Z) \text{ for all } y \in Y \text{ with } g(y) \in W. \tag{A.7}$$

Now we assume $f \pitchfork g^{-1}(W)$, i.e.,

$$\operatorname{Im}(df_x) + T_{f(x)}(g^{-1}(W)) = T_{f(x)}(Y) \text{ for } x \in X,\ f(x) \in g^{-1}(W). \tag{A.8}$$

We need to show $(g \circ f) \pitchfork W$. So let $x \in X$ be a point such that $g(f(x)) \in W$ and let $c \in T_{g(f(x))}(Z)$. By (A.7), there are vectors $b_1 \in T_{f(x)}(Y)$ and $b_2 \in T_{g(f(x))}(W)$ such that $dg_{f(x)}(b_1) + b_2 = c$. By (A.8), there are vectors $a \in T_x(X)$ and $b_3 \in T_{f(x)}(g^{-1}(W))$ such that $df_x(a) + b_3 = b_1$. Putting these two equations together we get

$$c = dg_{f(x)}(df_x(a) + b_3) + b_2 = dg_{f(x)}(df_x(a_1)) + dg_{f(x)}(b_3) + b_2.$$

By the chain rule, we have $d(g \circ f)_x = dg_{f(x)} \circ df_x$. By Proposition 4.13, we know $T_{f(x)}(g^{-1}(W)) = (dg_{f(x)})^{-1}(T_{g(f(x))}(W)$ in $T_{f(x)}(Y)$. In particular, $dg_{f(x)}(b_3) \in T_{g(f(x))}(W)$ and thus $dg_{f(x)}(b_3) + b_2 \in T_{g(f(x))}(W)$. Hence we have $c = d(g \circ f)_x(a) + dg_{f(x)}(b_3) + b_2$ lies in $[\operatorname{Im}(d(g \circ f)_x) + T_{g(f(x))}(W)] \subset T_{g(f(x))}(Z)$. Since c was an arbitrary element in $T_{g(f(x))}(Y)$, we have proven

$$\operatorname{Im}(d(g \circ f)_x) + T_{g(f(x))}(W) = T_{g(f(x))}(Z) \text{ for all } x \in X \text{ with } g(f(x)) \in W.$$

In other words, $(g \circ f) \pitchfork W$. Now we assume $(g \circ f) \pitchfork W$, i.e.,

$$\operatorname{Im}(d(g \circ f)_x) + T_{g(f(x))}(W) = T_{g(f(x))}(Z) \tag{A.9}$$

for $x \in X$ and $g(f(x)) \in W$. Let $x \in X$ be a point such that $f(x) \in g^{-1}(W)$ and let $b \in T_{f(x)}(Y)$. Since $g(f(x)) \in W$, we can use (A.9) to find vectors $a \in T_x(X)$ and $c \in T_{g(f(x))}(W)$ such that $d(g \circ f)_x(a) + c = dg_{f(x)}(b)$. By the chain rule, we have $d(g \circ f)_x(a) = dg_{f(x)}(df_x(a))$. Thus, $dg_{f(x)}(b - df_x(a)) = c \in T_{g(f(x))}(W)$. In other words,

$$b - df_x(a) \in (dg_{f(x)})^{-1}(T_{g(f(x))}(W)) = T_{f(x)}(g^{-1}(W)).$$

Since b was an arbitrary element, we have proven $\operatorname{Im}(df_x) + T_{f(x)}(g^{-1}(W)) = T_{f(x)}(Y)$ for all $x \in X$ with $f(x) \in g^{-1}(W)$. In other words, $f \pitchfork g^{-1}(W)$.

Solution (to Exercise 5.6) By definition, $\Gamma(A) \pitchfork \Delta$ if and only if $\Gamma(A) + \Delta(V) = V \times V$ since A is a linear map and $\Delta(V)$ is a vector space. See also Exercises 2.11 and 2.14. Let (v_1, v_2) be an arbitrary element of $V \times V$. We need to check under which conditions we can find $v, w \in V$ such that

$$(v_1, v_2) = (v, v) + (w, Aw) = (v + w, v + Aw),$$

i.e., $v_1 = v + w$ and $v_2 = v + Aw$. If we can find a suitable w, then we just set $v := v_1 - w$. Hence, by taking the difference of the two equations, we see that the question is reduced to checking whether we can find a w such that $v_2 - v_1 = Aw - w = (A - I)w$ where I is the identity map of V. But such a w exists for any choice of v_1 and v_2 if and only if $A - I$ is invertible, i.e., if and only if $\det(A - I) \neq 0$ which happens if and only if $+1$ is not an eigenvalue of A because the eigenvalues are the λ such that $\det(A - \lambda I) = 0$.

Solution (to Exercise 5.7) Let $\Delta_X = \{(x, x) : x \in X\} \subseteq X \times X$ be the diagonal of X and $\Gamma(f) = \{(x, f(x)) : x \in X\} \subseteq X \times X$ be the graph of f. Then the set of fixed points of f in X is the intersection $\Delta_X \cap \Gamma(f)$. Recall that a zero-dimensional manifold is just a discrete set of points. Since X is compact, $X \times X$ is also compact by Proposition 1.15. Hence a zero-dimensional submanifold of $X \times X$ is a discrete subset. By Lemma 1.17 and Exercise 1.3, this implies that the set is finite. Thus, in order to prove that f has only finitely many fixed points, it suffices to show that $\Delta_X \cap \Gamma(f)$ is a zero-dimensional submanifold of $X \times X$. Hence we would like to show that the intersection of Δ_X and $\Gamma(f)$ is transverse in $X \times X$. Since both have codimension equal to $\dim X$ in $X \times X$, the Transversality Theorem 5.2 then implies that $\Delta_X \cap \Gamma(f)$ is a zero-dimensional submanifold of $X \times X$. By definition, $\Delta_X \pitchfork \Gamma(f)$ means $T_{(x,x)}(\Gamma(f)) + T_{(x,x)}(\Delta_X) = T_{(x,x)}(X \times X)$ for every point $(x, x) \in \Delta_X \cap \Gamma(f)$. We know $T_{(x,x)}(\Gamma(f)) = \Gamma(df_x)$ and $T_{(x,x)}(\Delta_X) = \Delta_{T_x(X)}$ by Exercise 2.14. Moreover, we know $T_{(x,x)}(X \times X) = T_x(X) \times T_x(X)$. Hence we need to show $\Gamma(df_x) + \Delta_{T_x(X)} = T_x(X) \times T_x(X)$ for every point $(x, x) \in \Delta_X \cap \Gamma(f)$, i.e., $\Gamma(df_x) \pitchfork \Delta_{T_x(X)}$. Now, we have shown $\Gamma(df_x) \pitchfork \Delta_{T_x(X)}$ if $+1$ is not an eigenvalue of df_x in Exercise 5.6. Thus, since f is a Lefschetz map by assumption, this proves the assertion.

Solution (to Exercise 5.8) We can represent every element $x + iy \in \mathbb{C}$ by the real 2×2-matrix $\begin{pmatrix} x & -y \\ y & x \end{pmatrix}$, and we will freely use our terminology for smooth maps. We define the smooth map $f : \mathbb{C}^5 \setminus \{0\} \to \mathbb{C}$ by $f(z_1, \ldots, z_5) = z_1^2 + z_2^2 + z_3^2 + z_4^3 + z_5^{6k-1}$. Setting $Z := f^{-1}(0)$, we need to show that the intersection of Z and $\mathbb{S}^9$ is transverse in $\mathbb{C}^5 \setminus \{0\} \cong \mathbb{R}^{10} \setminus \{0\}$. The tangent space to Z at a point $z \in Z$ is the kernel of the derivative df_z. Since f is a polynomial in the variables $z_1, \ldots, z_5$, we can use our usual rules for partial differentiation to get the following matrix for $df_z\colon \mathbb{C}^5 \to \mathbb{C}$ with respect to the standard bases $df_z = (2z_1, 2z_2, 2z_3, 3z_4^2, (6k-1)z_5^{6k-2})$. The linear map df_z is surjective whenever df_z is not the zero map. Hence, for every $z \neq 0$, df_z is surjective. Thus, 0 is a regular value for f, and the tangent space $T_z(Z)$ is the kernel of df_z. This is a space of complex codimension 1 in $\mathbb{C}^5$ and real codimension 2 in $\mathbb{R}^{10}$. The tangent space $T_z(\mathbb{S}^9)$ consists of all vectors w in $\mathbb{R}^{10}$ which are orthogonal to z with respect to the inner product in $\mathbb{R}^{10}$, i.e., all vectors w which satisfy $w \cdot z = 0$. This is a vector subspace of $\mathbb{R}^{10}$ of dimension 9. Hence in order to show that the intersection of Z and $\mathbb{S}^9$ is transverse in $\mathbb{R}^{10}$ we need to show: for every $z \in Z \cap \mathbb{S}^9$, there is at least one vector w in $T_z(Z)$ which does not belong to $T_z\mathbb{S}^9$. Then we have $T_z(Z) + T_z(\mathbb{S}^9) \subseteq T_z(\mathbb{R}^{10})$, and it is a vector subspace of the same dimension as $T_z(\mathbb{R}^{10})$ and therefore equals $T_z(\mathbb{R}^{10})$. So let $z = (z_1, \ldots, z_5)$ be a fixed point in $Z \cap \mathbb{S}^9$. The tangent space $T_z(Z)$ is the kernel of df_z by Proposition 4.13. Hence we need to find at least one vector $w \in \mathbb{C}^5 = \mathbb{R}^{10}$ with $df_z(w) = 0$ and $w \cdot z \neq 0$. Set $m := 2\cdot 3\cdot(6k-1)$ and $w := \left(\frac{m}{2}z_1, \frac{m}{2}z_2, \frac{m}{2}z_3, \frac{m}{3}z_4, \frac{m}{6k-1}z_5\right)$. Then we have

$$\begin{aligned} df_z(w) &= \left(2z_1, 2z_2, 2z_3, 3z_4^2, (6k-1)z_5^{6k-2}\right) \cdot \begin{pmatrix} \frac{m}{2}z_1 \\ \frac{m}{2}z_2 \\ \frac{m}{2}z_3 \\ \frac{m}{3}z_4 \\ \frac{m}{6k-1}z_5 \end{pmatrix} \\ &= m\left(z_1^2 + z_2^2 + z_3^2 + z_4^3 + z_5^{6k-1}\right) = 0, \end{aligned}$$

since z is by assumption a point on Z. Hence $w \in T_z(Z)$. On the other hand, we can calculate the complex inner product of w and z as vectors in $\mathbb{C}^5$,

$$\begin{aligned} w \cdot z &= \left(\frac{m}{2}z_1, \frac{m}{2}z_2, \frac{m}{2}z_3, \frac{m}{3}z_4, \frac{m}{6k-1}z_5\right) \cdot \begin{pmatrix} \overline{z}_1 \\ \overline{z}_2 \\ \vdots \\ \overline{z}_5 \end{pmatrix} \\ &= \frac{m}{2}|z_1|^2 + \frac{m}{2}|z_2|^2 + \frac{m}{2}|z_3|^2 + \frac{m}{3}|z_4|^2 + \frac{m}{6k-1}|z_5|^2 > 0, \end{aligned}$$

which is bigger than zero, since z is a point on $\mathbb{S}^9$. Since the orthogonal complement of z in $\mathbb{C}^5$ is contained in the orthogonal complement of z in $\mathbb{R}^{10}$, this computation shows that z is not in the orthogonal complement of z in $\mathbb{R}^{10}$. Thus, w is a vector in $T_z(\mathbb{C}^5) = T_z(\mathbb{R}^{10})$ which is in $T_z(Z)$, but not in $T_z(\mathbb{S}^9)$. Hence the intersection of Z and $\mathbb{S}^9$ is transverse in $\mathbb{R}^{10}$. The codimension of $Z \cap \mathbb{S}^9$ in $\mathbb{R}^{10}$ is $2 + 1$ by the codimension formula. Thus $\dim(Z \cap \mathbb{S}^9) = 10 - 3 = 7$.

Solution (to Exercise 5.9) We need to show that the derivative $dg_x = df_x + A$ of g has full rank n for every x. To do so we define the map $F_k\colon \mathrm{M}(p,n;k) \times U \to \mathrm{M}(p,n)$ by $F_k(Q,x) := Q - df_x$. Since f is smooth, F_k is a smooth map, and, by Theorem 2.23, the domain of F_k is a smooth manifold of dimension $k(p+n-k)+n$. For $k \leq k' < n$, we have $k(p+n-k)+n \leq k'(p+n-k')+n$. Hence for $k = 0, 1, \ldots, n-1$, $\dim_{\mathbb{R}}(\mathrm{M}(p,n;k) \times U)$ equals

$$k(p+n-k)+n \leq (n-1)(p+n-(n-1))+n = (2n-p)+pn-1.$$

Since we assume $p \geq 2n$, we get $\dim_{\mathbb{R}}(\mathrm{M}(p,n;k) \times U) < pn = \dim_{\mathbb{R}} (\mathrm{M}(p,n))$. This means the derivative of F_k cannot be surjective for $k = 0, 1, \ldots, n-1$, and the only regular values for F_k with $k = 0, 1, \ldots, n-1$ are matrices in $\mathrm{M}(p,n)$ which are not in the image of F_k. By Sard's Theorem 5.17, the set of regular values for F_k is dense in $\mathrm{M}(p,n)$. Since the finite intersection of dense subsets is again a dense subset, the set of matrices which are not in the image of F_k for $k = 0, 1, \ldots, n-1$ is dense in $\mathrm{M}(p,n)$. Hence there is a matrix A in $\mathrm{M}(p,n)$ arbitrarily close to the zero matrix which is not in the image of the F_k for $k = 0, 1, \ldots, n-1$. Thus, the matrix $df_x + A$ has rank n since otherwise A had been in the image of one of the F_k with $k < n$. Hence we have found the desired matrix A such that $dg_x = df_x + A \cdot x$ has rank n for all x, and thus such that g is an immersion.

Solution (to Exercise 5.10) A point x is regular if and only if we have $df_x = \left(\frac{\partial f}{\partial x_1}(x), \ldots, \frac{\partial f}{\partial x_k}(x)\right) \neq 0$, and x is a non-degenerate critical point if $df_x = 0$ and $\det(H(f)_x) \neq 0$. Hence f is Morse if and only if $(\det(H(f)_x))^2 + \sum_{i=1}^{k}\left(\frac{\partial f}{\partial x_i}(x)\right)^2 > 0$ for all $x \in U$.

A.6 Abstract Smooth Manifolds

Solution (to Exercise 6.1) (a) A line in $\mathbb{R}^2$ is determined by an equation of the form $ax + by + c = 0$ with fixed $(a,b,c) \in \mathbb{R}^3$. Since an equation of the form $ax + by + c = 0$ with $a = b = 0$ does not define a line, we have to exclude triples of the form $(0,0,c)$. Moreover, the equations $ax + by + c = 0$ and $(\lambda a)x(\lambda b)y + (\lambda c) = 0$ with $\lambda \neq 0$ determine the same line. Let $\sim$ denote the equivalence relation defined by $(a,b,c) \sim (\lambda a, \lambda b, \lambda c)$ if there is a $\lambda \neq 0$. Then X can be identified with the set of equivalence classes $X = (\mathbb{R}^3 \setminus \{(0,0,0),(0,0,1)\})/\sim$. This is the subspace of $\mathbb{RP}^2$ given by removing the point $[0:0:1]$. Since any subspace consisting of just one point is closed in $\mathbb{RP}^2$, we have shown that X can be identified with the open subset $\mathbb{RP}^2 \setminus \{[0:0:1]\}$ of $\mathbb{RP}^2$.

(b) Every line in $\mathbb{R}^2$ is determined by the point where it crosses the x-axis and a direction which can be expressed by an angle $\in [0, 2\pi]$. Since we have not specified a direction for the line, two angles which differ by adding π determine the same line. Any angle between 0 and 2π can be described by a point on the unit circle, where the points s and $-s$ on $\mathbb{S}^1$ correspond to angles which differ by adding π. Hence any line in $\mathbb{R}^2$ is determined by a $(s,x) \in \mathbb{S}^1 \times \mathbb{R}$ where s is uniquely determined up to multiplying with ± 1.

Solution (to Exercise 6.2) (a) To show that $\mathbb{CP}^1$ is second-countable and Hausdorff we can show as for $\mathbb{RP}^n$ that the quotient map is open and the graph of the equivalence relation is closed. Then we apply Lemma 6.5 and Exercise 1.7. It remains to construct an atlas. We define $V_1 := \{[z:w] \in \mathbb{CP}^1 : z \neq 0\}$ and $V_2 := \{[z:w] \in \mathbb{CP}^1 : w \neq 0\}$. The preimage of V_1 in $\mathbb{C}^2$ is the open subset $\{(z,w) \in \mathbb{C}^2 : z \neq 0\}$, and the preimage of V_2 in $\mathbb{C}^2$ is the open subset $\{(z,w) \in \mathbb{C}^2 : w \neq 0\}$. Hence V_1 and V_2 are open in $\mathbb{CP}^1$. Since either z or w must be non-zero for every point $[z:w] \in \mathbb{CP}^1$, $\{V_1, V_2\}$ provides an open cover of $\mathbb{CP}^1$. We define maps $\phi_1 \colon \mathbb{R}^2 \to V_1$ and $\phi_2 \colon \mathbb{R}^2 \to V_2$ by $\phi_1 \colon (x,y) \mapsto [1 : (x+iy)]$ and $\phi_2 \colon (x,y) \mapsto [(x+iy) : 1]$. The inverses are $\phi_1^{-1} \colon V_1 \to \mathbb{R}^2$ and $\phi_2^{-1} \colon V_2 \to \mathbb{R}^2$ defined by

$$\phi_1^{-1} \colon [(x_1+iy_1) : (x_2+iy_2)] \mapsto \frac{1}{x_1^2+y_1^2}(x_1x_2+y_1y_2, x_1y_2-x_2y_1) \text{ and}$$
$$\phi_2^{-1} \colon [(x_1+iy_1) : (x_2+iy_2)] \mapsto \frac{1}{x_2^2+y_2^2}(x_1x_2+y_1y_2, x_2y_1-x_1y_2),$$

where the right-hand sides arise from calculating the quotients of complex numbers. As in the proof of Theorem 6.36, we can check that these maps do not depend on the chosen representatives and that ϕ and ϕ_i^{-1} are mutual inverses which are both continuous. We see that the change-of-coordinate maps are smooth, since the composites

$$\phi_1^{-1}(V_1 \cap V_2) \xrightarrow{\phi_1} V_1 \cap V_2 \xrightarrow{\phi_2^{-1}} \phi_2^{-1}(V_1 \cap V_2) \text{ and}$$

$$\phi_2^{-1}(V_1 \cap V_2) \xrightarrow{\phi_2} V_1 \cap V_2 \xrightarrow{\phi_1^{-1}} \phi_2^{-1}(V_1 \cap V_2)$$

are given by $(x, y) \mapsto \frac{1}{x^2+y^2}(x, -y)$ and are therefore smooth maps.

(b) We copy the relation we used before and just add a condition to make sure that the norms are respected. We define $\sim_s$ by $(z_0, w_0) \sim_s (z_1, w_1)$ if and only if $z_1 = \lambda z_0$ and $w_1 = \lambda w_0$ for some $\lambda \in \mathbb{C} \setminus \{0\}$ with $|\lambda| = 1$. Then we can check that $\mathbb{S}^3/\sim_s \cong (\mathbb{C}^2 \setminus \{(0,0)\})/\sim$.

(c) For $w \neq 0$, we consider the points $[z : w] = [z/w : 1]$ in $\mathbb{CP}^1$ as points in $\mathbb{C}$ by identifying them with z/w. This misses only the point $[1 : 0]$ in $\mathbb{CP}^1$. Now we mimic the stereographic projection and define $h\colon \mathbb{CP}^1 \to \mathbb{S}^2$ by

$$[z : w] \mapsto \begin{cases} \frac{1}{|z/w|^2+1}(2z/w, |z/w|^2 - 1) & \text{if } w \neq 0 \\ (0,1) & \text{if } w = 0, \end{cases}$$

where we consider $\mathbb{S}^2$ as a subset of $\mathbb{C} \times \mathbb{R}$ as before. We need to check that $h([z : w])$ actually is a point on $\mathbb{S}^2$. This is true for $h([z : 0]) = (0, 1)$, and for the other points we check

$$\begin{aligned} |h([z : w])|^2 &= \frac{2z/w}{|z/w|^2+1} \cdot \frac{2\bar{z}/\bar{w}}{|z/w|^2+1} + \frac{(|z/w|^2-1)^2}{(|z/w|^2+1)^2} \\ &= \frac{4|z/w|^2 + |z/w|^4 - 2|z/w|^2 + 1}{(|z/w|^2+1)^2} \\ &= \frac{|z/w|^4 + 2|z/w|^2 + 1}{(|z/w|^2+1)^2} = 1. \end{aligned}$$

Now we check that h is well defined, i.e., if $\lambda \neq 0 \in \mathbb{C}$, we need to check that h sends $[z : w]$ and $[\lambda z : \lambda w]$ to the same point. For the latter, it suffices to observe $|z/w| = |(\lambda z)/(\lambda w)|$. Finally, for $(z, 0) \in \mathbb{S}^3$, we compute $h(\varphi([z : 0])) = (0, 1) = \hbar(z, 0)$ where we use that $|z|^2 = 1$ as $(z, 0)$ is a point on $\mathbb{S}^3$. And for $(z, w) \in \mathbb{S}^3$ with $w \neq 0$, we get

$$\begin{aligned} h(\varphi([z : w])) &= \frac{1}{|z/w|^2+1}(2z/w, |z/w|^2 - 1) \\ &= \frac{|w|^2}{|z|^2+|w|^2}(2z/w, |z/w|^2 - 1) \\ &= (2z/w \cdot (w\bar{w}), |z|^2 - |w|^2) = (2z \cdot \bar{w}, |z|^2 - |w|^2) \\ &= \hbar(z, w), \end{aligned}$$

where we used $|z|^2 + |w|^2 = 1$. Hence we have $h \circ \varphi = \hbar$.

(d) The fibre consists of all points (z, w) in $\mathbb{S}^3$ with $[z : w] = [z_0 : w_0]$. By the definition of $\mathbb{C}\mathrm{P}^1$, that means (z, w) is in the fibre if and only if there is a $\lambda \in \mathbb{C} \setminus \{0\}$ such that $(z, w) = (\lambda z_0, \lambda w_0)$. The additional feature we need to remember is that (z, w) is in $\mathbb{S}^3$. Hence λ needs to satisfy $|\lambda| = 1$. Summarising, we have $(z, w) \in \varphi^{-1}([z_0 : w_0])$ if and only if $(z, w) = (\lambda z_0, \lambda w_0)$ for some $\lambda \in \mathbb{C} \setminus \{0\}$ with $|\lambda| = 1$. In other words, the points in the fibre $\varphi^{-1}([z_0 : w_0])$ are in one-to-one correspondence to points λ on the circle $\mathbb{S}^1 \subset \mathbb{C}$. This shows again that the fibre of the Hopf map at any point is diffeomorphic to $\mathbb{S}^1$.

Solution (to Exercise 6.3) (a) For $z \in \mathbb{C} \setminus \{0\}$, let $[z]$ be its equivalence class in H^2. We pick a point $[z_0] \in H^2$. Choosing $\varepsilon > 0$ small enough, i.e., in our case $0 < \varepsilon < \lambda/2$ is enough; the open ball $\mathbb{B}^2_\varepsilon(z_0) \subset \mathbb{C} \setminus \{0\}$ does not contain any point z with $[z] = [z_0]$. Hence the restriction $\mathbb{B}^2_\varepsilon(z_0) \to H^2$ of the quotient map to $\mathbb{B}^2_\varepsilon(z_0)$ is a homeomorphism onto its image $\bar{\mathbb{B}}^2_\varepsilon([z_0]) \subset H^2$. The inverse is given by sending a point $[z] \in \bar{\mathbb{B}}^2_\varepsilon([z_0])$ to the point $z \in \mathbb{B}^2_\varepsilon(z_0)$. Since ε is small enough, there is a unique such lift in $\mathbb{B}^2_\varepsilon(z_0)$. This defines a homeomorphism $\phi \colon \mathbb{B}^2_\varepsilon(z_0) \to \bar{\mathbb{B}}^2_\varepsilon([z_0])$ whose inverse ψ is a local chart around $[z_0] \in H^2$. If ψ_1 and ψ_1 are charts around $[z_1]$ and $[z_2]$, respectively, then, since we chose ε small enough, the change of coordinate map $\psi_2 \circ \psi_1^{-1} \colon \mathbb{B}^2_\varepsilon(z_1) \cap \mathbb{B}^2_\varepsilon(z_2) \to \mathbb{B}^2_\varepsilon(z_1) \cap \mathbb{B}^2_\varepsilon(z_2)$ is just the identity. Since this is a smooth map, we have shown that H^2 is an abstract smooth 2-manifold.

(b) We only construct charts. We write $[z]$ for the image of $z \in A$ in $A/\mathbb{Z}$. We define the map $h \colon A/\mathbb{Z} \to H^2, [z]_A \mapsto [z]_H$ where $[z]_A$ and $[z]_H$ denote the equivalence classes of $z \in \mathbb{C}$ in $A/\mathbb{Z}$ and H^2, respectively. We claim that h is a homeomorphism. To prove this claim, we check first that h is injective. Assume $w, z \in A$ with $h([w]_A) = h([z)_A]$. After possibly switching w and z, we can assume $|w| \leq |z|$. Then there is a $k \in \mathbb{Z}$, $k \leq 0$, such that $w = \lambda^k z$. If $k = 0$, then $w = z$, and if $k < 0$, then not both z and w can be in A, unless one of w and z has absolute value 1 and the other one has absolute value $1/\lambda$. So let us say $|w| = 1$ and $|z| = 1/\lambda$. In this case, their images in the quotient $A/\mathbb{Z}$ are the same. Hence h is injective. Next we check that f is surjective. Let $z \in \mathbb{C} \setminus \{0\}$. Then there is a $k \in \mathbb{Z}$ such that $\lambda^k z \in A$. Hence $h([\lambda^k z]_A) = [z]_H \in H^2$, and h is surjective. Since A is a closed subset of $\mathbb{C}$, the map h is continuous. Since A is compact and the quotient map $A \to A/\mathbb{Z}$ is continuous, $A/\mathbb{Z}$ is compact by Lemma 1.13 and Exercise 1.1. Hence h is a continuous bijection with compact domain and Hausdorff codomain. It now follows from Corollary 1.23 that h is a homeomorphism.

(c) Again, we only construct charts. For $(z, w) \in \mathbb{C}^2 \setminus \{0\}$, let $[z, w]$ be its equivalence class in H^4. We pick a point $[z_0, w_0] \in H^4$. Choosing $\varepsilon > 0$ small enough, the open ball $\mathbb{B}^4_\varepsilon(z_0) \subset \mathbb{C}^2 \setminus \{0\}$ does not contain any point (z, w) with $[z, w] = [z_0, w_0]$. Hence the restriction $\mathbb{B}^4_\varepsilon(z_0, w_0) \to H^4$ to $\mathbb{B}^2_\varepsilon(z_0)$ is a homeomorphism onto its image $\bar{\mathbb{B}}^4_\varepsilon([z_0, w_0]) \subset H^4$. The inverse is given by sending a point $[z, w] \in \bar{\mathbb{B}}^4_\varepsilon([z_0, w_0])$ to the point $(z, w) \in \mathbb{B}^4_\varepsilon(z_0, w_0)$. Since ε is small enough, there is a unique such lift in $\mathbb{B}^4_\varepsilon(z_0, w_0)$. This defines a homeomorphism $\phi\colon \mathbb{B}^4_\varepsilon(z_0, w_0) \to \bar{\mathbb{B}}^4_\varepsilon([z_0, w_0])$, whose inverse ψ is a local chart around $[z_0, w_0] \in H^4$. If ψ_1 and ψ_1 are charts around $[z_1, w_1]$ and $[z_2, w_2]$, respectively, then, by choosing ε small enough, the change of coordinate map $\psi_2 \circ \psi_1^{-1}\colon \mathbb{B}^4_\varepsilon(z_1, w_1) \cap \mathbb{B}^4_\varepsilon(z_2, w_2) \to \mathbb{B}^4_\varepsilon(z_1, w_1) \cap \mathbb{B}^4_\varepsilon(z_2, w_2)$ is just the identity.

(d) We consider $\mathbb{S}^3$ as a subset in $\mathbb{C}^2 \setminus \{0\}$ and $\mathbb{S}^1$ as the quotient $[1, 1/\lambda]/(1 \sim 1/\lambda)$, i.e., the closed interval $[1, 1/\lambda] \subset \mathbb{R}$ where we identify the endpoints. We define a map

$$f\colon \mathbb{S}^3 \times \mathbb{S}^1 = \mathbb{S}^3 \times [1, 1/\lambda]/(1 \sim 1/\lambda) \to H^4, (z, w, s) \mapsto [sz, sw].$$

To check that this is well defined, we compute $f(z, w, 1) = [z, w] = [1/\lambda z, 1/\lambda w] = f(z, w, 1/\lambda)$. Now we show that f is a homeomorphism. Recall that every point in $\mathbb{C}^2 \setminus \{0\}$ is determined by a direction, i.e., a point in $\mathbb{S}^3$, and the distance from the origin. Since we identify points p_1 and p_2 in H^4 if $p_2 = \lambda^k p_1$ for some $k \in \mathbb{Z}$, it suffices to specify the distance from the origin up to a multiple of λ^k. This shows that f is injective. Moreover, f is surjective since every point p in $\mathbb{C}^2 \setminus \{0\}$ determines a point in $\mathbb{S}^3$, i.e., the point (z, w) where the line from the origin to p meets $\mathbb{S}^3$. Then there is a unique $k \in \mathbb{Z}$ such that $1 \le \lambda^k |p| < 1/\lambda$. This real number $\lambda^k |p|$ is the coordinate s. Since f is continuous and has a compact domain and a Hausdorff codomain, f is a homeomorphism by Corollary 1.23. We now show that f is a diffeomorphism. Our copy of $\mathbb{S}^1$ equals the quotient $\mathbb{R}/\mathbb{Z}$, where we identify two real numbers s_1 and s_2 if $s_1 - s_2 \in \mathbb{Z}$. This makes it easy to provide local charts. Around any $x \in \mathbb{R}$, we can look at the open neighbourhood $(-\varepsilon + x, x + \varepsilon)$ which maps homeomorphically onto its image in $\mathbb{S}^1 = \mathbb{R}/\mathbb{Z}$ under the quotient map $\mathbb{R} \to \mathbb{R}/\mathbb{Z}$. On points in $\mathbb{C}^2 \setminus \{0\}$ and $\mathbb{S}^3$, f and its inverse are just scaling by a real number, which is some integer power of λ. Hence composition with local charts results in a smooth map since the local charts on $\mathbb{S}^1 = \mathbb{R}/\mathbb{Z}$ are just identity maps and the local charts on $\mathbb{S}^3$ are diffeomorphisms. Hence f is a diffeomorphism as it is a smooth homeomorphism with a smooth inverse.

Solution (to Exercise 6.4) Since $H(m,n)$ is a closed subspace in $\mathbb{RP}^m \times \mathbb{RP}^n$, it is Hausdorff and second-countable. It remains to construct charts. Recall the subsets $V_i := \{[x_0 : \ldots : x_n] \in \mathbb{RP}^n : x_i \neq 0\}$ which are open in $\mathbb{RP}^n$. The product spaces $V_i \times V_j$ are open in $\mathbb{RP}^m \times \mathbb{RP}^n$, and hence the subsets $V_{ij} = H(m,n) \cap (V_i \times V_j)$ are open in $H(m,n)$. The union of all $V_i \times V_j$ for all i, j covers $\mathbb{RP}^m \times \mathbb{RP}^n$. Hence the union of all $V_{ij} = H(m,n) \cap (V_i \times V_j)$ for all i, j covers $H(m,n)$. In fact, it suffices to consider all pairs (i, j) with $i \neq j$, since if x_i is the only coordinate in $[x]$ with $x_i \neq 0$, then we must have $y_i = 0$ in order to satisfy the condition $\sum_{i=0}^{m} x_i y_i = 0$. Thus we cannot have that x_i and y_i are the only non-zero coordinates in $[x]$ and $[y]$, respectively. Now we define maps $\phi_{ij}\colon \mathbb{R}^{m+n-1} \to V_{ij}$ by sending the $(m+n-1)$-tuple $((x_1, \ldots, x_m), (y_1, \ldots, y_{n-1}))$ to

$$\left([x_1 : \ldots : 1 : \ldots : x_m], [y_1 : \ldots : -\sum_{s=1, s\neq i}^{m} x_s y_s : \ldots : 1 : \ldots : y_{n-1}]\right),$$

where the first 1 and the sum are at position $i+1$, and the second 1 is at position $j+1$. Note that this actually yields an element in $H(m,n)$, and not just $\mathbb{RP}^m \times \mathbb{RP}^n$, since the defining condition $H(m,n)$ is satisfied. Their inverses $\phi_{ij}^{-1}\colon V_{ij} \to \mathbb{R}^{m+n-1}$ are given by sending $([x_0 : \ldots : x_m], [y_0 : \ldots : y_n])$ to

$$\left(\frac{1}{x_i}(x_0, \ldots, \widehat{x_i}, \ldots, x_m), \frac{1}{y_j}(y_0, \ldots, \widehat{y_j}, \ldots, y_n)\right).$$

We can check that the change of coordinate maps are smooth just as we did for real projective space since the sum $\sum_{i=0}^{m} x_i y_i$ is a polynomial and hence smooth.

A.7 Smooth Homotopy

Solution (to Exercise 8.1) (a) Let f and g be two smooth maps $Y \to X$. Let F be a homotopy from the identity map of X to the constant map $X \to \{x_0\}$ for some $x_0 \in X$. Then we can use F to define a homotopy from f to $Y \to \{x_0\}$ and a homotopy from $Y \to \{x_0\}$ to g. Putting these two homotopies together yields a homotopy H from f to g. It only remains to make sure that H is smooth. To achieve this we apply the technique used in the main text. After composing with a smooth bump function, we can assume $F(x,t) = x$ for all $(x,t) \in X \times [0,1/4]$ and $F(x,t) = x_0$ for all $(x,t) \in X \times [3/4,1]$. Then we can define H by

$$H\colon Y \times [0,1] \to X,\ (y,t) \mapsto \begin{cases} F(f(y),2t) & t \in [0,1/2] \\ F(g(y),2(1-t)) & t \in [1/2,1]. \end{cases}$$

(b) Let $Y = X$ and let $f\colon X \to X$ be the identity and $g\colon X \to \{x_0\} \subset X$ be the constant map for some point $x_0 \in X$. By the assumption, f and g are homotopic. Hence X is contractible.

(c) The map $F\colon \mathbb{R}^k \times [0,1] \to \mathbb{R}^k$, $(x,t) \mapsto (1-t)x$ is a smooth homotopy from the identity map to the constant map $\mathbb{R}^k \to \{0\}$.

Solution (to Exercise 8.2) By Sard's Theorem 5.17, there is a regular value $y \in \mathbb{S}^n$ for f. Assume there is a point $x \in f^{-1}(y)$. Since $\dim T_xX = k$ and $\dim T_y\mathbb{S}^n = n$, df_x cannot be surjective if $k < n$. Thus, if $k < n$, then $f^{-1}(y)$ must be empty. Hence we can assume that the image of f is contained in $U := \mathbb{S}^n \setminus \{y\}$. Now we can use stereographic projection from y to define a diffeomorphism $\psi\colon U \to \mathbb{R}^n$. Thus, $\psi \circ f$ is homotopic to a constant map by point (c) of Exercise 8.1. Composing the homotopy with the inverse of ψ defines a homotopy from f to a constant map.

Solution (to Exercise 8.4) For $k = 1$, the antipodal map is $(x,y) \mapsto (-x, -y)$. The map

$$F_1\colon \mathbb{S}^1 \times [0,1] \to \mathbb{S}^1,\ ((x,y),t) \mapsto \begin{pmatrix} \cos(\pi t) & -\sin(\pi t) \\ \sin(\pi t) & \cos(\pi t) \end{pmatrix} \begin{pmatrix} x \\ y \end{pmatrix}$$

is a smooth homotopy from the identity of $\mathbb{S}^1$ to the antipodal map. To convince ourselves that $F_1(x,y,t)$ is an element in $\mathbb{S}^1$, we can either just calculate its norm or observe that the matrix $\begin{pmatrix} \cos(\pi t) & -\sin(\pi t) \\ \sin(\pi t) & \cos(\pi t) \end{pmatrix}$ is an element in $\mathrm{O}(2)$ for every t. Elements in $\mathrm{O}(2)$ preserve the scalar product and hence the norm of vectors in $\mathbb{R}^2$. For an arbitrary odd k, we have $\mathbb{S}^k \subset \mathbb{R}^{k+1}$ and $k+1$ is even. Then we define a smooth homotopy $F_k\colon \mathbb{S}^k \times [0,1] \to \mathbb{S}^k$ from the identity in $\mathbb{S}^k$ to the antipodal map by sending $((x_1,y_1),\ldots,(x_{(k+1)/2},y_{(k+1)/2}),t)$ to $(F_1(x_1,y_1,t),\ldots,F_1((x_{(k+1)/2},y_{(k+1)/2},t)$. Then, for every t, the linear map $F_k(-,-,t)\colon \mathbb{R}^{k+1} \to \mathbb{R}^{k+1}$ is an element in $\mathrm{O}(k+1)$ and preserves the norm on $\mathbb{R}^{k+1}$.

Solution (to Exercise 8.5) Given two points $x, y \in X$, we define the relation $x \sim y$, and say x and y are path-connected, by $x \sim y$ if and only if there is a smooth path $\gamma\colon [0,1] \to X$ with $\gamma(0) = x$ and $\gamma(1) = y$. We now show that $\sim$ is an equivalence relation by linking it to the homotopy relation. Let $f\colon \{x\} \to X$, $f(x) = x$, and $g\colon \{x\} \to X$, $g(x) = y$. If $F\colon \{x\} \times [0,1] \to X$ is a smooth homotopy from f to g, then $\gamma(t) := F(x,t)$ is a smooth path from x to y. Conversely, if γ is a smooth path from x to y, then $F(x,t) := \gamma(t)$ is a smooth homotopy from f to g. Thus, $x \sim y$ if and only if $f \sim g$. Since homotopy defines an equivalence relation, we see that path-connectedness $x \sim y$ is also an equivalence relation. The equivalence class $[x]$ of a point $x \in X$ under the relation $\sim$ is the set $[x] = \{y \in X\colon x \sim y\}$. An important feature of an equivalence relation is that equivalence classes are either equal or disjoint, i.e., for any x and y in X, we have either $[x] = [y]$ or $[x] \cap [y] = \emptyset$. We are going to use this fact in the following way. If we can show that every equivalence class $[x]$ is an open subset of X, then we know that every $[x]$ is also a closed subset, since $X \setminus [x]$ is the union of all the other equivalence classes and is therefore open itself as arbitrary unions of open sets are open.

Let $x_0 \in X$ be an arbitrary point. We now show that $[x_0]$ is open. Let $x \in X$ be a point in $[x_0]$. Since X is a smooth manifold, there is a local parametrisation $\phi\colon \mathbb{B}_\varepsilon(0) \to U$ with $\phi(0) = x$, where U is open in X and $\mathbb{B}_\varepsilon(0)$ is the open ball of radius ε in $\mathbb{R}^{\dim X}$. Given any $y \in U$, let $\phi^{-1}(y)$ be its preimage in $\mathbb{B}_\varepsilon(0)$. In $\mathbb{B}_\varepsilon(0)$, all points are path-connected to 0. Hence there is a smooth path $\gamma\colon [0,1] \to \mathbb{B}_\varepsilon(0)$ defined by $\gamma(t) := t \cdot \phi^{-1}(y)$ with $\gamma(0) = 0$ and $\gamma(1) = \phi^{-1}(y)$. Since ϕ is a diffeomorphism, the composite $\phi \circ \gamma$ is a smooth path from x to y in X, i.e., $x \sim y$. This shows that U is contained in $[x_0]$. Thus $[x_0]$ is an open subset in X, since every point $x \in [x_0]$ has an open neighbourhood in X which is completely contained in $[x_0]$. This shows that $[x_0]$ is a non-empty, open, and closed subset of X. Since X is connected, this implies $[x_0] = X$. Thus X is path-connected.

Solution (to Exercise 8.6) Given two points $x, y \in X$, we define the relation $x \sim y$, if and only if there is a smooth path $\gamma\colon [0,1] \to X \setminus \{x_0\}$ with $\gamma(0) = x$ and $\gamma(1) = y$. As in the solution to Exercise 8.5, we show that this is again an equivalence relation and that every equivalence class $[x]$ is open. The difference is that it is not clear a priori that $X \setminus \{x_0\}$ is connected. Thus, we cannot conclude $[x] = X \setminus \{x_0\}$ yet. For example, $\mathbb{R}$ is path-connected, but $\mathbb{R} \setminus \{0\}$ is not connected, since there are two open and closed equivalence classes $[1] = \mathbb{R}^{>0}$ and $[-1] = \mathbb{R}^{<0}$. Another example is the space $H_0 = \{(x,y,z) \in \mathbb{R}^3\colon x^2 + y^2 - z^2 = 0\}$ for which we showed in Exercise 2.7 that H_0 is not a smooth manifold. While H_0 is connected, the subspace $H_0 \setminus \{0\}$, where 0 denotes the origin in $\mathbb{R}^3$, is not connected. Now we show that $X \setminus \{x\}$ is still path-connected when X is a connected smooth manifold of dimension $\dim X = n > 1$. To do so, it suffices to show that every point in X

has an open neighbourhood that is still path-connected when we remove x_0. In other words, we need to show that every equivalence class $[x]$ remains path-connected when we remove x_0. This is true for all points in X which are not x_0 by Exercise 8.5. So let V be an open neighbourhood of x_0 in X. After possibly shrinking V, we can assume that there is a diffeomorphism $\phi\colon \mathbb{B}^n_\varepsilon(0) \to V$ such that $\phi(0) = x_0$ where $\mathbb{B}^n_\varepsilon(0)$ is the open ball around the origin in $\mathbb{R}^n$ with radius $\varepsilon > 0$. Since $n \geq 2$, any two points in $\mathbb{B}^n_\varepsilon(0)$ distinct from the origin are connected by a smooth path which does not pass through the origin. Note that the latter conclusion would not be true for $n = 1$. Since ϕ is a diffeomorphism, this shows that any two points distinct from x_0 are connected by a smooth path which does not pass through x_0. This shows that all points in X are connected by a smooth path which does not pass through x_0. Thus, $X \setminus \{x\}$ is path-connected.

Solution (to Exercise 8.7) For simplicity, we assume $x = -1$ in $\mathbb{S}^1 \subset \mathbb{C}$. By Example 3.6, the restriction of the map $p\colon \mathbb{R} \to \mathbb{S}^1 \subset \mathbb{C}$, $t \mapsto e^{2\pi i t}$, to $(-1/2, 1/2) \subset \mathbb{R}$ is a diffeomorphism. Let $(p_{|(-1/2,1/2)})^{-1}\colon \mathbb{S}^1 \setminus \{-1\} \to (-1/2, 1/2)$ be the inverse map. We can then define a smooth map $(\mathbb{S}^1 \setminus \{-1\}) \times [0,1] \to \mathbb{S}^1 \setminus \{-1\}$ by $(t,s) \mapsto p(s \cdot (p_{|(-1/2,1/2)})^{-1}(t))$. This defines a homotopy between the constant map at 1 for $s = 0$ and the identity of $\mathbb{S}^1 \setminus \{-1\}$ for $s = 1$.

Solution (to Exercise 8.8) By Exercise 8.7, $\mathbb{S}^1 \setminus \{-1\}$ is contractible. By Exercise 8.2, there is a homotopy h_t between the constant map at 1 and f/g, since $f(x)/g(x) \neq -1$ for all $x \in \mathbb{S}^1$. Then the map $h_t \cdot g$ defined by $x \mapsto h_t(x) \cdot g(x)$ is a homotopy between f and g.

Solution (to Exercise 8.9) The assumption $|f(x) - g(x)| < 2$ implies that $f(x)$ and $g(x)$ are never antipodal points. In particular, the straight line segment from $f(x)$ to $g(x)$ in $\mathbb{R}^{k+1}$ does not pass through the origin for all $x \in X$. Hence the vector $(1-t)f(x) + tg(x)$ is non-zero for all $x \in X$ and all $t \in [0,1]$. Thus, we can form the well defined map $H\colon X \times [0,1] \to \mathbb{S}^k$ by $x \mapsto \frac{(1-t)f(x)+tg(x)}{|(1-t)f(x)+tg(x)|}$. Since f and g are continuous, H is continuous and defines a homotopy from f to g.

Solution (to Exercise 8.10) (a) If k is odd, then $k+1$ is even, and we can define the map $s\colon \mathbb{S}^k \to \mathbb{R}^{k+1}$, $(x_1, \ldots, x_{k+1}) \mapsto (-x_2, x_1, -x_3, x_4, \ldots, -x_{k+1}, x_k)$. This map can be extended to a linear map $\mathbb{R}^{k+1} \to \mathbb{R}^{k+1}$ and therefore s is smooth. For each $x \in \mathbb{S}^k$, $s(x)$ is non-zero and satisfies $x \perp s(x)$. Thus $s(x)$ is a tangent vector at x, i.e., $s(x) \in T_x(\mathbb{S}^k) \setminus \{0\}$.

Hence $\mathbf{v}\colon \mathbb{S}^k \to T(\mathbb{S}^k)$, $\mathbf{v}(x) := (x, s(x))$, is the desired non-vanishing vector field on $\mathbb{S}^k$.

(b) Given a vector field $\mathbf{v}\colon \mathbb{S}^k \to T(\mathbb{S}^k)$ which has no zeros. Let $\mathbf{v}(x) = (x, s(x))$. Since $s(x) \neq 0$ for every $x \in \mathbb{S}^k$, we can define a new vector field by $x \mapsto s(x)/|s(x)|$. By replacing s with this new non-vanishing vector field, we can assume $|s(x)| = 1$. Hence we can assume $s(x) \in \mathbb{S}^k$ and $s(x) \cdot x = 0$ for every $x \in \mathbb{S}^k$. Now we define the map $F\colon \mathbb{S}^k \times [0,1] \to \mathbb{S}^k$ by $F(x,t) := \cos(\pi t)x + \sin(\pi t)s(x)$. We need to check that $F(x,t)$ is in fact an element in $\mathbb{S}^k$ for every $x \in \mathbb{S}^k$:

$$\begin{aligned}
&F(x,t) \cdot F(x,t) \\
&= (\cos(\pi t)x + \sin(\pi t)s(x)) \cdot (\cos(\pi t)x + \sin(\pi t)s(x)) \\
&= \cos^2(\pi t)(x \cdot x) + 2\cos(\pi t)\sin(\pi t)(x \cdot s(x)) + \sin^2(\pi t)(s(x) \cdot s(x)) \\
&= \cos^2(\pi t) + \sin^2(\pi t) = 1,
\end{aligned}$$

where we use $x \cdot x = 1 = s(x) \cdot s(x)$ and $x \cdot s(x) = 0$. Thus $F(x,t)$ is a vector of norm 1 for every x and every t. Moreover, F is a smooth map with $F(x,0) = x$ and $F(x,1) = -x$, i.e., F is a smooth homotopy from the identity to the antipodal map on $\mathbb{S}^k$.

Solution (to Exercise 8.11) For $1 \leq i \leq k+1$, let $r_i\colon \mathbb{S}^k \to \mathbb{S}^k$ be the reflection map on the ith coordinate, i.e., $r_i(x_1, \ldots, x_{k+1}) = (x_1, \ldots, -x_i, \ldots, x_{k+1})$. Then the map $\mathbb{S}^k \times [0,1] \to \mathbb{S}^k$ defined by sending $(x_1, \ldots, x_{k+1}, t)$ to the point $(x_1, \ldots, x_{i-1}, \cos(\pi t)x_i - \sin(\pi t)x_{i+1}, \sin(\pi t)x_i + \cos(\pi t)x_{i+1}, x_{i+2}, \ldots, x_{k+1})$ is a homotopy from the identity on $\mathbb{S}^k$ to the map $r_i \circ r_{i+1}\colon \mathbb{S}^k \to \mathbb{S}^k$. The antipodal map is equal to the composition of reflections $r_1 \circ r_2 \circ \cdots \circ r_{k+1}$. Since k is even, $r_2 \circ \cdots \circ r_{k+1}$ is homotopic to the identity. Thus, the antipodal map is homotopic to the reflection r_1.

Solution (to Exercise 8.12) (a) That f_0 is an immersion means that $d(f_0)_x$ is injective for all $x \in X$. We need to show that there is an $\varepsilon > 0$ such that $d(f_t)_x$ is injective for all points (x,t) in $X \times [0,\varepsilon) \subset X \times I$. Given a point $x_0 \in X$, that $d(f_0)_{x_0}$ is injective implies that the matrix representing $d(f_0)_{x_0}$ has an $m \times m$-submatrix $A(x_0, 0)$ with non-vanishing determinant where $m = \dim X$. Since the determinant is continuous, this submatrix has non-vanishing determinant in an open neighbourhood of $(x_0, 0)$ in $X \times [0,1]$. Since X is compact, finitely many such neighbourhoods suffice to cover all of $X \times \{0\}$. Hence the minimum of the ε_i for the open intervals $[0, \varepsilon_i)$ covering $\{0\}$ defines an $\varepsilon > 0$ such that the intersection of these finitely many neighbourhoods contains $X \times [0,\varepsilon)$. Thus $d(f_t)_x$ is injective for all $(x,t) \in X \times [0,\varepsilon)$.

(b) Since X is compact, f_0 and each f_t are automatically proper maps. Hence we need to show that when f_0 is a one-to-one immersion, then so is f_t in a small neighbourhood. By the previous point, it remains to show that f_t is still one-to-one if t is small enough. We define a smooth map

$$G\colon X \times I \to Y \times I,\ G(x,t) := (f_t(x), t).$$

If f_t is not one-to-one in some small neighbourhood of 0, then, for every $\varepsilon > 0$, we can find a t with $0 < t < \varepsilon$ and $x, y \in X$ such that $f_t(x) = f_t(y)$. For example, for every $\varepsilon_i = 1/i$, we could find such a t_i, x_i, and y_i. Thus there is an infinite sequence $t_i \to 0$, and an infinite sequence of points $x_i \neq y_i \in X$ where f_{t_i} fails to be injective, i.e., such that $f_{t_i}(x_i) = G(x_i,t_i) = G(y_i,t_i) = f_{t_i}(y_i)$. Since X is compact, we may pass to subsequences which converge $x_i \to x_0$ and $y_i \to y_0$. Since G is continuous, this implies

$$G(x_0,0) = \lim_i G(x_i,t_i) = \lim_i G(y_i,t_i) = G(y_0,0).$$

But $G(x_0,0) = f_0(x_0)$ and $G(y_0,0) = f_0(x_0)$. By assumption, f_0 is injective, and hence $x_0 = y_0$. Now, after choosing local coordinates, we can express the derivative of G at $(x_0,0)$ by the matrix

$$dG_{(x_0,0)} = \begin{pmatrix} & * \\ d(f_0)_{x_0} & \vdots \\ & * \\ 0\cdots 0 & 1 \end{pmatrix},$$ where the 0s in the lowest row arise from the fact that the first coordinates do not depend on t, and the 1 in the bottom right-hand corner is the derivative of the function $t \mapsto t$. Since f_0 is an immersion, $d(f_0)_{x_0}$ has $k = \dim X$ many independent rows. Thus the matrix of $dG_{(x_0,0)}$ has $k+1$ independent rows, and hence $dG_{(x_0,0)}$ is an injective linear map. In other words, G is an immersion around $(x_0,0)$ and hence G must be one-to-one on some neighbourhood of $(x_0,0)$. But we have shown above that the sequences (x_i,t_i) and (y_i,t_i) both converge to $(x_0,0)$. This means that, for large i, both (x_i,t_i) and (y_i,t_i) belong to this neighbourhood. This contradicts the injectivity of G.

Solution (to Exercise 8.13) Let $\rho\colon \mathbb{R} \to \mathbb{R}$ be a smooth function with $\rho(s) = 1$ for $|s| < 1$ and $\rho(s) = 0$ for $|s| > 2$. Then we define $f_t\colon \mathbb{R} \to \mathbb{R}$ by $f_t(x) = x\rho(tx)$. For $t = 0$, we have $f_0(x) = x$ for all x, i.e., $f_0 = \mathrm{id}$. Hence f_0 is an immersion and an embedding. However, for any fixed $t > 0$, we have $|tx| > 2$

when $x > 2/|t|$. Hence, for this fixed t, $f_t(x) = 0$ for all $x > 2/|t|$. Thus, f_t is neither an immersion nor an embedding. To emphasise what is going wrong, let us replace the domain with a compact subspace of R, say $X := [a, b]$ with $b > 0$. Then we can choose $\varepsilon > 0$ such that $1/\varepsilon > \max\{|a|, |b|\}$. This implies that $x \in X$ is not bigger than $1/|t|$. Then, for $f_t(x) = x\rho(tx)$, we have $f_t(x) = x$ for all x and all $t < \varepsilon$.

Solution (to Exercise 8.14) We define the map

$$F: U \times [0,1] \to \mathbb{R},\ (x,t) \mapsto (\det(H(f_t)_x))^2 + \sum_{i=1}^{k} \left(\frac{\partial f}{\partial x_i}(x)\right)^2 .$$

Since f is smooth, F depends smoothly on both variables. By Exercise 5.10 and the assumption, we know $F(x,0) > 0$ for all $x \in U \times \{0\}$. Since $K \subset U$ is compact, F has a minimum on $K \times \{0\}$, i.e., there is a $\delta > 0$ such that $F(x,0) \geq 2\delta$ for all $x \in K$. Since F is continuous, there is an open neighbourhood $W \subset U \times [0,1]$ containing $K \times \{0\}$ such that $F(x,t) > \delta$ for all $(x,t) \in W$. In fact, we can cover $K \times \{0\}$ by open subsets $W_i \subset U \times [0,1]$ such that $F(x,t) > \delta$ for all $(x,t) \in W_i$. Each such open subset W_i has the form $V_i \times [0, \varepsilon_i)$ for some open $V_i \subset U$ and $\varepsilon_i > 0$. Since K is compact, finitely many such open W_i suffice to cover $K \times \{0\}$. Let ε be the minimum of the finitely many ε_i. Then we have $F(x,t) > \delta$ for all $(x,t) \in K \times [0,\varepsilon)$. Since F is continuous, for any fixed $t \in [0,\varepsilon)$, there is again an open subset $V \subset \mathbb{R}^k$ containing K such that $F(x,t) > 0$ for all $(x,t) \in V \times \{t\}$. Thus f_t is Morse in a neighbourhood of K for all sufficiently small t.

Solution (to Exercise 8.15) For $x \in X$, let $\phi_x: U_x \to X$ be a local parametrisation around x. Then $f_0 \circ \phi_x$ is a Morse function on U. Since $\{0\}$ is a compact subset of U, Exercise 8.14 implies that there is an open subset $V_x \subset U_x$ containing $\{0\}$ and an $\varepsilon(x) > 0$ such that f_t is Morse on V_x for all $t \in [0, \varepsilon(x))$. The images $\phi_x(V_x)$ are open in X and cover X. Since X is compact, finitely many suffice to cover X, say $X = \phi_{x_1}(V_{x_1}) \cup \cdots \cup \phi_{x_n}(V_{x_n})$. We set $\varepsilon := \min\{\varepsilon(x_1), \ldots, \varepsilon(x_n)\}$. Then $f_t: X \to \mathbb{R}$ is a Morse function for all $t \in [0, \varepsilon)$.

A.8 Manifolds with Boundary

Solution (to Exercise 9.1) (a) The boundary of X is $\partial X = \{(x,y) \in \mathbb{R}^2 : x = -1\}$. The derivative of f is given by the 1×2-matrix $df_{(x,y)} =$

$(2x \quad 2y)$. Hence $df_{(x,y)}$ is a surjective linear map for all $(x, y) \neq (0,0)$. Since $f(0,0) = 0 \neq 1$, $df_{(x,y)}$ is surjective for all $(x, y) \in f^{-1}(1)$ and 1 is a regular value of f. The restriction of f to the boundary of X is the map $\partial f : \partial X \to Y$ given by $(-1, y) \mapsto 1 + y^2$. Hence the derivative of ∂f is given by the 1×1-matrix $(\partial f)_{(-1,y)} = 2y$. This is a linear map which is surjective if and only if $y \neq 0$. Since $(-1,0) \in \partial X$ and $\partial f(-1,0) = 1$, we see that 1 is not a regular value of ∂f.

(b) The preimage $f^{-1}(1)$ is just the unit sphere $\mathbb{S}^1$. Hence the boundary $\partial(f^{-1}(1))$ is empty. However,

$$f^{-1}(1) \cap \partial X = \{(x, y) \in \mathbb{R}^2 : x^2 + y^2 = 1\} \cap \{(x, y) \in \mathbb{R}^2 : x = -1\},$$

and thus $f^{-1}(1) \cap \partial X = \{(-1,0)\} \neq \emptyset$. Hence $\partial(f^{-1}(1)) \neq f^{-1}(1) \cap \partial X$. This is not a contradiction to the Preimage Theorem for manifolds with boundary, since the conclusion of the theorem required that 1 is a regular of both f and ∂f. But we showed in the first part that 1 is not a regular value of ∂f.

Solution (to Exercise 9.2) Let $U \subset \mathbb{R}^k$ and $V \subset \mathbb{H}^k$ be open neighbourhoods of 0. Suppose there is a diffeomorphism $\theta : U \to V$. After composing with a translation if necessary, we can assume that 0 is sent to a boundary point of V. In fact, we can assume that $\theta(0) = 0$. The linear map $d\theta_0 : \mathbb{R}^k \to \mathbb{R}^k$ is an isomorphism. By the Inverse Function Theorem 3.4, there are subsets W_1 and W_2 in $\mathbb{R}^k$ containing 0 which are both open in $\mathbb{R}^k$ such that θ maps W_1 diffeomorphically onto W_2. Since W_2 is open in $\mathbb{R}^k$ and contained in the image of θ, we get that V must be open in $\mathbb{R}^k$, not just in $\mathbb{H}^k$. However, since V contains 0, it satisfies $V \cap \partial\mathbb{H}^k \neq \emptyset$ and cannot be open in $\mathbb{R}^k$.

Solution (to Exercise 9.3) Let $f : X \to Y$ be a diffeomorphism of manifolds with boundary. Let $\phi : U \to X$ and $\psi : V \to Y$ be local parametrisations, where U and V are open subsets of $\mathbb{H}^k$. Let $\theta : U \to V$ be the induced map. By shrinking U and V if necessary, we can assume that θ is a diffeomorphism with $f \circ \phi = \psi \circ \theta$. Boundary points of X are those which are in the image $\phi(\partial U) = \phi(U \cap \partial\mathbb{H}^k)$. Similarly, boundary points of Y are those which are in the image $\psi(\partial V) = \psi(V \cap \partial\mathbb{H}^k)$. Hence we need to show $\theta(\partial U) \subset \partial V$, for then we get that $f(\phi(\partial U)) = \psi(\theta(\partial U)) \subset \psi(\partial V) \subset \partial Y$. The argument is again based on the Inverse Function Theorem. Suppose there is a point $u \in \partial U$ which is mapped to an interior point $v = \theta(u)$ in V. Since θ is a diffeomorphism, the derivative $d(\theta^{-1})_v : \mathbb{R}^k \to \mathbb{R}^k$ of its inverse is an isomorphism. But, since $v \in \mathrm{Int}(V)$, V contains a neighbourhood W of v that is open in $\mathbb{R}^k$. Thus the Inverse Function Theorem implies that $\theta^{-1}(W)$ contains a neighbourhood of

u that is open in $\mathbb{R}^k$. Hence u is also an interior point in U, which contradicts the assumption $u \in \partial U$.

Solution (to Exercise 9.4) (a) The image of F is the product $\mathbb{S}^1 \times [-1/2, 1/2]$. This is a product of a manifold without a boundary $\mathbb{S}^1$ and the manifold $[-1/2, 1/2]$ with boundary. The boundary of $[-1/2, 1/2]$ consists of the disjoint union of $\{-1/2\}$ and $\{1/2\}$. By Proposition 9.8, we obtain the boundary $\partial X = \partial(\mathbb{S}^1 \times [-1/2, 1/2]) = (\mathbb{S}^1 \times \{-1/2\}) \cup (\mathbb{S}^1 \times \{1/2\})$.

(b) We define the maps $\phi_+ \colon (-\pi, \pi) \times [0, 3/4) \to Y$ by

$$\begin{aligned}\phi_+(t,s) :=& ((1 + (-1/2 + s)\cos(t/2))\cos t,\\ &(1 + (-1/2 + s)\cos(t/2))\sin t, (-1/2 + s)\sin(t/2)),\end{aligned}$$

$\phi_- \colon (-\pi, \pi) \times [0, 3/4) \to Y$ by

$$\begin{aligned}\phi_-(t,s) :=& ((1 + (1/2 - s)\cos(t/2))\cos t,\\ &(1 + (1/2 - s)\cos(t/2))\sin t, (1/2 - s)\sin(t/2)),\end{aligned}$$

$\psi_+ \colon (0, 2\pi) \times [0, 3/4) \to Y$ by

$$\begin{aligned}\psi_+(t,s) :=& ((1 + (-1/2 + s)\cos(t/2))\cos t,\\ &(1 + (-1/2 + s)\cos(t/2))\sin t, (-1/2 + s)\sin(t/2)),\end{aligned}$$

and $\psi_- \colon (0, 2\pi) \times [0, 3/4) \to Y$ by

$$\begin{aligned}\psi_-(t,s) :=& ((1 + (1/2 - s)\cos(t/2))\cos t,\\ &(1 + (1/2 - s)\cos(t/2))\sin t, (1/2 - s)\sin(t/2)).\end{aligned}$$

As one can check by calculating the partial derivatives that each of these maps are diffeomorphisms, and the union of their images covers Y. Hence we can use these four maps as local parametrisations of Y. The boundary of Y is then given by the union of the points

$$\begin{aligned}\partial Y = &\phi_+((-\pi, \pi) \times \{0\}) \cup \phi_-((-\pi, \pi) \times \{0\})\\ &\cup \psi_+((0, 2\pi) \times \{0\}) \cup \psi_-((0, 2\pi) \times \{0\}).\end{aligned}$$

Setting $s = 0$ in the formulae for those maps gives

$$\begin{aligned}\partial Y = &\left\{\left(\left(1 - \frac{1}{2}\cos(t/2)\right)\cos t, \left(1 - \frac{1}{2}\cos(t/2)\right)\sin t, -\frac{1}{2}\sin(t/2)\right) \in \mathbb{R}^3 : t \in \mathbb{R}\right\}\\ &\cup \left\{\left(\left(1 + \frac{1}{2}\cos(t/2)\right)\cos t, \left(1 + \frac{1}{2}\cos(t/2)\right)\sin t, \frac{1}{2}\sin(t/2)\right) \in \mathbb{R}^3 : t \in \mathbb{R}\right\}.\end{aligned}$$

In fact, the two sets describing ∂Y are the same which we see when we replace t with $t + 2\pi$ and use some trigonometric identities:

$$\begin{cases} \left(1 - \frac{1}{2}\cos\left(\frac{t+2\pi}{2}\right)\right)\cos(t+2\pi) & = \left(1 + \frac{1}{2}\cos(t/2)\right)\cos t, \\ \left(1 - \frac{1}{2}\cos\left(\frac{t+2\pi}{2}\right)\right)\sin(t+2\pi) & = \left(1 + \frac{1}{2}\cos(t/2)\right)\sin t \\ -\frac{1}{2}\sin\left(\frac{t+2\pi}{2}\right) & = \frac{1}{2}\sin(t/2). \end{cases}$$

Hence

$$\partial Y = \left\{\left(\left(1 + \frac{1}{2}\cos(t/2)\right)\cos t, \left(1 + \frac{1}{2}\cos(t/2)\right)\sin t, \frac{1}{2}\sin(t/2)\right) \in \mathbb{R}^3 : t \in \mathbb{R}\right\}.$$

Now we would like to show that ∂Y is diffeomorphic to $\mathbb{S}^1$. Remembering the trigonometric identities $\sin t = 2\sin(t/2)\cos(t/2)$ and $\cos t = \cos^2(t/2) - \sin^2(t/2)$, we see that the map

$$\varphi\colon \mathbb{R}^2 \to \mathbb{R}^3,\ (x,y) \mapsto \left(\left(1 + \frac{1}{2}x\right)\left(x^2 - y^2\right), \left(1 + \frac{1}{2}x\right)2xy, \frac{1}{2}y\right)$$

restricts to a bijection from $\mathbb{S}^1$ onto ∂Y. For injectivity, note that the last coordinate determines y uniquely, then the circle equation determines x up to sign, and the first and/or second coordinate determine the sign of x. It remains to check that $\varphi_{|\mathbb{S}^1}$ is a local diffeomorphism. Since $\varphi_{|\mathbb{S}^1}$ is a bijection onto its image, this will show that it is a diffeomorphism. First we observe that φ is smooth, since the three functions in each coordinate are just polynomials and hence smooth. To see that $\varphi_{|\mathbb{S}^1}$ is a local diffeomorphism, we use the maps $\phi\colon U \to \mathbb{S}^1$, $t \mapsto (\cos(t/2), \sin(t/2))$, and

$$\psi\colon U \to \partial Y, t \mapsto \left(\left(1 + \frac{1}{2}\cos(t/2)\right)\cos t, \left(1 + \frac{1}{2}\cos(t/2)\right)\sin t, \frac{1}{2}\sin(t/2)\right),$$

where $U \subset \mathbb{R}$ is some sufficiently small open subset. Then ϕ and ψ serve as local parametrisations of $\mathbb{S}^1$ and ∂Y, respectively, for suitable choices of U. The induced map $\theta\colon U \to U$, which arises as the composite $\psi^{-1} \circ \varphi_{|\mathbb{S}^1} \circ \phi$ is just the identity $t \mapsto t$. Hence $\varphi_{|\mathbb{S}^1}$ is a local diffeomorphism.

Solution (to Exercise 9.5) (a) We proceed as before when we showed that tangent spaces are well defined. Let $\psi\colon V \to X$ be another local parametrisation around x with $\psi(0) = x$, where V is an open subset of $\mathbb{H}^k$. After replacing U by $\phi^{-1}(\phi(U) \cap \psi(V)) \subset U$ and V by $\psi^{-1}(\phi(U) \cap \psi(V)) \subset V$, we can assume $\phi(U) = \psi(V)$. Then the map

θ defined by $\theta := \psi^{-1} \circ \phi \colon U \to V$ is a diffeomorphism since it is the composite of two diffeomorphisms. By definition of θ, we have $\phi = \psi \circ \theta$. Differentiating yields via the chain rule $d\phi_0 = d\psi_0 \circ d\theta_0$. It follows from the proof of Proposition 9.6 that $d\theta_0(\mathbb{H}^k) \subseteq \mathbb{H}^k$. This implies $d\phi_0(\mathbb{H}^k) \subseteq d\psi_0(\mathbb{H}^k)$ in $\mathbb{R}^N$. By switching the roles of ϕ and ψ in the argument, we also get $d\psi_0(\mathbb{H}^k) \subseteq d\phi_0(\mathbb{H}^k)$ in $\mathbb{R}^N$. Hence, the image of the upper half-plane $H_x(X)$ is well defined in $T_x(X)$.

(b) The codimension of $T_x(\partial X)$ in $T_x(X)$ is one. Thus the orthogonal complement of $T_x(\partial X)$ is one-dimensional and is spanned by one vector. By definition of ∂X as the image of the points in $\partial\mathbb{H}^k$ under local parametrisations, we know that $d\phi_0(e_k)$ spans the complement of $T_x(\partial X)$ in $T_x(X)$, since $d\phi_0$ is an isomorphism and $e_k = (0, \ldots, 0, 1)$ is non-zero and not contained in $d\phi_0(\mathbb{H}^k)$. We also know by the definition of $H_x(X)$ that $d\phi_0(e_k) \in H_x(X)$, and therefore $d\phi_0(-e_k) \notin H_x(X)$. But we do not know whether $d\phi_0(-e_k)$ is orthogonal to $T_x(\partial X)$ in $T_x(X)$. To make $d\phi_0(-e_k)$ into a vector which is orthogonal to $T_x(\partial X)$, we apply the Gram–Schmidt process. It produces a unit vector which is orthogonal to $T_x(\partial X)$. We denote this vector $n(x)$, this is the outward unit normal vector to ∂X. Note that $-n(x)$ is a unit vector contained in $H_x(X)$ and orthogonal to $T_x(\partial X)$, and this is the inward unit normal vector to ∂X.

(c) From what we have learned in the previous point, we can construct $n(x)$ by applying the Gram–Schmidt orthonormalisation process to $d\phi_0(-e_k)$. This process depends smoothly on the coefficients in the matrix representing $d\phi_0$. Since the derivative $d\phi_u$ depends smoothly on u, $d\phi_u(-e_k)$ depends smoothly on u. By the independence of the choice of local parametrisation, we see that $n(y) = d\phi_u(-e_k)$ for all $y \in \phi(\partial U)$, which is an open neighbourhood of x in ∂X, where $\phi(u) = y$. Thus, in total, we see that $n(x)$ depends smoothly on x in ∂X.

Solution (to Exercise 9.6) (a) The fact that F is a diffeomorphism follows from the same argument as in Exercise 2.14, that is, the inverse of F is the projection π onto the first factor since we have $\pi \circ F = \mathrm{Id}_X$ and $F \circ \pi = \mathrm{Id}_{\Gamma(f)}$. Let $X \subset \mathbb{R}^N$, $Y \subset \mathbb{R}^M$. Since f is smooth, for every point $x \in X$, there is an open subset $U \subset \mathbb{R}^N$ and a smooth map $\tilde{f} \colon U \to \mathbb{R}^M$ with $\tilde{f}_{X \cap U} = f_{X \cap U}$. Then $(\tilde{f} \times \mathrm{Id}) \colon U \to \mathbb{R}^N \times \mathbb{R}^M$ is a smooth extension of $F_{X \cap U}$. Hence F is smooth. The inverse π is a smooth map since it extends to the smooth projection on all of $\mathbb{R}^N \times \mathbb{R}^M$. Hence F is a diffeomorphism. If X has dimension n, then any local parametrisation $\phi \colon V \to X$ with $V \subset \mathbb{H}^n$ can be extended to a local parametrisation $F \circ \phi \colon V \to \Gamma(f)$. Thus the graph $\Gamma(f)$ is a smooth manifold with boundary.

(b) Since F is a diffeomorphism, we get from Exercise 9.3 that
$F(\partial X) = \partial\Gamma(f)$, i.e., $\partial\Gamma(F) = \{(x, f(x)) \in \Gamma(f) : x \in \partial X\}$.

A.9 Brouwer Fixed Point Theorem

Solution (to Exercise 10.1) If A is not invertible, then 0 is an eigenvalue, and we are done. So assume A is non-singular. For any vector $v \in \mathbb{S}^{n-1} \subset \mathbb{R}^n$, the vector $Av/|Av|$ has norm one and lies on $\mathbb{S}^{n-1}$. Note that this map is not defined for $v = 0$, and we cannot continuously extend it to 0. Hence we cannot just consider this as a map $\mathbb{D}^n \to \mathbb{D}^n$. Let $g\colon \mathbb{S}^{n-1} \to \mathbb{S}^{n-1}$ be the map $v \mapsto Av/|Av|$. Now we use the assumption on A. If $v \in Q := \{(x_1, \ldots, x_n) \in \mathbb{S}^{n-1} : \text{ all } x_i \geq 0\}$, then Av has only non-negative entries, since the entries in v are all non-negative and all the entries in A are, by assumption, non-negative. Since $|Av| > 0$ is non-negative as well, we know that $g(v)$ is an element in Q. Thus we can restrict g to a map $g\colon Q \to Q$. Now we can compose with a homeomorphism $\varphi\colon Q \xrightarrow{\cong} \mathbb{D}^{n-1}$ to get a continuous map

$$f\colon \mathbb{D}^{n-1} \xrightarrow{\varphi} Q \xrightarrow{g} Q \xrightarrow{\varphi^{-1}} \mathbb{D}^{n-1}.$$

By the Brouwer Fixed Point Theorem 10.7 for continuous maps, f must have a fixed point $y \in \mathbb{D}^{n-1}$ with $f(y) = y$. Hence the image of $w := \varphi^{-1}(y)$ is a vector in $Q \subset \mathbb{S}^{n-1} \subset \mathbb{R}^n$ with $Aw/|Aw| = w$, i.e., $Aw = |Aw| \cdot w$. Since w is non-zero, being a point on $\mathbb{S}^{n-1}$, w is an eigenvector with real non-negative eigenvalue $|Aw|$.

Solution (to Exercise 10.2) (a) Assuming that X is simply connected, f can be extended to a map $F\colon \mathbb{D}^2 \to X$. Let $r\colon X \to \mathbb{S}^1$ be the map defined by $p \mapsto p/|p|$. We compose these maps to get $g\colon \mathbb{D}^2 \xrightarrow{F} X \xrightarrow{r} \mathbb{S}^1 \hookrightarrow \mathbb{D}^2$.

(b) The image of g is contained in $\mathbb{S}^1$ by definition of g as a composition with the inclusion $\mathbb{S}^1 \hookrightarrow \mathbb{D}^2$. Hence the only possible fixed points of g are contained in $\mathbb{S}^1$. But g sends points in $\mathbb{S}^1 \subset \mathbb{D}^2$ to their antipodal points. Hence g does not have any fixed points.

(c) If $X = \mathbb{R}^2 \setminus \{(0, 0)\}$ is simply connected we can construct a continuous map $g\colon \mathbb{D}^2 \to \mathbb{D}^2$ without a fixed point. This contradicts Brouwer's Fixed Point Theorem 10.7. Hence $X = \mathbb{R}^2 \setminus \{(0, 0)\}$ cannot be simply connected.

Solution (to Exercise 10.3) Assume that there was such a homotopy $F: \mathbb{S}^1 \times [0,1] \to X = \mathbb{R}^2 \setminus \{(0,0)\}$. Then we get a homotopy H between the identity map of $\mathbb{R}^2 \setminus \{(0,0)\}$ and a constant map as follows. Let $G: X \times [0,1] \to X$ be a homotopy between the identity of X, i.e., $G(p,0) = p$, and the map r, i.e., $G(p,1) = p/|p|$. Then we define a homotopy $H: \mathbb{R}^2 \setminus \{(0,0)\} \times [0,1] \to \mathbb{R}^2 \setminus \{(0,0)\}$ by

$$(p,t) \mapsto \begin{cases} G(p,2t) & \text{for } 0 \le t \le 1/2 \\ F(r(p), 1-2t) & \text{for } 1/2 \le t \le 1. \end{cases}$$

This would imply that $\mathbb{R}^2 \setminus \{(0,0)\}$ is contractible. This contradicts our result that $\mathbb{R}^2 \setminus \{(0,0)\}$ is not simply connected.

A.10 Brouwer Degree Mod 2 and Borsuk–Ulam Theorem

Solution (to Exercise 11.2) Let z be a regular value for g. If z is not in the image of g, then $\deg_2(g)$ and $\deg_2(g \circ f)$ both vanish, and the claim is true. So let us assume that z is in the image of g. By Exercise 5.5, z then is a regular value for $g \circ f$ if and only if every $y \in g^{-1}(z)$ is a regular value for f. Hence we can use z to compute $\deg_2(g \circ f)$ if and only if we can use $y \in g^{-1}(z)$ to compute $\deg_2(f)$. If z is not in the image of $g \circ f$, then y is not in the image of f. In this case, both $\deg_2(g \circ f)$ and $\deg_2(f)$ vanish, and the claim is true. So assume there are points $x \in X$ with $g(f(x)) = z$. Then it remains to apply the definition of $\deg_2$. By our assumptions, we have

$$\begin{aligned} \deg_2(g \circ f) &= \#(g \circ f)^{-1}(z) = \# f^{-1}(g^{-1}(z)) \\ &= (\# g^{-1}(z)) \cdot (\# f^{-1}(y)) = \deg_2(g) \cdot \deg_2(f), \end{aligned}$$

where we use that the number of points in the fibre $f^{-1}(y)$ is the same for every $y \in g^{-1}(z)$ which is a regular value of f.

Solution (to Exercise 11.3) (a) Assume f is not surjective. Then there is a $y \in Y$ which is not in the image of f. Hence y is a regular value for f and we can use it to compute the degree of f as $\deg_2(f) = \# f^{-1}(y) = 0$. This contradicts the assumption that $\deg_2(f) \neq 0$. Hence f must be surjective.

(b) Let us assume $\deg_2(f) \neq 0$ and derive a contradiction. By the previous point, if $\deg_2(f) \neq 0$, then f is surjective, i.e., $Y = f(X)$. Since f is continuous and X is compact, Y would be compact by Lemma 1.13 and Exercise 1.1. This contradicts the assumption. Hence we must have $\deg_2(f) = 0$.

Solution (to Exercise 11.4) Let $f\colon \mathbb{S}^1 \to \mathbb{S}^1$ be a smooth map without fixed points, and let $\alpha\colon \mathbb{S}^1 \to \mathbb{S}^1$ be the antipodal map. Since $\alpha^{-1}(x) = -x$ for all $x \in \mathbb{S}^1$, there is exactly one preimage point for each x and all $x \in \mathbb{S}^1$ are regular values. Hence $\deg_2(\alpha) = 1$. If we can show that f and α are homotopic, then the invariance of $\deg_2$ under homotopy implies $\deg_2(f) = 1$. By Exercise 11.3, $\deg_2(f) = 1$ implies that f is surjective. Now we construct a smooth homotopy between f and α. First we consider the straight-line homotopy $G(x,t)\colon \mathbb{S}^1 \times [0,1] \to \mathbb{R}^2$ given by $G(x,t) = f(x)(1-t) - tx$. We would like to turn G into a homotopy between f and α. Hence we need to manipulate G such that its image is contained in $\mathbb{S}^1 \subset \mathbb{R}^2$. We can arrange this if $G(x,t) \neq 0$. For then, we can consider $\frac{G(x,t)}{|G(x,t)|}$ in $\mathbb{S}^1$. Hence we need to check $G(x,t) \neq 0$ for all $(x,t) \in \mathbb{S}^1 \times [0,1]$. For a fixed x and varying t, $f(x)(1-t) - tx$ describes the line segment in $\mathbb{R}^2$ between the two points $f(x)$ and $-x$ on $\mathbb{S}^1$. The only way this line segment can pass $0 \in \mathbb{R}^2$ is when $f(x)$ and $-x$ are antipodal points, i.e., when $f(x) = x$. But, by the assumption on f, $f(x) \neq x$ for all $x \in \mathbb{S}^1$. Thus, we can define a smooth map $F(x,t)\colon \mathbb{S}^1 \times [0,1] \to \mathbb{S}^1$ by $F(x,t) := \frac{f(x)(1-t)-tx}{|f(x)(1-t)-tx|}$, and F is a smooth homotopy between f and α.

Solution (to Exercise 11.6) We define a homotopy from $p_0(z) = z^m$ to $p_1(z) = p(z)$ by $p_t(z) = tp(z) + (1-t)z^m = z^m + t(a_1 z^{m-1} + \cdots + a_m)$. For $z \neq 0$, we consider $p_t(z)/z^m = 1 + t\cdot(a_1/z + \cdots + a_m/z^m)$. As $z \to \infty$ moves towards infinity, the term $\frac{a_1}{z} + \cdots + \frac{a_m}{z^m} \to 0$ converges to zero. Hence, if W is a closed ball around the origin in $\mathbb{C}$ with sufficiently large radius, none of the p_t has a zero on ∂W. Thus, the homotopy $p_t/|p_t|\colon \partial W \to \mathbb{S}^1$ is defined for all $t \in [0,1]$. This implies $\deg_2(p/|p|) = \deg_2(p_0/|p_0|)$. Since $p_0(z) = z^m$ and $\#\{z \in \partial W\colon z^m/|z|^m = 1\} = m$, we have $\deg_2(p_0/|p_0|) = m$ modulo 2. Hence, if m is odd, then $\deg_2(p/|p|) \neq 0$, and there must be $w \in W$ with $p(w) = 0$ by Example 11.16.

Solution (to Exercise 11.7) Assume that $f_1, \ldots, f_k$ do not have a common zero. Then we can form the smooth odd map $f\colon \mathbb{S}^k \to \mathbb{R}^{k+1} \setminus \{0\}$ defined by $f := (f_1, \ldots, f_k, 0)$. Now we can apply Corollary 11.24 to f and L being the x_{k+1}-axis. Hence f intersects L at least once. But x with $f(x) \in L$ is a common zero of the $f_1, \ldots, f_k$. Hence the $f_1, \ldots, f_k$ must have a common zero after all.

Solution (to Exercise 11.8) We let $f_1, \ldots, f_k$ be smooth functions $\mathbb{S}^k \to \mathbb{R}$ defined by $f_i(x) := g_i(x) - g_i(-x)$. Then each f_i is smooth and odd. The functions $f_1, \ldots, f_k$ satisfy the assumption of the previous exercise. Hence there is a common zero of the $f_1, \ldots, f_k$, which is the desired point $p \in \mathbb{S}^k$.

Solution (to Exercise 11.9) Since the p_i are all homogeneous of odd order, they satisfy $p_i(-x) = (-1)^{m_i} p_i(x) = -p(x_i)$. Moreover, for any $x \in \mathbb{R}^{n+1} \setminus \{0\}$, we can consider the associated map $q_i \colon \mathbb{S}^n \to \mathbb{R}$ given by $x \mapsto p_i(x/|x|)$. Since $n + 1 \geq 2$, this is a smooth map. Hence we have n smooth real-valued functions $q_1, \ldots, q_n$ on $\mathbb{S}^n$, with $q_i(-x) = p_i(-x/|x|) = -p_i(x/|x|) = -q_i(x)$. By Exercise 11.7, we know that the maps $q_1, \ldots, q_n$ have a common zero on $\mathbb{S}^n$, say $x_0 \in \mathbb{S}^n$. Since the p_i are homogeneous, $q_i(x_0) = p_i(x_0) = 0$ implies $p_1(x) = \cdots = p_n(x) = 0$ for all $x = \lambda x_0$ for some $\lambda \in \mathbb{R}$. Hence the line spanned by x_0 in $\mathbb{R}^{n+1}$ is the desired line on which all p_i vanish simultaneously.

Solution (to Exercise 11.10) Assume that such a map f exists. Then we can define the map

$$g \colon \mathbb{B}^2 = \{(x, y) \in \mathbb{R}^2 : x^2 + y^2 \leq 1\} \to \mathbb{S}^1,\ g(x, y) := f\left(x, y, \sqrt{1 - x^2 - y^2}\right).$$

Note that g is continuous, since g is the composite of f with the inverse of the projection from the upper hemisphere of the sphere to $\mathbb{B}^2$. Then, we have $g(-x, -y) = f(-x, -y, 0) = -f(x, y, 0) = -g(x, y)$ for any $(x, y) \in \mathbb{S}^1 = \partial\mathbb{B}^2$, i.e., for (x, y) such that $1 - x^2 - y^2 = 0$, by the assumption on f. In particular, $g_{\partial\mathbb{B}^2}$ is smooth since f is smooth, and hence $g_{|\partial\mathbb{B}^2} \colon \mathbb{S}^1 \to \mathbb{S}^1$ satisfies the assumptions of the Borsuk–Ulam Theorem. Thus, $\deg_2(g_{\partial\mathbb{B}^2}) = 1$. However, $g_{|\partial\mathbb{B}^2}$ can be extended to a smooth map on all of $\mathbb{B}^2$. Thus, by the Boundary Theorem 11.14 for degrees, $\deg_2(g_{\partial\mathbb{B}^2}) = 0$. Hence the assumption that f exists leads to a contradiction.

Solution (to Exercise 11.11) Assume that there is no such point p. Then we can consider the map $g \colon \mathbb{S}^2 \to \mathbb{S}^1$ given by $g(p) = \frac{f(p)-f(-p)}{|f(p)-f(-p)|}$. By the assumption, g is smooth and satisfies $g(-p) = -g(p)$ for all $p \in \mathbb{S}^2$. This contradicts the result of Exercise 11.10.

Solution (to Exercise 11.12) Let $U \subset \mathbb{R}^2$ and $V \subset \mathbb{R}^n$ be open subsets. After possibly translating, we can assume that V contains the origin. Moreover, since V is open, there is an $\varepsilon > 0$ such that $\bar{\mathbb{B}}^n_\varepsilon(0) \subset V$, where $\bar{\mathbb{B}}^n_\varepsilon(0)$ denotes the closed ball of radius ε centred at the origin. We can then consider the subset $\mathbb{S}^2_\varepsilon \subset \bar{\mathbb{B}}^3_\varepsilon(0) \subset \bar{\mathbb{B}}^n_\varepsilon(0) \subset V$. Assume now there is a homeomorphism

$\varphi\colon V \to U$. Then $\varphi_{|\mathbb{S}^2_\varepsilon}$ is still a homeomorphism. Since scaling defines a homeomorphism $\mathbb{S}^2 \to \mathbb{S}^2_\varepsilon$, composition with $\varphi_{|\mathbb{S}^2_\varepsilon}$ would yield a continuous map $\mathbb{S}^2 \to \mathbb{R}^2$ with $f(p) \neq f(-p)$ for all $p \in \mathbb{S}^2$. This contradicts the assumed continuous version of Exercise 11.11.

A.11 Tubular Neighbourhoods and Transversality

Solution (to Exercise 12.1) Let X and Y be submanifolds of $\mathbb{R}^N$. We define the map $F\colon X \times \mathbb{R}^N \to \mathbb{R}^N, (x,a) \mapsto x + a$. The derivative of F is the linear map $dF_{(x,a)}\colon T_x(X) \times \mathbb{R}^N \to \mathbb{R}^N, (v,w) \mapsto v+w$. Thus $dF_{(x,a)}$ is surjective at every point (x,a). This shows that F is a submersion, and therefore transverse to every submanifold of $\mathbb{R}^N$. In particular, F is transverse to both boundaryless submanifolds $\mathrm{Int}(Y)$ and ∂Y. By the Transversality Theorem 12.42, the set of $a \in R^N$ such that the map $t_a\colon X \to \mathbb{R}^N,\ x \mapsto x + a$ is transverse to each of $\mathrm{Int}(Y)$ and ∂Y is a dense subset of $\mathbb{R}^N$. Hence the set of $a \in R^N$ such that t_a is transverse to both $\mathrm{Int}(Y)$ and ∂Y simultaneously is a dense subset of $\mathbb{R}^N$. The derivative $d(t_a)_x\colon T_x(X) \to \mathbb{R}^N$ is just $v \mapsto v$. Moreover, the tangent spaces of $X+a$ and X are equal since any local parametrisation ϕ of X defines a local parametrisation $\phi + a$ of $X + a$. Since also the derivatives of ϕ and $\phi + a$ are equal, we have $T_x(X) = T_{x+a}(X + a)$. Hence $t_a \pitchfork Y$ implies

$$\mathbb{R}^N = \mathrm{Im}\,(d(t_a)_x) + T_{t_a(x)}(Y) = T_x(X) + T_{x+a}(Y) = T_{x+a}(X + a) + T_{x+a}(Y).$$

If $y = x + a \in Y$, we get that $X + a$ and Y meet transversely in $y = x + a$. If $x + a \notin Y$, then $x + a \notin (X + a) \cap Y$, and $X + a$ and Y meet transversely in $x + a$ automatically.

Solution (to Exercise 12.2) (a) Let X be a compact submanifold of $\mathbb{R}^n$, and let $w \in \mathbb{R}^n$. Since X is compact, the continuous function $\delta\colon X \to \mathbb{R}$ defined by $\delta(x) = |w - x|^2$ has a minimum. Let $x \in X$ be one of the points where δ has its minimum. Hence x is a point of X which is closest to w. Now let $c\colon (-a,a) \to X$ be any smooth curve on Y with $c(0) = x$. The smooth function $f\colon (-a,a) \to \mathbb{R}$ defined by $f(t) := |w - c(t)|^2$ then has a minimum at $t = 0$. Thus its derivative df_0 at 0 vanishes. Writing $f(t) = |w - c(t)|^2 = (w - c(t)) \cdot (w - c(t))$ using the scalar product, we see that df_0 is given by $df_0 = (-dc_0) \cdot 2(w - c(0))$. In particular, we get $w - x = w - c(0)$ is orthogonal to dc_0 in $\mathbb{R}^n$. Since every tangent vector in $T_x(X)$ is the velocity vector dc_0 for some smooth curve c on X with $c(0) = x$ by Exercise 2.16, this shows $w - x \in N_x(X)$ by definition of $N_x(X)$ as the orthogonal complement of $T_x(X)$ in $\mathbb{R}^n$.

(b) Let $N \subset N(X)$ be the open neighbourhood of $X \times \{0\}$ in $N(X)$ which is mapped diffeomorphically onto $X^{\varepsilon} \subset \mathbb{R}^n$ by h. Given $w \in X^{\varepsilon}$, there is a unique element $n \in N$ with $h(n) = w$. Since elements in $N(X)$ are pairs (x, v) with $v \in N_x(X)$, there is a unique $x \in X$ and $v \in N_x(X)$ such that $n = (x, v)$ and $\sigma(n) = x$. Since $h(x, v) = x + v$ by definition, and $h(n) = w$ by the choice of n, we must have $v = w - x \in N_x(X)$. Hence the pair (x, v) is uniquely determined by $w \in X^{\varepsilon}$. Since we have the commutative diagram

$$\begin{array}{ccc} N(X) & \xrightarrow{\ h\ } & X^{\varepsilon} \\ & \overset{\sigma}{\searrow} & \downarrow \pi \\ & & X \end{array}, \tag{A.10}$$

we know $\pi(w) = \sigma(n) = x$. By point (a), we know that any $x_0 \in X$ with minimal distance to w must satisfy $w - x_0 \in N_{x_0}(X)$. We just learned that $\sigma(n) = x \in X$ is the unique element in X with this property. Hence $\pi(w)$ is the unique point of X closest to w.

Solution (to Exercise 12.3) Let X be a submanifold of $\mathbb{R}^N$. Every k-dimensional vector subspace V of $\mathbb{R}^N$ is the span of a k-tuple of linearly independent vectors in $\mathbb{R}^N$. So let $S \subset (\mathbb{R}^N)^k$ be the set consisting of all k-tuples of linearly independent vectors in $\mathbb{R}^N$. For a k-tuple of vectors $[v] := v_1, \ldots, v_k$ in $\mathbb{R}^N$, let $A_{[v]}$ be the $N \times k$-matrix with the v_i as column vectors. Then the k-tuple $v_1, \ldots, v_k$ is linearly independent if and only if the $k \times k$-matrix $(A_{[v]})^{\top} A_{[v]}$ is invertible, i.e., $\det((A_{[v]})^{\top} A_{[v]}) \neq 0$. Hence S is the inverse image of the open subset $\mathbb{R} \setminus \{0\}$ under the continuous map $\mathbb{R}^{Nk} \to \mathbb{R}$ defined by $[v] \mapsto \det((A_{[v]})^{\top} A_{[v]})$. Thus, S is an open subset in $\mathbb{R}^{Nk}$. We define the map $\varphi \colon \mathbb{R}^k \times S \to \mathbb{R}^N$ by

$$([t], [v]) := ((t_1, \ldots, t_k), v_1, \ldots, v_k) \mapsto t_1 v_1 + \cdots + t_k v_k.$$

Since S is open in $\mathbb{R}^{Nk}$, the tangent space to S at any $[v]$ is just $\mathbb{R}^{Nk}$. Moreover, φ is linear in each coordinate. Thus the derivative of φ at any point $([t], [v])$ is just φ. Since φ is surjective, $d\varphi_{([t],[v])}$ is surjective. Thus φ is a submersion. Hence, the Transversality Theorem 12.42 implies that the set of $s = [v] \in S$ such that the map $\varphi_{[v]} \colon \mathbb{R}^k \to \mathbb{R}^N$ given by $(t_1, \ldots, t_k) \mapsto t_1 v_1 + \cdots + t_k v_k$ is transverse to every submanifold in $\mathbb{R}^N$ is dense in S. In particular, the set of $s = [v]$ such that $\varphi_{[v]} \pitchfork X$ is dense in S. This means

$$\mathbb{R}^N = \operatorname{Im}(d\varphi_{[v]}) + T_x(X) = \operatorname{Im}(\varphi_{[v]}) + T_x(X)$$

for every $x \in X$. However, by definition of φ, $\operatorname{Im}(\varphi_{[v]})$ is just the span of the k-tuple $[v] = \{v_1, \ldots, v_k\}$ in $\mathbb{R}^N$. Thus we have shown that almost every V is transverse to X.

Solution (to Exercise 12.4) Suppose that $f\colon \mathbb{R}^n \to \mathbb{R}^n$ is a smooth map with $n > 1$, and let $K \subset \mathbb{R}^n$ be compact and $\varepsilon > 0$. If $df_x \neq 0$ for all $x \in \mathbb{R}^n$, then we can take $g = f$. Now assume there is an $x \in X$ such that $df_x = 0$. We would like to replace f with a suitable smooth function $g\colon \mathbb{R}^n \to \mathbb{R}^n$ satisfying the two conditions

(1) $dg_x \neq 0$ for all $x \in X$.
(2) $|f(x) - g(x)| < \varepsilon$ for all $x \in K$.

The idea for the solution is to replace f with $f + A$ for a suitable matrix $A \in \mathrm{M}(n)$. Since K is compact, for any given $A \in \mathrm{M}(n) \setminus \{0\}$, the set of norms $\{|Ax| \in \mathbb{R} : x \in K\}$ has a maximum $\mu_A > 0$. Hence if $\mu_A < \varepsilon$, then $|Ax| < \varepsilon$ for all $x \in K$, and we can define $g\colon \mathbb{R}^n \to \mathbb{R}^n$ by $g(x) = f(x) + Ax$. This map is smooth and satisfies condition (2). Since $\frac{\varepsilon}{2\mu_A}A$ is a linear map, the derivative of g at x is $dg_x = df_x + \frac{\varepsilon}{2\mu_A}A$. In order to prove the assertion, it remains to show that we can find an $A \in \mathrm{M}(n)$ such that $df_x + A \neq 0$ and $\mu_A < \varepsilon$. To do this, we define the map $F\colon \mathbb{R}^n \times \mathrm{M}(n) \to \mathrm{M}(n)$ by $F(x, A) := df_x + A$. The derivative $dF_{(x,A)}$ of F at a point (x, A) is the sum of the derivative of df_x and the identity map on $\mathrm{M}(n)$. In particular, $dF_{(x,A)}\colon \mathbb{R}^n \times \mathrm{M}(n) \to \mathrm{M}(n)$ is always surjective. Hence F is a submersion, and thus transverse to every submanifold of $\mathrm{M}(n)$. By the Transversality Theorem 12.42, this implies that the subset of $A \in \mathrm{M}(n)$ such that the map $F_A\colon \mathbb{R}^n \to \mathrm{M}(n)$ given by $F_A(x) := F(x, A)$ is transverse to the submanifold $\{0\}$ of $\mathrm{M}(n)$ is dense in $\mathrm{M}(n)$. However, for $n > 1$, $\dim \mathbb{R}^n = n$ is strictly less than $n^2 = \dim \mathrm{M}(n)$. Thus, since $\{0\}$ is a zero-dimensional submanifold of $\mathrm{M}(n)$, F_A is transverse to $\{0\}$ if and only if the intersection $\operatorname{Im}(F_A) \cap \{0\}$ is empty, i.e., $F_A(x) \neq 0$ for all $x \in X$. Finally, the subset of matrices in $\mathrm{M}(n)$ with $\max\{|Ax| : x \in K\} < \varepsilon$ is open in $\mathrm{M}(n)$. Thus, by the Transversality Theorem 12.42, we can choose an $A \in \mathrm{M}(n)$ with $F_A(x) = df_x + A \neq 0$ for all $x \in X$ and $\max\{|Ax| : x \in K\} < \varepsilon$.

Solution (to Exercise 12.5) For $n = 1$, we construct a counter-example as follows. Let $f\colon \mathbb{R} \to \mathbb{R}$ be defined by $f(x) = x^2$, let $K = [-2, 2] \subset \mathbb{R}$, and let $\varepsilon = 1$. Let $g\colon \mathbb{R} \to \mathbb{R}$ be a smooth function with $|f(x) - g(x)| < 1$ for all $x \in K$. In particular, this implies $3 < g(-2) < 5$, $3 < g(2) < 5$, and $1 < g(0) < 1$. This shows $(g(0) - g(-2))/2 < 0$ and $(g(2) - g(0))/2 > 0$.

By the Mean Value Theorem in calculus, there are real numbers $c \in (-2, 0)$ and $e \in (0, 2)$ such that $g'(c) = (g(0) - g(-2))/2 < 0$ and $g'(e) = (g(2) - g(0))/2 > 0$. Since g is smooth, we can apply the Intermediate Value Theorem to g' and get a number $e \in (c, e)$ with $g'(d) = 0$. Hence we cannot find g with both $g'(x) \neq 0$ for all x and $|f - g| < \varepsilon$ on K.

A.12 Intersection Theory Modulo 2

Solution (to Exercise 13.2) (a) Let $Z \subset Y$ be any closed submanifold with $\dim X + \dim Z = \dim Y$. For $\dim X \geq 1$, this implies $\dim Y > 0$ and $\dim Z = \dim Y - \dim X < \dim Y$. In particular, Z is not all of Y, and there exist points in Y which are not in Z. We can assume that Y is connected, since f being homotopic to a constant map implies that the image of f lies in only one connected component of Y. Hence any part of Z in a different connected component satisfies $f^{-1}(Z) = \emptyset$. Assume that f is homotopic to the constant map $g_0 \colon X \to \{y_0\} \subset Y$. If $y_0 \in Z$, then g and Z are not transverse, since $\dim Z < \dim Y$ and $\dim\{y_0\} = 0$. Since Y is connected and a smooth manifold, Y is path-connected by Exercise 8.5. Hence there is a path γ from y_0 to a point y_1 with $y_1 \notin Z$. Then g_0 is homotopic to the constant map $g_1 \colon X \to \{y_1\} \subset Y$ by composing g_0 with the path γ. Since being smoothly homotopic is a transitive relation, we have $f \simeq g_1$. Since $y_1 \notin Z$, we have $g_1 \pitchfork Z$. This means $g_1^{-1}(Z) = \emptyset$, i.e., $I_2(g_1, Z) = \#g_1^{-1}(Z) = 0$. By the homotopy invariance of $I_2(-, Z)$, we have $I_2(f, Z) = I_2(g_1, Z)$, and hence $I_2(f, Z) = 0$.

(b) Assume that $X = \{x\}$ is a one-point space. In particular, we have $\dim X = 0$. Thus any closed submanifold $Z \subset Y$ with $\dim X + \dim Z = \dim Y$ is of dimension $\dim Z = \dim Y$. This implies that every such Z is transverse to every other submanifold of Y. In particular, $\{x\} \pitchfork Z$, i.e., $f \pitchfork Z$ for every map $f \colon X = \{x\} \to Y$. Hence, by definition of intersection numbers, we can calculate $I_2(f, Z)$ by assuming $f(x) = y \in Z$. Then we get that $I_2(f, Z) = \#f^{-1}(Z) = \#\{y\} = 1 \neq 0$.

Solution (to Exercise 13.3) Since $\mathbb{S}^1$ is compact, the intersection number $I_2(\mathrm{id}_{\mathbb{S}^1}, \{x\})$ is defined for every point $x \in \mathbb{S}^1$ where we consider intersections in $\mathbb{S}^1$, i.e., $Y = X = \mathbb{S}^1$. Since the $\mathrm{id}_{\mathbb{S}^1}^{-1}(x) = \{x\}$, we have $I_2(\mathrm{id}_{\mathbb{S}^1}, \{x\}) = 1 \neq 0$. But we just learned in Exercise 13.2 that if $\mathrm{id}_{\mathbb{S}^1}$ was homotopic to a constant map, then $I_2(\mathrm{Id}_{\mathbb{S}^1}, \{x\})$ would have to be zero, since $\dim \mathbb{S}^1 = 1$. Thus there is at least one map $\mathbb{S}^1 \to \mathbb{S}^1$ which is not homotopic to a constant map.

Solution (to Exercise 13.4) Let Y be contractible and $\dim Y > 0$. Let $f: X \to Y$ be a smooth map with X compact and $Z \subset Y$ closed, and $\dim X + \dim Z = \dim Y$. Since Y is contractible, the identity map id_Y is homotopic to a constant map $Y \to \{y\} \subset Y$. Thus f is homotopic to the constant map $X \to \{y\} \subset Y$. Since $\dim Y > 0$, we have $\{y\} \neq Y$ and there are points in Y different from y. Hence, since $\dim X \geq 1$, Exercise 13.2 implies $I_2(f, Z) = 0$. Finally, we recall that Euclidean space $\mathbb{R}^k$ is contractible for all k.

Solution (to Exercise 13.5) If X is compact, then the intersection number $I_2(\mathrm{id}_X, \{x\})$ is defined for every point $x \in X$, where we consider intersections in X. Moreover, we have $I_2(\mathrm{id}_X, \{x\}) = \#\{x\} = 1$. If X is contractible and $\dim X \geq 1$, we can apply Exercise 13.4. Or we note that we have $\mathrm{id}_X \sim g$ where $g: X \to \{y\}$ is some constant map. Since $\dim X \geq 1$, we have $g \pitchfork \{x\}$ for $x \in X$ if and only if $x \neq y$. Hence $I_2(g, \{x\}) = \#g^{-1}(x) = 0$. However, since $\mathrm{id}_X \simeq g$, we have $I_2(\mathrm{id}_X, \{x\}) = I_2(g, \{x\})$. This is impossible. Hence, if $\dim X \geq 1$, X cannot be both compact and contractible. If $\dim X = 0$, then the only way X can be contractible is that it consists of just one point. This is a compact space.

Solution (to Exercise 13.6) Let $f: X \to \mathbb{S}^k$ be a smooth map with X compact and $0 < \dim X < k$. Let Z be a closed submanifold $Z \subset \mathbb{S}^k$ of dimension complementary to X, i.e., $\dim X + \dim Z = \dim \mathbb{S}^k$. The image of the derivative df_x at any point has at most dimension $\dim X = \dim T_x(X)$ in $T_{f(x)}(\mathbb{S}^k)$. Hence, since $\dim X < k$, df_x cannot be surjective. Hence the only points in $\mathbb{S}^k$ which are regular values of f are the points which are not in the image of f. By Sard's Theorem, however, we know that regular values exist. Hence there must be a point $p \in \mathbb{S}^k$ with $p \notin f(X) \cap Z$. Now stereographic projection provides a diffeomorphism $\phi: \mathbb{S}^k \setminus \{p\} \to \mathbb{R}^k$. Since p is not in the image of f, the composition $\phi \circ f: X \to \mathbb{S}^k \setminus \{p\} \to \mathbb{R}^k$ is defined. Similarly, since p is not in Z, $\phi_{|Z}$ is a diffeomorphism. Hence we can consider $\phi(Z)$ as a submanifold of $\mathbb{R}^k$ of dimension $\dim Z$. This implies that $I_2(f, Z) = I_2(\phi \circ f, \phi(Z))$, where the latter is the intersection number mod 2 in $\mathbb{R}^k$. However, since $\mathbb{R}^k$ is contractible, Exercise 13.4 implies that $I_2(\phi \circ f, \phi(Z)) = 0$. Thus $I_2(f, Z) = 0$.

Solution (to Exercise 13.7) Considering $\mathbb{S}^1 \subset \mathbb{C}$, we can define two submanifolds X and Z as $X = \mathbb{S}^1 \times \{1\}$ and $Z = \{1\} \times \mathbb{S}^1$ in $\mathbb{S}^1 \times \mathbb{S}^1$. They intersect in the single point q given by $q := \{1\} \times \{1\} \in S^1 \times S^1$. The tangent space to $\mathbb{S}^1 \times \mathbb{S}^1$ at q is $T_q(\mathbb{S}^1 \times \mathbb{S}^1) = T_1(\mathbb{S}^1) \times T_1(\mathbb{S}^1) = T_1(X) \times T_1(Z)$. Hence X and Z are transverse in $\mathbb{S}^1 \times \mathbb{S}^1$. Thus their intersection number in $\mathbb{S}^1 \times \mathbb{S}^1$ is $I_2(X, Z) = 1$. Now assume there is a diffeomorphism $\varphi\colon \mathbb{S}^1 \times \mathbb{S}^1 \to \mathbb{S}^2$. The map $\iota\colon \mathbb{S}^1 \hookrightarrow \mathbb{S}^1 \times \mathbb{S}^1$, $z \mapsto (z, 1)$, maps $\mathbb{S}^1$ diffeomorphically onto X. Composition with φ gives us a map $\varphi \circ \iota\colon \mathbb{S}^1 \to \mathbb{S}^2$. Since φ is a diffeomorphism, $\varphi(Z)$ is the diffeomorphic image of Z, and we would obtain the equality $I_2(X, Z) = I_2(\iota, Z) = I_2(\varphi \circ \iota, \varphi(Z))$ where the latter is the intersection number in $\mathbb{S}^2$. However, by Exercise 13.6, $I_2(\varphi \circ \iota, \varphi(Z)) = 0$ since $\dim \mathbb{S}^1 = 1 < 2$.

Solution (to Exercise 13.8) (a) Let $i_t\colon X \times [0,1] \to Y$ be a smooth homotopy with i_0 being the embedding of X in Y, $i_1(X) = Z \subset Y$, and i_t is an embedding for all $t \in [0,1]$. We define a new smooth map $F\colon X \times [0,1] \to Y \times [0,1]$ by $F(x,t) := (i_t(x), t)$. Now we define $W := F(X \times [0,1]) \subset Y \times [0,1]$ to be the image of $X \times [0,1]$ under F. We claim that W is a compact smooth manifold with boundary. First, since X is compact, $X \times [0,1]$ is compact, and hence $F(X \times [0,1])$ is compact by Lemma 1.13 and Exercise 1.1. Since i_t is an embedding for every $t \in [0,1]$, and the identity map on the second component is obviously an embedding as well, F is also an embedding. Thus the image of F is a smooth manifold. The boundary of W is then given by $\partial W = W \cap \partial(Y \times [0,1]) = X \times \{0\} \cup Z \times \{1\}$. Hence W is a cobordism between X and Z.

(b) Let X and Z be cobordant in Y, and let C be a compact submanifold C in Y with dimension complementary to X and Z. Let f denote the restriction to W of the projection map $Y \times [0,1] \to Y$. Then $\partial f = f_{|\partial W}$ is a smooth map which can be extended to a map $f\colon W \to Y$. Hence, by the Boundary Theorem 13.11, $I_2(\partial f, V) = 0$ for any closed submanifold V of Y of dimension $\dim V = \dim Y - \dim \partial W$. Since $\dim \partial W = \dim X = \dim Z$, we have, in particular, $I_2(\partial f, C) = 0$. By the Transversality Homotopy Theorem 12.44, we can assume $f \pitchfork C$ and $\partial f \pitchfork C$. In particular, we can assume $X \pitchfork C$ and $Z \pitchfork C$. Hence $I_2(X, C) = \#(X \cap C)$ and $I_2(Z, C) = \#(Z \cap C)$. By definition of f, $(\partial f)^{-1}(C)$ is given by

$$\begin{aligned}(\partial f)^{-1}(C) &= \partial W \cap (C \times [0,1]) \\ &= (X \times \{0\}) \cap (C \times [0,1]) \cup (Z \times \{1\}) \cap (C \times [0,1]) \\ &= ((X \cap C) \times \{0\}) \cup ((Z \cap C) \times \{1\}).\end{aligned}$$

Thus, we get

$$\begin{aligned} 0 = I_2(\partial f, C) &= \#(\partial f)^{-1}(C) \\ &= \#((X \cap C) \times \{0\}) + \#((Z \cap C) \times \{1\}) \\ &= \#(X \cap C) + \#(Z \cap C) = I_2(X, C) + I_2(Z, C). \end{aligned}$$

Since we are working modulo 2, this implies $I_2(C, X) = I_2(C, Z)$.

Solution (to Exercise 13.9) By Theorem 12.21, there is an open neighbourhood $Z^{\varepsilon} \subset X$ of Z and a diffeomorphism $Z^{\varepsilon} \cong N^{\varepsilon}(Z, X)$ to an open subset $N^{\varepsilon}(Z, X) \subset N(Z, X)$. By Corollary 12.22, the assumption implies that there is a diffeomorphism $Z \times \mathbb{R}^k \to N(Z, X)$. Hence we may identify $N(Z, X) = Z \times \mathbb{R}^k$ and $N^{\varepsilon}(Z, X) = Z \times (-\varepsilon, \varepsilon)$. However, for any $v \neq 0$ in $\mathbb{R}^k$, the subspaces $Z = Z \times \{0\}$ and the deformation $Z \times \{v\}$ do not intersect in $Z \times \mathbb{R}^k$. Thus, the existence of the diffeomorphism $Z^{\varepsilon} \cong Z \times (-\varepsilon, \varepsilon)$ implies that there is a deformation Z' of Z in X such that $Z \cap Z' = \emptyset$. Hence we have $I_2(Z, Z) = 0$.

Solution (to Exercise 13.10) (a) We need to find a deformation Z' of Z such that Z and Z' are transverse and then count the intersection points. For example, let us look at the map $f\colon [-1, 1] \to [-1, 1] \times (-1, 1)$ defined by $f(t) := (t, \frac{1}{2}\sin(t \cdot \pi/2))$. Let Z' be the image of f in X. There is a smooth homotopy $f_s\colon [-1, 1] \times [0, 1] \to X$, $(t, s) \mapsto (t, s \cdot (1/2)\sin(t\pi/2))$ with f_0 being the embedding of Z into X, i.e., the map $f_0\colon [-1, 1] \to X$ defined by $t \mapsto (t, 0)$, and f_1 being the embedding of Z' into X. (Note that, by abuse of notation, we have not distinguished between a point in $[-1, 1] \times (-1, 1)$ and its equivalence class in X). Moreover, there is only one point where Z and Z' meet, namely the image of $f(0)$ in X, i.e., the point with coordinates $p = (0, 0)$. Finally, the intersection of Z and Z' is transverse. We check this locally for the induced maps on the codomains of local charts: the tangent space to X at p is the x-axis in $\mathbb{R}^2$ and the tangent space to Z' at p is the line spanned by the vector $\begin{pmatrix} 1 \\ 1/2 \end{pmatrix} \in \mathbb{R}^2$. Hence $T_p(Z)$ and $T_p(Z')$ span a two-dimensional subspace in the two-dimensional vector space $T_p(X)$, i.e., they span all of $T_p(X)$. Thus, $I_2(Z, Z) = I_2(Z, Z') = \#(Z \cap Z') = 1$.

(b) By Exercise 13.9, the existence of a submersion $g\colon X \to \mathbb{R}$ such that $Z = g^{-1}(0)$ would imply $I_2(Z, Z) = 0$. This contradicts the previous point. Hence such a g cannot exist.

Solution (to Exercise 13.11) (a) We can show that f is smooth by checking it for the induced map on local charts. It is proper, since its domain $\mathbb{S}^1$ is compact. It is injective, since $\cos(t_1/2) = \pm\cos(t_2/2)$ implies $t_2 = t_1 + 2\pi$, and similarly for sin.

(b) We pick a point on $\mathbb{RP}^2$, say $Q = [0 : 0 : 1]$, and let $q : \mathbb{S}^1 \to \mathbb{RP}^2$ be the constant map with value Q. Let $\varphi : \mathbb{R} \to \mathbb{R}$ be a smooth bump function such that

$$\varphi(t) = \begin{cases} 0 & x \leq 1/4 \\ 1 & x \geq 3/4. \end{cases}$$

We define a smooth homotopy by $H : [0, 2\pi] \times [0, 1] \to \mathbb{RP}^2$ by

$$(t, s) \mapsto \left[\sqrt{1 - \varphi(s)}\cos(t) : \sqrt{1 - \varphi(s)}\sin(t) : \sqrt{\varphi(s)}\right],$$

such that $H(t, 0) = 2f(t)$ and $H(t, 1) = q(t)$ for all t. Finally, we note that the map $\tilde{H} : [0, 2\pi] \times [0, 1] \to \mathbb{RP}^2$ defined by

$$(t, s) \mapsto \left[\sqrt{1 - \varphi(s)}\cos(t/2) : \sqrt{1 - \varphi(s)}\sin(t/2) : \sqrt{\varphi(s)}\right]$$

does not provide a homotopy between f and the constant map c. One issue is that it does not induce a map on $\mathbb{S}^1 \times [0, 1]$, since we can compute $\tilde{H}(0, s) = \left[\sqrt{1 - \varphi(s)} : 0 : \sqrt{\varphi(s)}\right]$, whereas $\tilde{H}(2\pi, s) = \left[-\sqrt{1 - \varphi(s)} : 0 : \sqrt{\varphi(s)}\right]$. There is a minus sign occurring in the first coordinate because we have $\cos(2\pi/2) = -\cos(0)$. Hence $\tilde{H}(0, s) \neq \tilde{H}(2\pi, s)$ in $\mathbb{RP}^2$ when $0 < \varphi(s) < 1$. This indicates that we cannot continuously deform the loop corresponding to f to a constant map without cutting it open at some point. And the above calculation of the intersection number for f shows that it is impossible to remedy this defect.

A.13 Orientation

Solution (to Exercise 14.2) Let $(e_1, \ldots, e_k)$ be the ordered basis of $\mathbb{R}^k$ which defines the standard orientation of $\mathbb{R}^k$. The orientation of $\mathbb{H}^k$ is given by the standard orientation of $\mathbb{R}^k$ restricted to the subspace $\mathbb{H}^k \subset \mathbb{R}^k$. The boundary orientation of $\partial\mathbb{H}^k$ is given by requiring that, at any point $x \in \partial\mathbb{H}^k$, the outward pointing unit normal vector $n_x = -e_k$ fits into a positively oriented basis $(n_x, e_1, \ldots, e_{k-1}) = (-e_k, e_1, \ldots, e_{k-1})$ for $\mathbb{R}^k$. The matrix which sends $(e_1, \ldots, e_k)$ to $(-e_k, e_1, \ldots, e_{k-1})$ is given by

$$A = \begin{pmatrix} 0 & 1 & & \dots & 0 \\ 0 & 0 & 1 & \dots & 0 \\ \vdots & \vdots & & \ddots & \vdots \\ 0 & 0 & \dots & & 1 \\ -1 & 0 & \dots & & 0 \end{pmatrix}$$

which can be transformed into the diagonal matrix

$$D = \begin{pmatrix} 1 & 0 & \dots & 0 \\ 0 & 1 & \dots & 0 \\ \vdots & & \ddots & \vdots \\ 0 & 0 & \dots & -1 \end{pmatrix}$$

by interchanging two columns exactly $k-1$ times. Hence we get that $\det(D) = (-1)^{k-1}\det(A)$. Since $\det(D) = -1$, we get $\det(A) = 1 > 0$ if and only if $(-1)^k = 1$, i.e., if k is even.

Solution (to Exercise 14.3) Assume that $df_{x_0}\colon T_{x_0}(X) \to T_{f(x_0)}(Y)$ preserves orientation at some point $x_0 \in X$. Since f is a diffeomorphism, df_x is an isomorphism for all $x \in X$. Hence $\det(df_x) \neq 0$ for all $x \in X$. In particular, the two subsets $U := \{x \in X : \det(df_x) > 0\}$ and $V := \{x \in X : \det(df_x) < 0\}$ are open and cover X. By assumption $x_0 \in U$, and hence U is non-empty. Since X is connected, this implies $U = X$.

Solution (to Exercise 14.4) (a) At $x = (a,b,c) \in \mathbb{S}^2$, the tangent space $T_x(\mathbb{S}^2)$ is the two-dimensional vector subspace of $\mathbb{R}^3$ which is orthogonal to x. Since $(a,b,c) \neq (0,0,0)$, let us assume that, say, $b \neq 0$. A basis for $T_x(\mathbb{S}^2)$ is given by, for example, $v = (-b,a,0)$ and $v = (0,c,-b)$. The outward pointing normal vector is given by $n_x = (a,b,c)$. The boundary orientation of $\mathbb{S}^2$, is the orientation of $T_x(\mathbb{S}^2)$ determined by the basis (n_x, v, w). This basis is positively oriented in $T_x(\mathbb{R}^3) = \mathbb{R}^3$ if and only if the matrix

$$A = \begin{pmatrix} a & -b & 0 \\ b & a & c \\ c & 0 & -b \end{pmatrix}$$

has positive determinant, since this is the matrix that transforms the standard basis of $\mathbb{R}^3$ into the basis (n_x, v, w). The determinant of A is $\det(A) = -a^2b - b^3 - bc^2 = -b(a^2+b^2+c^2) = -b$. Thus, if $b < 0$, (n_x, v, w) is a positively oriented basis of $T_x(\mathbb{S}^2)$. If $b > 0$, we take the basis (n_x, w, v). And if $b = 0$, we start over with either a or c taking the role of b.

(b) The boundary orientation of $\mathbb{S}^k$ is, at any point $x \in \mathbb{S}^k$, given on $T_x(\mathbb{S}^k)$ by choosing the ordered basis $(n_x, v_1, \dots, v_k)$ to be positively oriented where n_x is the outward pointing unit normal vector in $T_x(\mathbb{R}^{k+1}) = \mathbb{R}^{k+1}$ and $(v_1, \ldots, v_k)$ is an ordered basis of $T_x(\mathbb{S}^k)$. Since $\mathbb{S}^k \subset \mathbb{R}^{k+1}$ is of codimension one, n_x spans the orthogonal complement $N_x(\mathbb{S}^k, \mathbb{R}^{k+1})$ of $T_x(\mathbb{S}^k)$ in $\mathbb{R}^{k+1}$. Hence the orientation of $T_x(\mathbb{S}^k)$ induced by the direct sum $N_x(\mathbb{S}^k, \mathbb{R}^{k+1}) \oplus T_x(\mathbb{S}^k) = T_x(\mathbb{R}^{k+1}) = \mathbb{R}^{k+1}$ equals the orientation of $\mathbb{S}^k$ as the preimage under g.

Solution (to Exercise 14.5) (a) If E reverses orientation, then $r_1 \circ E$ preserves orientations. Then if there is a homotopy F between $r_1 \circ E$ and the identity, then, after composing all maps with r_1, $r_1 \circ F$ is a homotopy between $E = r_1 \circ r_1 \circ E$ and r_1. So let E be a linear isomorphism of $\mathbb{R}^k$ that preserves orientations. The proof now proceeds by induction on the dimension k. We need to check two initial cases.

(b) Let $k = 1$. Then $E\colon \mathbb{R} \to \mathbb{R}$ is given by multiplication by a real number $\lambda > 0$, and $E_t = t \cdot 1 + (1 - t) \cdot \lambda$ is a homotopy between $E = \lambda$ and $\mathrm{Id} = 1$. Note that each E_t is non-zero and therefore a linear isomorphism.

(c) Now let $k = 2$ and assume that E has only complex eigenvalues. Then $E_t = tE + (1 - t)\mathrm{Id}$ is a linear homotopy between Id and E. Moreover, each E_t is a linear isomorphism. To show this, we show that $\det(E_t) \neq 0$ for all $t \in [0, 1]$. If $E = \begin{pmatrix} a & b \\ c & d \end{pmatrix}$, then we get

$$\begin{aligned}\det(E_t) &= (t(a-1)+1)(t(d-1)+1) - t^2 bc \\ &= t^2(a-1)(d-1) + t(a+d-2) + 1 - t^2 bc \\ &= t^2(ad - bc - a - d + 1) + t(a + d - 2) + 1.\end{aligned}$$

The discriminant of this quadratic equation in t is

$$\begin{aligned}&(a+d-2)^2 - 4(ad - bc - a - d + 1) \\ &\quad = (a+d)^2 - 4(a+d) + 4 - 4(ad - bc) + 4(a+d) - 4 \\ &\quad = (a+d)^2 - 4(ad - bc).\end{aligned}$$

This is also the discriminant of the equation $t^2 + t(a+d) - (ad - bc) = 0$, which is the characteristic polynomial in t of E. By assumption, this polynomial has only complex roots, i.e., its discriminant is negative. Hence there is no real t such that $\det(E_t) = 0$.

(d) We assume $k \geq 2$ and the assertion to be true in all dimensions $< k$. Then E has either at least one real eigenvalue or at least one complex eigenvalue. Let $V \subset \mathbb{R}^k$ be the corresponding eigenspace, which is either one- or two-dimensional. Then E maps V into itself. Hence $\mathbb{R}^k$ splits into a direct sum $\mathbb{R}^k = V \oplus W$. By choosing a basis of $\mathbb{R}^k$ consisting of a basis of V and one for W, we can represent E as a matrix of the form $E = \begin{pmatrix} A & B \\ 0 & C \end{pmatrix}$. Here A is either a (1×1)- or a $2(\times 2)$-matrix determined by the type of the eigenvalues. Then we can define a linear homotopy E_t by $E_t = \begin{pmatrix} A & tB \\ 0 & C \end{pmatrix}$. Since E is a linear isomorphism and the determinant is multiplicative, we have $0 \neq \det(E) = \det(A)\det(C) = \det(E_t)$. Thus E_t is also a linear isomorphism for every t. For $t = 0$, we see that E_0 maps V to V by A and W to W by C. Since $\dim W$ is strictly less than k, we can apply the induction hypothesis to C and W and the initial cases to A and V, respectively. Hence we have a homotopy C_t consisting of linear isomorphisms between C and the identity and a homotopy A_t between A and the identity. Then $t \mapsto \begin{pmatrix} A_t & tB \\ 0 & C_t \end{pmatrix}$ is a linear homotopy between E and the identity map of $\mathbb{R}^k$.

Solution (to Exercise 14.6) Let X and Z be transverse submanifolds in Y and assume X, Z, and Y are oriented. Let $i\colon X \hookrightarrow Y$ be the inclusion of X into Y. The intersection $X \cap Z$ equals the preimage $i^{-1}(Z)$. By Section 14.4, at any $y \in X \cap Z$, the preimage orientation on $S := i^{-1}(Z)$ is induced by the direct sum $N_y(S,X) \oplus T_y(S) = T_y(X)$, where $N_y(S,X)$ is the orthogonal complement of $T_y(S)$ in $T_y(X)$. The orientation on $N_y(S,X)$ is induced by the direct sum $di_y(N_y(S,X)) \oplus T_y(Z) = T_y(Y)$, and the fact that $d(i_y)_{|N_y(S,X)}$ is an isomorphism onto its image. Since all these vector spaces are subspaces in $T_y(Y)$, and are oriented as subspaces of $T_y(Y)$, we can identify $N_y(S,X)$ with its image under di_y in $T_y(Y)$ and can rewrite this equation as $N_y(S,X) \oplus T_y(Z) = T_y(Y)$. Now let $N_y(S,Z)$ be the orthogonal complement of $T_y(S)$ in $T_y(Z)$. Then the orientation of $T_y(S)$ is determined by the direct sum $N_y(S,X) \oplus N_y(S,Z) \oplus T_y(S) = T_y(Y)$. Now if we start with the inclusion $j\colon Z \hookrightarrow Y$ of Z in Y instead, we get that the orientation of S considered as the preimage $j^{-1}(X)$ in Z, is determined by the direct sum $N_y(S,Z) \oplus N_y(S,X) \oplus T_y(S) = T_y(Y)$. We learned that the signs of the orientations of $N_y(S,X) \oplus N_y(S,Z)$ and $N_y(S,Z) \oplus N_y(S,X)$ differ by $(-1)^{\dim N_y(S,X)\dim N_y(S,Z)}$. Now it remains to remark that, by definition of the normal spaces as orthogonal complements, we have $\dim N_y(S,X) = \operatorname{codim}_X X \cap Z = \operatorname{codim}_Y Z$, and $\dim N_y(S,Z) = \operatorname{codim}_Z X \cap Z = \operatorname{codim}_Y X$, where the subscript denotes the ambient space for taking the codimension.

Solution (to Exercise 14.7) First, we assume that Z is orientable. At any point $z \in Z$, let $(v_1, \ldots, v_n)$ be an oriented basis of $T_z Z$. Since $T_z Z$ is a vector subspace of codimension one in $T_z X$, we can choose a single vector $n_z \in (T_z Z)^{\perp} = N_z(Z, X) \subset T_z X$ such that $(n_z, v_1, \ldots, v_n)$ is an oriented basis of $T_z X$. Since both X and Z are oriented, we can make this choice of n_z smoothly for every $z \in Z$. Since n_z has to be non-trivial in order to be part of a basis, we can define a diffeomorphism $\varphi \colon Z \times \mathbb{R} \to N(Z, X)$ by $\varphi(z, t) := (z, t \cdot n_z)$. Hence $N(Z, X)$ is a trivial line bundle over Z. Now we assume that $N(Z, X)$ is trivial and let $\varphi \colon Z \times \mathbb{R} \to N(Z, X)$ be a trivialising diffeomorphism. Let n_z denote the vector in $(T_z Z)^{\perp} = N_z(Z, X) \subset T_z X$ such that $\varphi(z, 1) = (z, n_z)$. We then define a basis $(v_1, \ldots, v_n)$ of $T_z Z$ to be positively oriented if $(n_z, v_1, \ldots, v_n)$ is a positively oriented basis of $T_z X$. Since φ is a diffeomorphism, this defines a smooth choice of an orientation for all $z \in Z$. Hence Z is orientable.

Solution (to Exercise 14.8) (a) Any basis of $V \times V$ consists of the product $(\alpha \times 0, 0 \times \beta)$ where α and β are ordered bases of V. The sign of this basis satisfies $\operatorname{sign}(\alpha \times 0, 0 \times \beta) = \operatorname{sign}(\alpha) \cdot \operatorname{sign}(\beta)$. Switching the orientation of V changes both $\operatorname{sign}(\alpha)$ and $\operatorname{sign}(\beta)$. Changing both signs simultaneously results in multiplying with $(-1)^2 = 1$. Hence the sign of the basis of $V \times V$ is independent of the choice of orientation for V.

(b) Let X be an orientable manifold. The orientation of $X \times X$ is given by a smooth choice of orientation of each $T_{(x,y)}(X \times X) = T_x(X) \times T_y(X)$. Changing the orientation of X means changing the orientation of both $T_x(X)$ and $T_y(X)$. As in the previous point, this means multiplying the sign of any ordered basis of $T_{(x,y)}(X \times X)$ by $+1$. Hence the product orientation on $X \times X$ is the same for all choices of orientation on X.

(c) Let X be a smooth manifold which is not orientable. Any Euclidean space $\mathbb{R}^m$ is oriented as a manifold by the choice of the standard orientation of the tangent space $T_z(\mathbb{R}^m) = \mathbb{R}^m$ for any $z \in \mathbb{R}^m$. For any points $x \in X$ and $z \in \mathbb{R}^m$, the tangent space $T_{(x,z)}(X \times \mathbb{R}^m)$ is just $T_x(X) \times \mathbb{R}^m$. If there was a smooth choice for an orientation of $X \times \mathbb{R}^m$, then each tangent space $T_x(X)$ of X would inherit a smooth choice of orientation from the product $T_x(X) \times \mathbb{R}^m$. This contradicts the non-orientability of X. Now let Y be any smooth manifold. If $X \times Y$ is orientable, then also $X \times U$ for an open subspace $U \subset Y$ which is diffeomorphic to some $\mathbb{R}^m$. But then $X \times \mathbb{R}^m$ would also inherit an orientation, which is not possible. Applied to $Y = X$, we see that $X \times X$ is not orientable.

(d) We can cover X by local parametrisations $\phi \colon U \to X$. The union of the images of the maps $\phi \times \phi \colon U \times U \to X \times X$ is then an open subspace V

of $X \times X$ which includes Δ. We orient each individual $\phi(U)$ by requiring the diffeomorphism $\phi\colon U \to \phi(U)$ to be orientation preserving. This induces an orientation on $(\phi \times \phi)(U \times U) = \phi(U) \times \phi(U)$. As we argued before, there is a well-defined orientation on $\phi(U) \times \phi(U)$ which is independent of the local parametrisations chosen. Thus, $\phi(U) \times \phi(U)$, which is an open neighbourhood of Δ in $X \times X$, is orientable. However, this does not mean that Δ is always orientable. For, the tangent space to Δ at any point (x,x) is the diagonal of $T_x(X) \times T_x(X)$. This diagonal is isomorphic to $T_x(X)$. Thus if we had a smooth choice of orientations for all diagonals in $T_x(X) \times T_x(X)$, then we would have a smooth choice of orientations for all $T_x(X)$. In other words, Δ is orientable if and only if X is orientable.

Solution (to Exercise 14.9) To determine the fibre over b, we write $z_0 = x_0 + iy_0$ and $z_1 = x_1 + iy_1$. Then we get $\hbar(z_0, z_1) = (0,1,0)$ which implies $2z_0\bar{z}_1 = i$ and $|z_0|^2 = |z_1|^2 = \frac{1}{2}$, i.e., $y_0 = x_1$, $y_1 = -x_0$, and $x_0^2 + x_1^2 = \frac{1}{2}$. Thus the fibre over b has the form

$$\hbar^{-1}(b) = \left\{(z_0, z_1) \in \mathbb{S}^3 : \bar{z}_1 = \frac{i}{2z_0}\right\} = \{(x_0, y_0, x_1, y_1) \in \mathbb{S}^3 : y_0 = x_1, y_1 = -x_0\}.$$

Let $q = (x_0, x_1, x_1, -x_0) \in \hbar^{-1}(b)$ be a point in the fibre over b. Since not both x_0 and x_1 can be zero, we assume that $x_0 \neq 0$. The tangent space $T_q\mathbb{S}^3$ is the vector space

$$T_q\mathbb{S}^3 = \{\mathbf{u} \in \mathbb{R}^4 : \mathbf{u} \perp q\} = \operatorname{Span}\left\{ q_1^\perp = \begin{pmatrix} -x_1 \\ x_0 \\ 0 \\ 0 \end{pmatrix}, q_2^\perp = \begin{pmatrix} -x_1 \\ 0 \\ x_0 \\ 0 \end{pmatrix}, q_3^\perp = \begin{pmatrix} x_0 \\ 0 \\ 0 \\ x_0 \end{pmatrix} \right\}.$$

The orientation of $T_q\mathbb{S}^3$ as a boundary of the unit ball is such that the outward pointing vector q together with the basis vectors of $T_q\mathbb{S}^3$ form a positively oriented basis of $\mathbb{R}^4$. The matrix expressing the basis $(q, q_1^\perp, q_2^\perp, q_3^\perp)$ in the standard basis of $\mathbb{R}^4$ is

$$\begin{pmatrix} x_0 & -x_1 & -x_1 & x_0 \\ x_1 & x_0 & 0 & 0 \\ x_1 & 0 & x_0 & 0 \\ -x_0 & 0 & 0 & x_0 \end{pmatrix}.$$

The determinant of this matrix is $2x_0^4 + 2x_0^2x_1^2 = 2x_0^2(x_0^2 + x_1^2) = x_0^2 > 0$. In particular, it is positive, and the basis $(q_1^\perp, q_2^\perp, q_3^\perp)$ is a positively oriented basis of $T_q\mathbb{S}^3$. By Proposition 4.13, the tangent space $T_q\hbar^{-1}(b)$ equals the kernel of $d\tilde{\hbar}_q$. In Exercise 4.12 , we computed this map as represented by the matrix

$$d\tilde{\hbar}_q = 2 \cdot \begin{pmatrix} x_1 & -x_0 & x_0 & x_1 \\ x_0 & x_1 & x_1 & -x_0 \\ x_0 & x_1 & -x_1 & x_0 \end{pmatrix}.$$

The kernel of this map is the span of the vector $q_0^\perp = (-x_1, x_0, x_0, x_1)$. The normal space $N_q(\hbar^{-1}(b); \mathbb{S}^3) \subset T_q\mathbb{S}^3$ of vectors which are orthogonal to $T_q\hbar^{-1}(b)$ is the span of $(q_1^\perp - q_2^\perp, q_3^\perp)$. The map $d\tilde{\hbar}_q$ sends $q_1^\perp - q_2^\perp$ and $q_3^\perp$ to, respectively,

$$d\tilde{\hbar}_q(q_1^\perp - q_2^\perp) = 2\begin{pmatrix} -2x_0^2 \\ 0 \\ -2x_0x_1 \end{pmatrix}, \; d\tilde{\hbar}_q(q_3^\perp) = 2\begin{pmatrix} -2x_0x_1 \\ 0 \\ 2x_0^2 \end{pmatrix}.$$

These two vectors form a basis $(d\tilde{\hbar}_q(q_1^\perp - q_2^\perp), d\tilde{\hbar}_q(q_3^\perp))$ of $T_b\mathbb{S}^2$. We need to check the orientation of this basis. The tangent space $T_b\mathbb{S}^2$ has a basis $(\mathbf{e}_1^3, \mathbf{e}_3^3)$ as a subspace in $\mathbb{R}^3$. This basis is negatively oriented, since, together with the outward pointing vector $b = \mathbf{e}_2^3$, the basis $(\mathbf{e}_2^3, \mathbf{e}_1^3, \mathbf{e}_3^3)$ is a negatively oriented basis of $\mathbb{R}^3$, as we need to make one permutation to get the standard basis. The matrix B which expresses $(d\tilde{\hbar}_q(q_1^\perp - q_2^\perp), d\tilde{\hbar}_q(q_3^\perp))$ in terms of the basis $(\mathbf{e}_1^3, \mathbf{e}_3^3)$ is given by $B = 4 \cdot \begin{pmatrix} -x_0^2 & -x_0x_1 \\ -x_0x_1 & x_0^2 \end{pmatrix}$. We see that $\det B = 16(-x_0^4 - x_0^2x_1^2) = -16x_0^2(x_0^2 + x_1^2) = -8x_0^2 < 0$ is negative. Hence the basis $(d\tilde{\hbar}_q(q_1^\perp - q_2^\perp), d\tilde{\hbar}_q(q_3^\perp))$ is a negatively oriented basis of $T_b\mathbb{S}^2$. This defines an orientation on the normal space $N_q(\hbar^{-1}(b); \mathbb{S}^3)$ by declaring the orientation of the basis $(q_1^\perp - q_2^\perp, q_3^\perp)$ to be negative. Finally, the orientation of $T_q\hbar^{-1}(b)$ is such that the direct sum $N_q(\hbar^{-1}(b); \mathbb{S}^3) \oplus T_q\hbar^{-1}(b) = T_q\mathbb{S}^3$ induces the given orientation on $T_q\mathbb{S}^3$. We check this by looking at the basis $(q_1^\perp - q_2^\perp, q_3^\perp, q_0^\perp)$ of $N_q(\hbar^{-1}(b); \mathbb{S}^3) \oplus T_q\hbar^{-1}(b)$. The transition matrix from the basis $(q_1^\perp, q_2^\perp, q_3^\perp)$ to the basis $(q_1^\perp - q_2^\perp, q_3^\perp, q_0^\perp)$ is

$$\begin{pmatrix} 1 & 0 & 1 \\ -1 & 0 & 1 \\ 0 & 1 & x_1/x_0 \end{pmatrix}.$$

The determinant of this matrix is -2. In particular, it is negative. Since we checked that the basis $(q_1^\perp, q_2^\perp, q_3^\perp)$ is a negatively oriented basis, we see that the orientation of the basis $(q_1^\perp - q_2^\perp, q_3^\perp, q_0^\perp)$ of $T_q\mathbb{S}^3$ is positive. Since the sign of $(q_1^\perp - q_2^\perp, q_3^\perp)$ is positive as a basis of $N_q(\hbar^{-1}(a); \mathbb{S}^3)$, we need that the basis $q_0^\perp$ also has positive sign. Hence the vector $q_0^\perp$ provides a positively oriented basis of $T_q\hbar^{-1}(b)$.

A.14 Brouwer Degree and Hopf Invariant

Solution (to Exercise 15.1) (a) For $1 \leq i \leq k+1$, let $r_i : \mathbb{S}^k \to \mathbb{S}^k$ be the ith reflection map, i.e., $r_i(x_1, \ldots, x_{k+1}) = (x_1, \ldots, -x_i, \ldots, x_{k+1})$. The antipodal map is equal to the composition of reflections $r_1 \circ r_2 \circ \cdots \circ r_{k+1}$. Each reflection r_i is a diffeomorphism which reverses the orientation on $\mathbb{S}^k$. The composition of two such reflections $r_i \circ r_{i+1}$, however, is then a diffeomorphism which preserves the orientation on $\mathbb{S}^k$. Hence, if $k = 2n$ is even, $a = r_1 \circ r_2 \circ \cdots \circ r_{2n+1}$ has degree -1. Whereas if $k = 2n-1$ is odd, $a = r_1 \circ r_2 \circ \cdots \circ r_{2n}$ has degree $+1$. In other words, $\deg(a) = (-1)^{k+1}$.

(b) By Exercise 8.4, we know that a is homotopic to the identity if k is odd. By the previous point, we have $\deg(a) = -1$ if k is even. Since deg is homotopy invariant, $\deg(a) = -1 \neq 1 = \deg(\mathrm{id})$ implies that the antipodal map is not homotopic to the identity if k is even.

Solution (to Exercise 15.2) Recall from Exercise 8.10 that if a vector field $\mathbf{v}$ on $\mathbb{S}^n$ does not have a zero, then the antipodal map of $\mathbb{S}^n$ is homotopic to the identity. By Exercise 15.1, the latter is the case if and only if n is odd.

Solution (to Exercise 15.3) We begin as in the solution to Exercise 11.6. We use the homotopy $p_t(z) = tp(z) + (1-t)z^m = z^m + t(a_1 z^{m-1} + \cdots + a_m)$ from $p_0(z) = z^m$ to $p_1(z) = p(z)$. We then observe that, if W is a closed ball around the origin in $\mathbb{C}$ with sufficiently large radius, none of the p_t has a zero on ∂W. Hence the homotopy $p_t/|p_t| \colon \partial W \to \mathbb{S}^1$ is defined for all $t \in [0,1]$. Thus, $\deg(p/|p|) = \deg(p_0/|p_0|)$. Since $p_0(z) = z^m$, the degree of $p_0/|p_0|$ is the same as $\deg(z^m) = m$, i.e., $\deg(p/|p|) = m$. Thus, if $m > 0$, $p/|p|$ does not extend to all of W, since otherwise its degree had to be zero. Hence p must have a zero inside W. Note that the final argument is not available for $\deg_2$ if m is even.

Solution (to Exercise 15.4) Let us try to set up the argument we used in Exercise 15.3: Let $p(x) = x^m + a_1 x^{m-1} + \cdots + a_m$ be a monic real polynomial. Define a homotopy $p_t(x) = tp(x) + (1-t)x^m = x^m + t(a_1 x^{m-1} + \cdots + a_m)$ from $p_0(x) = x^m$ to $p_1(x) = p(x)$. If $W = [-a, a]$ is a large enough closed interval in $\mathbb{R}$ containing the origin, none of the p_t has a zero on $\partial W = \{-a, a\}$. Hence the homotopy $p_t/|p_t| \colon \partial W = \{-a, a\} \to \{-1, +1\}$ is defined for all $t \in [0,1]$. Thus, $\deg(p/|p|) = \deg(p_0/|p_0|)$. However, for each $t \in [0,1]$, the map $p_t/|p_t|(x)$ is constant with value either -1 or $+1$. A constant map has degree 0. Hence we get $\deg(p/|p|) = 0$. But this does not imply that we cannot extend $p/|p|$ to all of W. In particular, we do not get a contradiction to the assumption that p does not have a zero in W.

Solution (to Exercise 15.5) (a) First, g is in fact a map $\mathbb{S}^1 \to \partial\mathbb{D}_0$, since any point z with $|z| = 1$ is sent to a point with $|g(z) - z_0| = z_0 + rz - z_0 = |rz| = r$ since $|z| = 1$. Moreover, g is smooth and has an inverse $g^{-1}\colon \partial\mathbb{D}_0 \to \mathbb{S}^1$ given by $g^{-1}(z) = \frac{z-z_0}{|z-z_0|}$. Note that g^{-1} is also smooth since taking norms and dividing by $|z - z_0|$ are smooth operations in $\mathbb{C} \setminus \{z_0\}$. Hence we know that g either preserves or reverses orientations. To check that g preserves orientations, it suffices to note that the derivative dg_x of g at any point x, for example dg_1, is given by multiplication with $r > 0$.

(b) Let $y \in \mathbb{S}^1$. Since g is a diffeomorphism, y is a regular value for p if and only if it is a regular value for $p \circ g$. Moreover, g defines a bijection between the finite sets $\{z \in \partial D_0\colon p(z) = y\}$ and $\{z \in \mathbb{S}^1\colon p(g(z)) = y\}$. Since g preserves orientations, the orientation numbers at each point in these sets agree. Hence the degrees of the above maps are the same.

(c) Since $p(z) \neq 0$ for all $z \in \mathbb{D}_0 \setminus \{z_0\}$ by our choice of $\mathbb{D}_0$, we have $|q(z_0 + trz)| \neq 0$ for all $|z| = 1$. Hence we can define a smooth map $H\colon \mathbb{S}^1 \times [0,1] \to \mathbb{S}^1$ by $H(z,t) := \frac{z^m q(z_0+trz)}{|q(z_0+trz)|}$. Write $h_t(z) = H(z,t)$. For all $z \in \mathbb{S}^1$, we have $h_0(z) = \frac{z^m q(z_0)}{q(z_0)|} = \frac{q(z_0)}{q(z_0)|} \cdot z^m$ and

$$h_1(z) = \frac{z^m q(z_0 + rz)}{|q(z_0 + rz)|} = \frac{(z_0 + rz - z_0)^m q(z_0 + rz)}{|z_0 + rz - z_0|^m |q(z_0 + rz)|} = \frac{p(g(z))}{|p(g(z))|}.$$

Hence h_t is the desired homotopy.

(d) We know from the calculation in Example 15.11 that the map $\mathbb{S}^1 \to \mathbb{S}^1$, $z \mapsto z^m$, has degree m. Since multiplying with a constant c does not change the degree, we also have $\deg(z \mapsto c \cdot z^m) = \deg(h_0) = m$. Since degrees are homotopy invariant, the previous point shows $\deg(h_1) = \deg(h_0)$. As we observed before, $\deg(p/|p|) = \deg(h_1)$. Hence we can conclude $\deg(p/|p|) = m$.

Solution (to Exercise 15.6) Let z be a regular value for g. If z is not in the image of g, then $\deg(g)$ and $\deg(g \circ f)$ both vanish, and the claim is true. So let us assume that z is in the image of g. We showed in Exercise 5.5 that z then is a regular value for $g \circ f$ if and only if every $y \in g^{-1}(z)$ is a regular value for f. Hence we can use z to compute $\deg(g \circ f)$ if and only if we can use $y \in g^{-1}(z)$ to compute $\deg(f)$. If z is not in the image of $g \circ f$, then y is not in the image of f. In this case, both $\deg(g \circ f)$ and $\deg(f)$ vanish, and the claim is true. So, assume there are points $x \in X$ with $g(f(x)) = z$. Given such an x, the chain rule gives us $d(g \circ f)_x = dg_{f(x)} \circ df_x$. Since x and $f(x)$ are regular points by assumption, all three derivatives in this equation are isomorphisms. Recall that we equip x with orientation number $+1$ if df_x preserves orientation and with orientation number -1 if df_x reverses orientation. Moreover, whether df_x preserves or reverses orientation is determined by the sign of its determinant. Similarly, for $f(x)$ with respect to $dg_{f(x)}$ and x with respect to $d(g \circ f)_x$. By the chain rule and since the determinant is a multiplicative function, we get the formula for signs: $\operatorname{sign}(\det(d(g \circ f)_x)) = \operatorname{sign}(\det(dg_{f(x)})) \cdot \operatorname{sign}(\det(df_x))$. Since the degree is the sum of the orientation numbers at all points in the fibre, we get

$$\begin{aligned}
\deg(g \circ f) &= \sum_{x \in (g \circ f)^{-1}(z)} \operatorname{sign}(\det(d(g \circ f)_x)) \\
&= \sum_{x \in (g \circ f)^{-1}(z)} [\operatorname{sign}(\det(dg_{f(x)})) \cdot \operatorname{sign}(\det(df_x))] \\
&= \left(\sum_{y \in g^{-1}(z)} \operatorname{sign}(\det(dg_y)) \right) \cdot \left(\sum_{x \in f^{-1}(y)} \operatorname{sign}(\det(df_x)) \right) \\
&= \deg(g) \cdot \deg(f).
\end{aligned}$$

Note that we used here that we can compute $\deg f$ using any regular value $y \in Y$ for f with $g(y) = z$.

Solution (to Exercise 15.7) Let $a \colon \mathbb{S}^k \to \mathbb{S}^k$ denote the antipodal map. We assume that f had no fixed point. Then x and $a(f(x))$ are never antipodal points, and we get $|x - (a \circ f)(x)| < 2$ for all $x \in \mathbb{S}^k$. By Exercise 8.9, this implies that $\mathrm{id}_{\mathbb{S}^k}$ and $a \circ f$ are homotopic. By Exercise 15.6, we have $\deg(a \circ f) = \deg(a) \cdot \deg(f)$ and by Exercise 15.1 we know $\deg(a) = (-1)^{k+1}$. Thus we obtain $1 = \deg(\mathrm{id}) = (-1)^{k+1} \deg(f)$, which implies $\deg(f) = (-1)^{k+1}$. Thus, if f does not have a fixed point, then $\deg(f) = (-1)^{k+1}$.

Solution (to Exercise 15.8) At every point $[x] \in \mathbb{RP}^k$, the fibre under q consists of two antipodal points $q^{-1}([x]) = \{x, -x\} \subset \mathbb{S}^k$. Let $a \colon \mathbb{S}^k \to \mathbb{S}^k$ be the antipodal map. We have $q(x) = q(-x)$ for all $x \in \mathbb{S}^k$, i.e., we have $q = q \circ a$. This implies $dq_{-x} \circ da_x = dq_x$ for all $x \in \mathbb{S}^k$. In particular, this implies $\det(dq_{-x}) \cdot \det(da_x) = \det(dq_x)$. If k is odd, then a preserves orientation, i.e., $\det(da_x) > 0$. Thus, $\det(dq_{-x})$ and $\det(dq_x)$ have the same sign. This implies that x and $-x$ contribute to the degree with the same orientation number. Hence

$\deg(q)$ is either $+2$ or -2. This also shows that, if $\det(dq_x) > 0$, i.e., if q preserves orientations, then $\deg(q) = +2$.

Solution (to Exercise 15.9) (a) If $f : \mathbb{S}^k \to \mathbb{S}^k$ sends no pair of antipodal points to antipodal points, then the line segment in $\mathbb{R}^{k+1}$ between $f(x)$ and $f(-x)$ does not pass through the origin. Thus the map

$$H : \mathbb{S}^k \times [0,1] \to \mathbb{S}^k, \; x \mapsto \frac{(1-t)f(x) + tf(-x)}{|(1-t)f(x) + tf(-x)|}$$

is well defined and smooth for every x and t. Hence H defines a homotopy between f and $f \circ a$. For $t = 1/2$, we have $H(x, 1/2) = \frac{f(x)+f(-x)}{|f(x)+f(-x)|} = g(x)$ for all $x \in \mathbb{S}^k$. Thus f and $f \circ a$ are both smoothly homotopic to the map g as well. The invariance of the degree under homotopy implies $\deg(f) = \deg(f \circ a) = \deg(g)$.

(b) The previous point gives us, using Exercise 15.6,

$$\deg(g) = \deg(f \circ a) = \deg(f) \cdot \deg(a) = \deg(f) = \deg(g).$$

Since we assume that k is even, we have $\deg(a) = -1$ by Exercise 15.1. Thus we get $\deg(g) = -\deg(g)$, which implies $\deg(g) = 0$.

(c) Since k is odd, the degree of the quotient map $q : \mathbb{S}^k \to \mathbb{RP}^k$ is defined. Since g satisfies $g(x) = g(-x)$ and q is a quotient map, g can be written as a composition $\mathbb{S}^k \xrightarrow{q} \mathbb{RP}^k \xrightarrow{[g]} \mathbb{S}^k$. By Exercise 15.8, we have $\deg(q) = 2$. Thus, by Exercise 15.6, we get $\deg(g) = \deg(q) \cdot \deg([g]) = 2 \cdot \deg([g])$. This shows that $\deg(g)$ is even.

(d) The previous points show that the assumption that $f : \mathbb{S}^k \to \mathbb{S}^k$ sends no pair of antipodal points to antipodal points implies that $\deg(f) = \deg(g)$ must be even. This contradicts the assumption that $\deg(f)$ is odd. This proves the theorem.

Solution (to Exercise 15.10) Let $z \in \mathbb{S}^k$ be a regular value for λ and let $(x, y) \in X \times Y$ with $\lambda(x, y) = z$. Consider the derivative

$$d\lambda_{(x,y)} : T_x(X) \times T_y(Y) = T_{(x,y)}(X \times Y) \to T_z(\mathbb{S}^k).$$

Then the degree of λ is defined as the sum $\deg(\lambda) = \sum_{(x,y)\in\lambda^{-1}(z)} \operatorname{sign}(d\lambda_{(x,y)})$. As we learned in Section 14.2.1, switching the order of the factors in the domain corresponds to making $\dim X \cdot \dim Y$ many flips of basis vectors. Each flip requires to multiply orientation numbers with a factor (-1). Hence the map $s : X \times Y \to Y \times X$ induces multiplication by $(-1)^{mn}$ on the degree. In addition, we change the value of λ by sending $x - y$ to $y - x = -(x - y)$. Hence we compose with the antipodal map a on $\mathbb{S}^k$. This map has degree $(-1)^{k+1}$ as we learned in Exercise 15.1. y Exercise 15.6, deg sends composition of maps to products, and hence we get, using $k = m + n$,

$$L(Y,X) = \deg(a \circ \lambda \circ s) = (-1)^{k+1} \cdot \deg(\lambda) \cdot (-1)^{mn}$$
$$= (-1)^{(m+1)(n+1)} \deg(\lambda) = (-1)^{(m+1)(n+1)} L(X,Y).$$

Solution (to Exercise 15.11) Since Y does not have a boundary, the product $W \times Y$ is a smooth manifold with boundary $\partial(W \times Y) = \partial W \times Y = X \times Y$. Since W and Y are disjoint, λ extends to a smooth map on $W \times Y$. Then the Boundary Theorem 15.2 for degrees implies that $\deg(\lambda) = 0$.

Solution (to Exercise 15.12) Since taking preimages preserves codimensions, the dimension of $f^{-1}(w)$ and $f^{-1}(z)$ for any regular values w and z for f is given by $\dim f^{-1}(w) = \dim f^{-1}(z) = \dim \mathbb{S}^{2n-1} - \dim \mathbb{S}^n = n-1$. Hence, if n is odd, then $\dim f^{-1}(w)+1 = \dim f^{-1}(z) = n$ is odd, and hence we have $H(f) = L(f^{-1}(w), f^{-1}(z)) = -L(f^{-1}(z), f^{-1}(w)) = -H(f)$ by Exercise 15.10. Thus we must have $H(f) = 0$.

Solution (to Exercise 15.13) By Sard's Theorem, we can find a point $a \in \mathbb{S}^n$ which is a regular value for both g and $g \circ f$. The fibre $g^{-1}(a)$ consists of a finite number of points, say $a_1, \ldots, a_r$ in $\mathbb{S}^n$. We can assume that these points are ordered such that the orientation numbers of g at $a_1, \ldots, a_p$ are positive, while the orientation numbers of g at $a_{p+1}, \ldots, a_r$ are negative. The fibre $f^{-1}(g^{-1}(a))$ then consists of a disjoint union of the oriented submanifolds $f^{-1}(a_i) \subset \mathbb{S}^{2n-1}$:

$$f^{-1}(g^{-1}(a)) = f^{-1}(a_1) \sqcup \ldots \sqcup f^{-1}(a_p) \sqcup (-f^{-1}(a_{p+1})) \sqcup \ldots \sqcup (-f^{-1}(a_r)).$$

Similarly, we can choose b such that the fibre $f^{-1}(g^{-1}(b))$ consists of a disjoint union of the submanifolds $f^{-1}(b_j) \subset \mathbb{S}^{2n-1}$ for $j = 1, \ldots, s$ and such that the $f^{-1}(a_i)$ and $f^{-1}(b_j)$ are mutually disjoint. Again we order these points such that the orientation numbers of g at $b_1, \ldots, b_s$ are positive, while those at $b_{q+1}, \ldots, b_s$ are negative. The fibre $f^{-1}(g^{-1}(b))$ then consists of a disjoint union of the oriented submanifolds $f^{-1}(a_i) \subset \mathbb{S}^{2n-1}$:

$$f^{-1}(g^{-1}(b)) = f^{-1}(b_1) \sqcup \ldots \sqcup f^{-1}(b_q) \sqcup (-f^{-1}(b_{q+1}) \sqcup \ldots \sqcup (-f^{-1}(b_s).$$

Since orientations behave multiplicatively on products of manifolds, the product fibre $f^{-1}(g^{-1}(a)) \times f^{-1}(g^{-1}(b))$ is the disjoint union of submanifolds in $\mathbb{S}^{2n-1}$

$$f^{-1}(g^{-1}(a)) \times f^{-1}(g^{-1}(b)) = \sqcup_{i,j} (\pm f^{-1}(a_i)) \times (\pm f^{-1}(b_j))$$
$$= \sqcup_{i=1,\, j=1}^{p,q} (f^{-1}(a_i) \times f^{-1}(b_j)) \sqcup \sqcup_{i=p+1,\, j=1}^{r,q} (-(f^{-1}(a_i) \times f^{-1}(b_j)))$$
$$\sqcup \sqcup_{i=1,\, j=q+1}^{p,s} (-(f^{-1}(a_i) \times f^{-1}(b_j))) \sqcup \sqcup_{i=p+1,\, j=q+1}^{r,s} (f^{-1}(a_i) \times f^{-1}(b_j)).$$

Now we write λ for the map $\lambda := (g \circ f)^{-1}(a) \times (g \circ f)^{-1}(b) \to \mathbb{S}^{2n-2}$ defined by $\lambda(x, y) = \frac{x-y}{|x-y|}$ and we write λ_{ij} for the map $\lambda_{ij} := f^{-1}(a_i) \times f^{-1}(b_j) \to \mathbb{S}^{2n-2}$ defined by $\lambda_{ij}(x, y) = \frac{x-y}{|x-y|}$. Note that $H(f) = \deg(\lambda_{ij})$ for every pair i, j, since $H(f)$ does not depend on the choice of regular values. Since the degree is additive on oriented connected components, the above decomposition implies

$$
\begin{aligned}
H(g \circ f) &= \deg(\lambda) \\
&= \sum_{i=1,\, j=1}^{p,q} \deg(\lambda_{ij}) - \sum_{i=p+1,\, j=1}^{r,q} \deg(\lambda_{ij}) - \sum_{i=1,\, j=q+1}^{p,s} \deg(\lambda_{ij}) + \sum_{i=p+1,\, j=q+1}^{r,s} \deg(\lambda_{ij}) \\
&= H(f) \cdot [pq + (p - \deg(g))(q - \deg(g)) - (q(p - \deg(g))) - (p(q - \deg(g)))] \\
&= H(f) \cdot (\deg(g))^2.
\end{aligned}
$$

References

[1] Adams, J. F. 1960. On the non-existence of elements of Hopf invariant one. *Ann. Math. (2)*, **72**, 20–104.

[2] Adams, J. F., and Atiyah, M. F. 1966. K-theory and the Hopf invariant. *Quart. J. Math. Oxford Ser. (2)*, **17**, 31–38.

[3] Barratt, M. G., Jones, J. D. S., and Mahowald, M. E. 1984. Relations amongst Toda brackets and the Kervaire invariant in dimension 62. *J. London Math. Soc. (2)*, **30**(3), 533–550.

[4] Barratt, M. G., Jones, J. D. S., and Mahowald, M. E. 1987. The Kervaire invariant and the Hopf invariant. Pages 135–173 of: *Algebraic Topology*. Edited by H. R. Miller and D. C. Ravenel. Lecture Notes in Math., vol. 1286. Springer, Berlin.

[5] Bott, R., and Milnor, J. 1958. On the parallelizability of the spheres. *Bull. Amer. Math. Soc.*, **64**, 87–89.

[6] Bourbaki, Nicolas. 1998. *General Topology. Chapters 1–4*. Elements of Mathematics (Berlin). Springer-Verlag, Berlin. Translated from the French, Reprint of the 1989 English translation.

[7] Brieskorn, Egbert. 1966. Beispiele zur Differentialtopologie von Singularitäten. *Invent. Math.*, **2**, 1–14.

[8] Brouwer, L. E. J. 1911. Beweis der invarianz des n-dimensionalen gebiets. *Math. Ann.*, **71**(3), 305–313.

[9] Brouwer, L. E. J. 1911. Über Abbildung von Mannigfaltigkeiten. *Math. Ann.*, **71**(1), 97–115.

[10] Dundas, Bjørn. 2018. *A Short Course in Differential Topology*. Cambridge University Press, Cambridge.

[11] Ehresmann, Charles. 1951. Les connexions infinitésimales dans un espace fibré différentiable. Pages 29–55 of: *Colloque de topologie (espaces fibrés), Bruxelles, 1950*. Georges Thone, Liège.

[12] Eisenbud, David. 1978. An algebraic approach to the topological degree of a smooth map. *Bull. Amer. Math. Soc.*, **84**(5), 751–764.

[13] Eisenbud, David, and Levine, Harold I. 1977. An algebraic formula for the degree of a C^∞ map germ. *Ann. Math.*, **106**(1), 19–44.

[14] Guillemin, Victor, and Pollack, Alan. 2010. *Differential Topology*. AMS Chelsea Publishing, Providence, RI. Reprint of the 1974 original.

[15] Hatcher, Allen. 2002. *Algebraic Topology*. Cambridge University Press, Cambridge.

[16] Hill, M. A., Hopkins, M. J., and Ravenel, D. C. 2016. On the nonexistence of elements of Kervaire invariant one. *Ann. Math. (2)*, **184**(1), 1–262.

[17] Hill, Michael A., Hopkins, Michael J., and Ravenel, Douglas C. 2021. *Equivariant Stable Homotopy Theory and the Kervaire Invariant Problem*. New Mathematical Monographs, vol. 40. Cambridge University Press, Cambridge.

[18] Hirsch, Morris W. 1994. *Differential Topology*. Graduate Texts in Mathematics, vol. 33. Springer-Verlag, New York. Corrected reprint of the 1976 original.

[19] Holm, Per. 1987. The theorem of Brown and Sard. *Enseign. Math. (2)*, **33**(3–4), 199–202.

[20] Hopf, Heinz. 1927. Abbildungsklassen n-dimensionaler Mannigfaltigkeiten. *Math. Ann.*, **96**(1), 209–224.

[21] Hopf, Heinz. 1927. Vektorfelder in n-dimensionalen Mannigfaltigkeiten. *Math. Ann.*, **96**(1), 225–249.

[22] Hopf, Heinz. 1931. Über die Abbildungen der dreidimensionalen Sphäre auf die Kugelfläche. *Math. Ann.*, **104**(1), 637–665.

[23] Huybrechts, Daniel. 2005. *Complex Geometry*. Universitext. Springer-Verlag, Berlin. An introduction.

[24] Kass, Jesse Leo, and Wickelgren, Kirsten. 2021. An arithmetic count of the lines on a smooth cubic surface. *Compos. Math.*, **157**(4), 677–709.

[25] Kervaire, Michel A. 1958. Non-parallelizability of the n-sphere for $n > 7$. *Proc. Natl. Acad. Sci. USA*, **44**(3), 280–283.

[26] Kervaire, Michel A. 1960. A manifold which does not admit any differentiable structure. *Comment. Math. Helv.*, **34**, 257–270.

[27] Kervaire, Michel A., and Milnor, John W. 1963. Groups of homotopy spheres. I. *Ann. Math. (2)*, **77**, 504–537.

[28] Kosinski, Antoni A. 1993. *Differential Manifolds*. Pure and Applied Mathematics, vol. 138. Academic Press, Inc., Boston, MA.

[29] Lee, John M. 2013. *Introduction to Smooth Manifolds*. Second edn. Graduate Texts in Mathematics, vol. 218. Springer, New York.

[30] Lin, Weinan, Wang, Guozhen, and Xu, Zhouli. 2025. *On the Last Kervaire Invariant Problem*. arXiv:2412.10879.

[31] May, J. P. 1999. *A Concise Course in Algebraic Topology*. Chicago Lectures in Mathematics. University of Chicago Press, Chicago, IL.

[32] Milnor, John. 2011. Differential topology forty-six years later. *Notices Amer. Math. Soc.*, **58**(6), 804–809.

[33] Milnor, John W. 1956. On manifolds homeomorphic to the 7-sphere. *Ann. Math. (2)*, **64**, 399–405.

[34] Milnor, John W. 1963. *Morse Theory*. Annals of Mathematics Studies, vol. 51. Princeton University Press, Princeton, NJ. Based on lecture notes by M. Spivak and R. Wells.

[35] Milnor, John W. 1997. *Topology from the Differentiable Viewpoint*. Princeton Landmarks in Mathematics. Princeton University Press, Princeton, NJ. Based on notes by David W. Weaver, Revised reprint of the 1965 original.

[36] Milnor, John W., and Stasheff, James D. 1974. *Characteristic Classes*. Annals of Mathematics Studies, vol. 76. Princeton University Press, Princeton, NJ; University of Tokyo Press, Tokyo.

[37] Morel, Fabien. 2012. $\mathbb{A}^1$*-Algebraic Topology over a Field*. Lecture Notes in Mathematics, vol. 2052. Springer, Heidelberg.

[38] Morel, Fabien, and Voevodsky, Vladimir. 1999. $\mathbb{A}^1$-homotopy theory of schemes. *Inst. Hautes Études Sci. Publ. Math.*, **90**, 45–143.

[39] Morse, Marston. 1996. *The Calculus of Variations in the Large*. American Mathematical Society Colloquium Publications, vol. 18. American Mathematical Society, Providence, RI. Reprint of the 1932 original.

[40] Munkres, James R. 2000. *Topology*. Second edn. Prentice Hall, Inc., Upper Saddle River, NJ.

[41] Pontryagin, L. S. 1959. *Smooth Manifolds and Their Applications in Homotopy Theory*. American Mathematical Society Translations, Ser. 2, vol. 11. American Mathematical Society, Providence, RI.

[42] Singh, Tej Bahadur. 2019. *Introduction to Topology*. Springer, Singapore.

[43] Spivak, Michael. 1965. *Calculus on Manifolds. A Modern Approach to Classical Theorems of Advanced Calculus*. W. A. Benjamin, Inc., New York.

[44] Thom, René. 1954. Quelques propriétés globales des variétés différentiables. *Comment. Math. Helv.*, **28**, 17–86.

[45] Thom, René. 1956. Un lemme sur les applications différentiables. *Bol. Soc. Mat. Mexicana (2)*, **1**, 59–71.

[46] Tu, Loring W. 2011. *An Introduction to Manifolds*. Second edn. Universitext. Springer, New York.

[47] Vick, James W. 1994. *Homology Theory*. Second edn. Graduate Texts in Mathematics, vol. 145. Springer-Verlag, New York. An introduction to algebraic topology.

[48] Voisin, Claire. 2007. *Hodge Theory and Complex Algebraic Geometry. I*. English edn. Cambridge Studies in Advanced Mathematics, vol. 76. Cambridge University Press, Cambridge. Translated from the French by Leila Schneps.

[49] Wall, C. T. C. 2016. *Differential Topology*. Cambridge Studies in Advanced Mathematics, vol. 156. Cambridge University Press, Cambridge.

[50] Whitney, Hassler. 1944. The self-intersections of a smooth n-manifold in $2n$-space. *Ann. Math. (2)*, **45**, 220–246.

[51] Whitney, Hassler. 1944. The singularities of a smooth n-manifold in $(2n - 1)$-space. *Ann. Math. (2)*, **45**, 247–293.

Index

For EU product safety concerns, contact us at Calle de José Abascal, 56–1°,
28003 Madrid, Spain or eugpsr@cambridge.org.

www.ingramcontent.com/pod-product-compliance
Ingram Content Group UK Ltd.
Pitfield, Milton Keynes, MK11 3LW, UK
UKHW020054260926
475953UK00009B/663